SAS® Certification Prep Guide
Base Programming for SAS®9
Third Edition

SAS® Documentation

The correct bibliographic citation for this manual is as follows: SAS Institute Inc. 2011. *SAS® Certification Prep Guide: Base Programming for SAS®9, Third Edition*. Cary, NC: SAS Institute Inc.

SAS® Certification Prep Guide: Base Programming for SAS®9, Third Edition

SAS® Publishing provides a complete selection of books and electronic products to help customers use SAS software to its fullest potential. For more information about our e-books, e-learning products, CDs, and hard-copy books, visit the SAS Publishing Web site at **support.sas.com/publishing** or call 1-800-727-3228.

Contents

About This Book

Audience

The *SAS Certification Prep Guide: Base Programming for SAS®9* is for new or experienced SAS programmers who want to prepare for the SAS Base Programming for SAS®9 exam.

Requirements and Details

Purpose

The *SAS Certification Prep Guide: Base Programming for SAS®9* helps prepare you to take the SAS Base Programming for SAS®9 exam. The book covers the objectives tested on the exam, including basic concepts, producing reports, creating and modifying SAS data sets, and reading various types of raw data. Before attempting the exam you should also have experience programming in the SAS®9 environment.

The book includes quizzes that enable you to test your understanding of material in each chapter. Additionally, solutions to all quizzes are included at the back of the book.

Note: Exam objectives are subject to change. Please view the current exam objectives at **support.sas.com/certify**.

Programming Environments

This book assumes you are running Base SAS or SAS Enterprise Guide software in the windowing environment. You will learn how to write and manage your SAS programs in either the SAS windowing environment workspace or in the SAS Enterprise Guide workspace.

If you are not sure which programming workspace you are using, select **Help** ⇨ **About** from the SAS software main menu. If you are using SAS Enterprise Guide, the About window displays the name "Enterprise Guide." If you are using the SAS windowing environment, the About window displays the name "SAS for Windows."

Because the two programming workspaces differ, you will occasionally see notes in this book that provide information specific to either SAS Enterprise Guide or to the SAS windowing environment.

Key to Icons Used in This Book

The following icons designate special information:

🗒	Specifies information that is additional to the previous information.
⚠	Specifies a caution.

How to Create Practice Data

If you are using the SAS 9.3 windowing environment, you can practice what you learn in this book by using sample data that you create from within the SAS®9 environment. To set up this practice data, select **Help** ⇨ **Learning SAS Programming** from the main SAS menu. When the SAS Online Training Sample Data window appears, click **OK** to create a permanent SAS library named sasuser, which contains the sample data.

You can access additional sample data by visiting the **SAS Certification** page on the SAS Training and Bookstore Web site at **http://support.sas.com/ publishing/cert/**. There you will find links to practice data as well as any updates to the guide.

Setting Result Formats in the SAS Windowing Environment

In the SAS windowing environment, you can use the Preferences window to specify whether you want your output in HTML or LISTING format, or both. Your preferences are saved until you modify them, and they apply to all output that you create. *SAS Certification Prep Guide: Base Programming for SAS®9* generally shows output in HTML format, but some sample programs in this book specify features that appear only in LISTING output. To create both HTML and LISTING output, do the following:

Start SAS and select **Tools** ⇨ **Options** ⇨ **Preferences**. Then click the **Results** tab and select the **Create listing** and **Create HTML** check boxes. If you want to store your HTML output in a folder other than the one shown, de-select the **Use WORK folder** check box and browse to the desired folder. HTML files are named sashtml.htm. Click **OK** to close the Preferences window.

Note: In SAS 9.3, HTML output in the SAS windowing environment is the default for Windows and UNIX, but not for other operating systems and not in batch mode. When you run SAS in batch mode or on other operating systems, the LISTING destination is open and is the default. Your actual defaults might be different because of your registry or configuration file settings.

SAS Certification Practice Exam: Base Programming for SAS®9

The SAS Certification Practice Exam: Base Programming for SAS®9 was designed to help you prepare for the SAS Base Programming for SAS®9 exam. This practice exam was constructed to give you a view of the type of questions on the official certification exam. You can get more information about this exam at **support.sas.com/ basepractice**.

SAS Base Programming for SAS®9

For information about how to register for the official SAS Base Programming for SAS®9 exam, see the SAS Global Certification Web site at `http://support.sas.com/certify`.

Additional Resources

Other resources might be helpful when you are learning SAS programming. You can refer to them as needed to enhance your understanding of the material covered in this book. You can access SAS Help, documentation, and other resources from your SAS software or on the Web.

From SAS Software	
Help	• SAS®9: Select **Help** ⇨ **SAS Help and Documentation**. • SAS Enterprise Guide: Select **Help** ⇨ **SAS Enterprise Guide Help**.
Documentation	• SAS®9: Select **Help** ⇨ **SAS Help and Documentation**. • SAS Enterprise Guide: Access online documentation on the Web. See "On the Web" below.

On the Web	
Bookstore	`http://support.sas.com/publishing/`
Training	`http://support.sas.com/training/`
Certification	`http://support.sas.com/certify/`
SAS Learning Edition	`http://support.sas.com/learn/le/`
SAS Global Academic Program	`http://support.sas.com/learn/ap/`
SAS OnDemand	`http://support.sas.com/ondemand/`
Knowledge Base	`http://support.sas.com/resources/`
Support	`http://support.sas.com/techsup/`
Learning Center	`http://support.sas.com/learn/`
Community	`http://support.sas.com/community/`

Syntax Conventions

The following example shows the general form of SAS code is shown in the book:

DATA *output-SAS-data-set*

      (**DROP**=*variables(s)* | **KEEP**=*variables(s)*);

   SET *SAS-data-set* <options>;

   BY *variable(s)*

RUN;

In the general form above:

- DATA, DROP=, KEEP=, SET, BY, and RUN are in uppercase bold because they must be spelled as shown.

- *output-SAS-data-set*, *variable(s)*, *SAS-data-set*, and *options* are in italics because each represents a value that you supply.

- *<options>* is enclosed in angle brackets because it is optional syntax.

- DROP= and KEEP= are separated by a vertical bar (|) to indicate that they are mutually exclusive.

The general forms of SAS statements and commands that are shown in this book include only the syntax that you need to know to prepare for the certification exam. For complete syntax, see the appropriate SAS reference guide.

Chapter 1
Base Programming

Overview

Introduction

To program effectively using SAS, you need to understand basic concepts about SAS programs and the SAS files that they process. In particular, you need to be familiar with SAS data sets.

In this chapter, you'll examine a simple SAS program and see how it works. You'll learn details about SAS data sets (which are files that contain data that is logically arranged in a form that SAS can understand). You'll see how SAS data sets are stored temporarily or permanently in SAS libraries. Finally, you'll learn how to use SAS windows to manage your SAS session and to process SAS programs.

Figure 1.1 *SAS Library with SAS Data Sets and Data Files*

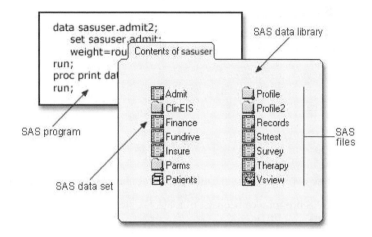

Objectives

In this chapter, you learn about

- the structure and components of SAS programs

- the steps involved in processing SAS programs

- SAS libraries and the types of SAS files that they contain

- temporary and permanent SAS libraries

- the structure and components of SAS data sets

- the SAS windowing environment.

SAS Programs

You can use SAS programs to access, manage, analyze, or present your data. Let's begin by looking at a simple SAS program.

A Simple SAS Program

This program uses an existing SAS data set to create a new SAS data set containing a subset of the original data set. It then prints a listing of the new data set using PROC PRINT. A SAS data set is a data file that is formatted in a way that SAS can understand.

```
data sasuser.admit2;
   set sasuser.admit;
   where age>39;
run;
proc print data=sasuser.admit2;
run;
```

Let's see how this program works.

Components of SAS Programs

The sample SAS program contains two steps: a DATA step and a PROC step.

```
data sasuser.admit2;
   set sasuser.admit;
   where age>39;
run;
proc print data=sasuser.admit2;
run;
```

These two types of steps, alone or combined, form most SAS programs.

A SAS program can consist of a DATA step or a PROC step or any combination of DATA and PROC steps.

Figure 1.2 *Components of a SAS Program*

DATA steps typically create or modify SAS data sets. They can also be used to produce custom-designed reports. For example, you can use DATA steps to

- put your data into a SAS data set

- compute values

- check for and correct errors in your data

- produce new SAS data sets by subsetting, supersetting, merging, and updating existing data sets.

In the previous example, the DATA step produced a new SAS data set containing a subset of the original data set. The new data set contains only those observations with an age value greater than 39.

PROC (procedure) steps invoke or call pre-written routines that enable you to analyze and process the data in a SAS data set. PROC steps typically present the data in the form of a report. They sometimes create new SAS data sets that contain the results of the procedure. PROC steps can list, sort, and summarize data. For example, you can use PROC steps to

- create a report that lists the data
- produce descriptive statistics
- create a summary report
- produce plots and charts.

Characteristics of SAS Programs

Next let's look at the individual statements in our sample program. SAS programs consist of SAS statements. A SAS statement has two important characteristics:

- It usually begins with a SAS keyword.
- It always ends with a semicolon.

As you've seen, a DATA step begins with a DATA statement, which begins with the keyword DATA. A PROC step begins with a PROC statement, which begins with the keyword PROC. Our sample program contains the following statements:

Table 1.1 SAS Program Statements

Statements	Sample Program Code
a DATA statement	`data sasuser.admit2;`
a SET statement	`set sasuser.admit;`
Additional programming statements	`where age>39;`
a RUN statement	`run;`
a PROC PRINT statement	`proc print data=sasuser.admit2;`
another RUN statement	`run;`

Layout for SAS Programs

SAS statements are free-format. This means that

- they can begin and end anywhere on a line
- one statement can continue over several lines
- several statements can be on the same line.

Blanks or special characters separate *words* in a SAS statement.

You can specify SAS statements in uppercase or lowercase. In most situations, text that is enclosed in quotation marks is case sensitive.

You've examined the general structure of our sample program. But what happens when you run the program?

Processing SAS Programs

When you submit a SAS program, SAS begins reading the statements and checking them for errors.

DATA and PROC statements signal the beginning of a new step. The RUN statement (for most procedures and the DATA step) and the QUIT statement (for some procedures) mark step boundaries. The beginning of a new step (DATA or PROC) also implies the end of the previous step. At a step boundary, SAS executes any statements that have not previously executed and ends the step. In our sample program, each step ends with a RUN statement.

```
data sasuser.admit2;
   set sasuser.admit;
   where age>39;
run;
proc print data=sasuser.admit2;
run;
```

Though the RUN statement is not always required between steps in a SAS program, using it can make the SAS program easier to read and debug, and it makes the SAS log easier to read.

Log Messages

Each time a step is executed, SAS generates a log of the processing activities and the results of the processing. The SAS log collects messages about the processing of SAS programs and about any errors that occur.

When SAS processes our sample program, you see the log messages shown below. Notice that you get separate sets of messages for each step in the program.

Figure 1.3 Log Messages

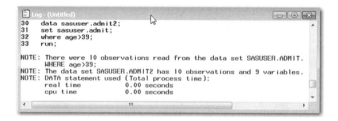

Results of Processing

*   DATA step Output

    Suppose you submit the sample program below:

    ```
    data sasuser.admit2;
    set sasuser.admit;
       where age>39;
       run;
    ```

When the program is processed, it creates a new SAS data set (sasuser.admit2) containing only those observations with age values greater than 39. The DATA step creates a new data set and produces messages in the SAS log, but it does not create a report or other output.

- Procedure Output

  If you add a PROC PRINT statement to this same example, the program produces the same new data set as before, but it also creates the following report, which is displayed in HTML:

```
data sasuser.admit2;
set sasuser.admit;
   where age>39;
run;
proc print data=sasuser.admit2;
run;
```

Figure 1.4 *PRINT Procedure Output*

The SAS System

Obs	ID	Name	Sex	Age	Date	Height	Weight	ActLevel	Fee
1	2523	Johnson, R	F	43	31	63	137	MOD	149.75
2	2539	LaMance, K	M	51	4	71	158	LOW	124.80
3	2568	Eberhardt, S	F	49	27	64	172	LOW	124.80
4	2571	Nunnelly, A	F	44	19	66	140	HIGH	149.75
5	2575	Quigley, M	F	40	8	69	163	HIGH	124.80
6	2578	Cameron, L	M	47	5	72	173	MOD	124.80
7	2579	Underwood, K	M	60	22	71	191	LOW	149.75
8	2584	Takahashi, Y	F	43	29	65	123	MOD	124.80
9	2589	Wilcox, E	F	41	16	67	141	HIGH	149.75
10	2595	Warren, C	M	54	7	71	183	MOD	149.75

Throughout this book, procedure output is shown in HTML in the style shown above unless otherwise noted.

You've seen the results of submitting our sample program. For other SAS programs, the results of processing might vary:

- Other Types of Procedural Output

  - Some SAS programs open an interactive window (a window that you can use to directly modify data), such as the REPORT window.

```
proc report data=sasuser.admit;
   columns id name sex age actlevel;
   run;
```

Figure 1.5 Interactive Report Window

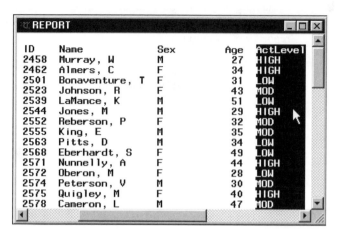

- SAS programs often invoke procedures that create output in the form of a report, as is the case with the TABULATE procedure:

```
proc tabulate data=sasuser.admit;
   class sex;
   var height weight;
   table sex*(height weight),mean;
run;
```

Figure 1.6 TABULATE Procedure Output

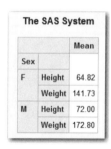

- Other SAS programs perform tasks such as sorting and managing data, which have no visible results except for messages in the log. (All SAS programs produce log messages, but some SAS programs produce only log messages.)

```
proc copy in=sasuser out=work;
   select admit;
run;
```

Figure 1.7 Log Output

```
36    proc copy in=sasuser out=work;
37       select admit;
38    run;

NOTE: Copying SASUSER.ADMIT to WORK.ADMIT (memtype=DATA).
NOTE: There were 21 observations read from the data set SASUSER.ADMIT.
NOTE: The data set WORK.ADMIT has 21 observations and 9 variables.
NOTE: PROCEDURE COPY used (Total process time):
      real time          0.15 seconds
      cpu time           0.04 seconds
```

SAS Libraries

So far you've learned about SAS programs. Now let's look at SAS libraries to see how SAS data sets and other SAS files are organized and stored.

How SAS Files Are Stored

Every SAS file is stored in a SAS library, which is a collection of SAS files. A SAS data library is the highest level of organization for information within SAS.

For example, in the Windows and UNIX environments, a library is typically a group of SAS files in the same folder or directory.

Figure 1.8   *SAS Data Library*

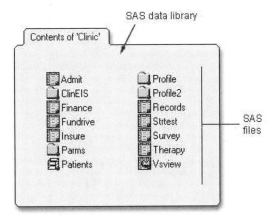

The table below summarizes the implementation of SAS libraries in various operating environments.

Table 1.2   *Environments and SAS Libraries*

In this environment...	A SAS library is...
Windows, UNIX, OpenVMS, OS/2 (directory based-systems)	a group of SAS files that are stored in the same directory. Other files can be stored in the directory, but only the files that have SAS file extensions are recognized as part of the SAS library. For more information, see the SAS documentation for your operating environment.
z/OS (OS/390)	a specially formatted host data set in which only SAS files are stored.

Storing Files Temporarily or Permanently

Depending on the library name that you use when you create a file, you can store SAS files temporarily or permanently.

Figure 1.9   *Temporary SAS Data Library*

Figure 1.10   *Permanent SAS Data Library*

Table 1.3   *Temporary and Permanent SAS Libraries*

Temporary SAS libraries last only for the current SAS session.	Storing files temporarily: If you don't specify a library name when you create a file (or if you specify the library name Work), the file is stored in the temporary SAS data library. When you end the session, the temporary library and all of its files are deleted.
Permanent SAS Libraries are available to you during subsequent SAS sessions.	Storing files permanently: To store files permanently in a SAS data library, you specify a library name other than the default library name Work. For example, by specifying the library name sasuser when you create a file, you specify that the file is to be stored in a permanent SAS data library until you delete it.

You can learn how to set up permanent SAS libraries in Chapter 2, "Referencing Files and Setting Options," on page 41.

Referencing SAS Files

Two-Level Names

To reference a permanent SAS data set in your SAS programs, you use a two-level name consisting of the library name and the filename, or data set name:

libref.filename

    In the two-level name, *libref* is the name of the SAS data library that contains the file, and *filename* is the name of the file, or data set. A period separates the libref and filename.

Figure 1.11   *Two-Level SAS Name*

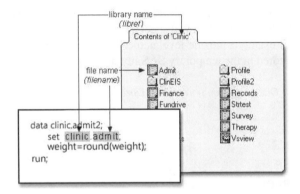

For example, suppose we want to create a new permanent sas library named Clinic. In our sample program, Clinic.Admit is the two-level name for the SAS data set Admit, which is stored in the library named Clinic. Notice that the LIBNAME statement is used to define the libref, Clinic, and to give SAS the physical location of the data files.

```
libname clinic 'c:\Users\Name\sasuser';
data clinic.admit2;
   set clinic.admit;
   weight =round(weight);
run;
```

Figure 1.12   *Two-Level Name Clinic.Admit*

Referencing Temporary SAS Files

To reference temporary SAS files, you can specify the default libref Work, a period, and the filename. For example, the two-level name Work.Test references the SAS data set named *Test* that is stored in the temporary SAS library Work.

Figure 1.13   *Two-Level Temporary SAS Library Work.Test*

Alternatively, you can use a one-level name (the *filename* only) to reference a file in a temporary SAS library. When you specify a one-level name, the default libref Work is assumed. For example, the one-level name Test also references the SAS data set named Test that is stored in the temporary SAS library Work.

Figure 1.14   *One-Level Temporary SAS Library Test*

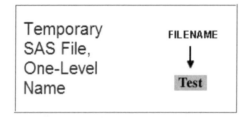

Referencing Permanent SAS Files

You can see that Clinic.Admit and Clinic.Admit2 are permanent SAS data sets because the library name is Clinic, not Work.

Figure 1.15   *Referencing Permanent SAS Files*

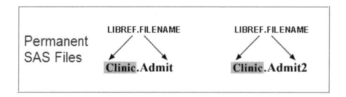

So, referencing a SAS file in any library except Work indicates that the SAS file is stored permanently. For example, when our sample program creates Clinic.Admit2, it stores the new Admit2 data set permanently in the SAS library Clinic.

Rules for SAS Names

These rules apply only to the filename portion of a SAS data set name. A libref can have a length of only eight characters.

SAS data set names and variable names

- can be 1 to 32 characters long

- must begin with a letter (A-Z, either uppercase or lowercase) or an underscore (_)

- can continue with any combination of numbers, letters, or underscores.

These are examples of valid data set names and variable names:

- Payroll

- LABDATA1995_1997

- _EstimatedTaxPayments3

SAS Data Sets

So far, you've seen the components and characteristics of SAS programs, including how they reference SAS data sets. Data sets are one type of SAS file. There are other types of SAS files (such as catalogs), but this chapter focuses on SAS data sets. For most procedures, data must be in the form of a SAS data set to be processed. Now let's take a closer look at SAS data sets.

Overview of Data Sets

As you saw in our sample program, for many of the data processing tasks that you perform with SAS, you

- access data in the form of a SAS data set

- analyze, manage, or present the data.

Conceptually, a SAS data set is a file that consists of two parts: a descriptor portion and a data portion. Sometimes a SAS data set also points to one or more indexes, which enable SAS to locate rows in the data set more efficiently. (The data sets that you work with in this chapter do not contain indexes.)

Figure 1.16 *Parts of a SAS Data Set*

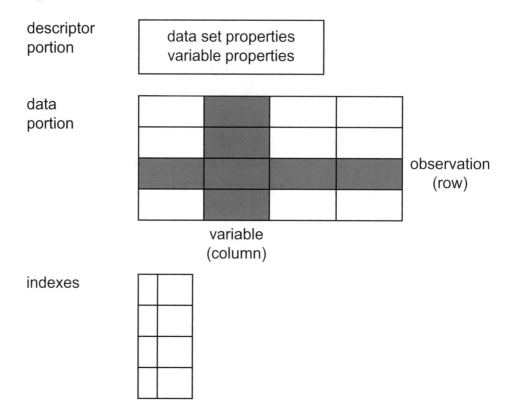

Descriptor Portion

The descriptor portion of a SAS data set contains information about the data set, including

- the name of the data set
- the date and time that the data set was created
- the number of observations
- the number of variables.

Let's look at another SAS data set. The table below lists part of the descriptor portion of the data set sasuser.insure, which contains insurance information for patients who are admitted to a wellness clinic. (It's a good idea to give your data set a name that is descriptive of the contents.)

Table 1.4 *Descriptor Portion of Attributes in a SAS Data Set sasuser.insure*

Data Set Name:	sasuser.INSURE
Member Type:	DATA
Engine:	V9
Created:	10:05 Tuesday, February 16, 2011
Observations:	21

Variables:	7
Indexes:	0
Observation Length:	64

Data Portion

The data portion of a SAS data set is a collection of data values that are arranged in a rectangular table. In the example below, the name **Murray** is a data value, Policy **32668** is a data value, and so on.

Figure 1.17 Parts of a SAS Data Set: Data: Data Portion

ID	Name	Policy	Company	PctInsured	Total	BalanceDue
2458	Murray, W	32668	MUTUALITY	100	98.64	0.00
2462	Almers, C	95824	RELIABLE	80	780.23	156.05
2501	Bonaventure, T	87795	A&R	80	47.38	9.48
2523	Johnson, R	39022	ACME	50	122.07	61.04

Observations (Rows)

Rows (called observations) in the data set are collections of data values that usually relate to a single object. The values **2458**, **Murray**, **32668**, **Mutuality**, **100**, **98.64** and **0.00** comprise a single observation in the data set shown below.

Figure 1.18 Parts of a SAS Data Set: Observations

Observation

ID	Name	Policy	Company	PctInsured	Total	BalanceDue
2458	Murray, W	32668	MUTUALITY	100	98.64	0.00
2462	Almers, C	95824	RELIABLE	80	780.23	156.05
2501	Bonaventure, T	87795	A&R	80	47.38	9.48
2523	Johnson, R	39022	ACME	50	122.07	61.04

This data set has seven observations, each containing information about an individual. A SAS data set can store any number of observations.

Variables (Columns)

Columns (called variables) in the data set are collections of values that describe a particular characteristic. The values **2458**, **2462**, **2501**, and **2523** comprise the variable ID in the data set shown below.

Figure 1.19 *Parts of a SAS Data Set: Variable*

	ID	Name	Policy	Company	PctInsured	Total	BalanceDue
Variables	2458	Murray, W	32668	MUTUALITY	100	98.64	0.00
	2462	Almers, C	95824	RELIABLE	80	780.23	156.05
	2501	Bonaventure, T	87795	A&R	80	47.38	9.48
	2523	Johnson, R	39022	ACME	50	122.07	61.04

This data set contains seven variables: ID, Name, Policy, Company, PctInsured, Total, and BalanceDue. A SAS data set can store thousands of variables.

Missing Values

Every variable and observation in a SAS data set must have a value. If a data value is unknown for a particular observation, a missing value is recorded in the SAS data set.

Figure 1.20 *Missing Data Values*

	ID	Name	Policy	Company	PctInsured	Total	BalanceDue
Missing Value	2458	Murray, W	32668	MUTUALITY	100	98.64	0.00
	2462	Almers, C	95824	RELIABLE	80	780.23	156.05
	2501	Bonaventure, T	87795	A&R	.	47.38	9.48
	2523	Johnson, R	39022	ACME	50	122.07	61.04

Variable Attributes

In addition to general information about the data set, the descriptor portion contains information about the properties of each variable in the data set. The properties information includes the variable's name, type, length, format, informat, and label.

When you write SAS programs, it's important to understand the attributes of the variables that you use. For example, you might need to combine SAS data sets that contain same-named variables. In this case, the variables must be the same type (character or numeric).

The following is a partial listing of the attribute information in the descriptor portion of the SAS data set insure.policy. First, let's look at the name, type, and length variable attributes.

Table 1.5 *Variable Attributes in the Descriptor Portion of a SAS Data Set insure.policy*

Variable	Type	Length	Format	Informat	Label
Policy	Char	8			Policy Number
Total	Num	8	DOLLAR8.2	COMMA10.	Total Balance
Name	Char	20			Patient Name

Name

Each variable has a name that conforms to SAS naming conventions. Variable names follow exactly the same rules as SAS data set names. Like data set names, variable names

- can be 1 to 32 characters long

- must begin with a letter (A-Z, either uppercase or lowercase) or an underscore (_)

- can continue with any combination of numbers, letters, or underscores.

Table 1.6 *Variable Name Attributes*

Variable	Type	Length	Format	Informat	Label
Policy	Char	8			Policy Number
Total	Num	8	DOLLAR8.2	COMMA10.	Total Balance
Name	Char	20			Patient Name

Your site may choose to restrict variable names to those valid in SAS 6, to uppercase variable names automatically, or to remove all restrictions on variable names.

Type

A variable's type is either character or numeric.

- Character variables, such as Name (shown below), can contain *any values*.

- Numeric variables, such as Total (shown below), can contain *only numeric values* (the numerals 0 through 9, +, -, ., and E for scientific notation).

Table 1.7 *Type Attribute*

Variable	Type	Length	Format	Informat	Label
Policy	Char	8			Policy Number
Total	Num	8	DOLLAR8.2	COMMA10.	Total Balance
Name	Char	20			Patient Name

A variable's type determines how missing values for a variable are displayed. In the following data set, Name and Sex are character variables, and Age and Weight are numeric variables.

- For character variables such as Name, a *blank* represents a missing value.

- For numeric variables such as Age, a *period* represents a missing value.

Figure 1.21   *Missing Values Represented Based on Variable Type*

Name	Sex	Age	Weight
	M	48	128.6
Laverne	M	58	158.3
Jaffe	F	.	115.5
Wilson	M	28	170.1

Length

A variable's length (the number of bytes used to store it) is related to its type.

- Character variables can be up to 32,767 bytes long. In the example below, Name has a length of 20 characters and uses 20 bytes of storage.

- All numeric variables have a default length of 8 bytes. Numeric values (no matter how many digits they contain) are stored as floating-point numbers in 8 bytes of storage.

Table 1.8   *Length Attribute*

Variable	Type	Length	Format	Informat	Label
Policy	Char	8			Policy Number
Total	Num	8	DOLLAR8.2	COMMA10.	Total Balance
Name	Char	20			Patient Name

You've seen that each SAS variable has a name, type, and length. In addition, you can optionally define format, informat, and label attributes for variables. Let's look briefly at these optional attributes—you'll learn more about them in later chapters as you need to use them.

Format

Formats are variable attributes that affect the way data values are written. SAS software offers a variety of character, numeric, and date and time formats. You can also create and store your own formats. To write values out using a particular form, you select the appropriate format.

Figure 1.22   *Formats*

For example, to display the value **1234** as $1,234.00 in a report, you can use the DOLLAR8.2 format, as shown for Total below.

Table 1.9 *Format Attribute*

Variable	Type	Length	Format	Informat	Label
Policy	Char	8			Policy Number
Total	Num	8	DOLLAR8.2	COMMA10.	Total Balance
Name	Char	20			Patient Name

Usually you have to specify the maximum width (*w*) of the value to be written. Depending on the particular format, you may also need to specify the number of decimal places (*d*) to be written. For example, to display the value **5678** as 5,678.00 in a report, you can use the COMMA8.2 format, which specifies a width of 8 including 2 decimal places.

You can permanently assign a format to a variable in a SAS data set, or you can temporarily specify a format in a PROC step to determine the way the data values appear in output.

Informat

Whereas formats write values out using some particular form, informats read data values in certain forms into standard SAS values. Informats determine how data values are read into a SAS data set. You *must* use informats to read numeric values that contain letters or other special characters.

Figure 1.23 *Informats*

For example, the numeric value **$12,345.00** contains two special characters, a dollar sign ($) and a comma (,). You can use an informat to read the value while removing the dollar sign and comma, and then store the resulting value as a standard numeric value. For Total below, the COMMA10. informat is specified.

Table 1.10 *Informat Attribute*

Variable	Type	Length	Format	Informat	Label
Policy	Char	8			Policy Number

Variable	Type	Length	Format	Informat	Label
Total	Num	8	DOLLAR8.2	COMMA10.	Total Balance
Name	Char	20			Patient Name

Label

A variable can have a label, which consists of descriptive text up to 256 characters long. By default, many reports identify variables by their names. You may want to display more descriptive information about the variable by assigning a label to the variable.

For example, you can label Policy as Policy Number, Total as Total Balance, and Name as Patient Name to display these labels in reports.

Table 1.11 Label Attribute

Variable	Type	Length	Format	Informat	Label
Policy	Char	8			Policy Number
Total	Num	8	DOLLAR8.2	COMMA10.	Total Balance
Name	Char	20			Patient Name

You may even want to use labels to shorten long variable names in your reports!

Using the Programming Workspace

Using the Main SAS Windows

When you start SAS, by default several primary windows are available to you. These include the Explorer, Log, Output, Results, and code editing window(s). The window you use to edit your SAS programs may vary, depending on your operating system and needs.

You use these SAS windows to explore and manage your files, to enter and submit SAS programs, to view messages, and to view and manage your output.

We'll tour each of these windows shortly.

Figure 1.24   *Using the Main SAS Windows*

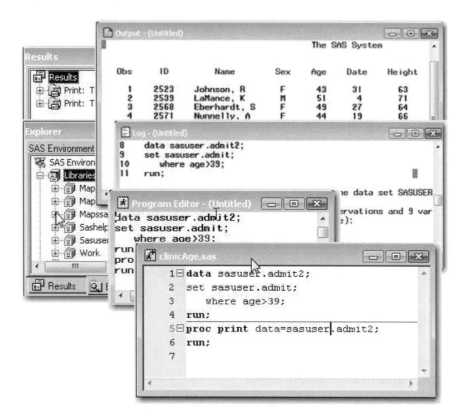

Your operating environment, and any options that you use when you start SAS, determine

- which of the main SAS windows are displayed by default

- their general appearance

- their position.

Features of SAS Windows

SAS windows have many features that help you get your work done. For example, you can

- maximize, minimize, and restore windows

- use pull-down menus, pop-up menus, and toolbars

- get more help.

This chapter and later chapters show SAS®9 windows in the Windows operating environment.

Figure 1.25 *Features of SAS Windows*

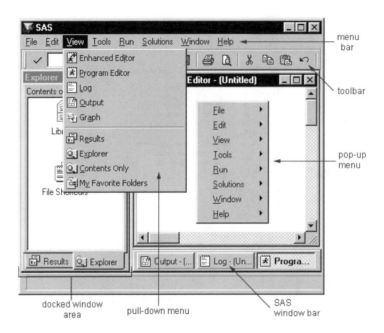

Minimizing and Restoring Windows

In the Windows environment, you can click the **Minimize** button to send a window that you aren't using to the SAS window bar. To restore the window to its former position, click the corresponding button on the SAS window bar.

In other operating environments, minimizing the window shrinks it to an icon.

Docking and Undocking Windows

In the Windows and OS/2 environments, the Explorer and Results windows are docked by default, so they can be resized but not minimized. If you prefer, you can select **Window** ➪ **Docked** to undock the active window, or you can turn docking off completely in the Preferences dialog box.

Issuing Commands

In SAS, you can issue commands by

- making selections from a menu bar

- by typing commands in a command box (or Toolbox) or on a command line.

Figure 1.26 *Using the SAS Command Box and Menu Bar to Issue Commands*

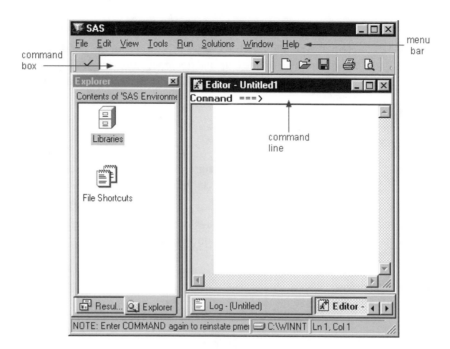

In most operating environments, SAS displays a menu bar by default. In the Windows environment, the menu bar selections correspond to the active window. To display a menu bar if it is not displayed, you can type **pmenus** in the command box (or ToolBox) or on the command line.

In all operating environments, SAS displays a command box (or ToolBox) or command line by default. You can display a command line in a particular window by activating that window and then using the Tools menu as indicated below:

- In the Windows environment, select **Tools** ⇨ **Options** ⇨ **Preferences**, and then select the **View** tab and select the **Command line** checkbox.

- In the UNIX and z/OS environments, select **Tools** ⇨ **Options** ⇨ **Turn Command Line On**.

- In the z/OS environment, you can display both a command line and a menu bar simultaneously in a window by selecting **Tools** ⇨ **Options** ⇨ **Command....**

In the Windows and UNIX operating environments, you can also display a command line in a window by activating the window, typing **command** in the command box (or ToolBox), and pressing Enter.

See the online help for a complete list of command-line commands.

Using Pop-Up Menus

Pop-up menus are context sensitive; they list actions that pertain only to a particular window. Generally, you display pop-up menus by clicking the right mouse button. If you like, you can specify a function key to open pop-up menus. Simply select **Tools** ⇨ **Options** ⇨ **Keys** and type **wpopup** as a function key setting.

To open a pop-up menu in the z/OS operating environment, type **?** in the selection field beside the item.

Getting Help

Help is available for all windows in SAS. From the **Help** menu, you can access comprehensive online help and documentation for SAS, or you can access task-oriented help for the active window. The Help menu is discussed in more detail later in this chapter.

Customizing Your SAS Environment

You can customize many features of the SAS workspace such as toolbars, pop-up menus, icons, and so on. Select the **Tools** menu to explore some of the customization options that are available.

You'll learn how to use features of SAS windows throughout this chapter. Now let's look at each of the main SAS windows individually.

The Explorer Window

In the Explorer window, you can view and manage your SAS files, which are stored in SAS data libraries. The library name is a logical name for the physical location of the files (such as a directory). You can think of the library name as a temporary nickname or shortcut.

You use the Explorer window to

- create new libraries and SAS files

- open any SAS file

- perform most file management tasks such as moving, copying, and deleting files

- create shortcuts to files that were not created with SAS.

Notice that the Explorer window displays a tree view of its contents.

Figure 1.27 *Tree View of the SAS Explorer Window*

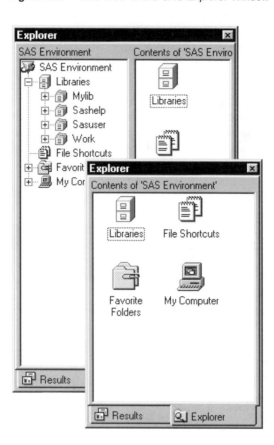

You can display the Explorer window by selecting **View** ⇨ **Explorer**. In the Windows and z/OS operating environments, if the Explorer window is docked, you can click the **Explorer** tab to display the window.

Navigating the Explorer Window

You can find your way around the Explorer window by double-clicking folders to open them and see their contents. You can also use pop-up menus to perform actions on a file (such as viewing its properties, or copying it). Pop-up menus contain different options for different file types.

To open a pop-up menu in the z/OS operating environment, type **?** in the selection field beside the item in the Explorer window. To simulate a double-click, type **s** in the selection field beside the item.

Code Editing Windows

You can use the following editors to write and edit SAS programs:

- the Enhanced Editor window

- the Program Editor window

- the host editor of your choice.

This training focuses on the two SAS code editing windows: the Enhanced Editor and the Program Editor windows. The Enhanced Editor is available only in the Windows operating environment.

The features of both editors are described below. In the remaining chapter and in future chapters, the general term code editing window will be used to refer to your preferred SAS code editing window.

Enhanced Editor Window

In the Windows operating environment, an Enhanced Editor window opens by default. You can use the Enhanced Editor window to enter, edit, and submit SAS programs. The initial window title is Editor - Untitled*n* until you open a file or save the contents of the editor to a file. Then the window title changes to reflect that filename. When the contents of the editor are modified, an asterisk is added to the title.

Figure 1.28   *Enhanced Editor Window*

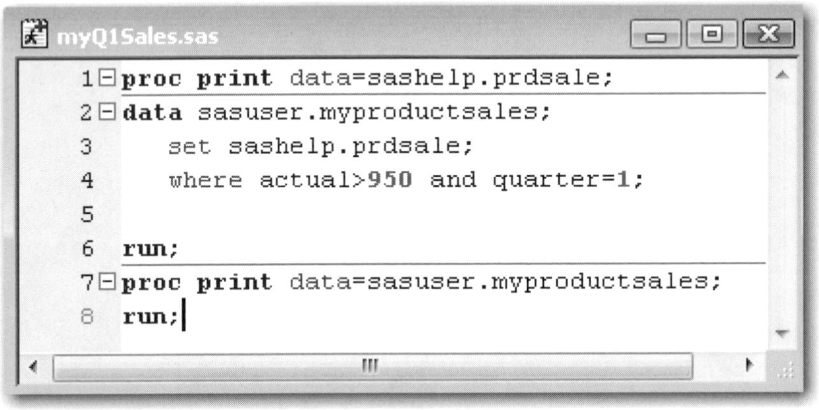

You can redisplay or open additional Enhanced Editor windows by selecting **View** ⇨ **Enhanced Editor**.

Enhanced Editor Features

In the Enhanced Editor, you can perform standard editing tasks such as

- opening SAS programs in various ways, including drag and drop

- entering, editing, and submitting SAS programs

- using the command line or menus

- saving SAS programs

- clearing contents.

In addition, the Enhanced Editor provides useful editing features, including

- color coding and syntax checking of the SAS programming language

- expandable and collapsible sections

- recordable macros

- support for keyboard shortcuts (Alt or Shift plus keystroke)

- multi-level undo and redo.

For more information about the Enhanced Editor, open or activate the Enhanced Editor window, and then select **Help** ⇨ **Using This Window**.

Clearing the Editor

In the Enhanced Editor, the code does not disappear when you submit it.

To clear any of these windows, you can activate the window and select **Edit** ⇨ **Clear All**.

Figure 1.29 *Cleared Editor Window*

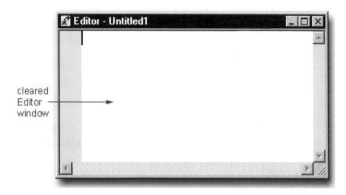

The Program Editor Window

As in the Enhanced Editor window, in the Program Editor window you enter, edit, and submit SAS programs. You can also open existing SAS programs. You can display the Program Editor window by selecting **View** ⇨ **Program Editor**.

Figure 1.30 *Program Editor Window*

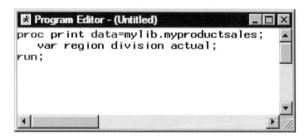

Features

As in the Enhanced Editor window, in the Program Editor window you can perform standard editing tasks such as

- opening SAS programs in various ways, including drag and drop

- entering, editing, and submitting SAS programs

- using the command line or menus

- saving SAS programs

- clearing contents

- recalling submitted statements.

However, the Program Editor does not provide some features of the enhanced Editor, such as syntax checking (this feature is available in SAS 9.2), expandable and collapsible sections, and recordable macros.

For more information about these features, open or activate the Program Editor window, and then select **Help** ⇨ **Using This Window**.

Clearing the Editor

At any time you can clear program code from the Program Editor window by activating the window and selecting **Edit** ⇨ **Clear All**.

When you submit SAS programs in the Program Editor window, the code in the window is automatically cleared.

The Log Window

The Log window displays messages about your SAS session and about any SAS programs that you submit. You can display the Log window by selecting **View** ⇨ **Log**.

Figure 1.31 *Log Window*

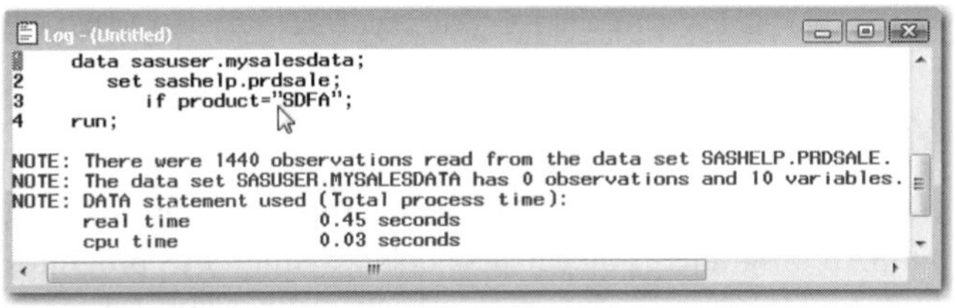

The Output Window

In the Output window, you browse LISTING output from SAS programs that you submit. (You can use a browser to view HTML output.)

By default, the Output window is positioned behind the code editing and Log windows. When you create output, the Output window automatically moves to the front of your display.

You can display the Output window at any time by selecting **View** ⇨ **Output**.

Figure 1.32   *Output Window*

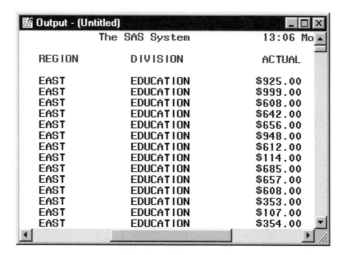

Not all SAS programs create output in the Output window. Some open interactive windows. Others produce only messages in the Log window.

In mainframe operating environments, when you create multiple pages of output, a message in the Output window border indicates that the procedure is suspended. In the example below, PROC PRINT output is suspended.

Figure 1.33   *Mainframe Window Showing Suspended Procedure*

To remove the message and view the remaining output, simply scroll to the bottom of the output.

The Results Window

The Results window helps you navigate and manage output from SAS programs that you submit. You can view, save, and print individual items of output. The Results window uses a tree structure to list various types of output that might be available after you run SAS.

On most operating systems, the Results window is positioned behind the Explorer window and is empty until you submit a SAS program that creates output. Then the Results window moves to the front of your display. The Results window displays separate icons for LISTING output and HTML output. In the example below, the first Print folder contains both types of output.

Figure 1.34 *Results Window*

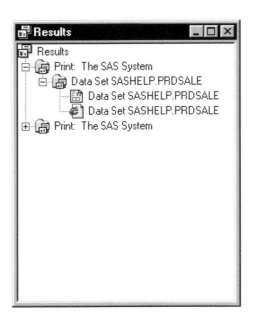

Viewing Output in the Results Window

You can display the Results window at any time by selecting **View** ⇨ **Results**. HTML is the default output type in the SAS windowing environment for UNIX and Windows. In these environments, when you submit a SAS program, the HTML output is automatically displayed in the Results Viewer and the file is listed in the Results window. In all other environments, LISTING is the default output.

The left pane of the following display shows the Results window, and the right pane shows the Results Viewer where the default HTML output is displayed. The Results window lists the files that were created when the SAS program executed.

Figure 1.35 *Results Window and Results Viewer*

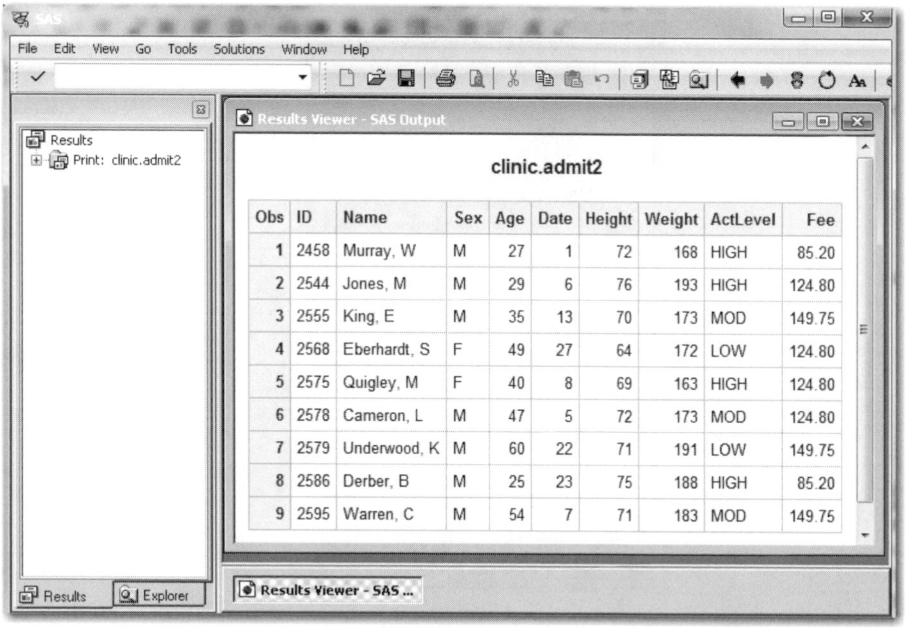

Creating SAS Libraries

Earlier in this chapter, you saw that SAS files are stored in libraries. By default, SAS defines several libraries for you (including Sashelp, Sasuser, and Work). You can also define additional libraries.

Sashelp
> a permanent library that contains sample data and other files that control how SAS works at your site. This is a read-only library.

Sasuser
> a permanent library that contains SAS files in the Profile catalog that store your personal settings. This is also a convenient place to store your own files.

Work
> a temporary library for files that do not need to be saved from session to session.

Figure 1.36 *Active SAS Libraries*

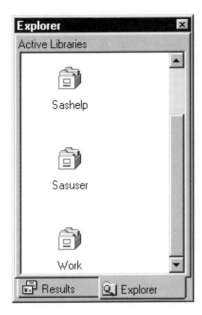

You can also *define* additional libraries. When you define a library, you indicate the location of your SAS files to SAS. Once you define a library, you can manage SAS files within it.

When you *delete* a SAS library, the pointer to the library is deleted, and SAS no longer has access to the library. However, the contents of the library still exist in your operating environment.

Defining Libraries

To define a library, you assign a library name to it and specify a path, such as a directory path. (In some operating environments you must create the directory or other storage location before defining the library.)

You can also specify an engine, which is a set of internal instructions that SAS uses for writing to and reading from files in a library.

Figure 1.37 *Defining LIbraries*

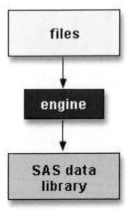

In this chapter, you learn about SAS libraries. You can define SAS libraries using programming statements. "Specifying Engines" on page 47 shows you how to write LIBNAME statements to define SAS libraries.

Depending on your operating environment and the SAS/ACCESS products that you license, you can create libraries with various engines. Each engine enables you to read a different file format, including file formats from other software vendors.

Creating and Using File Shortcuts

You've seen that the Explorer window gives you access to your SAS files. You can also create a file shortcut to an external file.

An external file is a file that is created and maintained in your host operating environment. External files contain data or text, such as

- SAS programming statements

- records of raw data

- procedure output.

SAS can use external files, but they are not managed by SAS.

A file shortcut (or *fileref*) is an optional name that is used to identify an external file to SAS. File shortcuts are stored in the File Shortcuts folder in the Explorer window. You can use a file shortcut to open, browse, and submit a file.

When you delete a file shortcut, the pointer to the file is deleted, and SAS no longer has access to the file. However, the file still exists in your operating environment.

Figure 1.38   *Creating and Using File Shortcuts*

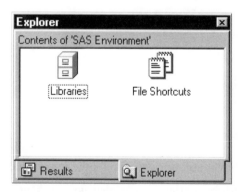

If you have used SAS before, a file shortcut is the same as a *file reference* or *fileref.*

Using SAS Solutions and Tools

Along with windows for working with your SAS files and SAS programs, SAS provides a set of ready-to-use solutions, applications, and tools. You can access many of these tools by using the Solutions and Tools menus.

Figure 1.39 *Using SAS Solutions and Tools*

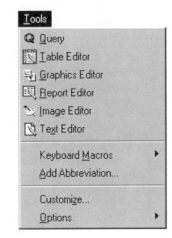

For example, you can use the table editor in the Tools menu to enter, browse, or edit data in a SAS data set.

Getting Help

You've learned to use SAS windows to perform common SAS tasks. As you begin working in SAS, be sure to take advantage of the different types of online help that are available from the Help menu.

Figure 1.40 *SAS Help*

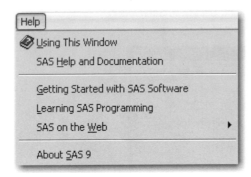

- **Using This Window** is task-oriented help for the active window.

- **SAS Help and Documentation** is a complete guide to syntax, examples, procedures, concepts, and what's new.

- **Getting Started tutorials** are listed under Help for products where they are available.

- Selecting **Learning SAS programming** enables you to create data that is used in online training courses, and displays SAS OnlineTutor if you have a site license.

- If you have Internet access, **SAS on the Web** provides links to information including Technical Support and Frequently Asked Questions.

To access SAS online help, documentation, and other resources from your SAS software, select **Help** ⇨ **SAS Documentation** from the SAS toolbar. You can also access SAS documentation in the SAS Knowledge Base at `support.sas.com`.

Chapter Summary

Text Summary

Components of SAS Programs

SAS programs consist of two types of steps: DATA steps and PROC (procedure) steps. These two steps, alone or combined, form most SAS programs. A SAS program can consist of a DATA step, a PROC step, or any combination of DATA and PROC steps. DATA steps typically create or modify SAS data sets, but they can also be used to produce custom-designed reports. PROC steps are pre-written routines that enable you to analyze and process the data in a SAS data set and to present the data in the form of a report. They sometimes create new SAS data sets that typically contain the results of the procedure.

Characteristics of SAS Programs

SAS programs consist of SAS statements. A SAS statement usually begins with a SAS keyword and always ends with a semicolon. A DATA step begins with the keyword DATA. A PROC step begins with the keyword PROC. SAS statements are free-format, so they can begin and end anywhere on a line. One statement can continue over several lines, and several statements can be on a line. Blanks or special characters separate "words" in a SAS statement.

Processing SAS Programs

When you submit a SAS program, SAS reads SAS statements and checks them for errors. When it encounters a subsequent DATA, PROC, or RUN statement, SAS executes the previous step in the program.

Each time a step is executed, SAS generates a log of the processing activities and the results of the processing. The SAS log collects messages about the processing of SAS programs and about any errors that occur.

The results of processing can vary. Some SAS programs open an interactive window or invoke procedures that create output in the form of a report. Other SAS programs perform tasks such as sorting and managing data, which have no visible results other than messages in the log.

SAS Libraries

Every SAS file is stored in a SAS library, which is a collection of SAS files such as SAS data sets and catalogs. In the Windows and UNIX environments, a SAS library is typically a group of SAS files in the same folder or directory.

Depending on the libref you use, you can store SAS files in a temporary SAS library or in permanent SAS libraries.

- Temporary SAS files that are created during the session are held in a special work space that is assigned the default libref Work. If you don't specify a libref when you create a file (or if you specify Work), the file is stored in the temporary SAS library. When you end the session, the temporary library is deleted.

- To store files permanently in a SAS library, you assign it a libref other than the default Work. For example, by assigning the libref sasuser to a SAS library, you specify that files within the library are to be stored until you delete them.

Referencing SAS Files

To reference a SAS file, you use a two-level name, *libref.filename*. In the two-level name, *libref* is the name for the SAS library that contains the file, and *filename* is the name of the file itself. A period separates the libref and filename.

To reference temporary SAS files, you specify the default libref Work, a period, and the filename. Alternatively, you can simply use a one-level name (the filename only) to reference a file in a temporary SAS library. Referencing a SAS file in any library *except Work* indicates that the SAS file is stored permanently.

SAS data set names can be 1 to 32 characters long, must begin with a letter (A-Z, either uppercase or lowercase) or an underscore (_), and can continue with any combination of numerals, letters, or underscores.

SAS Data Sets

For many of the data processing tasks that you perform with SAS, you access data in the form of a SAS data set and use SAS programs to analyze, manage, or present the data. Conceptually, a SAS data set is a file that consists of two parts: a descriptor portion and a data portion. Some SAS data sets also contain one or more indexes, which enable SAS to locate records in the data set more efficiently.

The descriptor portion of a SAS data set contains property information about the data set.

The data portion of a SAS data set is a collection of data values that are arranged in a rectangular table. Observations in the data set correspond to rows or data lines. Variables in the data set correspond to columns. If a data value is unknown for a particular observation, a missing value is recorded in the SAS data set.

Variable Attributes

In addition to general information about the data set, the descriptor portion contains property information for each variable in the data set. The property information includes the variable's name, type, and length. A variable's type determines how missing values for a variable are displayed by SAS. For character variables, a *blank* represents a missing value. For numeric variables, a *period* represents a missing value. You can also specify format, informat, and label properties for variables.

Using the Main SAS Windows

You use the following windows to explore and manage your files, to enter and submit SAS programs, to view messages, and to view and manage your output.

Table 1.12 *Windows and How They Are Used*

Use this window ...	To ...
Explorer	view your SAS files
	create new libraries and SAS files
	perform most file management tasks such as moving, copying, and deleting files
	create shortcuts to files that were not created with SAS

Use this window ...	To ...
Enhanced Editor (code editing window)	enter, edit, and submit SAS program The Enhanced Editor window is available only in the Windows operating environment.
Program Editor (code editing window)	enter, edit, and submit SAS programs
Log	view messages about your SAS session and about any SAS programs that you submit
Output	browse output from SAS programs
Results	navigate and manage output from SAS programs view, save, and print individual items of output

Points to Remember

- Before referencing SAS files, you must assign a name (libref, or library reference) to the library in which the files are stored (or specify that SAS is to assign the name automatically).

- You can store SAS files either temporarily or permanently.

- Variable names follow the same rules as SAS data set names. However, your site may choose to restrict variable names to those valid in Version 6 SAS, to uppercase variable names automatically, or to remove all restrictions on variable names.

Chapter Quiz

Select the best answer for each question. After completing the quiz, you can check your answers using the answer key in the appendix.

1. How many observations and variables does the data set below contain?

Name	Sex	Age
Picker	M	32
Fletcher		28
Romano	F	.
Choi	M	42

   a. 3 observations, 4 variables

   b. 3 observations, 3 variables

   c. 4 observations, 3 variables

   d. can't tell because some values are missing

2. How many program steps are executed when the program below is processed?

```
data user.tables;
    infile jobs;
    input date yyddmm8. name $ job $;
run;
proc sort data=user.tables;
    by name;
run;
proc print data=user.tables;
run;
```

   a. three

   b. four

   c. five

   d. six

3. What type of variable is the variable AcctNum in the data set below?

AcctNum	Gender
3456_1	M
2451_2	
Romano	F
Choi	M

   a. numeric

   b. character

   c. can be either character or numeric

   d. can't tell from the data shown

4. What type of variable is the variable Wear in the data set below?

Brand	Wear
Acme	43
Ajax	34
Atlas	.

   a. numeric

   b. character

   c. can be either character or numeric

   d. can't tell from the data shown

5. Which of the following variable names is valid?

   a. 4BirthDate

   b. $Cost

   c. _Items_

   d. Tax-Rate

6. Which of the following files is a permanent SAS file?

    a. Sashelp.PrdSale

    b. Sasuser.MySales

    c. Profits.Quarter1

    d. all of the above

7. In a DATA step, how can you reference a temporary SAS data set named Forecast?

    a. Forecast

    b. Work.Forecast

    c. Sales.Forecast (after assigning the libref Sales)

    d. only a and b above

8. What is the default length for the numeric variable Balance?

Name	Balance
Adams	105.73
Geller	107.89
Martinez	97.45
Noble	182.50

    a. 5

    b. 6

    c. 7

    d. 8

9. How many statements does the following SAS program contain?

    ```
    proc print data=new.prodsale
                   label double;
       var state day price1 price2; where state='NC';
       label state='Name of State'; run;
    ```

    a. three

    b. four

    c. five

    d. six

10. What is a SAS library?

    a. collection of SAS files, such as SAS data sets and catalogs

    b. in some operating environments, a physical collection of SAS files

    c. a group of SAS files in the same folder or directory

    d. all of the above

Chapter 2
Referencing Files and Setting Options

Overview

Introduction

When you begin a SAS session, it's often convenient to set up your environment first. For example, you may want to

- define libraries that contain the SAS data sets that you intend to use

- specify whether your procedure output is created as HTML (Hyper Text Markup Language) output, LISTING output, or another type of output.

- set features of your LISTING output, if you are creating any, such as whether the date and time appear

- specify how two-digit year values should be interpreted.

This chapter shows you how to define libraries, reference SAS files, and specify options for your SAS session. You also learn how to specify the form(s) of output to produce.

Figure 2.1 *SAS System Options Window*

Objectives

In this chapter, you learn to

- define new libraries by using programming statements

- reference SAS files to be used during your SAS session

- set results options to determine the type or types of output produced (HTML, LISTING output, or other) in desktop operating environments

- set system options to determine how date values are read and to control the appearance of any LISTING output that is created during your SAS session.

Defining Libraries

Often the first step in setting up your SAS session is to define the libraries. You can also use programming statements to assign library names.

Remember that to reference a permanent SAS file, you

1. assign a name (*libref*) to the SAS library in which the file is stored

2. use the libref as the first part of the file's two-level name (*libref.filename*) to reference the file within the library.

Figure 2.2 *Defining Libraries*

This example shows the two-level name for a SAS data set, **Admit**, which is stored in a SAS library to which the libref **Clinic** has been assigned.

LIBREF.FILENAME

Clinic.Admit

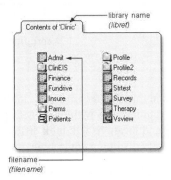

Assigning Librefs

To define libraries, you can use a LIBNAME statement. You can store the LIBNAME statement with any SAS program so that the SAS data library is assigned each time the program is submitted.

General form, basic LIBNAME statement:

LIBNAME *libref 'SAS-data-library'*;

where

- *libref* is 1 to 8 characters long, begins with a letter or underscore, and contains only letters, numbers, or underscores.
- *SAS-data-library* is the name of a SAS data library in which SAS data files are stored. The specification of the physical name of the library differs by operating environment.

The LIBNAME statement below assigns the libref Clinic to the SAS data library **D:\Users\Qtr\Reports** in the Windows environment.

```
libname clinic 'd:\users\qtr\reports';
```

Many of the examples in this book use the libref **sasuser**. The following LIBNAME statement assigns the libref sasuser to the c:\Users\name\sasuser folder in a Windows operating environment:

```
libname sasuser 'c:\Users\name\sasuser';
```

The table below gives some examples of physical names for SAS data libraries in various operating environments.

Table 2.1 *Physical Names for SAS Data Libraries*

Environment	Sample Physical Name
Windows	c:\fitness\data
UNIX	/users/april/fitness/sasdata
z/OS (OS/390)	april.fitness.sasdata

The code examples in this chapter are shown in the Windows operating environment. If you are running SAS within another operating environment, then the platform-specific names and locations will look different. Otherwise, SAS programming code will be the same across operating environments.

You can use multiple LIBNAME statements to assign as many librefs as needed.

Verifying Librefs

After assigning a libref, it is a good idea to check the Log window to verify that the libref has been assigned successfully.

Figure 2.3 *Log Output for Clinic libref*

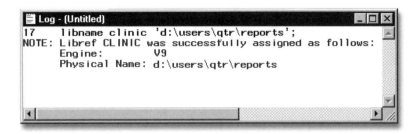

How Long Librefs Remain in Effect

The LIBNAME statement is global, which means that the librefs remain in effect until you modify them, cancel them, or end your SAS session.

Therefore, the LIBNAME statement assigns the libref for the current SAS session only. Each time you begin a SAS session, you must assign a libref to each permanent SAS data library that contains files that you want to access in that session. (Remember that Work is the default libref for a temporary SAS data library.)

Figure 2.4 *How Long Librefs Remain in Effect*

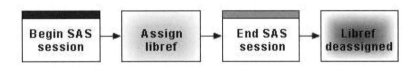

When you end your SAS session or delete a libref, SAS no longer has access to the files in the library. However, the contents of the library still exist on your operating system.

Remember that you can also assign a library from the SAS Explorer using the New Library window. Libraries that are created with the New Library window can be automatically assigned at startup by selecting **Enable at Startup**.

Specifying Two-Level Names

After you assign a libref, you specify it as the first element in the two-level name for a SAS file.

Figure 2.5 *Specifying Two-Level Names*

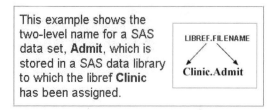

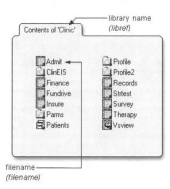

For example, in order for the PRINT procedure to read Clinic.Admit, you specify the two-level name of the file as follows:

```
proc print data=clinic.admit;
run;
```

Referencing Files in Other Formats

You can use the LIBNAME statement to reference not only SAS files but also files that were created with other software products, such as database management systems.

SAS can read or write these files by using the appropriate engine for that file type. For some file types, you need to tell SAS which engine to use. For others, SAS automatically chooses the appropriate engine.

A SAS engine is a set of internal instructions that SAS uses for writing to and reading from files in a SAS library.

Figure 2.6 *Referencing Files in Other Formats*

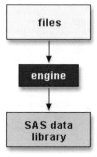

Specifying Engines

To indicate which engine to use, you specify the engine name in the LIBNAME statement, as shown below.

General form, LIBNAME statement for files in other formats:

LIBNAME *libref engine 'SAS-data-library'*;

where

- *libref* is 1 to 8 characters long, begins with a letter or underscore, and contains only letters, numbers, or underscores.
- *engine* is the name of a library engine that is supported in your operating environment.
- *SAS-data-library* is the name of a SAS library in which SAS data files are stored. The specification of the physical name of the library differs by operating environment.

Interface Library Engines

Interface library engines support read-only access to BMDP, OSIRIS, and SPSS files. With these engines, the physical filename that is associated with a libref is an actual filename, not a SAS library. This is an exception to the rules for librefs.

Table 2.2 *Engines and Their Descriptions*

Engine	Description
BMDP	allows read-only access to BMDP files
OSIRIS	allows read-only access to OSIRIS files
SPSS	allows read-only access to SPSS files

For example, the LIBNAME statement below specifies the libref Rptdata and the engine SPSS for the file **G:\Myspss.dat** in the Windows operating environment.

```
libname rptdata spss 'g:\myspss.spss';
```

For more information about interface library engines, see the SAS documentation for your operating environment.

SAS/ACCESS Engines

If your site licenses SAS/ACCESS software, you can use the LIBNAME statement to access data that is stored in a DBMS file. The types of data you can access depend on your operating environment and on which SAS/ACCESS products you have licensed.

Table 2.3 *Relational Databases and Their Associated Files*

Relational Databases	Nonrelational Files	PC Files
ORACLE	ADABAS	Excel (.xls)

Relational Databases	Nonrelational Files	PC Files
SYBASE	IMS/DL-I	Lotus (.wkn)
Informix	CA-IDMS	dBase
DB2 for z/OS DB2 for UNIX and PC	SYSTEM 2000	DIF
Oracle Rdb	Teradata	Access
ODBC	MySQL	SPSS
CA-OpenIngres	Netezza	Stata
	Ole DB	Paradox

Viewing SAS Libraries

Viewing the Contents of a SAS Library

You've seen that you can assign librefs in order to access different types of data. "Using the Programming Workspace" on page 20 explained that after you have assigned a libref, you can view

- details about the library that the libref references

- the library's contents

- contents and properties of files in the library.

The libraries that are currently defined for your SAS session are listed under Libraries in the Explorer window. To view details about a library, double-click **Libraries** (or select **Libraries** ⇨ **Open** from the pop-up menu). Then select **View** ⇨ **Details**.

Information for each library (name, engine, type, host pathname, and date modified) is listed under Active Libraries.

Figure 2.7 *Viewing Active Libraries*

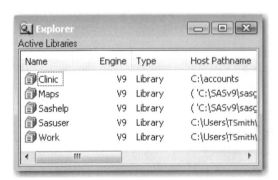

To view the contents of a library, double-click the library name (or select the library name, and then select **Open** from the pop-up menu). A list of the files contained in the library is displayed. If you have the details feature turned on, then information about each file (name, size, type, description, and date modified) is also listed.

Figure 2.8   *Viewing the Contents of a SAS Library*

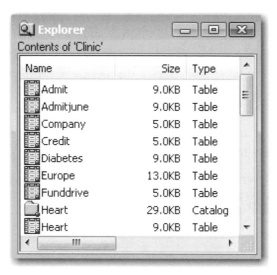

Viewing a File's Contents

To display a file's contents in a windowing environment, you can double-click the filename (or select the filename, and then select **Open** from the pop-up menu). If you select a SAS data set (type Table or View), its contents are displayed in the VIEWTABLE window.

Figure 2.9   *Viewing a File's Contents*

	ID	Name	Sex	Age
1	2458	Murray, W	M	27
2	2462	Almers, C	F	34
3	2501	Bonaventure, T	F	31
4	2523	Johnson, R	F	43
5	2539	LaMance, K	M	51
6	2544	Jones, M	M	29
7	2552	Reberson, P	F	32
8	2555	King, E	M	35
9	2563	Pitts, D	M	34
10	2568	Eberhardt, S	F	49
11	2571	Nunnelly, A	F	44
12	2572	Oberon, M	F	28
13	2574	Peterson, V	M	30
14	2575	Quigley, M	F	40
15	2578	Cameron, L	M	47

To display a file's properties, you can select the filename, and then select **Properties** from the pop-up menu.

If you are working in the z/OS operating environment, you can type **?** in the selection field next to a filename in the Explorer window to display a pop-up menu with a list of options for working with that file.

If you have installed SAS/FSP software, you can type **B** or **L** in the selection field next to a data set name to browse the data set observation by observation or to list the contents of the data set, respectively. For more information, see the documentation for SAS/FSP.

If SAS/FSP is not installed, you can view the contents of a SAS data set by using the PRINT procedure (PROC PRINT). You can learn how to use PROC PRINT in "Creating List Reports" on page 112.

Using PROC CONTENTS to View the Contents of a SAS Library

You've learned how to use SAS windows to view the contents of a SAS library or of a SAS file. Alternatively, you can use the CONTENTS procedure to create SAS output that describes either of the following:

- the contents of a library
- the descriptor information for an individual SAS data set.

General form, basic PROC CONTENTS step:

PROC CONTENTS DATA=*SAS-file-specification* **NODS;**

RUN;

where

- *SAS-file-specification* specifies an entire library or a specific SAS data set within a library. *SAS-file-specification* can take one of the following forms:
  - *<libref.>SAS-data-set* names one SAS data set to process.
  - *<libref.>_ALL_* requests a listing of all files in the library. (Use a period (.) to append _ALL_ to the libref.)
- **NODS** suppresses the printing of detailed information about each file when you specify _ALL_. (You can specify NODS *only* when you specify _ALL_.)

Examples

To view the contents of the entire clinic library, you can submit the following PROC CONTENTS step:

```
proc contents data=clinic._all_ nods;
run;
```

The output from this step lists only the names, types, sizes, and modification dates for the SAS files in the Clinic library.

Figure 2.10 *Using PROC CONTENTS to View the Contents of a Library*

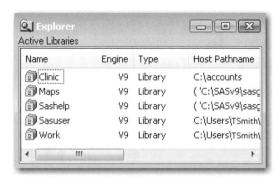

Figure 2.11 *Output from PROC CONTENTS on SAS Library Clinic*

#	Name	Member Type	File Size	Last Modified
1	ADMIT	DATA	9216	05Apr11:10:52:56
2	ADMIT2	DATA	9216	05Apr11:12:35:45
3	ADMITJUNE	DATA	9216	05Apr11:10:52:56
4	COMPANY	DATA	5120	05Apr11:10:52:56
5	CREDIT	DATA	5120	05Apr11:10:52:56
6	DIABETES	DATA	9216	05Apr11:10:52:57
7	EUROPE	DATA	13312	05Apr11:10:52:57

To view the descriptor information for only the clinic.Admit data set, you can submit the following PROC CONTENTS step:

```
proc contents data=clinic.admit;
run;
```

The output from this step lists information for clinic.admit, including an alphabetic list of the variables in the data set.

Figure 2.12 *PROC CONTENTS Output Showing the Descriptor Information for a Single Data Set in a Library*

Data Set Name	CLINIC.ADMIT	Observations	21
Member Type	DATA	Variables	9
Engine	V9	Indexes	0
Created	Tuesday, January 25, 2011 11:13:25 AM	Observation Length	64
Last Modified	Tuesday, January 25, 2011 11:13:25 AM	Deleted Observations	0
Protection		Compressed	NO
Data Set Type		Sorted	NO
Label			
Data Representation	WINDOWS_32		
Encoding	wlatin1 Western (Windows)		

Figure 2.13   PROC CONTENTS Output Showing the Engine/Host Dependent information for a Single Data Set in a Library

Engine/Host Dependent Information	
Data Set Page Size	8192
Number of Data Set Pages	1
First Data Page	1
Max Obs per Page	127
Obs in First Data Page	21
Number of Data Set Repairs	0
Filename	c:\clinic\admit.sas7bdat
Release Created	9.0301M0
Host Created	W32_7PRO

Figure 2.14   PROC CONTENTS Output Showing an Alphabetic List of the Variables in the Data Set

Alphabetic List of Variables and Attributes				
#	Variable	Type	Len	Format
8	ActLevel	Char	4	
4	Age	Num	8	
5	Date	Num	8	
9	Fee	Num	8	7.2
6	Height	Num	8	
1	ID	Char	4	
2	Name	Char	14	
3	Sex	Char	1	
7	Weight	Num	8	

Using PROC DATASETS

In addition to PROC CONTENTS, you can use PROC DATASETS to view the contents of a SAS library or a SAS data set. PROC DATASETS also enables you to perform a number of management tasks such as copying, deleting, or modifying SAS files.

PROC CONTENTS and PROC DATASETS overlap in terms of functionality. Generally, these two function the same:

- the CONTENTS procedure
- the CONTENTS statement in the DATASETS procedure.

PROC CONTENTS<*options*>;	PROC DATASETS<*options*>;
RUN;	CONTENTS<*options*>;
	QUIT;

The major difference between the CONTENTS procedure and the CONTENTS statement in PROC DATASETS is the default for *libref* in the DATA= option. For PROC CONTENTS, the default is either Work or User. For the CONTENTS statement, the default is the libref of the procedure input library. Notice also that PROC DATASETS supports RUN-group processing. It uses a QUIT statement to end the procedure. The QUIT statement and the RUN statement are not required.

However, the options for the PROC CONTENTS statement and the CONTENTS statement in the DATASETS procedure are the same. For example, the following PROC steps produce essentially the same output (with minor formatting differences):

```
proc datasets;
   contents data=clinic._all_ nods;


proc contents data=clinic._all_ nods;
run;
```

In addition to the CONTENTS statement, PROC DATASETS also uses several other statements. These statements enable you to perform tasks that PROC CONTENTS does not perform. For more information about PROC DATASETS, see the SAS documentation.

Viewing Descriptor Information for a SAS Data Set Using VARNUM

As with PROC CONTENTS, you can also use PROC DATASETS to display the descriptor information for a specific SAS data set.

By default, PROC CONTENTS and PROC DATASETS list variables *alphabetically*. To list variable names in the order of their *logical* position (or creation order) in the data set, you can specify the VARNUM option in PROC CONTENTS or in the CONTENTS statement in PROC DATASETS.

For example, either of these programs creates output that includes the list of variables shown below:

```
proc datasets;
   contents data=clinic.admit varnum;


proc contents data=clinic.admit varnum;
run;
```

Note: If you are using the sample data in the sasuser library, you may want to specify sasuser as the libref (instead of clinic):

```
contents data=sasuser.admit varnum;


proc contents data=sasuser.admit varnum;
run;
```

Figure 2.15 *Viewing Descriptor Information for a SAS Data Set Using VARNUM*

#	Variable	Type	Len	Format
	Variables in Creation Order			
1	ID	Char	4	
2	Name	Char	14	
3	Sex	Char	1	
4	Age	Num	8	
5	Date	Num	8	
6	Height	Num	8	
7	Weight	Num	8	
8	ActLevel	Char	4	
9	Fee	Num	8	7.2

Specifying Results Formats

Next, let's consider the appearance and format of your SAS output.

HTML and Listing Formats

In SAS 9.3 and later versions, when running SAS in windowing mode in the Windows and UNIX operating environments, HTML output is created by default. In other platforms, you can create HTML output using programming statements. When running SAS in batch mode, the default format is LISTING.

Let's look at these two result formats. The following PROC PRINT output is a listing of part of the SAS data set Clinic.Therapy:

Figure 2.16 *LISTING Output*

```
              The SAS System              11:02 Tuesday, April 12, 2011

                        Aer      Walk
     Obs     Date      Class    JogRun    Swim

       7    JUL2009     67       102       72
       9    SEP2009     78        77       54
      10    OCT2009     81        62       47
      11    NOV2009     84        31       52
      22    OCT2010     78        70       41
      23    NOV2010     82        44       58
      24    DEC2010     93        57       47
```

This is HTML output from the same program:

Figure 2.17 *HTML Output*

Obs	Date	AerClass	WalkJogRun	Swim
7	JUL2009	67	102	72
9	SEP2009	78	77	54
10	OCT2009	81	62	47
11	NOV2009	84	31	52
22	OCT2010	78	70	41
23	NOV2010	82	44	58
24	DEC2010	93	57	47

If you aren't running SAS in a desktop operating environment, you might want to skip this topic and go on to "Setting System Options" on page 58. For details on creating HTML output using programming statements on any SAS platform, see "Producing HTML Output " on page 280.

The Results Tab of the Preferences Window

You use the Preferences window to set the result format(s) that you prefer. Your preferences are saved until you modify them, and they apply to all output that you create.

To open this window in desktop operating environments, select **Tools** ⇨ **Options** ⇨ **Preferences**. Then click the **Results** tab. You can remember this sequence using the mnemonic TOPR (pronounced "topper"). You can choose **Listing**, **HTML**, or both. You can also specify options for displaying and storing your results.

Results tab options may differ somewhat, depending on your operating environment. The example below is from the Windows operating environment.

The following display shows the SAS **Results** tab with the new default settings specified.

Figure 2.18 *SAS Results Tab with the New Default Settings*

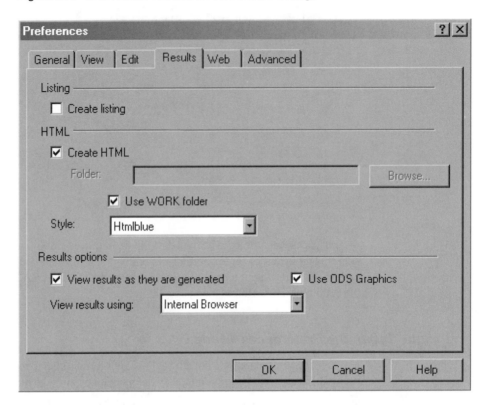

The default settings in the **Results** tab are as follows:

- The **Create listing** check box is not selected, so LISTING output is not created.

- The **Create HTML** check box is selected, so HTML output is created.

- The **Use WORK folder** check box is selected, so both HTML and graph image files are saved in the WORK folder (and not your current directory).

- The default style, HTMLBlue, is selected from the **Style** drop-down list.

- The **Use ODS Graphics** check box is selected, so ODS Graphics is enabled.

- **Internal browser** is selected from the **View results using:** drop-down list, so results are viewed in an internal SAS browser.

If you create HTML files, they are stored in the folder that you specify and are by default incrementally named sashtml.htm, sashtml1.htm, sashtml2.htm, and so on, throughout your SAS session.

Now look at the two choices for viewing HTML results: Internal browser and Preferred web browser. (These options appear in the Results tab only in the Windows operating environment.)

Internal Browser vs. Preferred Web Browser

To view HTML output, you can choose between these two options in the Results tab of the Preferences window in the Windows operating environment:

- the Internal browser, called the Results Viewer window. SAS provides this browser as part of your SAS installation.

- the Preferred web browser. If you select this option, SAS uses the browser that is specified in the Web tab of the Preferences window. By default, this is the default browser for your PC.

Internal Browser

The **Results Viewer** is displayed as a SAS window, as shown below.

Figure 2.19   *Internal Browser Showing HTML Output*

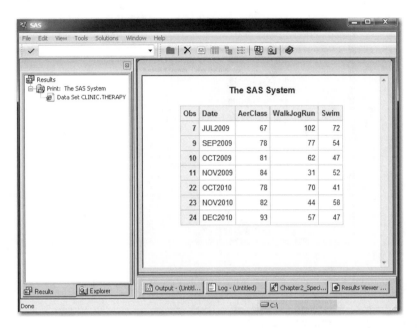

Preferred Web Browser

If you select the preferred web browser, your HTML output is displayed in a separate browser that is independent of SAS. For example, the HTML output below is displayed in Internet Explorer.

Figure 2.20 *Preferred Web Browser Showing HTML Output*

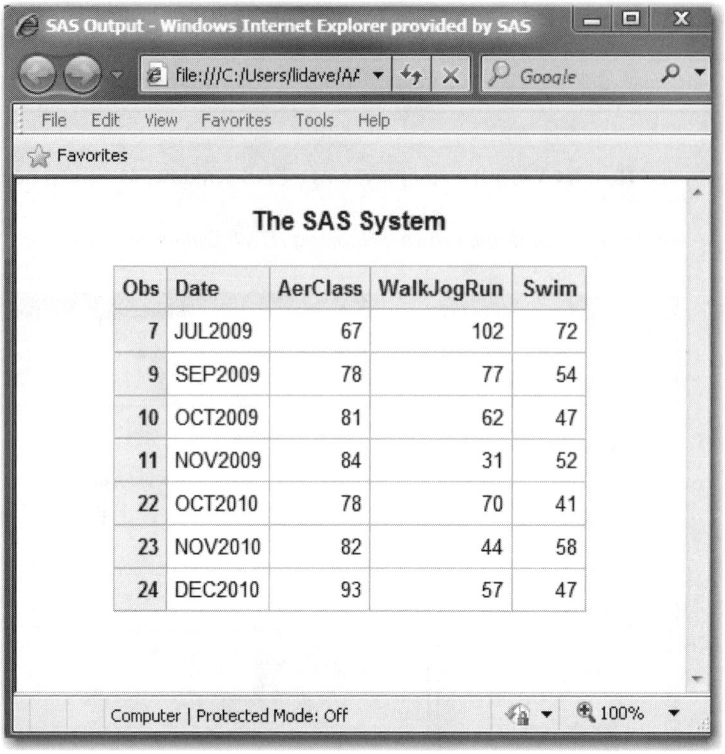

Obs	Date	AerClass	WalkJogRun	Swim
7	JUL2009	67	102	72
9	SEP2009	78	77	54
10	OCT2009	81	62	47
11	NOV2009	84	31	52
22	OCT2010	78	70	41
23	NOV2010	82	44	58
24	DEC2010	93	57	47

Setting System Options

Overview

If you create your procedure output as LISTING output, you can also control the appearance of your output by setting system options such as

- line size (the maximum width of the log and output)
- page size (the number of lines per printed page of output)
- the display of page numbers
- the display of date and time.

The above options do not affect the appearance of HTML output.

All SAS system options have default settings that are used unless you specify otherwise. For example, page numbers are automatically displayed in LISTING output (unless your site modifies this default).

Figure 2.21 *LISTING Output Showing Default Settings*

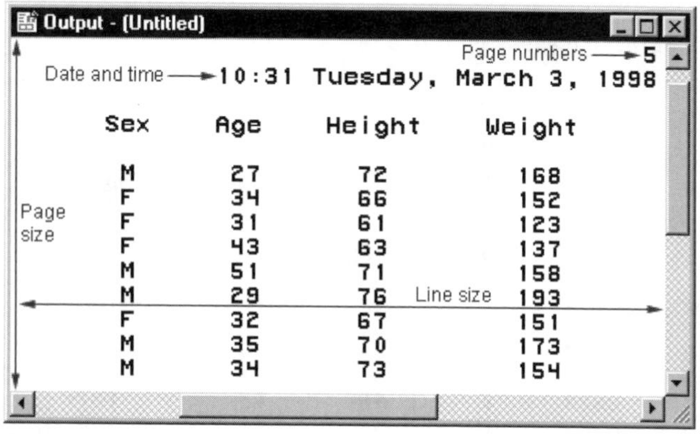

Sex	Age	Height	Weight
M	27	72	168
F	34	66	152
F	31	61	123
F	43	63	137
M	51	71	158
M	29	76	193
F	32	67	151
M	35	70	173
M	34	73	154

Changing System Options

To modify system options in your LISTING output, you submit an OPTIONS statement. You can place an OPTIONS statement anywhere in a SAS program to change the settings from that point onward. However, it is good programming practice to place OPTIONS statements outside of DATA or PROC steps so that your programs are easier to read and debug.

Because the OPTIONS statement is global, the settings remain in effect until you modify them, or until you end your SAS session.

General form, OPTIONS statement:

OPTIONS *options***;**

where *options* specifies one or more system options to be changed. The available system options depend on your host operating system.

Example: DATE | NODATE and NUMBER | NONUMBER Options

By default, page numbers and dates appear with LISTING output. The following OPTIONS statement suppresses the printing of both page numbers and the date and time in LISTING output.

```
options nonumber nodate;
```

In the following example, the NONUMBER and NODATE system options suppress the display of page numbers and the date in the PROC PRINT output. Page numbers are not displayed in the PROC FREQ output, either, but the date does appear at the top of the PROC FREQ output since the DATE option was specified.

```
ods listing;
options nonumber nodate;
proc print data=sasuser.admit;
   var id sex age height weight;
   where age>=30;
run;
options date;
proc freq data=sasuser.diabetes;
```

```
   where fastgluc>=300;
   tables sex;
run;
proc print data=sasuser.diabetes;
run;
ods listing close;
```

Figure 2.22 *PROC PRINT LISTING Output for sasuser.diabetes*

```
                   The SAS System

   Obs      ID     Sex     Age     Height     Weight

     2     2462     F      34       66         152
     3     2501     F      31       61         123
     4     2523     F      43       63         137
     5     2539     M      51       71         158
     7     2552     F      32       67         151
     8     2555     M      35       70         173
     9     2563     M      34       73         154
    10     2568     F      49       64         172
    11     2571     F      44       66         140
    13     2574     M      30       69         147
    14     2575     F      40       69         163
    15     2578     M      47       72         173
    16     2579     M      60       71         191
    17     2584     F      43       65         123
    20     2589     F      41       67         141
    21     2595     M      54       71         183
```

Figure 2.23 *PROC FREQ LISTING Output for sasuser.diabetes*

```
                 The SAS System          11:11 Saturday, May 14, 2011

                 The FREQ Procedure

                                    Cumulative     Cumulative
   Sex     Frequency     Percent    Frequency      Percent

   F           2         25.00          2           25.00
   M           6         75.00          8          100.00
```

Figure 2.24 *PROC PRINT LISTING Output for sasuser.diabetes*

```
                           The SAS System          11:03 Monday, May 16, 2011

                                                         Fast     Post
   Obs     ID     Sex   Age   Height   Weight   Pulse    Gluc     Gluc

     1    2304     F     16     61      102      100      568      625
     2    1128     M     43     71      218       76      156      208
     3    4425     F     48     66      162       80      244      322
     4    1387     F     57     64      142       70      177      206
     5    9012     F     39     63      157       68      257      318
     6    6312     M     52     72      240       77      362      413
     7    5438     F     42     62      168       83      247      304
     8    3788     M     38     73      234       71      486      544
     9    9125     F     56     64      159       70      166      215
    10    3438     M     15     66      140       67      492      547
    11    1274     F     50     65      153       70      193      271
    12    3347     M     53     70      193       78      271      313
    13    2486     F     63     65      157       70      152      224
    14    1129     F     48     61      137       69      267      319
    15    9723     M     52     75      219       65      348      403
    16    8653     M     49     68      185       79      259      311
    17    4451     M     54     71      196       81      373      431
    18    3279     M     40     70      213       82      447      504
    19    4759     F     60     68      164       71      155      215
    20    6488     F     59     64      154       75      362      409
```

Figure 2.25 *PROC FREQ LISTING Output with Date and Still No Page Number*

```
                 The SAS System          13:31 Monday, April 18, 2011

                 The FREQ Procedure

                                    Cumulative     Cumulative
   Sex     Frequency     Percent    Frequency      Percent

   F           2         25.00          2           25.00
   M           6         75.00          8          100.00
```

Example: PAGENO= Option

If you print page numbers, you can specify the beginning page number for your LISTING report by using the PAGENO= option. If you don't specify PAGENO=, output is numbered sequentially throughout your SAS session, starting with page 1.

In the following example, the output pages are numbered sequentially throughout the SAS session, beginning with number 3.

```
ods listing;
options nodate number pageno=3;
proc print data=hrd.funddrive;
run;
ods listing close;
```

Since SAS 9.3 does not creating LISTING output by default, the ODS LISTING statement was added to generate LISTING output.

Figure 2.26 *LISTING Output with the PAGENO= Option Set*

```
                    The SAS System                                 3

     Obs    LastName     Qtr1    Qtr2    Qtr3    Qtr4

       1    ADAMS          18      18      20      20
       2    ALEXANDER      15      18      15      10
       3    APPLE          25      25      25      25
       4    ARTHUR         10      25      20      30
       5    AVERY          15      15      15      15
       6    BAREFOOT       20      20      20      20
       7    BAUCOM         25      20      20      30
       8    BLAIR          10      10       5      10
       9    BLALOCK         5      10      10      15
      10    BOSTIC         20      25      30      25
      11    BRADLEY        12      16      14      18
      12    BRADY          20      20      20      20
      13    BROWN          18      18      18      18
      14    BRYANT         16      18      20      18
      15    BURNETTE       10      10      10      10
      16    CHEUNG         30      30      30      30
      17    LEHMAN         20      20      20      20
      18    VALADEZ        14      18      40      25
```

Example: PAGESIZE= Option

The PAGESIZE= option (alias PS=) specifies how many lines each page of output contains. In the following example, each page of the output that the PRINT procedure produces contains 15 lines (including those used by the title, date, and so on).

```
options number date pagesize=15;
proc print data=sasuser.admit;
run;
```

Figure 2.27 *LISTING Output with PAGESIZE= 15*

```
                    The SAS System           13:23 Tuesday, April 12, 2011    17

     Obs    LastName     Qtr1    Qtr2    Qtr3    Qtr4

      12    BRADY          20      20      20      20
      13    BROWN          18      18      18      18
      14    BRYANT         16      18      20      18
      15    BURNETTE       10      10      10      10
      16    CHEUNG         30      30      30      30
      17    LEHMAN         20      20      20      20
      18    VALADEZ        14      18      40      25
```

Example: LINESIZE= Option

The LINESIZE= option (alias LS=) specifies the width of the print line for your procedure output and log. Observations that do not fit within the line size continue on a different line.

In the following example, the observations are longer than 64 characters, so the observations continue on a subsequent page.

```
ODS listing;
options number linesize=64;
proc print data=flights.europe;
run;
ODS listing close;
```

Figure 2.28 LISTING Output with the LINESIZE= Option Set, Page 1

```
                         The SAS System                    55
                              13:23 Tuesday, April 12, 2011

                                          Day
                         Non                Of
  Obs  Transferred    Revenue   Deplaned   Capacity  Month   Revenue

   1        17           7         222        250       1     150634
   2         8           3         163        250       1     156804
   3        15           5         227        250       1     190098
   4        13           4         222        250       1     150634
   5        14           6         158        250       1     193930
   6        18           7         172        250       2     166470
   7         8           1         114        250       2     167772
   8         7           4         187        250       2     163248
```

Figure 2.29 LISTING Output with the LINESIZE= Option Set, Page 2

```
                         The SAS System                    56
                              13:23 Tuesday, April 12, 2011



  Obs  Flight    Date  Depart  Orig  Dest  Miles  Mail  Freight  Boarded

   9     821   09MAR99  14:56   LGA   LON   3442   219    368      203
  10     271   09MAR99  13:17   LGA   PAR   3635   357    282      159
  11     821   10MAR99   9:31   LGA   LON   3442   389    479      188
  12     271   10MAR99  11:40   LGA   PAR   3856   415    463      182
  13     622   03MAR99  12:19   LGA   FRA   3857   296    414      180
  14     821   03MAR99  14:56   LGA   LON   3442   448    282      151
  15     271   03MAR99  13:17   LGA   PAR   3635   352    351      147
  16     219   04MAR99   9:31   LGA   LON   3442   331    376      232
```

Handling Two-Digit Years: Year 2000 Compliance

If you use two-digit year values in your data lines, external files, or programming statements, you should consider another important system option, the YEARCUTOFF= option. This option specifies which 100-year span is used to interpret two-digit year values.

Figure 2.30 *The Default 100 Year Span in SAS*

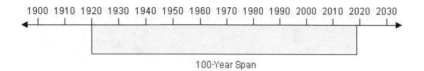

100-Year Span

All versions of SAS represent dates correctly from 1582 A.D. to 20,000 A.D. (Leap years, century, and fourth-century adjustments are made automatically. Leap seconds are ignored, and SAS does not adjust for Daylight Savings Time.) However, you should be aware of the YEARCUTOFF= value to ensure that you are properly interpreting two-digit years in data lines.

As with other system options, you specify the YEARCUTOFF= option in the OPTIONS statement:

```
options yearcutoff=1925;
```

How the YEARCUTOFF= Option Works

When a two-digit year value is read, SAS interprets it based on a 100-year span which starts with the YEARCUTOFF= value. The default value of YEARCUTOFF= is `1920`.

Figure 2.31 *Default YEARCUTOFF= Date (1920)*

1920 ◄—100 years—► 2019

Table 2.4 *Date Expressions and How They Are Interpreted*

Date Expression	Interpreted As
12/07/41	12/07/1941
18Dec15	18Dec2015
04/15/30	04/15/1930
15Apr95	15Apr1995

However, you can override the default and change the value of YEARCUTOFF= to the first year of another 100-year span. For example, if you specify YEARCUTOFF=1950, then the 100-year span will be from 1950 to 2049.

```
options yearcutoff=1950;
```

Using YEARCUTOFF=1950, dates are interpreted as shown below:

Figure 2.32 *Interpreting Dates When YEARCUTOFF=1950*

1950 ◄—100 years—► 2049

Table 2.5 *Date Expressions and How They Are Interpreted*

Date Expression	Interpreted As
12/07/41	12/07/2041
18Dec15	18Dec2015
04/15/30	04/15/2030
15Apr95	15Apr1995

How Four-Digit Years Are Handled

Remember, the value of the YEARCUTOFF= system option affects only two-digit year values. A date value that contains a four-digit year value will be interpreted correctly even if it does not fall within the 100-year span set by the YEARCUTOFF= system option.

You can learn more about reading date values in "Reading Date and Time Values" on page 571.

Using System Options to Specify Observations

You've seen how to use SAS system options to change the appearance of output and interpret two-digit year values. You can also use the FIRSTOBS= and OBS= system options to specify the observations to process from SAS data sets.

You can specify either or both of these options as needed. That is, you can use

- FIRSTOBS= to start processing at a specific observation

- OBS= to stop processing after a specific observation

- FIRSTOBS= and OBS= together to process a specific group of observations.

General form, FIRSTOBS= and OBS= options in an OPTIONS statement:

FIRSTOBS=*n*

OBS=*n*

where *n* is a positive integer. For FIRSTOBS=, *n* specifies the number of the *first* observation to process. For OBS=, *n* specifies the number of the *last* observation to process. By default, FIRSTOBS=1. The default value for OBS= is MAX, which is the largest signed, eight-byte integer that is representable in your operating environment. The number can vary depending on your operating system.

⚠ Each of these options applies to every input data set that is used in a program or a SAS process.

Examples: FIRSTOBS= and OBS= Options

The data set clinic.heart contains 20 observations. If you specify FIRSTOBS=10, SAS reads the 10th observation of the data set first and reads through the last observation (for a total of 11 observations).

```
options firstobs=10;
proc print data=sasuser.heart;
run;
```

The PROC PRINT step produces the following output:

Figure 2.33 PROC PRINT Output with FIRSTOBS=10

The SAS System

Obs	Patient	Sex	Survive	Shock	Arterial	Heart	Cardiac	Urinary
10	509	2	SURV	OTHER	79	84	256	90
11	742	1	DIED	HYPOVOL	100	54	135	0
12	609	2	DIED	NONSHOCK	93	101	260	90
13	318	2	DIED	OTHER	72	81	410	405
14	412	1	SURV	BACTER	61	87	296	44
15	601	1	DIED	BACTER	84	101	260	377
16	402	1	SURV	CARDIO	88	137	312	75
17	98	2	SURV	CARDIO	84	87	260	377
18	4	1	SURV	HYPOVOL	81	149	406	200
19	50	2	SURV	HYPOVOL	72	111	332	12
20	2	2	DIED	OTHER	101	114	424	97

If you specify OBS=10 instead, SAS reads through the 10th observation, in this case for a total of 10 observations. (Notice that FIRSTOBS= has been reset to the default value.)

```
options firstobs=1 obs=10;
proc print data=sasuser.heart;
run;
```

Now the PROC PRINT step produces this output:

Figure 2.34 *PROC PRINT Output with FIRSTOBS=1 and Obs=10*

The SAS System

Obs	Patient	Sex	Survive	Shock	Arterial	Heart	Cardiac	Urinary
1	203	1	SURV	NONSHOCK	88	95	66	110
2	54	1	DIED	HYPOVOL	83	183	95	0
3	664	2	SURV	CARDIO	72	111	332	12
4	210	2	DIED	BACTER	74	97	369	0
5	101	2	DIED	NEURO	80	130	291	0
6	102	2	SURV	OTHER	87	107	471	65
7	529	1	DIED	CARDIO	103	106	217	15
8	524	2	DIED	CARDIO	145	99	156	10
9	426	1	SURV	OTHER	68	77	410	75
10	509	2	SURV	OTHER	79	84	256	90

Combining FIRSTOBS= and OBS= processes observations in the middle of the data set. For example, the following program processes only observations 10 through 15, for a total of 6 observations:

```
options firstobs=10 obs=15;
proc print data=sasuser.heart;
run;
```

Here is the output:

Figure 2.35 *PROC PRINT Output with FIRSTOBS=10 and Obs=15*

The SAS System

Obs	Patient	Sex	Survive	Shock	Arterial	Heart	Cardiac	Urinary
10	509	2	SURV	OTHER	79	84	256	90
11	742	1	DIED	HYPOVOL	100	54	135	0
12	609	2	DIED	NONSHOCK	93	101	260	90
13	318	2	DIED	OTHER	72	81	410	405
14	412	1	SURV	BACTER	61	87	296	44
15	601	1	DIED	BACTER	84	101	260	377

To reset the number of the last observation to process, you can specify OBS=MAX in the OPTIONS statement.

```
options obs=max;
```

This instructs any subsequent SAS programs in the SAS session to process through the last observation in the data set being read.

Using FIRSTOBS= and OBS= for Specific Data Sets

As you saw above, using the FIRSTOBS= or OBS= system options determines the first or last observation, respectively, that is read for all steps for the duration of your current SAS session or until you change the setting. However, you may want to

- override these options for a given data set

- apply these options to a specific data set only.

To affect any single file, you can use FIRSTOBS= or OBS= as data set options instead of as system options. You specify the data set option in parentheses immediately following the input data set name.

A FIRSTOBS= or OBS= specification from a data set option overrides the corresponding FIRSTOBS= or OBS= system option.

Example: FIRSTOBS= and OBS= as Data Set Options

As shown in the last example below, this program processes only observations 10 through 15, for a total of 6 observations:

```
options firstobs=10 obs=15;
proc print data=sasuser.heart;
run;
```

You can create the same output by specifying FIRSTOBS= and OBS= as data set options, as follows. The data set options override the system options for this instance only.

```
options firstobs=10 obs=15;
proc print data=sasuser.heart(firstobs=10 obs=15);
run;
```

To specify FIRSTOBS= or OBS= for this program only, you could omit the OPTIONS statement altogether and simply use the data set options.

The SAS System Options Window

You can also set system options by using the SAS System Options window. The changed options are reset to the defaults at the end of your SAS session.

To view the SAS System Options window, select **Tools** ⇨ **Options** ⇨ **System**.

Figure 2.36 *The SAS System Options Window*

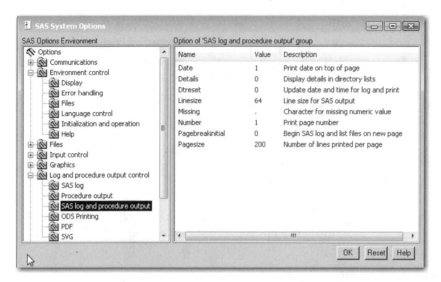

Changing Options

To change an option:

1. Expand the groups and subgroups under SAS Options Environment until you find the option that you want to change. (Options in subgroups are listed in the right pane of the window.)

2. Click the name of the option that you want to change, and display its pop-up menu. Then select one of the choices:

   - **Modify Value** opens a window in which you type or select a new value for the option.

   - **Set to Default** immediately resets the option to its default value.

For example, the SAS System Options window above shows options for the **SAS log and procedure output** subgroup under the group **Log and procedure output control**.

Finding Options Quickly

To locate an option in the SAS System Options window:

1. Place your cursor over the name of any option group or subgroup, and display its pop-up menu.

2. Click **Find Option**. The Find Option dialog box appears.

3. Type the name of the option that you want to locate, and click **OK**.

The SAS System Options window expands to the appropriate option subgroup. All subgroup options also appear, and the option that you located is highlighted.

Additional Features

When you set up your SAS session, you can set SAS system options that affect LISTING output, information written to the SAS log, and much more. Here are some additional system options that you are likely to use with SAS:

Table 2.6 *Selected System Options and Their Descriptions*

FORMCHAR= *'formatting-characters'*	specifies the formatting characters for your output device. Formatting characters are used to construct the outlines of tables as well as dividers for various procedures, such as the FREQ and TABULATE procedures. If you do not specify formatting characters as an option in the procedure, then the default specifications given in the FORMCHAR= system option are used.
FORMDLIM= *'delimit-character'*	specifies a character that is used to delimit page breaks in SAS System output. Normally, the delimit character is null. When the delimit character is null, a new physical page starts whenever a page break occurs.
LABEL|NOLABEL	permits SAS procedures to temporarily replace variable names with descriptive labels. The LABEL system option must be in effect before the LABEL option of any procedure can be used. If NOLABEL is specified, then the LABEL option of a procedure is ignored. The default setting is LABEL.
REPLACE|NOREPLACE	specifies whether permanently stored SAS data sets are replaced. If you specify NOREPLACE, a permanently stored SAS data set cannot be replaced with one that has the same name. This prevents you from inadvertently replacing existing SAS data sets. The default setting is REPLACE.
SOURCE|NOSOURCE	controls whether SAS source statements are written to the SAS log. NOSOURCE specifies not to write SAS source statements to the SAS log. The default setting is SOURCE.

You can also use programming statements to control the result format of each item of procedure output individually. For more information, see "Producing HTML Output" on page 280.

Chapter Summary

Text Summary

Referencing SAS Files in SAS Libraries

To reference a SAS file, you assign a libref (library reference) to the SAS library in which the file is stored. Then you use the libref as the first part of the two-level name (*libref.filename*) for the file. To assign a libref, you can submit a LIBNAME statement. You can store the LIBNAME statement with any SAS program to assign the libref automatically when you submit the program. The LIBNAME statement assigns the libref for the current SAS session only. You must assign a libref each time you begin a SAS session in order to access SAS files that are stored in a permanent SAS library other than clinic. (Work is the default libref for a temporary SAS library.)

You can also use the LIBNAME statement to reference data in files that were created with other software products, such as database management systems. SAS can write to or read from the files by using the appropriate engine for that file type. For some file types, you need to tell SAS which engine to use. For others, SAS automatically chooses the appropriate engine.

Viewing Librefs

The librefs that are in effect for your SAS session are listed under Libraries in the Explorer window. To view details about a library, double-click **Libraries** (or select **Libraries** ⇨ **Open** from the pop-up menu). Then select **View** ⇨ **Details**. The library's name, engine, host pathname, and date are listed under Active Libraries.

Viewing the Contents of a Library

To view the contents of a library, double-click the library name in the Explorer window (or select the library name and then select **Open** from the pop-up menu). Files contained in the library are listed under Contents.

Viewing the Contents of a File

If you are working in a windowing environment, you can display the contents of a file by double-clicking the filename (or selecting the filename and then selecting **Open** from the pop-up menu) under Contents in the Explorer window. If you select a SAS data set, its contents are displayed in the VIEWTABLE window.

If you are working in the z/OS operating environment, you can type **?** in the selection field next to a filename in the Explorer window to display a pop-up menu with a list of options for working with that file.

Listing the Contents of a Library

To list the contents of a library, use the CONTENTS procedure. Append a period and the _ALL_ option to the libref to get a listing of all files in the library. Add the NODS option to suppress detailed information about the files. As an alternative to PROC CONTENTS, you can use PROC DATASETS.

Specifying Result Formats

In desktop operating environments, you can choose to create your SAS procedure output as an HTML document, a listing (traditional SAS output), or both. You choose the results format(s) that you prefer in the Preferences window. Your preferences are saved until you modify them, and they apply to all output that is created during your SAS session. To open this window, select **Tools** ⇨ **Options** ⇨ **Preferences**. Then click the **Results** tab. Choose **Create listing**, **Create HTML**, or both.

If you choose **Create HTML**, then each HTML file is displayed in the browser that you specify (in the Windows operating environment, the internal browser is the Results Viewer window). HTML files are stored in the location that you specify and are by default incrementally named sashtml.htm, sashtml1.htm, sashtml2.htm, and so on throughout your SAS session. To specify where HTML files are stored, type a path in the **Folder** box (or click **Browse** to locate a pathname). If you prefer to store your HTML files temporarily and to delete them at the end of your SAS session, click **Use WORK** folder instead of specifying a folder. To specify the presentation style for HTML output, you can select an item in the **Style** box.

Setting System Options

For your LISTING output, you can also control the appearance of your output by setting system options such as line size, page size, the display of page numbers, and the display of the date and time. (These options do not affect the appearance of HTML output.)

All SAS system options have default settings that are used unless you specify otherwise. For example, page numbers are automatically displayed (unless your site modifies this default). To modify system options, you submit an OPTIONS statement. You can place an OPTIONS statement anywhere in a SAS program to change the current settings. Because the OPTIONS statement is global, the settings remain in effect until you modify them or until you end your SAS session.

If you use two-digit year values in your SAS data lines, you must be aware of the YEARCUTOFF= option to ensure that you are properly interpreting two-digit years in your SAS program. This option specifies which 100-year span is used to interpret two-digit year values.

To specify the observations to process from SAS data sets, you can use the FIRSTOBS= and OBS= options.

You can also use the SAS System Options window to set system options.

Additional Features

You can set a number of additional SAS system options that are commonly used.

Syntax

LIBNAME *libref* '*SAS-data-library*';

LIBNAME *libref engine* '*SAS-data-library*';

OPTIONS *options*;

PROC CONTENTS DATA= *libref.*_**ALL**_ **NODS;**

PROC DATASETS;

      CONTENTS DATA= *libref.*_**ALL**_ **NODS;**

QUIT;

Points to Remember

- LIBNAME and OPTIONS statements remain in effect for the current SAS session only.

- When you work with date values, check the default value of the YEARCUTOFF= system option and change it if necessary.

Chapter Quiz

Select the best answer for each question. After completing the quiz, check your answers using the answer key in the appendix.

1. If you submit the following program, how does the output look?

```
options pagesize=55 nonumber;
proc tabulate data=clinic.admit;
   class actlevel;
   var age height weight;
   table actlevel,(age height weight)*mean;
run;
options linesize=80;
proc means data=clinic.heart min max maxdec=1;
   var arterial heart cardiac urinary;
   class survive sex;
run;
```

   a. The PROC MEANS output has a print line width of 80 characters, but the PROC TABULATE output has no print line width.

   b. The PROC TABULATE output has no page numbers, but the PROC MEANS output has page numbers.

   c. Each page of output from both PROC steps is 55 lines long and has no page numbers, and the PROC MEANS output has a print line width of 80 characters.

   d. The date does not appear on output from either PROC step.

2. How can you create SAS output in HTML format on any SAS platform?

   a. by specifying system options

   b. by using programming statements

   c. by using SAS windows to specify the result format

   d. you can't create HTML output on all SAS platforms

3. In order for the date values **05May1955** and **04Mar2046** to be read correctly, what value must the YEARCUTOFF= option have?

   a. a value between 1947 and 1954, inclusive

   b. 1955 or higher

   c. 1946 or higher

   d. any value

4. When you specify an engine for a library, you are always specifying

   a. the file format for files that are stored in the library.

   b. the version of SAS that you are using.

   c. access to other software vendors' files.

   d. instructions for creating temporary SAS files.

5. Which statement prints a summary of all the files stored in the library named Area51?

   a. `proc contents data=area51._all_ nods;`

   b. `proc contents data=area51 _all_ nods;`

   c. `proc contents data=area51 _all_ noobs;`

   d. `proc contents data=area51 _all_.nods;`

6. The following PROC PRINT output was created immediately after PROC TABULATE output. Which system options were specified when the report was created?

   ```
                                                          1
                          10:03 Friday, March 17, 2000

                                       Act
          Obs  ID  Height Weight Level     Fee

           1  2458   72    168    HIGH    85.20
           2  2462   66    152    HIGH   124.80
           3  2501   61    123    LOW    149.75
           4  2523   63    137    MOD    149.75
           5  2539   71    158    LOW    124.80
           6  2544   76    193    HIGH   124.80
           7  2552   67    151    MOD    149.75
           8  2555   70    173    MOD    149.75
           9  2563   73    154    LOW    124.80
   ```

   a. OBS=, DATE, and NONUMBER

   b. NUMBER, PAGENO=1, and DATE

   c. NUMBER and DATE only

   d. none of the above

7. Which of the following programs correctly references a SAS data set named *SalesAnalysis* that is stored in a permanent SAS library?

   a.
   ```
   data saleslibrary.salesanalysis;
       set mydata.quarter1sales;
       if sales>100000;
   run;
   ```

   b.
   ```
   data mysales.totals;
       set sales_99.salesanalysis;
       if totalsales>50000;
   run;
   ```

   c.
   ```
   proc print data=salesanalysis.quarter1;
       var sales salesrep month;
   run;
   ```

   d.
   ```
   proc freq data=1999data.salesanalysis;
       tables quarter*sales;
   run;
   ```

8. Which time span is used to interpret two-digit year values if the YEARCUTOFF= option is set to 1950?

   a. 1950-2049

    b. 1950-2050

    c. 1949-2050

    d. 1950-2000

9. Assuming you are using SAS code and not special SAS windows, which one of the following statements is *false*?

    a. LIBNAME statements can be stored with a SAS program to reference the SAS library automatically when you submit the program.

    b. When you delete a libref, SAS no longer has access to the files in the library. However, the contents of the library still exist on your operating system.

    c. Librefs can last from one SAS session to another.

    d. You can access files that were created with other vendors' software by submitting a LIBNAME statement.

10. What does the following statement do?

```
libname osiris spss 'c:\myfiles\sasdata\data.spss';
```

    a. defines a library called Spss using the OSIRIS engine

    b. defines a library called Osiris using the SPSS engine

    c. defines two libraries called Osiris and Spss using the default engine

    d. defines the default library using the OSIRIS and SPSS engines

Chapter 3
Editing and Debugging SAS Programs

Overview

Introduction

Now that you're familiar with the basics, you can learn how to use the SAS programming windows to edit and debug programs effectively.

Figure 3.1   *SAS Programming Windows*

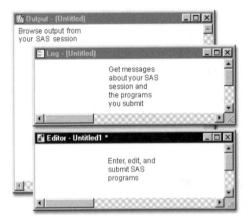

Objectives

In this chapter you learn to

* open a stored SAS program

* edit SAS programs

- clear SAS programming windows

- interpret error messages in the SAS log

- correct errors

- resolve common problems.

Opening a Stored SAS Program

Overview

You can open a stored program in the code editing window, where you can edit the program and submit it again.

You can open a program using

- file shortcuts

- My Favorite Folders

- the Open window

- the INCLUDE command.

In operating environments that provide drag-and-drop functionality, you can also open a SAS file by dragging and dropping it into the code editing window.

Figure 3.2 *Code Editor Window*

Using File Shortcuts

File shortcuts are available in the SAS Explorer window. To open a program using file shortcuts:

1. At the top level of the SAS Explorer window, double-click **File Shortcuts**.

2. Double-click the file shortcut that you want to open, or select **Open** from the pop-up menu for the file.

Figure 3.3   *File Short Cuts*

The file opens in a new code editing window, with the filename shown as the window title.

You can select **Submit** from the pop-up menu to execute the file directly from **File Shortcuts**.

Remember that you open **File Shortcuts** with a single-click in the z/OS (z/OS) or CMS operating environments. To open a file, you type

*   *?* beside the item to display pop-up menus

*   *S* beside the item to simulate a double-click.

Using My Favorite Folders

To view and manage any files in your operating environment, you can use the My Favorite Folders window.

To open a file that is stored in My Favorite Folders:

1.   Select **View** ⇨ **My Favorite Folders**.

2.   Double-click the file, or select **Open** from the pop-up menu for the file.

The file opens in a new code editing window, with the filename shown as the window title.

You can select **Submit** from the pop-up menu to execute the file directly from My Favorite Folders.

Using the Open Window

To use the Open window:

1.   With the code editing window active, select **File** ⇨ **Open** (or **File** ⇨ **Open Program**).

2.   In the Open window, click the file that you want to open (or type the path for the file).

3. Click **Open** or **OK**.

Figure 3.4 *The Open Window*

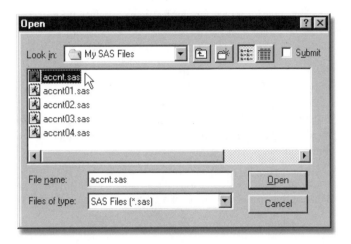

The file opens in a new code editing window, with the filename shown as the window title.

In the Windows environment, you can submit your file directly from the Open window by clicking the **Submit** check box before clicking **Open**.

The appearance of the Open window varies by operating environment. In mainframe operating environments, files are not listed, so you must type the path for the file.

If you're not sure of the name of the file in which your program is stored, you can do either of the following:

- use My Favorite Folders to locate the file in your operating environment

- issue the **X** command to temporarily suspend your SAS session.

  The X command enables you to use operating system facilities without ending your SAS session. When you issue the X command, you temporarily exit SAS and enter the host system.

To resume your SAS session after issuing the X command, issue the appropriate host system command, as shown below.

Table 3.1 *Resuming a SAS Session*

Operating Environment	Host command to resume SAS session
z/OS	RETURN or END
UNIX	exit
Windows	EXIT

Issuing an INCLUDE Command

You can also open a program by issuing an INCLUDE command.

General form, basic INCLUDE command:

INCLUDE *'file-specification'*

where *file-specification* is the physical name by which the host system recognizes the file.

Example

Suppose you want to include the program **D:\Programs\Sas\Myprog1.sas** in the Windows operating environment. To do so, you can issue the following INCLUDE command:

```
include 'd:\programs\sas\myprog1.sas'
```

Because this is a command (not a SAS statement), no semicolon is used at the end.

Editing SAS Programs

Now that you know how to open a SAS program, let's review the characteristics of SAS statements and look at enhancing the readability of your SAS programs.

SAS Program Structure

Remember that SAS programs consist of SAS statements. Consider the SAS program that is shown in a code editing window below.

Figure 3.5 *An Example SAS Program*

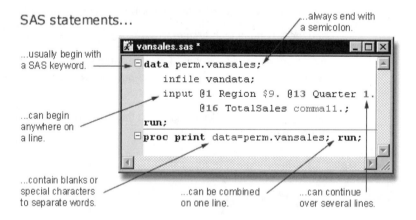

Although you can write SAS statements in almost any format, a consistent layout enhances readability and helps you understand the program's purpose. It's a good idea to

- begin DATA and PROC steps in column one
- indent statements within a step
- begin RUN statements in column one
- include a RUN statement after every DATA step or PROC step.

```
data sashelp.prdsale;
    infile jobs;
```

```
    input date name $ job $;
run;
proc print data=sashelp.prdsale;
run;
```

Now let's look at the Enhanced Editor window features that you can use to edit your SAS programs.

If you are running SAS in an operating environment other than Windows, you can skip theEnhanced Editor window section and see "Program Editor Features" on page 84 instead.

Using the Enhanced Editor

When you edit SAS programs in the Enhanced Editor, you can take advantage of a number of useful features. The following section shows you how to

- use color-coding to identify code elements
- automatically indent the next line when you press the **Enter** key
- collapse and expand sections of SAS procedures, DATA steps, and macros
- bookmark lines of code for easy access to different sections of your program
- open multiple views of a file.

Figure 3.6 *Enhanced Editor Window*

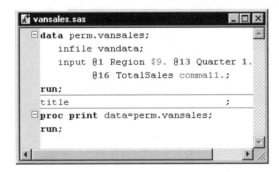

You can also

- save color-coding settings in a color scheme
- create macros that record and play back program editing commands
- create shortcuts for typing in text using abbreviations
- customize keyboard shortcuts for most Enhanced Editor commands
- create and format your own keywords
- automatically reload modified disk files
- access Help for SAS procedures by placing the insertion point within the procedure name and pressing F1.

Entering and Viewing Text

The Enhanced Editor uses visual aides such as color coding and code sections to help you write and debug your SAS programs.

You can use the margin on the left side of the Editor window to

- select one or more lines of text

- expand and collapse code sections

- display bookmarks.

Finding and Replacing Text

To find text in the Editor window, select **Edit** ⇨ **Find** or press Ctrl+F. You can set commonly used find options, including specifying that the text string is a regular expression. You can also specify whether to search in the code only, in the comments only, or in both the code and comments.

Figure 3.7 *The Find Window*

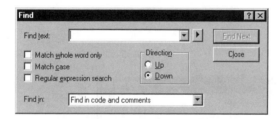

Using Abbreviations

You can define a character string so that when you type it and then press the Tab key or the Enter key, the string expands to a longer character string.

For example, you could define the abbreviation myv7sasfiles, which would expand to `'c:\winnt\profiles\myid\personal\mysasfiles\v7';`. Abbreviations are actually macros that insert one or more lines of text.

To create an abbreviation,

1. Select **Tools** ⇨ **Add Abbreviation**.

2. For **Abbreviation**, type the name of the abbreviation.

3. For **Text to insert for abbreviation**, type the text that the abbreviation will expand into.

4. Click **OK**.

To use an abbreviation, type the abbreviation. When an abbreviation is recognized, a tooltip displays the expanded text. Press the Tab key or Enter key to accept the abbreviation.

You can modify or delete abbreviations that you create.

Opening Multiple Views of a File

When you use the Enhanced Editor, you can see different parts of the same file simultaneously by opening multiple views of the same file. While you are working with multiple views, you are working with only *one file*, not multiple copies of the same file.

To open multiple views of the same file,

1. make the file the active window.

2. select **Window** ⇨ **New Window**.

The filename in the title bar is appended with a colon and a view number. Changes that you make to a file in one view, such as changing text or bookmarking a line, occur in all views simultaneously. Actions such as scroll bar movement, text selection, and expanding or contracting a section of code occur only in the active window.

Setting Enhanced Editor Options

You can customize the Enhanced Editor by setting Enhanced Editor options. To open the Enhanced Editor Options window, activate an Editor window and select **Tools** ⇨ **Options** ⇨ **Enhanced Editor**.

Click the tabs that are located along the top of the window to navigate to the settings that you want to change, and then select the options that you want.

For example, **Show line numbers** specifies whether to display line numbers in the margin. When line numbers are displayed, the current line number is red. **Insert spaces for tabs** specifies whether to insert the space character or the tab character when you press the Tab key. If it is selected, the space character is used. If it is not selected, the tab character is used.

When you are finished, click **OK**.

Figure 3.8 *Enhanced Editor Options Window*

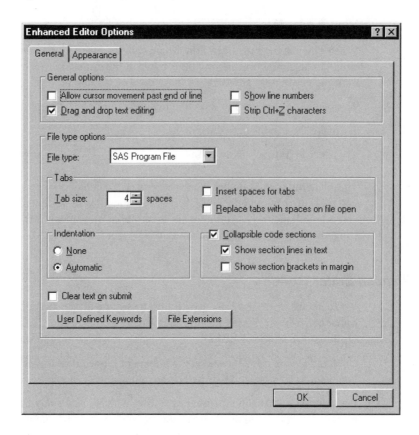

 For more information about setting options in the Enhanced Editor, select **Help** ⇨ **Using This Window** from an Enhanced Editor window.

Program Editor Features

You use the Program Editor window to submit SAS programs. You can also enter and edit SAS programs in this window, or you can open existing SAS programs that you created in another text editor.

To edit SAS programs in the Program Editor window, you can display line numbers. Then you can use text editor line commands and block text editor line commands within the line numbers to insert, delete, move, and copy lines within the Program Editor window.

In all operating environments, you can use line numbers, text editor line commands, and block text editor line commands. Depending on your operating environment, you can also use features such as dragging and dropping text, inserting lines with the Enter key, or marking and deleting text. For information about these features, see the SAS documentation for your operating environment.

Line Numbers

In some operating environments, line numbers appear in the Program Editor window by default. In other systems, the use of line numbers is optional. Activating line numbers can make it easier for you to edit your program regardless of your operating environment.

Figure 3.9 *Line Numbers*

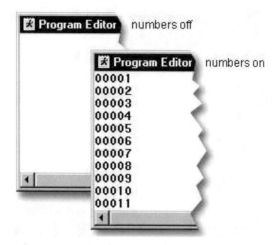

To turn on line numbers, use the NUMS command. Type nums on the command line or in the command box and press Enter. To turn line numbers off, enter the NUMS command again. Line numbers activated using the NUMS command will remain in effect for the duration of your SAS session.

To permanently display line numbers, select **Tools ⇨ Options ⇨ Program Editor**. Select the **Editing** tab. Select the **Display line numbers** option. Click **OK**.

Text Editor Line Commands

Text editor line commands enable you to delete, insert, move, copy, and repeat text. You enter these commands in the line number area in the Program Editor window.

The table below summarizes the basic text editing line commands.

Table 3.2 *Text Editor Line Commands*

Command	Action
Cn	copies n lines (where n = a number up to 9999)
Dn	deletes n lines
In	inserts n blank lines
Mn	moves n lines
Rn	repeats current line n times
A	after (used with C, I, and M)
B	before (used with C, I, and M)

You can use text editor line commands to perform actions like these:

Table 3.3 Results from Using Text Editor Line Commands

Command	Action
00001 i3002 00003	inserts 3 lines after line 00002
0ib01 00002 00003	inserts 1 line before line 00001
0ib41 00002 00003	inserts 4 lines before line 00001
000c2 00002 0a003	copies 2 lines (00001 and 00002) after line 00003
00001 0d302 00003	deletes 3 lines (00002, 00003, and 00004)
00b01 00002 00m03	moves 1 line (00003) before line 00001

Example 1

In the example below, a PROC PRINT statement and a RUN statement need to be inserted after line 00004.

Figure 3.10 Incomplete Program

```
Program Editor - Program 1.sas
00001 data clinic.admitfee;
00002    infile feedata;
00003    input name $ 1-20 fee 22-27;
00004 run;
00005 proc tabulate data=clinic.admitfee;
00006    var fee;
00007    table fee,mean;
00008 run;
```

To insert the PROC PRINT and RUN statements in the program,

1. Type **i2** anywhere in the line number area for line 00004 in the Program Editor window. Two blank lines are inserted, and the cursor is positioned on the first new line.

2. Type the PROC PRINT statement on line 00005.

3. Type a RUN statement on line 00006.

Figure 3.11 *Complete Program*

Block Text Editor Line Commands

Block text editor line commands enable you to delete, repeat, copy, and move multiple lines in the Program Editor window.

The block text editor line commands include the following:

Table 3.4 *Block Text Editor Line Commands*

Command	Action
DD	deletes a block of lines
CC	copies a block of lines
MM	moves a block of lines
RR	repeats multiple lines
A	after (used with CC and MM commands)
B	before (used with CC and MM commands)

To use a block command, specify the command on the first line affected and on the final line affected, and then press Enter.

Example 1

In the program below, the PROC TABULATE step needs to be deleted.

Figure 3.12 *Original Program*

```
Program Editor - Program 1.sas
00001 data clinic.admitfee;
00002    infile feedata;
00003    input name $ 1-20 fee 22-27;
00004 run;
00005 proc print data=clinic.admitfee;
00006 run;
00007 proc tabulate data=clinic.admitfee;
00008    var fee;
00009    table fee,mean;
00010 run;
```

You can use the DD block command to remove the step. To use the DD command:

1. Type **DD** anywhere in the line number area for line 00007 to mark the first line affected.

2. Type **DD** anywhere in the line number area for line 00010 to mark the final line affected.

3. Press Enter to delete lines 00007-00010.

Figure 3.13 *Edited Program*

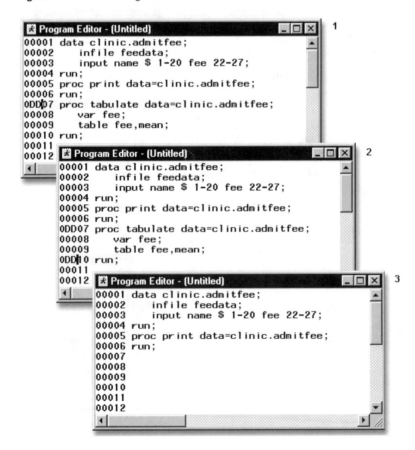

Example 2

In the example below, the PROC TABULATE step needs to be moved to the line after the PROC PRINT step.

Figure 3.14   *Original Program*

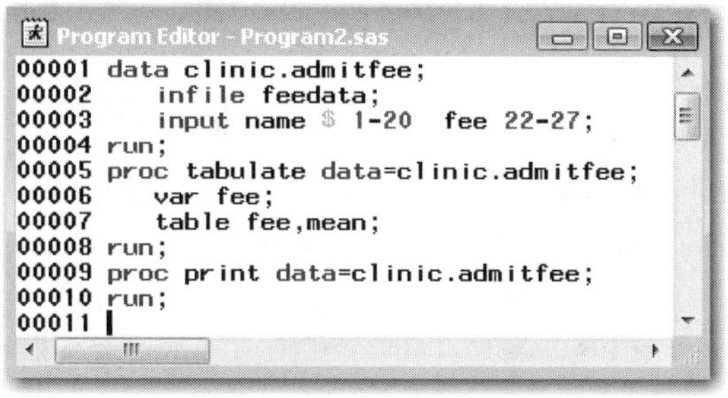

You can use the MM block command with the A or B command to move a block of code to a specific location. The MM is used to mark the block to be moved and the A or B is used to mark where you want the block to go. You use the A if you want to move the block after the line, and you use b if you want to move the block before the line.

To use the MM command:

1. Type **MM** anywhere in the line number area for line 00005 to mark the first line affected.

2. Type **MM** anywhere in the line number area for line 00008 to mark the last line affected.

3. Type **A** anywhere in the line number area for line 00010 to move the block after line 00010.

4. Press Enter to move lines 00005-00008 to line 00010.

Figure 3.15   *Edited Program*

Recalling SAS Programs

SAS statements disappear from the Program Editor window when you submit them. However, you can recall a program to the Program Editor by selecting **Run** ⇨ **Recall Last Submit**.

The area in which submitted SAS code is stored is known as the recall buffer. Program statements accumulate in the recall buffer each time you submit a program. The recall buffer is last-in/first-out (the most recently submitted statements are recalled first).

For example, if you submit two programs, you will need to select **Run** ⇨ **Recall Last Submit** two times to recall the first program to the Program Editor window. You can recall statements any time during your SAS session.

Figure 3.16 *Recalled Programs*

2nd recalled program

1st recalled program

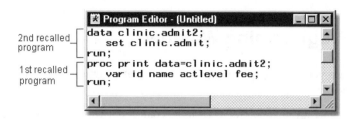

Saving SAS Programs

To save your SAS program to an external file, first activate the code editing window and select **File** ⇨ **Save As**. Then specify the name of your program in the Save As window. It's good practice to assign the extension .SAS to your SAS programs to distinguish them from other types of SAS files (such as .LOG for log files and .DAT for raw data files).

To save a SAS program as an external file in the z/OS operating environment, select **File** ⇨ **Save As** ⇨ **Write to File**, and then specify the name of your program in the Save As dialog box. You can specify SAS as the last level of the filename.

Figure 3.17 *Save as Window*

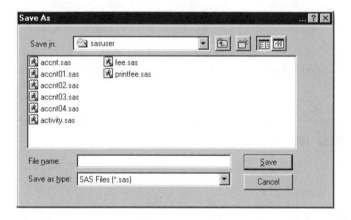

You can also save a SAS program by issuing a FILE command.

General form, basic FILE command:

FILE *'file-specification'*

where *file-specification* is the name of the file to be saved.

Example

Suppose you want to save a program as **D:\Programs\Sas\Newprog.Sas** in the Windows operating environment. To do so, you can issue the following FILE command:

```
file 'd:\programs\sas\newprog.sas'
```

Clearing SAS Programming Windows

When you run SAS programs, text accumulates in the Output window and in the Log window.

You might find it helpful to clear the contents of your SAS programming windows. To clear the Output window, Editor window, Program Editor window, or Log window, activate each window individually and select **Edit** ⇨ **Clear All**.

Figure 3.18 Cleared Log Window

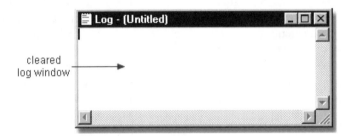

You can also clear the contents of a window by activating the window and then issuing the CLEAR command.

Text that has been cleared from windows cannot be recalled. However, in the Editor and Program Editor windows, you can select **Edit** ⇨ **Undo** to redisplay the text.

Interpreting Error Messages

Error Types

So far, the programs that are shown in this chapter have been error-free, but programming errors do occur. SAS can detect several types of errors. The most common are

- syntax errors that occur when program statements do not conform to the rules of the SAS language
- data errors that occur when some data values are not appropriate for the SAS statements that are specified in a program.

This chapter focuses on identifying and correcting syntax errors.

Syntax Errors

When you submit a program, SAS scans each statement for syntax errors, and then executes the step (if no syntax errors are found). SAS then goes to the next step and repeats the process. Syntax errors, such as misspelled keywords, generally prevent SAS from executing the step in which the error occurred.

You already know that notes are written to the Log window at the conclusion of execution of the program. When a program that contains an error is submitted, messages regarding the problem also appear in the Log window. When a syntax error is detected, the Log window

- displays the word ERROR

- identifies the possible location of the error

- gives an explanation of the error.

Example

The program below contains a syntax error. The DATA step copies the SAS data set Clinic.Admit into a new data set named Clinic.Admitfee. The PROC step should print the values for the variables ID, Name, Actlevel, and Fee in the new data set. However, print is misspelled in the PROC PRINT statement.

```
data clinic.admitfee;
   set clinic.admit;
run;
proc prin data=clinic.admitfee;
   var id name actlevel fee;
run;
```

When the program is submitted, messages in the Log window indicate that the procedure PRIN was not found and that SAS stopped processing the PRINT step due to errors. No output is produced by the PRINT procedure, because the second step fails to execute.

Figure 3.19 *Log Window Displaying Error Message*

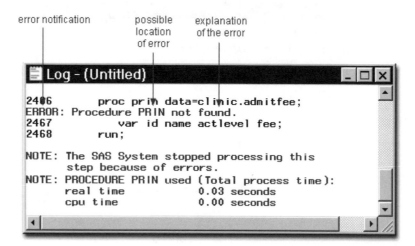

Problems with your statements or data might not be evident when you look at results in the Output window. Therefore, it is important to review the messages in the Log window each time you submit a SAS program.

Correcting Errors

Overview

To modify programs that contain errors, you can edit them in the code editing window. You can correct simple errors, such as the spelling error in the following program, by typing over the incorrect text, deleting text, or inserting text.

```
data clinic.admitfee;
   set clinic.admit;
run;
proc prin data=clinic.admitfee;
   var id name actlevel fee;
run;
```

If you use the Program Editor window, you usually need to recall the submitted statements from the recall buffer to the Program Editor window, where you can correct the problems. Remember that you can recall submitted statements by issuing the RECALL command or by selecting **Run** ⇨ **Recall Last Submit**.

In the program below, the missing t has been inserted into the PRINT keyword that is specified in the PROC PRINT statement.

Figure 3.20 Corrected Program

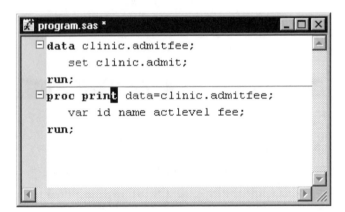

Some problems are relatively easy to diagnose and correct. But sometimes you might not know right away how to correct errors. The online Help provides information about individual procedures as well as help that is specific to your operating environment. From the **Help** menu, you can also select **SAS on the Web** for links to Technical Support and Frequently Asked Questions, if you have Internet access.

Resubmitting a Revised Program

After correcting your program, you can submit it again. However, before doing so, it is a good idea to clear the messages from the Log window so that you don't confuse the old error messages with the new messages. Then you can resubmit the program and view any resulting output.

Figure 3.21 *Correct PROC PRINT Output*

The SAS System

Obs	ID	Name	ActLevel	Fee
1	2458	Murray, W	HIGH	85.20
2	2462	Almers, C	HIGH	124.80
3	2501	Bonaventure, T	LOW	149.75
4	2523	Johnson, R	MOD	149.75
5	2539	LaMance, K	LOW	124.80
6	2544	Jones, M	HIGH	124.80
7	2552	Reberson, P	MOD	149.75
8	2555	King, E	MOD	149.75
9	2563	Pitts, D	LOW	124.80
10	2568	Eberhardt, S	LOW	124.80
11	2571	Nunnelly, A	HIGH	149.75
12	2572	Oberon, M	LOW	85.20
13	2574	Peterson, V	MOD	149.75
14	2575	Quigley, M	HIGH	124.80
15	2578	Cameron, L	MOD	124.80
16	2579	Underwood, K	LOW	149.75
17	2584	Takahashi, Y	MOD	124.80
18	2586	Derber, B	HIGH	85.20
19	2588	Ivan, H	LOW	85.20
20	2589	Wilcox, E	HIGH	149.75
21	2595	Warren, C	MOD	149.75

Remember to check the Log window again to verify that your program ran correctly.

Figure 3.22 *Log Message with No Errors*

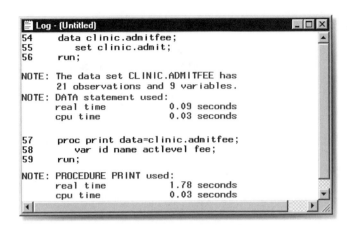

Because submitted steps remain in the recall buffer, resubmitting error-free steps wastes system resources. You can place SAS comment symbols before and after your error-free code until you debug the rest of your program. SAS ignores text between comment symbols during processing. When your entire program is error-free, remove the comment symbols and your entire SAS program is intact. See "Additional Features" on page 102 for instructions about creating a SAS comment statement.

To resubmit a section of code in the Windows operating environment, highlight the selected code in the code editing window. Then select **Run** ⇨ **Submit**.

This and other chapters show you the Enhanced Editor window only. If you are not using the Enhanced Editor as a code editing window, be sure to adapt the directions for the Program Editor window. For example, you might need to recall programs.

Resolving Common Problems

Overview

In addition to correcting spelling mistakes, you might need to resolve several other types of common syntax errors. These errors include

- omitting semicolons

- leaving quotation marks unbalanced

- specifying invalid options.

Another common problem is omitting a RUN statement at the end of a program. Although this is not technically an error, it can produce unexpected results. For the sake of convenience, we'll consider it together with syntax errors.

The table below lists these problems and their symptoms.

Table 3.5 *Identifying Problems in a SAS Program*

Problem	Symptom
missing RUN statement	"PROC (or DATA) step running" at top of active window
missing semicolon	log message indicating an error in a statement that seems to be valid
unbalanced quotation marks	log message indicating that a quoted string has become too long or that a statement is ambiguous
invalid option	log message indicating that an option is invalid or not recognized

Missing RUN Statement

Each step in a SAS program is compiled and executed independently from every other step. As a step is compiled, SAS recognizes the end of the current step when it encounters

- a DATA or PROC statement, which indicates the beginning of a new step

- a RUN or QUIT statement, which indicates the end of the current step.

When the program below is submitted, the DATA step executes, but the PROC step does not. The PROC step does not execute because there is no following DATA or PROC step to indicate the beginning of a new step, nor is there a following RUN statement to indicate the end of the step.

```
data clinic.admitfee;
   set clinic.admit;
run;
proc print data=clinic.admitfee;
   var id name actlevel fee;
```

Because there is nothing to indicate the end of the PROC step, the PRINT procedure waits before executing, and a "PROC PRINT running" message appears at the top of the active window.

Figure 3.23 *Program Window with Message*

Resolving the Problem

To correct the error, submit a RUN statement to complete the PROC step.

```
run;
```

If you are using the Program Editor window, you do *not* need to recall the program to the Program Editor window.

Missing Semicolon

One of the most common errors is the omission of a semicolon at the end of a statement. The program below is missing a semicolon at the end of the PROC PRINT statement.

```
data clinic.admitfee;
  set clinic.admit;
run;
proc print data=clinic.admitfee
    var id name actlevel fee;
run;
```

When you omit a semicolon, SAS reads the statement that lacks the semicolon, plus the following statement, as one long statement. The SAS log then lists errors that relate to the combined statement, not the actual mistake (the missing semicolon).

Figure 3.24 *Log Window with Error Message*

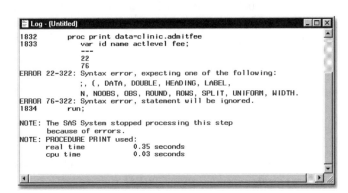

Resolving the Problem

To correct the error, do the following:

1. Find the statement that lacks a semicolon. You can usually locate the statement that lacks the semicolon by looking at the underscored keywords in the error message and working backwards.

2. Add a semicolon in the appropriate location.

3. Resubmit the corrected program.

4. Check the Log window again to make sure there are no other errors.

Unbalanced Quotation Marks

Some syntax errors, such as the missing quotation mark after HIGH in the program below, cause SAS to misinterpret the statements in your program.

```
data clinic.admitfee;
   set clinic.admit;
   where actlevel='HIGH;
run;
proc print data=clinic.admitfee;
   var id name actlevel fee;
run;
```

When the program is submitted, SAS is unable to resolve the DATA step, and a "DATA STEP running" message appears at the top of the active window.

Figure 3.25   *Program Editor with Message*

Sometimes a warning appears in the SAS log which indicates that

- a quoted string has become too long

- a statement that contains quotation marks (such as a TITLE or FOOTNOTE statement) is ambiguous due to invalid options or unquoted text.

Figure 3.26   *Log with Warning Message*

```
Log - (Untitled)   PROC PRINT running
93   proc print data=clinic.admitfee;
94      var id name actlevel fee;
95      title 'Patient Billing;
96      title2 'January 1998';
WARNING: The TITLE statement is ambiguous due to
         invalid options or unquoted text.
97   run;
```

When you have unbalanced quotation marks, SAS is often unable to detect the end of the statement in which it occurs. Simply adding a quotation mark and resubmitting your program usually does *not* solve the problem. SAS still considers the quotation marks to be unbalanced.

Therefore, you need to resolve the unbalanced quotation mark by canceling the submitted statements (in the Windows and UNIX operating environments) or by submitting a line of SAS code (in the z/OS operating environment) before you recall, correct, and resubmit the program.

If you do not resolve this problem when it occurs, it is likely that any subsequent programs that you submit in the current SAS session will generate errors.

Resolving the Error in the Windows Operating Environment

To resolve the error in the Windows operating environment:

1. Press the Ctrl and Break keys or click the Break Icon on the toolbar.

2. Select **1. Cancel Submitted Statements**, and then click **OK**.

   Figure 3.27 *Tasking Manager Window*

   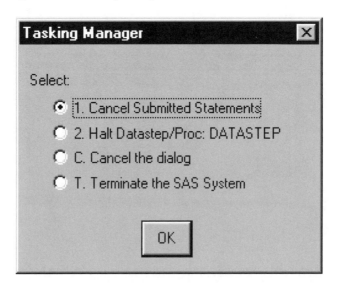

3. Select **Y to cancel submitted statements**, and then click **OK**.

   Figure 3.28 *Break Window*

   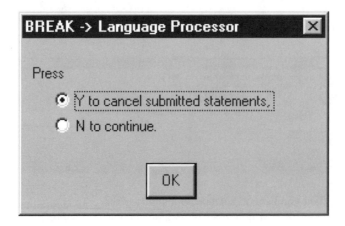

4. Correct the error and resubmit the program.

Resolving the Error in the UNIX Operating Environment

To resolve the error in the UNIX operating environment:

1. Open the Session Management window and click **Interrupt**.

Figure 3.29 *Session Management Window*

2. Select **1. Cancel Submitted Statements**, and then click **Y**.

Figure 3.30 *Tasking Management Window*

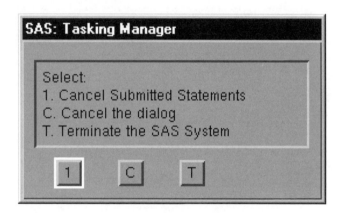

3. Correct the error and resubmit the program.

Resolving the Error in the z/OS Operating Environment

To resolve the error in the z/OS operating environment:

1. Submit an asterisk followed by a quotation mark, a semicolon, and a RUN statement.

```
*'; run;
```

2. Delete the line that contains the asterisk followed by the quotation mark, the semicolon, and the RUN statement.

3. Insert the missing quotation mark in the appropriate place.

4. Submit the corrected program.

You can also use the above method in the Windows and UNIX operating environments.

Invalid Option

An invalid option error occurs when you specify an option that is not valid in a particular statement. In the program below, the KEYLABEL option is not valid when used with the PROC PRINT statement.

```
data sasuser.admitfee;
   set clinic.admit;
   where weight>180 and (actlevel="MOD" or actlevel="LOW");
run;
proc print data=clinic.admitfee keylabel;
   label actlevel='Activity Level';
run;
```

When a SAS statement that contains an invalid option is submitted, a message appears in the Log window indicating that the option is not valid or not recognized.

Figure 3.31 *Log Window with Syntax Error Message*

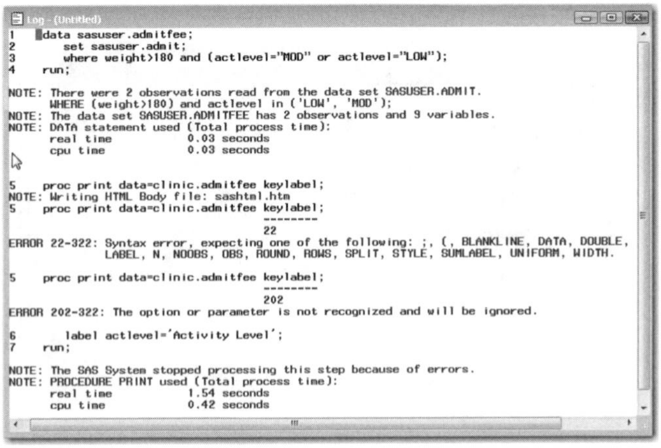

Resolving the Problem

To correct the error:

1. Remove or replace the invalid option, and check your statement syntax as needed.

2. Resubmit the corrected program.

3. Check the Log window again to make sure there are no other errors.

Additional Features

Comments in SAS Programs

You can insert comments into a SAS program to document the purpose of the program, to explain segments of the program, or to describe the steps in a complex program or calculation. A comment statement begins and ends with a comment symbol. There are two forms of comment statements:

```
*text;
```

or

```
/*text*/
```

SAS ignores text between comment symbols during processing.

The following program shows some of the ways comments can be used to describe a SAS program.

```
/* Read national sales data for vans */
/* from an external raw data file */
data perm.vansales;
   infile vandata;
   input @1 Region $9.
         @13 Quarter 1. /* Values are 1, 2, 3, or 4 */
         @16 TotalSales comma11.;
   /* Print the entire data set */
proc print data=perm.vansales;
run;
```

⚠ Avoid placing the /* comment symbols in columns 1 and 2. On some host operating systems, SAS might interpret a /* in columns 1 and 2 as a request to end the SAS job or session. For more information, see the SAS documentation for your operating environment.

System Options

SAS includes several system options that enable you to control error handling and Log window messages. The table shown below contains brief descriptions of some of these options. You can use either the OPTIONS statement or the SAS System Options window to specify these options.

Table 3.6   *Options for Controlling Error Handling and Log Window Messages*

Option	Descriptions
ERRORS=n	Specifies the maximum number of observations for which complete data error messages are printed.
FMTERR | NOFMTERR	Controls whether SAS generates an error message when a format of a variable cannot be found. NOFMTERR results in a warning instead of an error. FMTERR is the default.
SOURCE | NOSOURCE	Controls whether SAS writes source statements to the SAS log. SOURCE is the default.

Error Checking In the Enhanced Editor

If you are using the Enhanced Editor, you can use its color-coding for program elements, quoted strings, and comments to help you find coding errors.

You can also search for

- ending brackets or parentheses by pressing Ctrl+].

- matching DO-END pairs by pressing Alt+[. (You can learn about DO-END pairs in .)

See the following table for suggestions about finding syntax errors using the Enhanced Editor.

Table 3.7 *Finding Syntax Errors Using the Enhanced Editor*

To find...	Do this...
Undefined or misspelled keywords	In the Appearance tab of the Enhanced Editor Options dialog box, set the file elements **Defined keyword**, **User defined keyword**, and the **Undefined keyword** to unique color combinations.
	When SAS recognizes a keyword, the keyword changes to the defined colors. You'll be able to easily spot undefined keywords by looking for the color that you selected for undefined keywords.
Unmatched quoted strings	Look for one or more lines of the program that are the same color.
	Text following a quotation mark remains the same color until the string is closed with a matching quotation mark.
Unmatched comments	Look for one or more lines of the program that are the same color.
	Text that follows an open comment symbol (/*) remains the same color until the comment is closed with a closing comment symbol (*/).
Matching DO-END pairs	Place the cursor within a DO-END block and press Alt+[.
	The cursor moves first to the DO keyword. If one of the keywords is not found, the cursor remains as positioned.
	When both of the keywords exist, pressing Alt+[moves the cursor between the DO-END keywords.
Matching parentheses or brackets	Place the cursor on either side of the parenthesis or bracket. Press Ctrl+].
	The cursor moves to the matching parentheses or bracket. If one is not found, the cursor remains as positioned.
Missing semi-colons (;)	Look for keywords that appear in normal text.

Debugging with the DATA Step Debugger

Unlike most syntax errors, logic errors do not stop a program from running. Instead, they cause the program to produce unexpected results.

You can debug logic errors in DATA steps by using the DATA step debugger. This tool enables you to issue commands to execute DATA step statements one by one, and then to pause to display the resulting variables' values in a window. By observing the results that are displayed, you can determine where the logic error lies.

The debugger can be used only in interactive mode. Because the debugger is interactive, you can repeat the process of issuing commands and observing results as many times as needed in a single debugging session. To invoke the debugger, add the DEBUG option to the DATA statement, and submit the program.

```
data perm.publish / debug;
   infile pubdata;
   input BookID $ Publisher & $22. Year;
run;
proc print data=perm.publish;
run;
```

Figure 3.32 *Debugger Log Window*

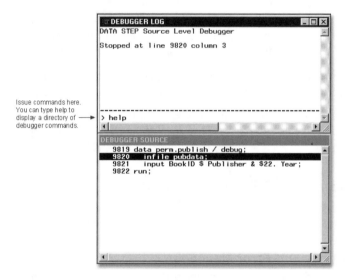

![icon] For detailed information about how to use the debugger, see the SAS documentation.

Chapter Summary

Text Summary

Opening a Stored SAS Program
A SAS program that is stored in an external file can be opened in the code editing window using

- file shortcuts
- My Favorite Folders
- the Open window
- the INCLUDE command.

Editing SAS Programs

SAS programs consist of SAS statements. Although you can write SAS statements in almost any format, a consistent layout enhances readability and enables you to understand the program's purpose.

In the Windows operating environment, the Enhanced Editor enables you to enter and view text and

- select one or more lines of text

- use color-coding to identify code elements

- automatically indent the next line when you press the Enter key

- collapse and expand sections of SAS procedures, DATA steps, and macros

- bookmark lines of code for easy access to different sections of your program.

Using the Editor window, you can also find and replace text, use abbreviations, open multiple views of a file, and set Enhanced Editor options.

The Program Editor window enables you to edit your programs just as you would with a word processing program. You can also use text editor commands and block text editor commands to edit SAS programs. Activating line numbers can make it easier for you to edit your program regardless of your operating environment.

Remember that SAS statements disappear from the Program Editor window when they are submitted. However, you can recall a program to the Program Editor window.

To save your SAS program to an external file, first activate the Program Editor window and select **File** ⇨ **Save As**. Then specify the name of your program in the Save As window.

Clearing SAS Programming Windows

Text and output accumulate in the Editor, Program Editor, Log, and Output windows throughout your SAS session. You can clear a window by selecting **Edit** ⇨ **Clear All**.

Interpreting Error Messages

When a SAS program that contains errors is submitted, error messages appear in the Log window. SAS can detect three types of errors: syntax, execution-time, and data. This chapter focuses on identifying and resolving common syntax errors.

Correcting Errors

To modify programs that contain errors, you can correct the errors in the code editing window. In the Program Editor window, you need to recall the submitted statements before editing them.

Before resubmitting a revised program, it is a good idea to clear the messages from the Log window so that you don't confuse old messages with the new. You can delete any error-free steps from a revised program before resubmitting it.

Resolving Common Problems

You might need to resolve several types of common problems: missing RUN statements, missing semicolons, unbalanced quotation marks, and invalid options.

Points to Remember

- It's a good idea to begin DATA steps, PROC steps, and RUN statements on the left and to indent statements within a step.

- End each step with a RUN statement.

- Review the messages in the Log window each time you submit a SAS program.

- You can delete any error-free steps from a revised program before resubmitting it, or you can submit only the revised steps in a program.

Chapter Quiz

Select the best answer for each question. After completing the quiz, check your answers using the answer key in the appendix.

1. As you write and edit SAS programs, it's a good idea to

    a. begin DATA and PROC steps in column one.

    b. indent statements within a step.

    c. begin RUN statements in column one.

    d. do all of the above.

2. Suppose you have submitted a SAS program that contains spelling errors. Which set of steps should you perform, in the order shown, to revise and resubmit the program?

    a. • Correct the errors.

       • Clear the Log window.

       • Resubmit the program.

       • Check the Log window.

    b. • Correct the errors.

       • Resubmit the program.

       • Check the Output window.

       • Check the Log window.

    c. • Correct the errors.

       • Clear the Log window.

       • Resubmit the program.

       • Check the Output window.

    d. • Correct the errors.

       • Clear the Output window.

       • Resubmit the program.

       • Check the Output window.

3. What happens if you submit the following program?

```
proc sort data=clinic.stress out=maxrates;
   by maxhr;
run;
proc print data=maxrates label double noobs;
   label rechr='Recovery Heart Rate;
   var resthr maxhr rechr date;
   where toler='I' and resthr>90;
```

```
      sum fee;
   run;
```

   a. Log messages indicate that the program ran successfully.

   b. A "PROC SORT running" message appears at the top of the active window, and
      a log message might indicate an error in a statement that seems to be valid.

   c. A log message indicates that an option is not valid or not recognized.

   d. A "PROC PRINT running" message appears at the top of the active window, and
      a log message might indicate that a quoted string has become too long or that the
      statement is ambiguous.

4. What generally happens when a syntax error is detected?

   a. SAS continues processing the step.

   b. SAS continues to process the step, and the Log window displays messages about
      the error.

   c. SAS stops processing the step in which the error occurred, and the Log window
      displays messages about the error.

   d. SAS stops processing the step in which the error occurred, and the Output
      window displays messages about the error.

5. A syntax error occurs when

   a. some data values are not appropriate for the SAS statements that are specified in
      a program.

   b. the form of the elements in a SAS statement is correct, but the elements are not
      valid for that usage.

   c. program statements do not conform to the rules of the SAS language.

   d. none of the above

6. How can you tell whether you have specified an invalid option in a SAS program?

   a. A log message indicates an error in a statement that seems to be valid.

   b. A log message indicates that an option is not valid or not recognized.

   c. The message "PROC running" or "DATA step running" appears at the top of the
      active window.

   d. You can't tell until you view the output from the program.

7. Which of the following programs contains a syntax error?

   a.
```
proc sort data=sasuser.mysales;
   by region;
run;
```

   b.
```
dat sasuser.mysales;
   set mydata.sales99;
   where sales<5000;
run;
```

   c.
```
proc print data=sasuser.mysales label;
   label region='Sales Region';
run;
```

   d. none of the above

8. What should you do after submitting the following program in the Windows or UNIX operating environment?

```
proc print data=mysales;
   where state='NC;
run;
```

   a. Submit a RUN statement to complete the PROC step.

   b. Recall the program. Then add a quotation mark and resubmit the corrected program.

   c. Cancel the submitted statements. Then recall the program, add a quotation mark, and resubmit the corrected program.

   d. Recall the program. Then replace the invalid option and resubmit the corrected program.

9. Which of the following commands opens a file in the code editing window?

   a. `file 'd:\programs\sas\newprog.sas'`

   b. `include 'd:\programs\sas\newprog.sas'`

   c. `open 'd:\programs\sas\newprog.sas'`

   d. all of the above

10. Suppose you submit a short, simple DATA step. If the active window displays the message "DATA step running" for a long time, what probably happened?

   a. You misspelled a keyword.

   b. You forgot to end the DATA step with a RUN statement.

   c. You specified an invalid data set option.

   d. Some data values weren't appropriate for the SAS statements that you specified.

Chapter 4
Creating List Reports

Overview

Introduction

To list the information in a data set, you can create a report with a PROC PRINT step. Then you can enhance the report with additional statements and options to create reports like those shown below.

Figure 4.1   *PROC PRINT*

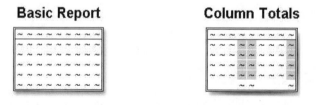

Objectives

In this chapter, you learn to

* specify SAS data sets to print

* select variables and observations to print

- sort data by the values of one or more variables

- specify column totals for numeric variables

- double space LISTING output

- add titles and footnotes to procedure output

- assign descriptive labels to variables

- apply formats to the values of variables.

Creating a Basic Report

To produce a simple list report, you first reference the library in which your SAS data set is stored. If you want, you can also set system options to control the appearance of your reports. Then you submit a basic PROC PRINT step.

General form, basic PROC PRINT step:

PROC PRINT DATA=*SAS-data-set***;**

RUN;

where *SAS-data-set* is the name of the SAS data set to be printed.

In the program below, the PROC PRINT statement invokes the PRINT procedure and specifies the data set Therapy in the SAS data library to which the libref Patients has been assigned.

```
libname patients 'c:\records\patients';
proc print data=patients.therapy;
run;
```

Notice the layout of the resulting report below. By default,

- all observations and variables in the data set are printed

- a column for observation numbers appears on the far left

- variables and observations appear in the order in which they occur in the data set.

Figure 4.2 *Patients.Therapy Data Set*

The SAS System

Obs	Date	AerClass	WalkJogRun	Swim
1	JAN1999	56	78	14
2	FEB1999	32	109	19
3	MAR1999	35	106	22
4	APR1999	47	115	24
5	MAY1999	55	121	31
6	JUN1999	61	114	67
7	JUL1999	67	102	72
8	AUG1999	64	76	77
9	SEP1999	78	77	54
10	OCT1999	81	62	47
11	NOV1999	84	31	52
12	DEC1999	2	44	55
13	JAN2000	37	91	83
14	FEB2000	41	102	27
15	MAR2000	52	98	19
16	APR2000	61	118	22
17	MAY2000	49	88	29
18	JUN2000	24	101	54
19	JUL2000	45	91	69
20	AUG2000	63	65	53
21	SEP2000	60	49	68
22	OCT2000	78	70	41
23	NOV2000	82	44	58
24	DEC2000	93	57	47

Be sure to specify the equal sign in the DATA= option in SAS procedures. If you omit the equal sign, your program produces an error similar to the following in the SAS log:

Figure 4.3 *Error Message*

```
35    proc print data patients.therapy;
                        -----------------
                        73
ERROR 73-322: Expecting an =.
36    run;

NOTE: The SAS System stopped processing this step because of errors.
```

Selecting Variables

Overview

By default, a PROC PRINT step lists all the variables in a data set. You can select variables and control the order in which they appear by using a VAR statement in your PROC PRINT step.

General form, VAR statement:

VAR *variable(s)*;

where *variable(s)* is one or more variable names, separated by blanks.

For example, the following VAR statement specifies that only the variables Age, Height, Weight, and Fee be printed, in that order:

```
proc print data=clinic.admit;
   var age height weight fee;
run;
```

The procedure output from the PROC PRINT step with the VAR statement lists only the values for the variables Age, Height, Weight, and Fee.

Figure 4.4 *Procedure Output*

The SAS System

Obs	Age	Height	Weight	Fee
1	27	72	168	85.20
2	34	66	152	124.80
3	31	61	123	149.75
4	43	63	137	149.75
5	51	71	158	124.80
6	29	76	193	124.80
7	32	67	151	149.75
8	35	70	173	149.75
9	34	73	154	124.80
10	49	64	172	124.80
11	44	66	140	149.75
12	28	62	118	85.20
13	30	69	147	149.75
14	40	69	163	124.80
15	47	72	173	124.80
16	60	71	191	149.75
17	43	65	123	124.80
18	25	75	188	85.20
19	22	63	139	85.20
20	41	67	141	149.75
21	54	71	183	149.75

In addition to selecting variables, you can control the default Obs column that PROC PRINT displays to list observation numbers. If you prefer, you can choose not to display observation numbers.

Figure 4.5 *Printing Observations*

Obs	Age	Height	Weight	Fee
1	27	72	168	85.20
2	34	66	152	124.80
3	31	61	123	149.75
4	43	63	137	149.75
5	51	71	158	124.80

Removing the OBS Column

To remove the Obs column, specify the NOOBS option in the PROC PRINT statement.

```
proc print data=work.example noobs;
   var age height weight fee;
run;
```

Identifying Observations

You've learned how to remove the Obs column altogether. As another alternative, you can use one or more variables to replace the Obs column in the output.

Using the ID Statement

To specify which variables should replace the Obs column, use the ID statement. This technique is particularly useful when observations are too long to print on one line.

General form, ID statement:

ID *variable(s)*;

where *variable(s)* specifies one or more variables to print instead of the observation number at the beginning of each row of the report.

Example

To replace the Obs column and identify observations based on an employee's ID number and last name, you can submit the following program.

```
proc print data=sales.reps;
   id idnum lastname;
run;
```

This is HTML output from the program:

Figure 4.6 *HTML Output*

IDnum	LastName	FirstName	City	State	Sex	JobCode	Salary	Birth	Hired	HomePhone
1269	CASTON	FRANKLIN	STAMFORD	CT	M	NA1	41690.00	06MAY60	01DEC80	203/781-3335
1935	FERNANDEZ	KATRINA	BRIDGEPORT	CT		NA2	51081.00	31MAR42	19OCT69	203/675-2962
1417	NEWKIRK	WILLIAM	PATERSON	NJ	,	NA2	52270.00	30JUN52	10MAR77	201/732-6611
1839	NORRIS	DIANE	NEW YORK	YN	F	NA1	43433.00	02DEC58	06JUL81	718/384-1767
1111	RHODES	JEREMY	PRINCETON	NJ	M	NA1	40586.00	17JUL61	03NOV80	201/812-1837
1352	RIVERS	SIMON	NEW YORK	NY	M	NA2	5379.80	05DEC48	19OCT74	718/383-3345
1332	STEPHENSON	ADAM	BRIDGEPORT	CT	M	NA1	42178.00	20SEP58	07JUN79	203/675-1497
1443	WELLS	AGNES	STAMFORD	CT	F	NA1	422.74	20NOV56	01SEP79	203/781-5546

In LISTING output, the IDnum and LastName columns are repeated for each observation that is printed on more than one line.

Figure 4.7 LISTING Output

```
IDnum  LastName   FirstName    City      State   Sex    JobCode

1269   CASTON     FRANKLIN    STAMFORD    CT      M      NA1
1935   FERNANDEZ  KATRINA     BRIDGEPO    CT             NA2
1417   NEWKIRK    WILLIAM     PATERSON    NJ      ,      NA2
1839   NORRIS     DIANE       NEW YORK    NY      F      NA1
1111   RHODES     JEREMY      PRINCETO    NJ      M      NA1
1352   RIVERS     SIMON       NEW YORK    NY      M      NA2
1332   STEPHENS   ADAM        BRIDGEPO    CT      M      NA1
1443   WELLS      AGNES       STAMFORD    CT      F      NA1


IDnum  LastName   Salary    Birth     Hired      HomePhone

1269   CASTON     41690.00  06MAY60  01DEC80    203/781-3335
1935   FERNANDEZ  51081.00  31MAR42  19OCT69    203/675-2962
1417   NEWKIRK    52270.00  30JUN52  10MAR77    201/732-6611
1839   NORRIS     43433.00  02DEC58  06JUL81    718/384-1767
1111   RHODES     40586.00  17JUL61  03NOV80    201/812-1837
1352   RIVERS      5379.80  05DEC48  19OCT74    718/383-3345
1332   STEPHENS   42178.00  20SEP58  07JUN79    203/675-1497
1443   WELLS        422.74  20NOV56  01SEP79    203/781-5546
```

If a variable in the ID statement also appears in the VAR statement, the output contains two columns for that variable. In the example below, the variable IDnum appears twice.

```
proc print data=sales.reps;
    id idnum lastname;
    var idnum sex jobcode salary;
run;
```

Figure 4.8 IDNUM Output

IDnum	LastName	IDnum	Sex	JobCode	Salary
1269	CASTON	1269	M	NA1	41690.00
1935	FERNANDEZ	1935		NA2	51081.00
1417	NEWKIRK	1417	,	NA2	52270.00
1839	NORRIS	1839	F	NA1	43433.00
1111	RHODES	1111	M	NA1	40586.00
1352	RIVERS	1352	M	NA2	5379.80
1332	STEPHENSON	1332	M	NA1	42178.00
1443	WELLS	1443	F	NA1	422.74

Selecting Observations

By default, a PROC PRINT step lists all the observations in a data set. You can control which observations are printed by adding a WHERE statement to your PROC PRINT step. There should be only one WHERE statement in a step. If multiple WHERE statements are issued, only the last statement is processed.

General form, WHERE statement:

WHERE *where-expression*;

where *where-expression* specifies a condition for selecting observations. The *where-expression* can be any valid SAS expression.

For example, the following WHERE statement selects only observations for which the value of Age is greater than 30:

```
proc print data=clinic.admit;
   var age height weight fee;
   where age>30;
run;
```

Here is the procedure output from the PROC PRINT step with the WHERE statement:

Figure 4.9 *PROC PRINT Output with WHERE Statement*

Obs	Age	Height	Weight	Fee
2	34	66	152	124.80
3	31	61	123	149.75
4	43	63	137	149.75
5	51	71	158	124.80
7	32	67	151	149.75
8	35	70	173	149.75
9	34	73	154	124.80
10	49	64	172	124.80
11	44	66	140	149.75
14	40	69	163	124.80
15	47	72	173	124.80
16	60	71	191	149.75
17	43	65	123	124.80
20	41	67	141	149.75
21	54	71	183	149.75

Specifying WHERE Expressions

In the WHERE statement you can specify any variable in the SAS data set, not just the variables that are specified in the VAR statement. The WHERE statement works for both character and numeric variables. To specify a condition based on the value of a character variable, you must

- enclose the value in quotation marks

- write the value with lowercase, uppercase, or mixed case letters exactly as it appears in the data set.

You use the following comparison operators to express a condition in the WHERE statement:

Table 4.1 *Comparison Operators in a WHERE Statement*

Symbol	Meaning	Example
= or eq	equal to	`where name='Jones, C.';`
^= or ne	not equal to	`where temp ne 212;`
> or gt	greater than	`where income>20000;`
< or lt	less than	`where partno lt "BG05";`
>= or ge	greater than or equal to	`where id>='1543';`
<= or le	less than or equal to	`where pulse le 85;`

You can learn more about valid SAS expressions in Chapter 5, "Creating SAS Data Sets from External Files," on page 151.

Using the CONTAINS Operator

The CONTAINS operator selects observations that include the specified substring. The symbol for the CONTAINS operator is **?**. You can use either the CONTAINS keyword or the symbol in your code, as shown below.

```
where firstname CONTAINS 'Jon';
where firstname ? 'Jon';
```

Specifying Compound WHERE Expressions

You can also use WHERE statements to select observations that meet multiple conditions. To link a sequence of expressions into compound expressions, you use logical operators, including the following:

Table 4.2 *Compound WHERE Expression Operators*

Operator		Meaning
AND	&	and, both. If both expressions are true, then the compound expression is true.
OR	|	or, either. If either expression is true, then the compound expression is true.

Examples of WHERE Statements

- You can use compound expressions like these in your WHERE statements:

```
where age<=55 and pulse>75;
where area='A' or region='S';
where ID>'1050' and state='NC';
```

- When you test for multiple values of the same variable, you specify the variable name in each expression:

```
where actlevel='LOW' or actlevel='MOD';
where fee=124.80 or fee=178.20;
```

- You can use the IN operator as a convenient alternative:

```
where actlevel in ('LOW','MOD');
where fee in (124.80,178.20);
```

- To control the way compound expressions are evaluated, you can use parentheses (expressions in parentheses are evaluated first):

```
where (age<=55 and pulse>75) or area='A';
where age<=55 and (pulse>75 or area='A');
```

Sorting Data

Overview

By default, PROC PRINT lists observations in the order in which they appear in your data set. To sort your report based on values of a variable, you must use PROC SORT to sort your data before using the PRINT procedure to create reports from the data.

The SORT procedure

- rearranges the observations in a SAS data set
- creates a new SAS data set that contains the rearranged observations
- replaces the original SAS data set by default
- can sort on multiple variables
- can sort in ascending or descending order
- does not generate printed output
- treats missing values as the smallest possible values.

General form, simple PROC SORT step:

PROC SORT DATA=*SAS-data-set*<OUT=*SAS-data-set*>;

    BY <DESCENDING> *BY-variable(s)*;

RUN;

where

- the **DATA=** option specifies the data set to be read

- the **OUT=** option creates an output data set that contains the data in sorted order

- *BY-variable(s)* in the required **BY** statement specifies one or more variables whose values are used to sort the data

- the **DESCENDING** option in the BY statement sorts observations in descending order. If you have more that one variable in the BY statement, DESCENDING applies only to the variable that immediately follows it.

⚠ If you don't use the OUT= option, PROC SORT overwrites the data set specified in the DATA= option.

Example

In the following program, the PROC SORT step sorts the permanent SAS data set Clinic.Admit by the values of the variable Age within the values of the variable Weight and creates the temporary SAS data set Wgtadmit. Then the PROC PRINT step prints a subset of the Wgtadmit data set.

```
proc sort data=clinic.admit out=work.wgtadmit;
   by weight age;
run;
proc print data=work.wgtadmit;
   var weight age height fee;
   where age>30;
run;
```

The report displays observations in ascending order of age within weight.

Figure 4.10 *Observations Displayed in Ascending Order of Age Within Weight*

Obs	Weight	Age	Height	Fee
2	123	31	61	149.75
3	123	43	65	124.80
4	137	43	63	149.75
6	140	44	66	149.75
7	141	41	67	149.75
9	151	32	67	149.75
10	152	34	66	124.80
11	154	34	73	124.80
12	158	51	71	124.80
13	163	40	69	124.80
15	172	49	64	124.80
16	173	35	70	149.75
17	173	47	72	124.80
18	183	54	71	149.75
20	191	60	71	149.75

Adding the DESCENDING option to the BY statement sorts observations in ascending order of age within descending order of weight. Notice that DESCENDING applies only to the variable Weight.

```
proc sort data=clinic.admit out=work.wgtadmit;
   by descending weight age;
run;
proc print data=work.wgtadmit;
   var weight age height fee;
   where age>30;
run;
```

Figure 4.11 *Observations Displayed in Descending Order*

Obs	Weight	Age	Height	Fee
2	191	60	71	149.75
4	183	54	71	149.75
5	173	35	70	149.75
6	173	47	72	124.80
7	172	49	64	124.80
9	163	40	69	124.80
10	158	51	71	124.80
11	154	34	73	124.80
12	152	34	66	124.80
13	151	32	67	149.75
15	141	41	67	149.75
16	140	44	66	149.75
18	137	43	63	149.75
19	123	31	61	149.75
20	123	43	65	124.80

Generating Column Totals

Overview

To produce column totals for numeric variables, you can list the variables to be summed in a SUM statement in your PROC PRINT step.

General form, SUM statement:

SUM *variable(s)*;

where *variable(s)* is one or more numeric variable names, separated by blanks.

The SUM statement in the following PROC PRINT step requests column totals for the variable BalanceDue:

```
proc print data=clinic.insure;
   var name policy balancedue;
   where pctinsured < 100;
   sum balancedue;
run;
```

Column totals appear at the end of the report in the same format as the values of the variables.

Figure 4.12 Column Totals

Obs	Name	Policy	BalanceDue
2	Almers, C	95824	156.05
3	Bonaventure, T	87795	9.48
4	Johnson, R	39022	61.04
5	LaMance, K	63265	43.68
6	Jones, M	92478	52.42
7	Reberson, P	25530	207.41
8	King, E	18744	27.19
9	Pitts, D	60976	310.82
10	Eberhardt, S	81589	173.17
13	Peterson, V	75986	228.00
14	Quigley, M	97048	99.01
15	Cameron, L	42351	111.41
17	Takahashi, Y	54219	186.58
18	Derber, B	74653	236.11
20	Wilcox, E	94034	212.20
21	Warren, C	20347	164.44
			2279.0

Requesting Subtotals

You might also want to subtotal numeric variables. To produce subtotals, add both a
SUM statement and a BY statement to your PROC PRINT step.

General form, BY statement in the PRINT procedure:

BY <DESCENDING> *BY-variable-1*

                   <...<DESCENDING><BY-variable-n>>

            <NOTSORTED>**;**

where

- *BY-variable* specifies a variable that the procedure uses to form BY groups. You can specify more than one variable, separated by blanks.

- the DESCENDING option specifies that the data set is sorted in descending order by the variable that immediately follows.

- the NOTSORTED option specifies that observations are not necessarily sorted in alphabetic or numeric order. If observations that have the same values for the BY variables are not contiguous, the procedure treats each contiguous set as a separate BY group.

⚠️ If you do not use the NOTSORTED option in the BY statement, the observations in the data set must either be sorted by all the variables that you specify, or they must be indexed appropriately.

Example

The SUM statement in the following PROC PRINT step requests column totals for the variable Fee, and the BY statement produces a subtotal for each value of ActLevel.

```
proc sort data=clinic.admit out=work.activity;
   by actlevel;
run;
proc print data=work.activity;
   var age height weight fee;
   where age>30;
   sum fee;
   by actlevel;
run;
```

In the output, the BY variable name and value appear before each BY group. The BY variable name and the subtotal appear at the end of each BY group.

Figure 4.13 *BY Group Output: High*

ActLevel=HIGH			
Obs	**Age**	**Height**	**Weight**	**Fee**
2	34	66	152	124.80
4	44	66	140	149.75
5	40	69	163	124.80
7	41	67	141	149.75
ActLevel				**549.10**

Figure 4.14 *BY Group Output: Low*

ActLevel=LOW

Obs	Age	Height	Weight	Fee
8	31	61	123	149.75
9	51	71	158	124.80
10	34	73	154	124.80
11	49	64	172	124.80
13	60	71	191	149.75
ActLevel				673.90

Figure 4.15 *BY Group Output: Mod*

ActLevel=MOD

Obs	Age	Height	Weight	Fee
15	43	63	137	149.75
16	32	67	151	149.75
17	35	70	173	149.75
19	47	72	173	124.80
20	43	65	123	124.80
21	54	71	183	149.75
ActLevel				848.60
				2071.60

Creating a Customized Layout with BY Groups and ID Variables

In the previous example, you may have noticed the redundant information for the BY variable. For example, in the partial PROC PRINT output below, the BY variable ActLevel is identified both before the BY group and for the subtotal.

Figure 4.16 *Creating a Customized Layout with BY Groups and ID Variables*

ActLevel=HIGH

Obs	Age	Height	Weight	Fee
2	34	66	152	124.80
4	44	66	140	149.75
5	40	69	163	124.80
7	41	67	141	149.75
ActLevel				549.10

To show the BY variable heading only once, you can use an ID statement and a BY statement together with the SUM statement. When an ID statement specifies the same variable as the BY statement,

- the Obs column is suppressed
- the ID/BY variable is printed in the left-most column

• each ID/BY value is printed only at the start of each BY group and on the line that contains that group's subtotal.

Example

The ID, BY, and SUM statements work together to produce the output shown below. The ID variable is listed only once for each BY group and once for each sum. The BY lines are suppressed. Instead, the value of the ID variable, ActLevel, identifies each BY group.

```
proc sort data=clinic.admit out=work.activity;
   by actlevel;
run;
proc print data=work.activity;
   var age height weight fee;
   where age>30;
   sum fee;
   by actlevel;
   id actlevel;
run;
```

Figure 4.17 *Creating Custom Output Example Output*

ActLevel	Age	Height	Weight	Fee
HIGH	34	66	152	124.80
	44	66	140	149.75
	40	69	163	124.80
	41	67	141	149.75
HIGH				549.10
LOW	31	61	123	149.75
	51	71	158	124.80
	34	73	154	124.80
	49	64	172	124.80
	60	71	191	149.75
LOW				673.90
MOD	43	63	137	149.75
	32	67	151	149.75
	35	70	173	149.75
	47	72	173	124.80
	43	65	123	124.80
	54	71	183	149.75
MOD				848.60
				2071.60

Requesting Subtotals on Separate Pages

As another enhancement to your PROC PRINT report, you can request that each BY group be printed on a separate page by using the PAGEBY statement.

General form, PAGEBY statement:

PAGEBY *BY-variable*:

where *BY-variable* identifies a variable that appears in the BY statement in the PROC PRINT step. PROC PRINT begins printing a new page if the value of the BY variable changes, or if the value of any BY variable that precedes it in the BY statement changes.

⚠️ The variable specified in the PAGEBY statement must also be specified in the BY statement in the PROC PRINT step.

Example

The PAGEBY statement in the program below prints BY groups for the variable ActLevel separately. The BY groups appear separated by horizontal lines in the HTML output.

```
proc sort data=clinic.admit out=work.activity;
   by actlevel;
run;
proc print data=work.activity;
   var age height weight fee;
   where age>30;
   sum fee;        ·
   by actlevel;
   id actlevel;
   pageby actlevel;
run;
```

Figure 4.18 *PAGEBY Example: High*

ActLevel	Age	Height	Weight	Fee
HIGH	34	66	152	124.80
	44	66	140	149.75
	40	69	163	124.80
	41	67	141	149.75
HIGH				**549.10**

Figure 4.19 *PAGEBY Example: Low*

ActLevel	Age	Height	Weight	Fee
LOW	31	61	123	149.75
	51	71	158	124.80
	34	73	154	124.80
	49	64	172	124.80
	60	71	191	149.75
LOW				**673.90**

Figure 4.20 *PAGEBY Example: Mod*

ActLevel	Age	Height	Weight	Fee
MOD	43	63	137	149.75
	32	67	151	149.75
	35	70	173	149.75
	47	72	173	124.80
	43	65	123	124.80
	54	71	183	149.75
MOD				848.60
				2071.60

Double Spacing LISTING Output

If you are generating SAS LISTING output, one way to control the layout is to double space it. To do so, specify the DOUBLE option in the PROC PRINT statement. For example,

```
proc print data=clinic.stress double;
   var resthr maxhr rechr;
   where tolerance='I';
run;
```

Double spacing does not apply to HTML output.

To generate SAS LISTING output, you must select **Tools ⇨ Options ⇨ Preferences**. Select the **Results** tab. Select the **Create listing** option.

Figure 4.21 *Double-Spaced LISTING Output*

SAS Output

OBS	RestHR	MaxHR	RecHR
2	68	171	133
3	78	177	139
8	70	167	122
11	65	181	141
14	74	152	113
15	75	158	108
20	78	189	138

Specifying Titles and Footnotes

Overview

Now you've learned how to structure your PRINT procedure output. However, you might also want to make your reports easy to interpret by

- adding titles and footnotes

- replacing variable names with descriptive labels

- formatting variable values.

Although this chapter focuses on PROC PRINT, you can apply these enhancements to most SAS procedure output.

TITLE and FOOTNOTE Statements

To make your report more meaningful and self-explanatory, you can associate up to 10 titles with procedure output by using TITLE statements before the PROC step. Likewise, you can specify up to 10 footnotes by using FOOTNOTE statements before the PROC step.

 Because TITLE and FOOTNOTE statements are global statements, place them anywhere within or before the PRINT procedure. Titles and footnotes are assigned as soon as TITLE or FOOTNOTE statements are read; they apply to all subsequent output.

General form, TITLE and FOOTNOTE statements:

TITLE<*n*> *'text'*;

FOOTNOTE<*n*> *'text'*;

where *n* is a number from 1 to 10 that specifies the title or footnote line, and *'text'* is the actual title or footnote to be displayed. The maximum title or footnote length depends on your operating environment and on the value of the LINESIZE= option.

The keyword **title** is equivalent to title1. Likewise, **footnote** is equivalent to footnote1. If you don't specify a title, the default title is The SAS System. No footnote is printed unless you specify one.

Be sure to match quotation marks that enclose the title or footnote text.

Using the TITLES and FOOTNOTES Windows

You can also specify titles in the TITLES window and footnotes in the FOOTNOTES window. Titles and footnotes that you specify in these windows are not stored with your program, and they remain in effect only during your SAS session.

To open the TITLES window, issue the TITLES command. To open the FOOTNOTES window, issue the FOOTNOTES command.

To specify a title or footnote, type in the text you want next to the number of the line where the text should appear. To cancel a title or footnote, erase the existing text. Notice that you do not enclose text in quotation marks in these windows.

Figure 4.22 *Titles Window*

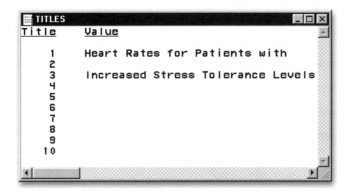

Example: Titles

The two TITLE statements below, specified for lines 1 and 3, define titles for the PROC PRINT output.

```
title1 'Heart Rates for Patients with';
title3 'Increased Stress Tolerance Levels';
proc print data=clinic.stress;
   var resthr maxhr rechr;
   where tolerance='I';
run;
```

Title lines for HTML output appear differently depending on the version of SAS that you use. In SAS Version 8, title lines simply appear consecutively, without extra spacing to indicate skipped title numbers. In SAS®9 HTML output, title line 2 is blank.

Figure 4.23 *HTML Output with Titles: SAS®8*

SAS Version 8 HTML Output:

Heart Rates for Patients with Increased Stress Tolerance Levels

Obs	RestHR	MaxHR	RecHR
2	68	171	133
3	78	177	139
8	70	167	122
11	65	181	141
14	74	152	113
15	75	158	108
20	78	189	138

Figure 4.24 *HTML Output with Titles: SAS®9*

SAS®9 HTML Output:

Heart Rates for Patients with

Increased Stress Tolerance Levels

Obs	RestHR	MaxHR	RecHR
2	68	171	133
3	78	177	139
8	70	167	122
11	65	181	141
14	74	152	113
15	75	158	108
20	78	189	138

In SAS LISTING output for all versions of SAS, title line 2 is blank, as shown below. Titles are centered by default.

Figure 4.25 *LISTING Output with Titles: All Versions*

```
        Heart Rates for Patients with

     Increased Stress Tolerance Levels

     OBS       RestHR      MaxHR       RecHR

      2          68         171         133
      3          78         177         139
      8          70         167         122
     11          65         181         141
     14          74         152         113
     15          75         158         108
     20          78         189         138
```

Example: Footnotes

The two FOOTNOTE statements below, specified for lines 1 and 3, define footnotes for the PROC PRINT output. Since there is no footnote2, a blank line is inserted between footnotes 1 and 2 in the output.

```
footnote1 'Data from Treadmill Tests';
footnote3 '1st Quarter Admissions';
proc print data=clinic.stress;
   var resthr maxhr rechr;
   where tolerance='I';
run;
```

Footnotes appear at the bottom of each page of procedure output. Notice that footnote lines are *pushed up* from the bottom. The FOOTNOTE statement that has the largest number appears on the bottom line.

Figure 4.26 HTML Output with Footnotes

Obs	RestHR	MaxHR	RecHR
2	68	171	133
3	78	177	139
8	70	167	122
11	65	181	141
14	74	152	113
15	75	158	108
20	78	189	138

Data from Treadmill Tests
1st Quarter Admissions

In SAS LISTING output, footnote line 2 is blank, as shown below. Footnotes are centered by default.

Figure 4.27 LISTING Output with Footnotes

```
  OBS      RestHR      MaxHR      RecHR

    2         68         171        133
    3         78         177        139
    8         70         167        122
   11         65         181        141
   14         74         152        113
   15         75         158        108
   20         78         189        138

       Data from Treadmill Tests

          1st Quarter Admissions
```

Modifying and Canceling Titles and Footnotes

TITLE and FOOTNOTE statements are global statements. That is, after you define a title or footnote, it remains in effect until you modify it, cancel it, or end your SAS session.

For example, the footnotes that are assigned in the PROC PRINT step below also appear in the output from the PROC TABULATE step.

```
footnote1 'Data from Treadmill Tests';
footnote3 '1st Quarter Admissions';
proc print data=clinic.stress;
   var resthr maxhr rechr;
   where tolerance='I';
run;
proc tabulate data=clinic.stress;
   where tolerance='I';
   var resthr maxhr;
   table mean*(resthr maxhr);
run;
```

Redefining a title or footnote line cancels any higher-numbered title or footnote lines, respectively. In the example below, defining a title for line 2 in the second report automatically cancels title line 3.

```
title3 'Participation in Exercise Therapy';
proc print data=clinic.therapy;
   var swim walkjogrun aerclass;
run;
title2 'Report for March';
proc print data=clinic.therapy;
run;
```

To cancel all previous titles or footnotes, specify a null TITLE or FOOTNOTE statement (a TITLE or FOOTNOTE statement with no number or text) or a TITLE1 or FOOTNOTE1 statement with no text. This will also cancel the default title The SAS System.

For example, in the program below, the null TITLE1 statement cancels all titles that are in effect before either PROC step executes. The null FOOTNOTE statement cancels all footnotes that are in effect after the PROC PRINT step executes. The PROC TABULATE output appears without a title or a footnote.

```
title1;
footnote1 'Data from Treadmill Tests';
footnote3 '1st Quarter Admissions';
proc print data=clinic.stress;
   var resthr maxhr rechr;
   where tolerance='I';
run;
footnote;
proc tabulate data=clinic.stress;
   var timemin timesec;
   table max*(timemin timesec);
run;
```

Assigning Descriptive Labels

Temporarily Assigning Labels to Variables

You can also enhance your PROC PRINT report by labeling columns with more descriptive text. To label columns, you use

- the LABEL statement to assign a descriptive label to a variable

- the LABEL option in the PROC PRINT statement to specify that the labels be displayed.

General form, LABEL statement:

LABEL *variable1='label1'*

variable2='label2'

... ;

Labels can be up to 256 characters long. Enclose the label in quotation marks.

 The LABEL statement applies only to the PROC step in which it appears.

Example

In the PROC PRINT step below, the variable name WalkJogRun is displayed with the label Walk/Jog/Run. Note the LABEL option in the PROC PRINT statement. Without the LABEL option in the PROC PRINT statement, PROC PRINT would use the name of the column heading `walkjogrun` even though you specified a value for the variable.

```
proc print data=clinic.therapy label;
   label walkjogrun=Walk/Jog/Run;
run;
```

Figure 4.28   *Output Created Without the LABEL Option*

Obs	Date	AerClass	Walk/Jog/Run	Swim
6	JUN1999	61	114	67
7	JUL1999	67	102	72
8	AUG1999	64	76	77
9	SEP1999	78	77	54
10	OCT1999	81	62	47
11	NOV1999	84	31	52
16	APR2000	61	118	22
20	AUG2000	63	65	53
22	OCT2000	78	70	41
23	NOV2000	82	44	58
24	DEC2000	93	57	47

Using Single or Multiple LABEL Statements

You can assign labels in separate LABEL statements . . .

```
proc print data=clinic.admit label;
   var age height;
   label age='Age of Patient';
   label height='Height in Inches';
run;
```

. . . or you can assign any number of labels in a single LABEL statement.

```
proc print data=clinic.admit label;
   var actlevel height weight;
   label actlevel='Activity Level'
         height='Height in Inches'
         weight='Weight in Pounds';
run;
```

Formatting Data Values

Temporarily Assigning Formats to Variables

In your SAS reports, formats control how the data values are displayed. To make data values more understandable when they are displayed in your procedure output, you can use the FORMAT statement, which associates formats with variables.

Formats affect only how the data values appear in output, *not* the actual data values as they are stored in the SAS data set.

General form, FORMAT statement:

FORMAT *variable(s) format-name*;

where

- *variable(s)* is the name of one or more variables whose values are to be written according to a particular pattern

- *format-name* specifies a SAS format or a user-defined format that is used to write out the values.

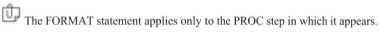 The FORMAT statement applies only to the PROC step in which it appears.

You can use a separate FORMAT statement for each variable, or you can format several variables (using either the same format or different formats) in a single FORMAT statement.

Table 4.3 *Formats That Are Used to Format Data*

This FORMAT statement ...	Associates ...	To display values as ...
`format date mmddyy8.;`	the format MMDDYY8. with the variable Date	06/05/03
`format net comma5.0` `      gross comma8.2;`	the format COMMA5.0 with the variable Net and the format COMMA8.2 with the variable Gross	1,234 5,678.90
`format net gross dollar9.2;`	the format DOLLAR9.2 with both variables, Net and Gross	$1,234.00 $5,678.90

For example, the FORMAT statement below writes values of the variable Fee using dollar signs, commas, and no decimal places.

```
proc print data=clinic.admit;
   var actlevel fee;
   where actlevel='HIGH';
   format fee dollar4.;
run;
```

Figure 4.29   *FORMAT Statement Example*

Obs	ActLevel	Fee
1	HIGH	$85
2	HIGH	$125
6	HIGH	$125
11	HIGH	$150
14	HIGH	$125
18	HIGH	$85
20	HIGH	$150

Specifying SAS Formats

The table below describes some SAS formats that are commonly used in reports.

Table 4.4   *Commonly Used SAS Formats*

Format	Specifies values ...	Example
COMMAw.d	that contain commas and decimal places	comma8.2
DOLLAR w.d	that contain dollar signs, commas, and decimal places	dollar6.2
MMDDYYw.	as date values of the form 09/12/97 (MMDDYY8.) or 09/12/1997 (MMDDYY10.)	mmddyy10.
w.	rounded to the nearest integer in w spaces	7.
w.d	rounded to d decimal places in w spaces	8.2
$w.	as character values in w spaces	$12.
DATEw.	as date values of the form 16OCT99 (DATE7.) or 16OCT1999 (DATE9.)	date9.

Field Widths

All SAS formats specify the total field width (**w**) that is used for displaying the values in the output. For example, suppose the longest value for the variable Net is a four-digit number, such as **5400**. To specify the COMMA**w.d** format for Net, you specify a field width of 5 or more. You must count the comma, because it occupies a position in the output.

⚠ When you use a SAS format, be sure to specify a field width (w) that is wide enough for the largest possible value. Otherwise, values might not be displayed properly.

Figure 4.30 *Specifying a Field Width (w) with the FORMAT Statement*

```
format net comma5.0;
   5  ,  4   0   0
   1  2  3   4   5
```

Decimal Places

For numeric variables you can also specify the number of decimal places (**d**), if any, to be displayed in the output. Numbers are rounded to the specified number of decimal places. In the example above, no decimal places are displayed.

Writing the whole number 2030 as 2,030.00 requires eight print positions, including two decimal places and the decimal point.

Figure 4.31 *Whole Number Decimal Places*

```
format qtr3tax comma8.2;
   2  ,  0  3  0  .  0  0
   1  2  3  4  5  6  7  8
```

Formatting 15374 with a dollar sign, commas, and two decimal places requires ten print positions.

Figure 4.32 *Specifying 10 Decimal Places*

```
format totsales dollar10.2;
  $  1  5  ,  3  7  4  .  0   0
  1  2  3  4  5  6  7  8  9  10
```

Examples

This table shows you how data values are displayed when different format, field width, and decimal place specifications are used.

Table 4.5 *Displaying Data Values with Formats*

Stored Value	Format	Displayed Value
38245.3975	COMMA12.2	38,245.40
38245.3975	12.2	38245.40
38245.3975	DOLLAR12.2	$38,245.40
38245.3975	DOLLAR9.2	$38245.40
38245.3975	DOLLAR8.2	38245.40
0	MMDDYY8.	01/01/60
0	MMDDYY10.	01/01/1960
0	DATE7.	01JAN60
0	DATE9.	01JAN1960

If a format is too small, the following message is written to the SAS log: "NOTE: At least one W.D format was too small for the number to be printed. The decimal may be shifted by the 'BEST' format."

Using Permanently Assigned Labels and Formats

You have seen how to *temporarily* assign labels and formats to variables. When you use a LABEL or FORMAT statement within a PROC step, the label or format applies only to the output from that step.

However, in your PROC steps, you can also take advantage of *permanently* assigned labels or formats. Permanent labels and formats can be assigned in the DATA step. These labels and formats are saved with the data set, and they can later be used by procedures that reference the data set.

For example, the DATA step below creates Flights.March and defines a format and label for the variable Date. Because the LABEL and FORMAT statements are inside the DATA step, they are written to the Flights.March data set and are available to the subsequent PRINT procedure.

```
data sasuser.paris;
set sasuser.laguardia;
   where dest="PAR" and (boarded=155 or boarded=146);
   label date='Departure Date';
   format date date9.;
run;

proc print data=sasuser.paris;
   var date dest boarded;
run;
```

Figure 4.33 *Using Permanent Labels and Formats*

Obs	Departure Date	Dest	Boarded
1	04MAR1999	PAR	146
2	07MAR1999	PAR	155
3	04MAR1999	PAR	146
4	07MAR1999	PAR	155

Notice that the PROC PRINT statement still requires the LABEL option in order to display the permanent labels. Other SAS procedures display permanently assigned labels and formats without additional statements or options.

You can learn about permanently assigning labels and formats in "Creating and Managing Variables" on page 304.

Additional Features

When you create list reports, you can use several other features to enhance your procedure output. For example, you can

- control where text strings split in labels by using the SPLIT= option.

```
proc print data=reps split='*';
   var salesrep type unitsold net commission;
   label salesrep='Sales*Representative';
run;
```

- create your own formats, which are particularly useful for formatting character values.

```
proc format;
   value $repfmt
          'TFB'='Bynum'
          'MDC'='Crowley'
          'WKK'='King';
run;
proc print data=vcrsales;
   var salesrep type unitsold;
   format salesrep $repfmt.;
run;
```

You can learn more about user-defined formats in "Creating and Applying User-Defined Formats" on page 233 .

Chapter Summary

Text Summary

Creating a Basic Report

To list the information in a SAS data set, you can use PROC PRINT. You use the PROC PRINT statement to invoke the PRINT procedure and to specify the data set that you are listing. Include the DATA= option to specify the data set that you are using. By default, PROC PRINT displays all observations and variables in the data set, includes a column for observation numbers on the far left, and displays observations and variables in the order in which they occur in the data set. If you use a LABEL statement with PROC PRINT, you must specify the LABEL option in the PROC PRINT statement.

To refine a basic report, you can

- select which variables and observations are processed
- sort the data
- generate column totals for numeric variables.

Selecting Variables

You can select variables and control the order in which they appear by using a VAR statement in your PROC PRINT step. To remove the Obs column, you can specify the NOOBS option in the PROC PRINT statement. As an alternative, you can replace the Obs column with one or more variables by using the ID statement.

Selecting Observations

The WHERE statement enables you to select observations that meet a particular condition in the SAS data set. You use comparison operators to express a condition in the WHERE statement. You can also use the CONTAINS operator to express a condition in the WHERE statement. To specify a condition based on the value of a character variable, you must enclose the value in quotation marks, and you must write the value with lower and uppercase letters exactly as it appears in the data set. You can also use the WHERE statement to select a subset of observations based on multiple conditions. To link a sequence of expressions into compound expressions, you use logical operators. When you test for multiple values of the same variable, you specify the variable name in each expression. You can use the IN operator as a convenient alternative. To control how compound expressions are evaluated, you can use parentheses.

Sorting Data

To display your data in sorted order, you use PROC SORT to sort your data before using PROC PRINT to create reports. By default, PROC SORT sorts the data set specified in the DATA= option permanently. If you do not want your data to be sorted permanently, you must create an output data set that contains the data in sorted order. The OUT= option in the PROC SORT statement specifies an output data set. If you need sorted data to produce output for only one SAS session, you should specify a temporary SAS data set as the output data set. The BY statement, which is required with PROC SORT, specifies the variable(s) whose values are used to sort the data.

Generating Column Totals

To total the values of numeric variables, use the SUM statement in the PROC PRINT step. You do not need to specify the variables in a VAR statement if you specify them in the SUM statement. Column totals appear at the end of the report in the same format as the values of the variables. To produce subtotals, add both the SUM statement and the BY statement to your PROC PRINT step. To show BY variable headings only once, use an ID and BY statement together with the SUM statement. As another enhancement to your report, you can request that each BY group be printed on a separate page by using the PAGEBY statement.

Double Spacing Output

To double space your SAS LISTING output, you can specify the DOUBLE option in the PROC PRINT statement.

Specifying Titles

To make your report more meaningful and self-explanatory, you can associate up to 10 titles with procedure output by using TITLE statements anywhere within or preceding the PROC step. After you define a title, it remains in effect until you modify it, cancel it, or end your SAS session. Redefining a title line cancels any higher-numbered title lines. To cancel all previous titles, specify a null TITLE statement (a TITLE statement with no number or text).

Specifying Footnotes

To add footnotes to your output, you can use the FOOTNOTE statement. Like TITLE statements, FOOTNOTE statements are global. Footnotes appear at the bottom of each page of procedure output, and footnote lines are *pushed up* from the bottom. The FOOTNOTE statement that has the largest number appears on the bottom line. After you define a footnote, it remains in effect until you modify it, cancel it, or end your SAS session. Redefining a footnote line cancels any higher-numbered footnote lines. To cancel all previous footnotes, specify a null FOOTNOTE statement (a FOOTNOTE statement with no number or text).

Assigning Descriptive Labels

To label the columns in your report with more descriptive text, you use the LABEL statement, which assigns a descriptive label to a variable. To display the labels that were assigned in a LABEL statement, you must specify the LABEL option in the PROC PRINT statement.

Formatting Data Values

To make data values more understandable when they are displayed in your procedure output, you can use the FORMAT statement, which associates formats with variables. The FORMAT statement remains in effect only for the PROC step in which it appears. Formats affect only how the data values appear in output, not the actual data values as they are stored in the SAS data set. All SAS formats can specify the total field width (w) that is used for displaying the values in the output. For numeric variables you can also specify the number of decimal places (d), if any, to be displayed in the output.

Using Permanently Assigned Labels and Formats

You can take advantage of permanently assigned labels or formats without adding LABEL or FORMAT statements to your PROC step.

Syntax

LIBNAME *libref 'SAS-data-library'*;

OPTIONS *options*;

PROC SORT DATA=*SAS-data-set* **OUT**=*SAS-data-set*;
  BY *variable(s)*;

RUN;

TITLE<*n* > *'text'*;

FOOTNOTE<*n*>*'text'*;

PROC PRINT DATA=*SAS-data-set*

  BY<DESCENDING>*BY-variable-1*<...<DESCENDING><*BY-variable-n*>>
    <NOTSORTED>;

  PAGEBY*BY-variable*;

  NOOBS LABEL DOUBLE;

  ID *variable(s)*;

  VAR *variable(s)*;

  WHERE *where-expression*;

  SUM *variable(s)*;

 LABEL *variable1='label1' variable2='label2' ...*;

 FORMAT *variable(s) format-name;*

RUN;

Sample Program

```
libname clinic 'c:\stress\labdata';
options nodate number pageno=15;
proc sort data=clinic.stress out=work.maxrates;
   by maxhr;
   where tolerance='I' and resthr>60;
run;
title 'August Admission Fees';
footnote 'For High Activity Patients';
proc print data=work.maxrates label double noobs;
   id name;
   var resthr maxhr rechr;
   label rechr='Recovery HR';
run;

proc print data=clinic.admit label;
   var actlevel fee;
   where actlevel='HIGH';
   label fee='Admission Fee';
   sum fee;
   format fee dollar4.;
run;
```

Points to Remember

- VAR, WHERE, SUM, FORMAT and LABEL statements remain in effect only for the PROC step in which they appear.

- If you don't use the OUT= option, PROC SORT permanently sorts the data set specified in the DATA= option.

- TITLE and FOOTNOTE statements remain in effect until you modify them, cancel them, or end your SAS session.

- Be sure to match the quotation marks that enclose the text in TITLE, FOOTNOTE, and LABEL statements.

- To display labels in PRINT procedure output, remember to add the LABEL option to the PROC PRINT statement.

- To permanently assign labels or formats to data set variables, place the LABEL or FORMAT statement inside the DATA step.

Chapter Quiz

Select the best answer for each question. After completing the quiz, you can check your answers using the answer key in the appendix.

1. Which PROC PRINT step below creates the following output?

Date	On	Changed	Flight
04MAR99	232	18	219
05MAR99	160	4	219
06MAR99	163	14	219
07MAR99	241	9	219
08MAR99	183	11	219
09MAR99	211	18	219
10MAR99	167	7	219

   a.
```
proc print data=flights.laguardia noobs;
    var on changed flight;
    where on>=160;
run;
```

   b.
```
proc print data=flights.laguardia;
    var date on changed flight;
    where changed>3;
run;
```

   c.
```
proc print data=flights.laguardia label;
    id date;
    var boarded transferred flight;
    label boarded='On' transferred='Changed';
    where flight='219';
run;
```

    d.
```
proc print flights.laguardia noobs;
   id date;
   var date on changed flight;
   where flight='219';
run;
```

2. Which of the following PROC PRINT steps is correct if labels are not stored with the data set?

    a.
```
proc print data=allsales.totals label;
   label region8='Region 8 Yearly Totals';
run;
```

    b.
```
proc print data=allsales.totals;
   label region8='Region 8 Yearly Totals';
run;
```

    c.
```
proc print data allsales.totals label noobs;
run;
```

    d.
```
proc print allsales.totals label;
run;
```

3. Which of the following statements selects from a data set only those observations for which the value of the variable Style is **RANCH**, **SPLIT**, or **TWOSTORY**?

    a. `where style='RANCH' or 'SPLIT' or 'TWOSTORY';`

    b. `where style in 'RANCH' or 'SPLIT' or 'TWOSTORY';`

    c. `where style in (RANCH, SPLIT, TWOSTORY);`

    d. `where style in ('RANCH','SPLIT','TWOSTORY');`

4. If you want to sort your data and create a temporary data set named Calc to store the sorted data, which of the following steps should you submit?

    a.
```
proc sort data=work.calc out=finance.dividend;
run;
```

    b.
```
proc sort dividend out=calc;
   by account;
run;
```

    c.
```
proc sort data=finance.dividend out=work.calc;
   by account;
run;
```

    d.
```
proc sort from finance.dividend to calc;
   by account;
run;
```

5. Which options are used to create the following PROC PRINT output?

```
                                    13:27 Monday, March 22, 1999

        Patient      Arterial      Heart      Cardiac      Urinary

          203           88          95          66           110

           54           83         183          95             0

          664           72         111         332            12

          210           74          97         369             0

          101           80         130         291             0
```

    a. the DATE system option and the LABEL option in PROC PRINT

    b. the DATE and NONUMBER system options and the DOUBLE and NOOBS options in PROC PRINT

    c. the DATE and NONUMBER system options and the DOUBLE option in PROC PRINT

    d. the DATE and NONUMBER system options and the NOOBS option in PROC PRINT

6. Which of the following statements can you use in a PROC PRINT step to create this output?

Month	Instructors	AerClass	WalkJogRun	Swim
01	1	37	91	83
02	2	41	102	27
03	1	52	98	19
04	1	61	118	22
05	3	49	88	29
	8	240	497	180

    a.
```
var month instructors;
sum instructors aerclass walkjogrun swim;
```

    b.
```
var month;
sum instructors aerclass walkjogrun swim;
```

    c.
```
var month instructors aerclass;
sum instructors aerclass walkjogrun swim;
```

    d. all of the above

7. What happens if you submit the following program?

```
proc sort data=clinic.diabetes;
run;
proc print data=clinic.diabetes;
   var age height weight pulse;
   where sex='F';
run;
```

    a. The PROC PRINT step runs successfully, printing observations in their sorted order.

    b. The PROC SORT step permanently sorts the input data set.

   c. The PROC SORT step generates errors and stops processing, but the PROC PRINT step runs successfully, printing observations in their original (unsorted) order.

   d. The PROC SORT step runs successfully, but the PROC PRINT step generates errors and stops processing.

8. If you submit the following program, which output does it create?

```
proc sort data=finance.loans out=work.loans;
   by months amount;
run;
proc print data=work.loans noobs;
   var months;
   sum amount payment;
   where months<360;
run;
```

   a.

Months	Amount	Payment
12	$3,500	$308.52
24	$8,700	$403.47
36	$10,000	$325.02
48	$5,000	$128.02
	$27,200	$1,165.03

   b.

Months	Amount	Payment
12	$3,500	$308.52
24	$8,700	$403.47
36	$10,000	$325.02
48	$5,000	$128.02
	27,200	1,165.03

   c.

Months	Amount	Payment
12	$3,500	$308.52
48	$5,000	$128.02
24	$8,700	$403.47
36	$10,000	$325.02
	$27,200	$1,165.03

   d.

Months	Amount	Payment
12	$3,500	$308.52
24	$8,700	$403.47
36	$10,000	$325.02
48	$5,000	$128.02
		$1,165.03

9. Choose the statement below that selects rows in which

   - the amount is less than or equal to $5000

   - the account is 101-1092 or the rate equals 0.095.

   a. ```
      where amount <= 5000 and
           account='101-1092' or rate = 0.095;
      ```

   b. ```
      where (amount le 5000 and account='101-1092')
           or rate = 0.095;
      ```

   c. ```
      where amount <= 5000 and
           (account='101-1092' or rate eq 0.095);
      ```

   d. ```
      where amount <= 5000 or account='101-1092'
           and rate = 0.095;
      ```

10. What does PROC PRINT display by default?

    a. PROC PRINT does not create a default report; you must specify the rows and columns to be displayed.

    b. PROC PRINT displays all observations and variables in the data set. If you want an additional column for observation numbers, you can request it.

    c. PROC PRINT displays columns in the following order: a column for observation numbers, all character variables, and all numeric variables.

    d. PROC PRINT displays all observations and variables in the data set, a column for observation numbers on the far left, and variables in the order in which they occur in the data set.

Chapter 5
Creating SAS Data Sets from External Files

Overview

Introduction

In order to create reports with SAS procedures, your data must be in the form of a SAS data set. If your data is not stored in the form of a SAS data set, you need to create a SAS data set by entering data, by reading raw data, or by accessing files that were created by other software.

This chapter shows you how to design and write DATA step programs to create SAS data sets from raw data that is stored in an external file and from data stored in Microsoft Excel worksheets. It also shows you how to read data from a SAS data set and write observations out to these destinations.

Regardless of the input data source — raw files or Excel worksheets — you use the DATA step to read in the data and create the SAS data set.

Figure 5.1 *Using the DATA Step to Create SAS Data Sets*

	ID	Age	Activity Level	Sex
1	2462	38	HIGH	F
2	2523	26	HIGH	M
3	2544	29	HIGH	M
4	2568	28	HIGH	M

SAS Data Set

Objectives

In this chapter, you learn to

- reference a SAS library

- reference a raw data file

- name a SAS data set to be created

- specify a raw data file to be read

- read standard character and numeric values in fixed fields

- create new variables and assign values

- select observations based on conditions

- read instream data

- submit and verify a DATA step program

- read a SAS data set and write the observations out to a raw data file.

- use the DATA step to create a SAS data set from an Excel worksheet

- use the SAS/ACCESS LIBNAME statement to read from an Excel worksheet

- create an Excel worksheet from a SAS data set

- use the IMPORT procedure to read external files

Raw Data Files

A raw data file is an external text file whose records contain data values that are organized in fields. Raw data files are non-proprietary and can be read by a variety of software programs. The sample raw data files in this chapter are shown with a ruler to help you identify where individual fields begin and end. The ruler is not part of the raw data file.

Figure 5.2 *Raw Data File*

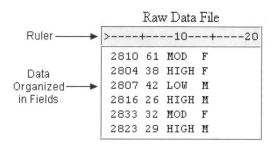

The table below describes the record layout for a raw data file that contains readings from exercise stress tests that have been performed on patients at a health clinic. Exercise physiologists in the clinic use the test results to prescribe various exercise therapies. The file contains fixed fields. That is, values for each variable are in the same location in all records.

Table 5.1 *Record Layout for Raw Data*

Field Name	Starting Column	Ending Column	Description of Field
ID	1	4	patient ID number
Name	6	25	patient name
RestHR	27	29	resting heart rate
MaxHR	31	33	maximum heart rate during test
RecHR	35	37	recovery heart rate after test
TimeMin	39	40	time, complete minutes
TimeSec	42	43	time, seconds
Tolerance	45	45	comparison of stress test tolerance between this test and the last test (I=increased, D=decreased, S=same, N=no previous test)

Steps to Create a SAS Data Set from a Raw Data File

Let's look at the steps for creating a SAS data set from a raw data file. In the first part of this chapter, you will follow these steps to create a SAS data set from a raw data file that contains fixed fields.

Before reading raw data from a file, you might need to reference the SAS library in which you will store the data set. Then you can write a DATA step program to read the raw data file and create a SAS data set.

To read the raw data file, the DATA step must provide the following instructions to SAS:

- the location or name of the external text file

- a name for the new SAS data set

- a reference that identifies the external file

- a description of the data values to be read.

After using the DATA step to read the raw data, you can use a PROC PRINT step to produce a report that displays the data values that are in the new data set.

The table below outlines the basic statements that are used in a program that reads raw data in fixed fields. Throughout this chapter, you'll see similar tables that show sample SAS statements.

Table 5.2   *Basic Statements for Reading Raw Data*

To do this...	Use this SAS statement...
Reference SAS data library	LIBNAME statement
Reference external file	FILENAME statement
Name SAS data set	DATA statement
Identify external file	INFILE statement
Describe data	INPUT statement
Execute DATA step	RUN statement
Print the data set	PROC PRINT statement
Execute final program step	RUN statement

You can also use additional SAS statements to perform tasks that customize your data for your needs. For example, you may want to create new variables from the values of existing variables.

Referencing a SAS Library

Using a LIBNAME Statement

As you begin to write the program, remember that you might need to use a LIBNAME statement to reference the permanent SAS library in which the data set will be stored.

To do this...	Use this SAS statement...	Example

Reference a SAS library	LIBNAME statement	`libname taxes 'c:`
		`\users\name`
		`\sasuser';`

For example, the LIBNAME statement below assigns the libref Taxes to the SAS library in the folder `C:\Acct\Qtr1\Report` in the Windows environment.

```
libname taxes 'c:\acct\qtr1\report';
```

You do not need to use a LIBNAME statement in all situations. For example, if you are storing the data set in a temporary SAS data set or if SAS has automatically assigned the libref for the permanent library that you are using.

Many of the examples in this chapter use the libref Sasuser, which SAS automatically assigns.

Referencing a Raw Data File

Using a FILENAME Statement

When reading raw data, you can use the FILENAME statement to point to the location of the external file that contains the data. Just as you assign a libref by using a LIBNAME statement, you assign a fileref by using a FILENAME statement.

Table 5.3 *Referencing a Raw Data File*

To do this...	Use this SAS statement...	Example
Reference a SAS library	LIBNAME statement	`libname libref` `'SAS-data-library';`
Reference an external file	FILENAME statement	`filename tests 'c:` `\users\tmill.dat';`

Filerefs perform the same function as librefs: they temporarily point to a storage location for data. However, librefs reference SAS data libraries, whereas filerefs reference external files.

Figure 5.3 *Filerefs and Librefs*

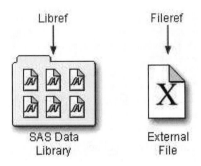

General form, FILENAME statement:

FILENAME *fileref* '*filename*';

where

- *fileref* is a name that you associate with an external file. The name must be 1 to 8 characters long, begin with a letter or underscore, and contain only letters, numerals, or underscores.

- '*filename*' is the fully qualified name or location of the file.

Defining a Fully Qualified Filename

The following FILENAME statement temporarily associates the fileref Tests with the external file that contains the data from the exercise stress tests. The complete filename is specified as **C:\Users\Tmill.dat** in the Windows environment.

```
filename tests 'c:\users\tmill.dat';
```

Figure 5.4 *File Location*

```
                   Raw Data File Tests
1---+----10---+----20---+----30---+----40---+--
2458 Murray, W          72    185 128 12 38 D
2462 Almers, C          68    171 133 10 5  I
2501 Bonaventure, T     78    177 139 11 13 I
2523 Johnson, R         69    162 114 9  42 S
2539 LaMance, K         75    168 141 11 46 D
2552 Reberson, P        69    158 139 15 41 D
```

Defining an Aggregate Storage Location

You can also use a FILENAME statement to associate a fileref with an aggregate storage location, such as a directory that contains multiple external files.

Figure 5.5 Aggregate Storage Location

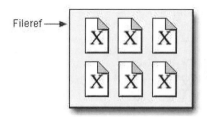

This FILENAME statement temporarily associates the fileref Finance with the aggregate storage directory `C:\Users\Personal\Finances`:

```
filename finance 'c:\users\personal\finances';
```

Viewing Active Filerefs

Like librefs, the filerefs currently defined for your SAS session are listed in the SAS Explorer window.

To view details about a referenced file, double-click **File Shortcuts** (or select **File Shortcuts** and then **Open** from the pop-up menu). Then select **View** ⇨ **Details**. Information for each file (name, size, type, and host path name) is listed.

Both the LIBNAME and FILENAME statements are global. In other words, they remain in effect until you change them, cancel them, or end your SAS session.

Referencing a Fully Qualified Filename

When you associate a fileref with an individual external file, you specify the fileref in subsequent SAS statements and commands.

Figure 5.6 Referencing a Fully Qualified Filename

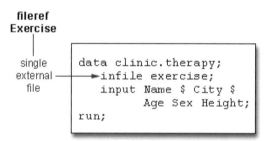

Referencing a File in an Aggregate Storage Location

To reference an external file with a fileref that points to an aggregate storage location, you specify the fileref followed by the individual filename in parentheses:

Figure 5.7 *Referencing a File in an Aggregate Storage Location*

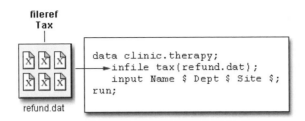

  In the Windows environment, you can omit the filename extension but you will need to add quotation marks when referencing an external file, as in

```
infile tax('refund');
```

For details on referencing external files stored in aggregate storage locations, see the SAS documentation for your operating environment.

Writing a DATA Step Program

Naming the Data Set

The DATA statement indicates the beginning of the DATA step and names the SAS data set to be created.

Table 5.4 *Naming the Data Set*

To do this...	Use this SAS statement...	Example
Reference a SAS library	LIBNAME statement	`libname libref 'SAS-data-library';`
Reference an external file	FILENAME statement	`filename tests 'c:\users\tmill.dat';`
Name a SAS data set	DATA statement	`data clinic.stress;`

General form, basic DATA statement:

DATA *SAS-data-set-1 <...SAS-data-set-n>*;

where *SAS-data-set* names (in the format *libref.filename*) the data set or data sets to be created.

Remember that the SAS data set name is a two-level name. For example, the two-level name Clinic.Admit specifies that the data set Admit is stored in the permanent SAS library to which the libref Clinic has been assigned.

Figure 5.8 *Two-Level Names*

Specifying the Raw Data File

When reading raw data, use the INFILE statement to indicate which file the data is in.

Table 5.5 *Specifying the Raw Data File*

To do this...	Use this SAS statement...	Example
Reference a SAS library	LIBNAME statement	`libname libref 'SAS-data-library';`
Reference an external file	FILENAME statement	`filename tests 'c:\users\tmill.dat';`
Name a SAS data set	DATA statement	`data clinic.stress;`
Identify an external file	INFILE statement	`infile tests;`

General form, INFILE statement:

INFILE *file-specification <options>*;

where

- *file-specification* can take the form *fileref* to name a previously defined file reference or *'filename'* to point to the actual name and location of the file

- *options* describe the input file's characteristics and specify how it is to be read with the INFILE statement.

To read the raw data file to which the fileref Tests has been assigned, you write the following INFILE statement:

```
infile tests;
```

Instead of using a FILENAME statement, you can choose to identify the raw data file by specifying the entire filename and location in the INFILE statement. For example, the following statement points directly to the `C:\Irs\Personal\Refund.dat` file:

```
infile 'c:\irs\personal\refund.dat';
```

Column Input

Column input specifies actual column locations for values. However, column input is appropriate only in certain situations. When you use column input, your data *must* be

- standard character or numeric values

- in fixed fields.

Standard and Nonstandard Numeric Data

Standard numeric data values can contain only

- numerals

- decimal points

- numbers in scientific or E-notation (2.3E4, for example)

- plus or minus signs.

Nonstandard numeric data includes

- values that contain special characters, such as percent signs (%), dollar signs ($), and commas (,)

- date and time values

- data in fraction, integer binary, real binary, and hexadecimal forms.

The external file that is referenced by the fileref Staff contains the personnel information for a technical writing department of a small computer manufacturer. The fields contain values for each employee's last name, first name, job code, and annual salary.

Notice that the values for Salary contain commas. So, the values for Salary are considered to be nonstandard numeric values. You cannot use column input to read these values.

Figure 5.9   *Raw Data File*

Raw Data File Staff

```
1---+----10---+----20---+---
EVANS    DONNY 112 29,996.63
HELMS    LISA  105 18,567.23
HIGGINS  JOHN  111 25,309.00
LARSON   AMY   113 32,696.78
MOORE    MARY  112 28,945.89
```

Fixed-Field Data

Raw data can be organized in several different ways.

The following external delimited file contains data that is not arranged in columns, meaning data that is not arranged in columns. Notice that the values for a particular field do not begin and end in the same columns. You cannot use column input to read this file.

Figure 5.10   *Fixed Field Data*

```
1---+----10---+----20
BARNES NORTH 360.98
FARLSON WEST 243.94
LAWRENCE NORTH 195.04
NELSON EAST 169.30
STEWART SOUTH 238.45
TAYLOR WEST 318.87
```

The following external file contains data that is arranged in columns or fixed fields. You can specify a beginning and ending column for each field. Let's look at how column input can be used to read this data.

Figure 5.11   *External File with Columns*

```
1---+----10---+----20
2810 61 MOD  F
2804 38 HIGH F
2807 42 LOW  M
2816 26 HIGH M
2833 32 MOD  F
2823 29 HIGH M
```

If you are not familiar with the content and structure of your raw data files, you can use PROC FSLIST to view them.

Describing the Data

The INPUT statement describes the fields of raw data to be read and placed into the SAS data set.

Table 5.6   *Describing the Data*

To do this...	Use this SAS statement...	Example
Reference a SAS library	LIBNAME statement	`libname libref 'SAS-data-library';`
Reference an external file	FILENAME statement	`filename tests 'c:\users\tmill.dat';`
Name a SAS data set	DATA statement	`data clinic.stress;`
Identify an external file	INFILE statement	`infile tests;`
Describe data	INPUT statement	`input ID $ 1-4 Name $ 6-25 ...;`
Execute the DATA step	RUN statement	`run;`

General form, INPUT statement using column input:

INPUT *variable <$> startcol-endcol . . .*;

where

- *variable* is the SAS name that you assign to the field
- the dollar sign ($) identifies the variable type as character (if the variable is numeric, then nothing appears here)
- *startcol* represents the starting column for this variable
- *endcol* represents the ending column for this variable.

Look at the small data file shown below. For each field of raw data that you want to read into your SAS data set, you must specify the following information in the INPUT statement:

- a valid SAS variable name

- a type (character or numeric)

- a range (starting column and ending column).

Figure 5.12 *Raw Data File*

```
     Raw Data File Exercise
1---+----10---+----20
2810 61 MOD   F
2804 38 HIGH  F
2807 42 LOW   M
2816 26 HIGH  M
2833 32 MOD   F
2823 29 HIGH  M
```

The INPUT statement below assigns the character variable ID to the data in columns 1-4, the numeric variable Age to the data in columns 6-7, the character variable ActLevel to the data in columns 9-12, and the character variable Sex to the data in column 14.

```
filename exer 'c:\users\exer.dat';
data exercise;
   infile exer;
   input ID $ 1-4 Age 6-7 ActLevel $ 9-12 Sex $ 14;
run;
```

Figure 5.13 *Assigning Column Ranges to Variables*

SAS Data Set Work.Exercise

Obs	ID	Age	ActLevel	Sex
1	2810	61	MOD	F
2	2804	38	HIGH	F
3	2807	42	LOW	M
4	2816	26	HIGH	M
5	2833	32	MOD	F
6	2823	29	HIGH	M

When you use column input, you can

- read any or all fields from the raw data file

- read the fields in any order

- specify only the starting column for values that occupy only one column.

```
   input ActLevel $ 9-12 Sex $ 14 Age 6-7;
```

Remember that when you name a new variable, you must specify the name in the exact case that you want it stored, for example NewBalance. Thereafter, you can specify the name in uppercase, lowercase, or mixed case.

Specifying Variable Names

Each variable has a name that conforms to SAS naming conventions. Variable names

- must be 1 to 32 characters in length

- must begin with a letter (A-Z) or an underscore (_)

- can continue with any combination of numerals, letters, or underscores.

Let's look at an INPUT statement that uses column input to read the three data fields in the raw data file below.

Figure 5.14 *Raw Data File*

```
      Raw Data File Admit
1---+----10---+----20
58MOD M
29LOW F
34LOW M
41HIGHF
30MOD F
22HIGHM
```

The values for the variable Age are located in columns 1-2. Because Age is a numeric variable, you do not specify a dollar sign ($) after the variable name.

```
input Age 1-2
```

The values for the variable ActLevel are located in columns 3-6. You specify a $ to indicate that ActLevel is a character variable.

```
input Age 1-2 ActLevel $ 3-6
```

The values for the character variable Sex are located in column 7. Notice that you specify only a single column.

```
input Age 1-2 ActLevel $ 3-6 Sex $ 7;
```

Submitting the DATA Step Program

Verifying the Data

To verify your data, it is a good idea to use the OBS= option in the INFILE statement. Adding OBS=*n* to the INFILE statement enables you to process only records 1 through *n*, so you can verify that the correct fields are read before reading the entire data file.

The program below reads the first ten records in the raw data file referenced by the fileref Tests. The data is stored in a permanent SAS data set, named Sasuser.Stress. Don't forget a RUN statement, which tells SAS to execute the previous SAS statements.

```
data sasuser.stress;
   infile tests obs=10;
   input ID $ 1-4 Name $ 6-25
         RestHR 27-29 MaxHR 31-33
         RecHR 35-37 TimeMin 39-40
```

```
        TimeSec 42-43 Tolerance $ 45;
run;
```

Figure 5.15 *SAS Data Set sasuser.stress*

ID	Name	RestHR	MaxHR	RecHR	TimeMin	TimeSec	Tolerance
2458	Murray, W	72	185	128	12	38	D
2462	Almers, C	68	171	133	10	5	I
2501	Bonaventure, T	78	177	139	11	13	I
2523	Johnson, R	69	162	114	9	42	S
2539	LaMance, K	75	168	141	11	46	D
2544	Jones, M	79	187	136	12	26	N
2552	Reberson, P	69	158	139	15	41	D
2555	King, E	70	167	122	13	13	I
2563	Pitts, D	71	159	116	10	22	S
2568	Eberhardt, S	72	182	122	16	49	N

Checking DATA Step Processing

If you submit the DATA step below it will run successfully.

```
data sasuser.stress;
   infile tests obs=10;
   input ID $ 1-4 Name $ 6-25
        RestHR 27-29 MaxHR 31-33
        RecHR 35-37 TimeMin 39-40
        TimeSec 42-43 Tolerance $ 45;
run;
```

Messages in the log verified that the raw data file was read correctly. The notes in the log indicate that

- 10 records were read from the raw data file

- the SAS data set Sasuser.Stress was created with 10 observations and 8 variables.

Figure 5.16 *SAS Log*

SAS Log

```
NOTE: The infile TESTS is:
      File Name=C:\My SAS Files\tests.dat,
      RECFM=V,LRECL=256

NOTE: 10 records were read from the infile TESTS.
      The minimum record length was 80.
      The maximum record length was 80.
NOTE: The data set SASUSER.STRESS has 10 observations
      and 8 variables.
NOTE: DATA statement used 0.07 seconds
```

Printing the Data Set

The messages in the log seem to indicate that the DATA step program correctly accessed the raw data file. But it is a good idea to look at the ten observations in the new data set before reading the entire raw data file. You can submit a PROC PRINT step to view the data.

Table 5.7 *Printing the Data Set*

To do this...	Use this SAS statement...	Example
Reference a SAS library	LIBNAME statement	`libname libref 'SAS-data-library';`
Reference an external file	FILENAME statement	`filename tests 'c:\users\tmill.dat';`
Name a SAS data set	DATA statement	`data clinic.stress;`
Identify an external file	INFILE statement	`infile tests obs=10;`
Describe data	INPUT statement	`input ID $ 1-4 Name $ 6-25 ...;`
Execute the DATA step	RUN statement	`run;`
Print the data set	PROC PRINT statement	`proc print data=clinic.stress;`
Execute the final program step	RUN statement	`run;`

The following PROC PRINT step prints the Sasuser.Stress data set.

```
proc print data=sasuser.stress;
run;
```

The PROC PRINT output indicates that the variables in the Sasuser.Stress data set were read correctly for the first ten records.

Figure 5.17 *PROC Print Output*

Obs	ID	Name	RestHR	MaxHR	RecHR	TimeMin	TimeSec	Tolerance
1	2458	Murray, W	72	185	128	12	38	D
2	2462	Almers, C	68	171	133	10	5	I
3	2501	Bonaventure, T	78	177	139	11	13	I
4	2523	Johnson, R	69	162	114	9	42	S
5	2539	LaMance, K	75	168	141	11	46	D
6	2544	Jones, M	79	187	136	12	26	N
7	2552	Reberson, P	69	158	139	15	41	D
8	2555	King, E	70	167	122	13	13	I
9	2563	Pitts, D	71	159	116	10	22	S
10	2568	Eberhardt, S	72	182	122	16	49	N

Reading the Entire Raw Data File

Now that you've checked the log and verified your data, you can modify the DATA step to read the entire raw data file. To do so, remove the OBS= option from the INFILE statement and re-submit the program.

```
data sasuser.stress;
   infile tests;
   input ID $ 1-4 Name $ 6-25
         RestHR 27-29 MaxHR 31-33
         RecHR 35-37 TimeMin 39-40
         TimeSec 42-43 Tolerance $ 45;
run;
```

Invalid Data

When you submit the revised DATA step and check the log, you see a note indicating that invalid data appears for the variable RecHR in line 14 of the raw data file, columns 35-37.

This note is followed by a column ruler and the actual data line that contains the invalid value for RecHR.

Figure 5.18 *SAS Log*

SAS Log

```
NOTE: Invalid data for RecHR in line 14 35-37.
RULE:       ----+----1----+----2----+----3----+----4----+----5---
14          2575 Quigley, M            74  152 Q13 11 26 I 45
ID=2575 Name=Quigley, M RestHR=74 MaxHR=152 RecHR=. TimeMin=11
TimeSec=26 Tolerance=I _ERROR_=1
_N_=14
NOTE: 21 records were read from the infile TESTS.
      The minimum record length was 80.
      The maximum record length was 80.
NOTE: The data set SASUSER.STRESS has 21 observations
      and 8 variables.
NOTE: DATA statement used 0.13 seconds
```

The value Q13 is a data-entry error. It was entered incorrectly for the variable RecHR.

RecHR is a numeric variable, but Q13 is not a valid number. So RecHR is assigned a missing value, as indicated in the log. Because RecHR is numeric, the missing value is represented with a period.

Notice, though, that the DATA step does not fail as a result of the invalid data but continues to execute. Unlike syntax errors, invalid data errors do not cause SAS to stop processing a program.

Assuming that you have a way to edit the file and can justify a correction, you can correct the invalid value and rerun the DATA step. If you did this, the log would then show that the data set Sasuser.Stress was created with 21 observations, 8 variables, and no messages about invalid data.

Figure 5.19 *SAS Log*

SAS Log

```
NOTE: The infile TESTS2 is:
      File Name=C:\My SAS Files\tests2.dat,
      RECFM=V,LRECL=256

NOTE: 21 records were read from the infile TESTS2.
      The minimum record length was 80.
      The maximum record length was 80.
NOTE: The data set SASUSER.STRESS has 21 observations
      and 8 variables.
NOTE: DATA statement used 0.14 seconds
```

After correcting the raw data file, you can print the data again to verify that it is correct.

```
proc print data=sasuser.stress;
run;
```

Figure 5.20 PROC Print Output

Obs	ID	Name	RestHR	MaxHR	RecHR	TimeMin	TimeSec	Tolerance
1	2458	Murray, W	72	185	128	12	38	D
2	2462	Almers, C	68	171	133	10	5	I
3	2501	Bonaventure, T	78	177	139	11	13	I
4	2523	Johnson, R	69	162	114	9	42	S
5	2539	LaMance, K	75	168	141	11	46	D
6	2544	Jones, M	79	187	136	12	26	N
7	2552	Reberson, P	69	158	139	15	41	D
8	2555	King, E	70	167	122	13	13	I
9	2563	Pitts, D	71	159	116	10	22	S
10	2568	Eberhardt, S	72	182	122	16	49	N
11	2571	Nunnelly, A	65	181	141	15	2	I
12	2572	Oberon, M	74	177	138	12	11	D
13	2574	Peterson, V	80	164	137	14	9	D
14	2575	Quigley, M	74	152	113	11	26	I
15	2578	Cameron, L	75	158	108	14	27	I
16	2579	Underwood, K	72	165	127	13	19	S
17	2584	Takahashi, Y	76	163	135	16	7	D
18	2586	Derber, B	68	176	119	17	35	N
19	2588	Ivan, H	70	182	126	15	41	N
20	2589	Wilcox, E	78	189	138	14	57	I
21	2595	Warren, C	77	170	136	12	10	S

Whenever you use the DATA step to read raw data, remember the steps that you followed in this chapter, which help ensure that you don't waste resources when accessing data:

* write the DATA step using the OBS= option in the INFILE statement
* submit the DATA step
* check the log for messages
* view the resulting data set
* remove the OBS= option and resubmit the DATA step
* check the log again
* view the resulting data set again.

Creating and Modifying Variables

Overview

So far in this chapter, you've read existing data. But sometimes existing data doesn't provide the information you need. To modify existing values or to create new variables, you can use an assignment statement in any DATA step.

General form, assignment statement:

variable=expression;

where

- *variable* names a new or existing variable

- *expression* is any valid SAS expression

The assignment statement is one of the few SAS statements that doesn't begin with a keyword.

For example, here is an assignment statement that assigns the character value **Toby Witherspoon** to the variable Name:

```
Name='Toby Witherspoon';
```

SAS Expressions

You use SAS expressions in assignment statements and many other SAS programming statements to

- transform variables

- create new variables

- conditionally process variables

- calculate new values

- assign new values.

An expression is a sequence of operands and operators that form a set of instructions. The instructions are performed to produce a new value:

- Operands are variable names or constants. They can be numeric, character, or both.

- Operators are special-character operators, grouping parentheses, or functions. You can learn about functions in "Transforming Data with SAS Functions" on page 404 .

Using Operators in SAS Expressions

To perform a calculation, you use arithmetic operators. The table below lists arithmetic operators.

Table 5.8 *Using Operators in SAS Expressions*

Operator	Action	Example	Priority
-	negative prefix	`negative=-x;`	I
**	exponentiation	`raise=x**y;`	I
*	multiplication	`mult=x*y;`	II
/	division	`divide=x/y;`	II
+	addition	`sum=x+y;`	III
-	subtraction	`diff=x-y;`	III

When you use more than one arithmetic operator in an expression,

- operations of priority I are performed before operations of priority II, and so on
- consecutive operations that have the same priority are performed
  - from right to left within priority I
  - from left to right within priority II and III
- you can use parentheses to control the order of operations.

⚠ When a value that is used with an arithmetic operator is missing, the result of the expression is missing. The assignment statement assigns a missing value to a variable if the result of the expression is missing.

You use the following comparison operators to express a condition.

Table 5.9 *Comparison Operators*

Operator	Meaning	Example
= or eq	equal to	`name='Jones, C.'`
^= or ne	not equal to	`temp ne 212`
> or gt	greater than	`income>20000`
< or lt	less than	`partno lt "BG05"`
>= or ge	greater than or equal to	`id>='1543'`
<= or le	less than or equal to	`pulse le 85`

To link a sequence of expressions into compound expressions, you use logical operators, including the following:

Table 5.10 *Logical Operators*

Operator	Meaning
AND or &	and, both. If both expressions are true, then the compound expression is true.
OR or		or, either. If either expression is true, then the compound expression is true.

More Examples of Assignment Statements

The assignment statement in the DATA step below creates a new variable, TotalTime, by multiplying the values of TimeMin by 60 and then adding the values of TimeSec.

```
data sasuser.stress;
   infile tests;
   input ID $ 1-4 Name $ 6-25 RestHR 27-29 MaxHR 31-33
         RecHR 35-37 TimeMin 39-40 TimeSec 42-43
         Tolerance $ 45;
   TotalTime=(timemin*60)+timesec;
run;
```

Figure 5.21 *Assignment Statement Output*

SAS Data Set Sasuser.Stress (Partial Listing)

ID	Name	RestHR	MaxHR	RecHR	TimeMin	TimeSec	Tolerance	TotalTime
2458	Murray, W	72	185	128	12	38	D	758
2462	Almers, C	68	171	133	10	5	I	605
2501	Bonaventure, T	78	177	139	11	13	I	673
2523	Johnson, R	69	162	114	9	42	S	582
2539	LaMance, K	75	168	141	11	46	D	706

The expression can also contain the variable name that is on the left side of the equal sign, as the following assignment statement shows. This statement re-defines the values of the variable RestHR as 10 percent higher.

```
data sasuser.stress;
   infile tests;
   input ID $ 1-4 Name $ 6-25 RestHR 27-29 MaxHR 31-33
         RecHR 35-37 TimeMin 39-40 TimeSec 42-43
         Tolerance $ 45;
   resthr=resthr+(resthr*.10);
run;
```

When a variable name appears on both sides of the equal sign, the original value on the right side is used to evaluate the expression. The result is assigned to the variable on the left side of the equal sign.

```
data sasuser.stress;
   infile tests;
   input ID $ 1-4 Name $ 6-25 RestHR 27-29 MaxHR 31-33
         RecHR 35-37 TimeMin 39-40 TimeSec 42-43
         Tolerance $ 45;
```

```
    resthr=resthr+(resthr*.10);
run;    ^           ^


    result      original value
```

Date Constants

You can assign date values to variables in assignment statements by using date constants. To represent a constant in SAS date form, specify the date as *'ddmmmyy'* or *'ddmmmyyyy'*, immediately followed by a D.

General form, date constant:

'ddmmmyy'd

or

"ddmmmyy"d

where

- *dd* is a one- or two-digit value for the day

- *mmm* is a three-letter abbreviation for the month (JAN, FEB, and so on)

- *yy* or *yyyy* is a two- or four-digit value for the year, respectively.

Be sure to enclose the date in quotation marks.

Example

In the following program, the second assignment statement assigns a date value to the variable TestDate.

```
data sasuser.stress;
   infile tests;
   input ID $ 1-4 Name $ 6-25 RestHR 27-29 MaxHR 31-33
         RecHR 35-37 TimeMin 39-40 TimeSec 42-43
         Tolerance $ 45;
   TotalTime=(timemin*60)+timesec;
   TestDate='01jan2000'd;
run;
```

You can also use SAS time constants and SAS datetime constants in assignment statements.

```
   Time='9:25't;


   DateTime='18jan2005:9:27:05'dt;
```

Subsetting Data

As you read your data, you can subset it by processing only those observations that meet a specified condition. To do this, you can use a subsetting IF statement in any DATA step.

Using a Subsetting IF Statement

The subsetting IF statement causes the DATA step to continue processing only those observations that meet the condition of the expression specified in the IF statement. The resulting SAS data set or data sets contain a subset of the original external file or SAS data set.

General form, subsetting IF statement:

IF *expression*;

where *expression* is any valid SAS expression.

- If the expression is *true*, the DATA step continues to process that observation.
- If the expression is *false*, no further statements are processed for that observation, and control returns to the top of the DATA step.

Example

The subsetting IF statement below selects only observations whose values for Tolerance are **D**. It is positioned in the DATA step so that other statements do not need to process unwanted observations.

```
data sasuser.stress;
   infile tests;
   input ID $ 1-4 Name $ 6-25 RestHR 27-29 MaxHR 31-33
         RecHR 35-37 TimeMin 39-40 TimeSec 42-43
         Tolerance $ 45;
   if tolerance='D';
   TotalTime=(timemin*60)+timesec;
run;
```

Because Tolerance is a character variable, the value **D** must be enclosed in quotation marks, and it must be the same case as in the data set.

See the SAS documentation for a comparison of the WHERE and subsetting IF statements when they are used in the DATA step.

Reading Instream Data

Overview

Throughout this chapter, our program has contained an INFILE statement that identifies an external file to read.

```
data sasuser.stress;
   infile tests;
   input ID $ 1-4 Name $ 6-25 RestHR 27-29 MaxHR 31-33
         RecHR 35-37 TimeMin 39-40 TimeSec 42-43
         Tolerance $ 45;
    if tolerance='D';
```

```
     TotalTime=(timemin*60)+timesec;
  run;
```

However, you can also read instream data lines that you enter directly in your SAS program, rather than data that is stored in an external file. Reading instream data is extremely helpful if you want to create data and test your programming statements on a few observations that you can specify according to your needs.

To read instream data, you use

- a DATALINES statement as the last statement in the DATA step and immediately preceding the data lines

- a null statement (a single semicolon) to indicate the end of the input data.

```
data sasuser.stress;
    input ID $ 1-4 Name $ 6-25 RestHR 27-29 MaxHR 31-33
          RecHR 35-37 TimeMin 39-40 TimeSec 42-43
          Tolerance $ 45;
    datalines;
    .
    .
    .
data lines go here
    .
    .
    .
;
```

General form, DATALINES statement:

DATALINES;

- You can use only one DATALINES statement in a DATA step. Use separate DATA steps to enter multiple sets of data.

- You can also use LINES; or CARDS; as the last statement in a DATA step and immediately preceding the data lines. Both LINES and CARDS are aliases for the DATALINES statement.

- If your data contains semicolons, use the DATALINES4 statement plus a null statement that consists of four semicolons (;;;;).

Example

To read the data for the treadmill stress tests as instream data, you can submit the following program:

```
data sasuser.stress;
    input ID $ 1-4 Name $ 6-25 RestHR 27-29 MaxHR 31-33
          RecHR 35-37 TimeMin 39-40 TimeSec 42-43
          Tolerance $ 45;
    if tolerance='D';
    TotalTime=(timemin*60)+timesec;
    datalines;
2458 Murray, W          72  185 128 12 38 D
2462 Almers, C          68  171 133 10  5 I
```

```
2501 Bonaventure, T        78  177 139 11 13 I
2523 Johnson, R            69  162 114  9 42 S
2539 LaMance, K            75 .168 141 11 46 D
2544 Jones, M              79  187 136 12 26 N
2552 Reberson, P           69  158 139 15 41 D
2555 King, E               70  167 122 13 13 I
2563 Pitts, D              71  159 116 10 22 S
2568 Eberhardt, S          72  182 122 16 49 N
2571 Nunnelly, A           65  181 141 15  2 I
2572 Oberon, M             74  177 138 12 11 D
2574 Peterson, V           80  164 137 14  9 D
2575 Quigley, M            74  152 113 11 26 I
2578 Cameron, L            75  158 108 14 27 I
2579 Underwood, K          72  165 127 13 19 S
2584 Takahashi, Y          76  163 135 16  7 D
2586 Derber, B             68  176 119 17 35 N
2588 Ivan, H               70  182 126 15 41 N
2589 Wilcox, E             78  189 138 14 57 I
2595 Warren, C             77  170 136 12 10 S
;
```

⚠ Notice that you do not need a RUN statement following the null statement (the semicolon after the data lines). The DATALINES statement functions as a step boundary, so the DATA step is executed as soon as SAS encounters it.

Creating a Raw Data File

Overview

Look at the SAS program and SAS data set that you created earlier in this chapter.

```
data sasuser.stress;
   infile tests;
   input ID $ 1-4 Name $ 6-25 RestHR 27-29 MaxHR 31-33
         RecHR 35-37 TimeMin 39-40 TimeSec 42-43
         Tolerance $ 45;
   if tolerance='D';
   TotalTime=(timemin*60)+timesec;
run;
```

Figure 5.22 *SAS Data Set sasuser.stress Output*

SAS Data Set Sasuser.Stress

ID	Name	RestHR	MaxHR	RecHR	TimeMin	TimeSec	Tolerance	TotalTime
2458	Murray, W	72	185	128	12	38	D	758
2539	LaMance, K	75	168	141	11	46	D	706
2552	Reberson, P	69	158	139	15	41	D	941
2572	Oberon, M	74	177	138	12	11	D	731
2574	Peterson, V	80	164	137	14	9	D	849
2584	Takahashi, Y	76	163	135	16	7	D	967

As you can see, the data set has been modified with SAS statements. If you wanted to write the new observations to a raw data file, you could reverse the process that you've been following and write out the observations from a SAS data set as records or lines to a new raw data file.

Using the _NULL_ Keyword

Because the goal of your SAS program is to create a raw data file and not a SAS data set, it is inefficient to print a data set name in the DATA statement. Instead, use the keyword _NULL_, which enables you to use the DATA step without actually creating a SAS data set. A SET statement specifies the SAS data set that you want to read from.

```
data _null_;
   set sasuser.stress;
```

The next step is to specify the output file.

Specifying the Raw Data File

You use the FILE and PUT statements to write the observations from a SAS data set to a raw data file, just as you used the INFILE and INPUT statements to create a SAS data set. These two sets of statements work almost identically.

When writing observations to a raw data file, use the FILE statement to specify the output file.

General form, FILE statement:

FILE *file-specification* *<options>* *<operating-environment-options>*;

where

- *file-specification* can take the form *fileref* to name a previously defined file reference or *'filename'* to point to the actual name and location of the file

- *options* names options that are used in creating the output file

- *operating-environment-options* names options that are specific to an operating environment (for more information, see the SAS documentation for your operating environment).

For example, if you want to read the Sasuser.Stress data set and write it to a raw data file that is referenced by the fileref Newdat, you would begin your program with the following SAS statements.

```
data _null_;
   set sasuser.stress;
   file newdat;
```

Instead of identifying the raw data file with a SAS fileref, you can choose to specify the entire filename and location in the FILE statement. For example, the following FILE statement points directly to the **C:\Clinic\Patients\Stress.dat** file. Note that the path specifying the filename and location must be enclosed in quotation marks.

```
data _null_;
   set sasuser.stress;
   file 'c:\clinic\patients\stress.dat';
```

Describing the Data

Whereas the FILE statement specifies the output raw data file, the PUT statement describes the lines to write to the raw data file.

General form, PUT statement using column output:

PUT *variable startcol-endcol...*;

where

- *variable* is the name of the variable whose value is written
- *startcol* indicates where in the line to begin writing the value
- *endcol* indicates where in the line to end the value.

In general, the PUT statement mirrors the capabilities of the INPUT statement. In this case you are working with column output. Therefore, you need to specify the variable name, starting column, and ending column for each field that you want to create. Because you are creating raw data, you don't need to follow character variable names with a dollar sign ($).

```
data _null_;
   set sasuser.stress;
   file 'c:\clinic\patients\stress.dat';
   put id $ 1-4 name  6-25 resthr 27-29 maxhr 31-33
       rechr 35-37 timemin 39-40 timesec 42-43
       tolerance 45 totaltime 47-49;
run;
```

Figure 5.23 *SAS Data Set sasuser.stress Output with PUT Statement*

SAS Data Set Sasuser.Stress

ID	Name	RestHR	MaxHR	RecHR	TimeMin	TimeSec	Tolerance	TotalTime
2458	Murray, W	72	185	128	12	38	D	758
2539	LaMance, K	75	168	141	11	46	D	706
2552	Reberson, P	69	158	139	15	41	D	941
2572	Oberon, M	74	177	138	12	11	D	731
2574	Peterson, V	80	164	137	14	9	D	849
2584	Takahashi, Y	76	163	135	16	7	D	967

The resulting raw data file would look like this.

Figure 5.24 *Creating a Raw Data File*

Raw Data File Stress.Dat

```
1---+----10---+----20---+----30---+----40---+----50---+
2458 Murray,  W         72   185 128 12 38  D 758
2539 LaMance, K         75   168 141 11 46  D 706
2552 Reberson, P        69   158 139 15 41  D 941
2572 Oberon, M          74   177 138 12 11  D 731
2574 Peterson, V        80   164 137 14 9   D 849
2584 Takahashi, Y       76   163 135 16 7   D 967
```

In later chapters you'll learn how to use INPUT and PUT statements to read and write raw data in other forms and record types.

⬚ If you do not execute a FILE statement before a PUT statement in the current iteration of the DATA step, SAS writes the lines to the SAS log. If you specify the PRINT fileref in the FILE statement, before the PUT statement, SAS writes the lines to the procedure output file.

Additional Features

In this section, you learned to read raw data by using an INPUT statement that uses column input. You also learned how to write to a raw data file by using the FILE statement with column input. However, column input is appropriate only in certain situations. When you use column input, your data must be

- standard character and numeric values. If the raw data file contains nonstandard values, then you need to use formatted input, another style of input. To learn about formatted input, see "Reading SAS Data Sets" on page 334 .

- in fixed fields. That is, values for a particular variable must be in the same location in all records. If your raw data file contains values that are not in fixed fields, you need to use list input. To learn about list input, see "Reading Free-Format Data" on page 536 .

Other forms of the INPUT statement enable you to read

- nonstandard data values such as hexadecimal, packed decimal, dates, and monetary values that contain dollar signs and commas

- free-format data (data that is not in fixed fields)

- implied decimal points

- variable-length data values

- variable-length records

- different record types.

Reading Microsoft Excel Data

Overview

In addition to reading raw data files, SAS can also read Microsoft Excel data. Whether the input data source is a SAS data set, a raw data file, or a file from another application, you use the DATA step to create a SAS data set. The difference between reading these various types of input is in how you reference the data. To read in Excel data you use one of the following methods:

- SAS/ACCESS LIBNAME statement

- Import Wizard

Remember, the Base SAS LIBNAME statement associates a SAS name (libref) with a SAS DATA library by pointing to its physical location. But, the SAS/ACCESS

LIBNAME statement associates a SAS name with an Excel workbook file by pointing to its location.

In doing so, the Excel workbook becomes a new library in SAS, and the worksheets in the workbook become the individual SAS data sets in that library.

The figure below illustrates the difference between how the two LIBNAME statements treat the data.

Figure 5.25   *Comparing Libname Statements*

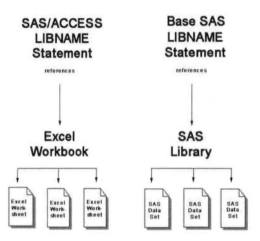

The next figure shows how the DATA step is used with three types of input data.

Figure 5.26   *Using the DATA step with Different Types of Output*

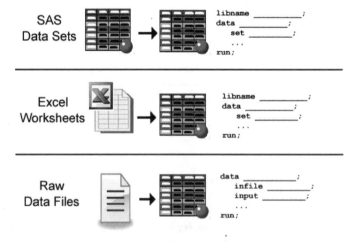

Notice how the INFILE and INPUT statements are used in the DATA step for reading raw data, but the SET statement is used in the DATA step for reading in the Excel worksheets.

Running SAS with Microsoft Excel

- You must have licensed SAS/ACCESS Interface to PC Files to use a SAS/ACCESS LIBNAME statement that references an Excel workbook.

- If you are running SAS version 9.1 or earlier and want to read in Microsoft Excel data, you must use Microsoft Excel 2003 or earlier.

- To read Microsoft Excel 2007 data you must be running SAS version 9.2 or later.

- The examples in this section are based on SAS version 9.2 running with Microsoft Excel 2007.

Steps for Reading Excel Data

Let's look at the steps for reading in an Excel workbook file.

To read the Excel workbook file, the DATA step must provide the following instructions to SAS:

- a libref to reference the Excel workbook to be read

- the name and location (using another libref) of the new SAS data set

- the name of the Excel worksheet that is to be read

The table below outlines the basic statements that are used in a program that reads Excel data and creates a SAS data set from an Excel worksheet. The PROC CONTENTS and PROC PRINT statements are not requirements for reading in Excel data and creating a SAS data set. However, these statements are useful for confirming that your Excel data has successfully been read into SAS.

Table 5.11   *Basic Steps for Reading Excel Data into a SAS Data Set*

To do this...	Use this SAS statement...	Example
Reference an Excel workbook file	SAS/ACCESS LIBNAME statement	libname results `'c:` `\users` `\exercise.xlsx';`
Output the contents of the SAS Library	PROC CONTENTS	proc contents data=results._all_;
Execute the PROC CONTENTS statement	RUN statement	run;
Name and create a new SAS data set	DATA statement	data work.stress;
Read in an Excel worksheet (as the input data for the new SAS data set)	SET statement	set results.'ActLevel$'n;
Execute the DATA step	RUN statement	run;
View the contents of a particular data set	PROC PRINT	proc print data=stress;

To do this...	Use this SAS statement...	Example
Execute the PROC PRINT statement	RUN statement	run;

The SAS/ACCESS LIBNAME Statement

The general form of the SAS/ACCESS LIBNAME statement is as follows:

General form, SAS/ACCESS LIBNAME statement:

LIBNAME *libref 'location-of-Excel-workbook ' <options>*;

where

- *libref* is a name that you associate with an Excel workbook.
- *'location-of-Excel-workbook'* is the physical location of the Excel workbook.

Example:

libname results **'c:\users\exercise.xlsx';**

Referencing an Excel Workbook

Overview

This example uses data similar to the scenario used for the raw data in the previous section. The data shows the readings from exercise stress tests that have been performed on patients at a health clinic.

The stress test data is located in an Excel workbook named exercise.xlsx (shown below), which is stored in the location **c:\users**.

Display 5.1 Excel Workbook

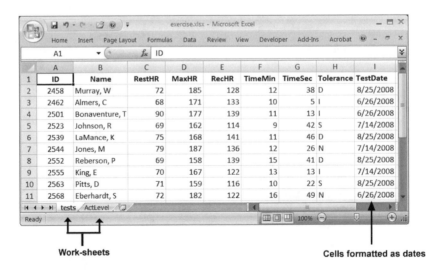

Work-sheets

Cells formatted as dates

Notice in the sample worksheet above that the date column is defined in Excel as dates. That is, if you right-click on the cells and select **Format Cells** (in Excel), the cells have a category of Date. SAS reads this data just as it is stored in Excel. If the date had been stored as text in Excel, then SAS would have read it as a character string.

To read in this workbook, you must first create a libref to point to the workbook's location:

```
libname results 'c:\users\exercise.xlsx';
```

The LIBNAME statement creates the libref **results**, which points to the Excel workbook exercise.xlsx. The workbook contains two worksheets, **tests** and **ActLevel**, which are now available in the new SAS library (results) as data sets.

After submitting the LIBNAME statement, you can look in the SAS Explorer window to see how SAS handles your Excel workbook. The Explorer window enables you to manage your files in the SAS windowing environment.

Display 5.2   *SAS Explorer Window*

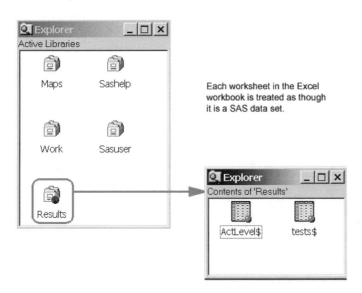

Name Literals

In the figure above, notice how the LIBNAME statement created a permanent library, **results**, which is the SAS name (libref) we gave to the workbook file and its location. The new library contains two SAS data sets, which accesses the data from the Excel worksheets. From this window you can browse the list of SAS libraries or display the descriptor portion of a SAS data set.

Notice that the Excel worksheet names have the special character ($) at the end. All Excel worksheets are designated this way. But remember, special characters such as these are not allowed in SAS data set names by default. So, in order for SAS to allow this character to be included in the data set name, you must assign a name literal to the data set name. A SAS name literal is a name token that is expressed as a string within quotation marks, followed by the uppercase or lowercase letter *n*. The name literal tells SAS to allow the special character ($) in the data set name.

Figure 5.27 Name Literal

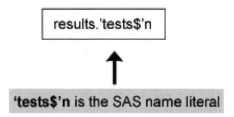

Named Ranges

A named range is a range of cells within a worksheet that you define in Excel and assign a name to. In the example below, the worksheet contains a named range, tests_week_1, which SAS recognizes as a data set.

The named range, tests_week_1, and its parent worksheet, tests, will appear in the SAS Explorer window as separate data sets, except that the data set created from the named range will have no dollar sign ($) appended to its name.

For more information on named ranges, see your Microsoft Excel documentation.

Figure 5.28 Named Range

Using PROC CONTENTS

In addition to using the SAS explorer window to view library data, you can also use the CONTENTS procedure with the _ALL_ keyword to produce information about a data library and its contents. In the example below, PROC CONTENTS outputs summary information for the SAS data set tests, including data set name, variables, data types, and other summary information. This statement is useful for making sure that SAS successfully read in your Excel data before moving on to the DATA step.

```
proc contents data=results._all_;
run;
```

Figure 5.29 *CONTENTS Procedure Output*

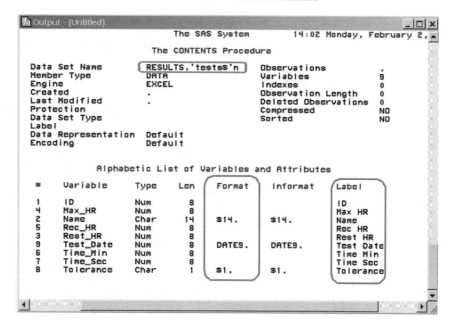

About the sample output above:
- The variables in the data set are pulled from the Excel column headings. SAS uses underscores to replace the spaces.

- The Excel dates in the Format column are converted to SAS dates with the default DATE9. format.

- This is partial output. The ACTLEVEL data set would also be included in the PROC CONTENTS report.

Creating the DATA Step

You use the DATA statement to indicate the beginning of the DATA step and name the SAS data set to be created. Remember that the SAS data set name is a two-level name. For example, the two-level name results.Admit specifies that the data set Admit is stored in the permanent SAS library to which the libref results has been assigned.

When reading Excel data, use the SET statement to indicate which worksheet in the Excel file that you want to read. To read in the Excel file you write the DATA and SET statements as follows:

```
data work.stress;
   set results.'ActLevel$'n;
run;
```

In this example, the DATA statement tells SAS to name the new data set, stress, and store it in the temporary library WORK. The SET statement in the DATA step specifies the libref (reference to the Excel file) and the worksheet name as the input data.

You can use several statements in the DATA step to subset your data as needed. Here, the WHERE statement is used with a variable to include only those participants whose activity level is HIGH.

```
data work.stress;
   set results.'ActLevel$'n;
   where ActLevel='HIGH';
run;
```

The figure below shows the partial output for this DATA step in table format.

Figure 5.30 *DATA Step Output*

Label changes column heading

	ID	Age	Activity Level	Sex
1	2462	38	HIGH	F
2	2523	26	HIGH	M
3	2544	29	HIGH	M
4	2568	28	HIGH	M

WHERE statement subsets data to only HIGH

Using PROC PRINT

After using the DATA step to read in the Excel data and create the SAS data set, you can use PROC PRINT to produce a report that displays the data set values.

You can also use the PRINT procedure to refer to a specific worksheet. Remember to use the name literal when referring to a specific Excel worksheet. In the example below, the first PRINT statement displays the data values for the new data set that was created in the DATA step. The second PRINT statement displays the contents of the Excel worksheet that was referenced by the LIBNAME statement.

```
proc print data=work.stress;
run;
proc print data=results.'ActLevel$'n;
run;
```

Disassociating a Libref

If SAS has a libref assigned to an Excel workbook, the workbook cannot be opened in Excel. To disassociate a libref, use a LIBNAME statement, specifying the libref and the CLEAR option.

```
libname results 'c:\users\exercise.xlsx';
   proc print data=results.'tests$'n;
run;
```

```
libname results clear;
```

SAS disconnects from the data source and closes any resources that are associated with that libref's connection.

LIBNAME Statement Options

There are several options that you can use with the LIBNAME statement to control how SAS interacts with the Excel data. The general form of the SAS/ACCESS LIBNAME statement (with options) is as follows:

```
libname libref 'location-of-Excel-workbook' <options>;
```

Example:

```
libname doctors 'c:\clinicNotes\addresses.xlsx' mixed=yes;
```

DBMAX_TEXT=n
> indicates the length of the longest character string where n is any integer between 256 and 32,767 inclusive. Any character string with a length greater than this value is truncated. The default is 1024.

GETNAMES=YES|NO
> determines whether SAS will use the first row of data in an Excel worksheet or range as column names.

> YES specifies to use the first row of data in an Excel worksheet or range as column names.

> NO specifies not to use the first row of data in an Excel worksheet or range as column names. SAS generates and uses the variable names F1, F2, F3, and so on.

> The default is YES.

MIXED=YES|NO
> Specifies whether to import data with both character and numeric values and convert all data to character.

> YES specifies that all data values will be converted to character.

> NO specifies that numeric data will be missing when a character type is assigned. Character data will be missing when a numeric data type is assigned.

> The default is NO.

SCANTEXT=YES|NO
> specifies whether to read the entire data column and use the length of the longest string found as the SAS column width.

YES scans the entire data column and uses the longest string value to determine the SAS column width.

NO does not scan the column and defaults to a width of 255.

The default is YES.

SCANTIME=YES|NO
specifies whether to scan all row values in a date/time column and automatically determine the TIME. format if only time values exist.

YES specifies that a column with only time values be assigned the TIME8. format.

NO specifies that a column with only time values be assigned the DATE9. format.

The default is NO.

USEDATE=YES|NO
specifies whether to use the DATE9. format for date/time values in Excel workbooks.

YES specifies that date/time values be assigned the DATE9. format.

NO specifies that date/time values be assigned the DATETIME. format.

The default is YES.

Creating Excel Worksheets

In addition to being able to read Excel data, SAS can also create Excel worksheets from SAS data sets.

To do this, you use the SAS/ACCESS LIBNAME statement. For example, to create a new worksheet named **high_stress** from the temporary SAS data set **work.high_stress** and save this worksheet in the new Excel file **newExcel.xlsx**, you would submit the following LIBNAME statement and DATA step:

```
libname clinic 'c:\Users\mylaptop\admitxl.xlsx' mixed=yes;
data clinic.admit;
   set work.admit;
run;
```

The IMPORT Wizard

Importing Data

As an alternative to using programming statements, you can use the Import Wizard to guide you through the process of creating a SAS data set from both raw data and from Excel worksheets. The Import Wizard enables you to create a SAS data set from different types of external files, such as

- dBase files (*.dbf)

- Excel 2007 (or earlier version) workbooks (*.xls, *.xlsx, *.xlsb, or *.xlsm)

- Microsoft Access tables (*.mdb, *.accdb)

- Delimited files (*.*)

- Comma-separated values (*.csv).

The data sources that are available to you depend on which SAS/ACCESS products you have licensed. If you do not have any SAS/ACCESS products licensed, the only type of data source files available to you are CSV files, TXT files, and delimited files.

To access the Import Wizard, select **File** ⇨ **Import Data** from the menu bar. The Import Wizard opens with the **Select import type** screen.

Figure 5.31 *Import Wizard*

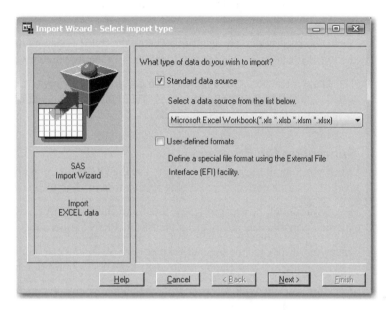

Follow the instructions on each screen of the Import Wizard to read in your data. If you need additional information, select the **Help** button at the bottom of each screen in the wizard.

Just as you can create a SAS data set from raw data by using the Import Wizard, you can use the Export Wizard to read data from a SAS data set and to write the data to an external data source. To access the Export Wizard, select **File** ⇨ **Export Data** from the menu bar.

Chapter Summary

Text Summary

Raw Data Files
A raw data file is an external file whose records contain data values that are organized in fields. The raw data files in this chapter contain fixed fields.

Steps to Create a SAS Data Set
You need to follow several steps to create a SAS data set using raw data. You need to

- reference the raw data file to be read

- name the SAS data set
- identify the location of the raw data
- describe the data values to be read.

Referencing a SAS Library

To begin your program, you might need to use a LIBNAME statement to reference the SAS library in which your data set will be stored.

Writing a DATA Step Program

The DATA statement indicates the beginning of the DATA step and names the SAS data set(s) to be created.

Next, you specify the raw data file by using the INFILE statement. The OBS= option in the INFILE statement enables you to process a specified number of observations.

This chapter teaches column input, the most simple input style. Column input specifies actual column locations for data values. The INPUT statement describes the raw data to be read and placed into the SAS data set.

Submitting the Program

When you submit the program, you can use the OBS= option with the INFILE statement to verify that the correct data is being read before reading the entire data file.

After you submit the program, view the log to check the DATA step processing. You can then print the data set by using the PROC PRINT procedure.

Once you've checked the log and verified your data, you can modify the DATA step to read the entire raw data file by removing the OBS= option from the INFILE statement.

If you are working with a raw data file that contains invalid data, the DATA step continues to execute. Unlike syntax errors, invalid data errors do not cause SAS to stop processing a program. If you have a way to edit the invalid data, it's best to correct the problem and rerun the DATA step.

Creating and Modifying Variables

To modify existing values or to create new variables, you can use an assignment statement in any DATA step. Within assignment statements, you can specify any SAS expression.

You can use date constants to assign dates in assignment statements. You can also use SAS time constants and SAS datetime constants in assignment statements.

Subsetting Data

To process only observations that meet a specified condition, use a subsetting IF statement in the DATA step.

Reading Instream Data

To read instream data lines instead of an external file, use a DATALINES statement, a CARDS statement, or a LINES statement and enter data directly in your SAS program. Omit the RUN at the end of the DATA step.

Creating a Raw Data File

When the goal of your SAS program is to create a raw data file and not a SAS data set, it is inefficient to list a data set name in the DATA statement. Instead use the keyword

NULL, which allows the power of the DATA step without actually creating a SAS data set. A SET statement specifies the SAS data set that you want to read from.

You can use the FILE and PUT statements to write out the observations from a SAS data set to a raw data file just as you used the INFILE and INPUT statements to create a SAS data set. These two sets of statements work almost identically.

Microsoft Excel Files

You can read Excel worksheets by using the SAS/ACCESS LIBNAME statement.

Steps to Create a SAS Data Set from Excel Data

You need to follow several steps to create a SAS data set using Excel. You need to

- provide a name for the new SAS data set

- provide the location or name of the libref and Excel worksheet

Referencing an Excel Workbook

To begin your program, you need to use a LIBNAME statement to reference the Excel workbook.

Writing a DATA Step Program

The DATA statement indicates the beginning of the DATA step and names the SAS data set(s) to be created.

Next, you specify the Excel worksheet to be read by using the SET statement. You must use a SAS name literal since SAS uses the special character ($) to name Excel worksheets.

Submitting the Program

When you submit the program, you can use the CONTENTS procedure to explore the new library and contents.

After you submit the program, view the log to check the DATA step processing. You can then print the data sets created from the Excel worksheets by using the PROC PRINT procedure.

Once you've checked the log and verified your data, you can modify the DATA step along with the WHERE statement to subset parts of the data as needed.

Syntax

Reading Data from a Raw File or Reading Instream Data

LIBNAME *libref 'SAS-data-library'*;

FILENAME *fileref 'filename'*;

DATA *SAS-data-set;*

      INFILE *file-specification<OBS=n>*;

      INPUT *variable <$> startcol-endcol...*;

      IF *expression*;

      variable=expression;

DATALINES;

      instream data goes here if used

;

RUN; /* *not used with the DATALINES statement* */

PROC PRINT DATA= *SAS-data set*;

RUN;

Creating a Raw Data File

LIBNAME *libref 'SAS-data-library'*;

DATA _NULL_;

SET *SAS-data-set;*

      FILE *file-specification;*

      PUT *variable startcol-endcol...*;

RUN;

Reading Data from an Excel Workbook

LIBNAME *libref '<location-of-Excel-workbook>'*;

PROC CONTENTS DATA=*libref._ALL_*;

DATA *SAS-data-set;*

    SET *libref.'worksheet_name$'n;*

      WHERE *where-expression;*

RUN;

PROC PRINT DATA= *SAS-data set*;

RUN;

Sample Programs

Reading Data from an External File

```
libname clinic 'c:\bethesda\patients\admit';
filename admit 'c:\clinic\patients\admit.dat';
data clinic.admittan;
   infile admit obs=5;
   input ID $ 1-4 Name $ 6-25 RestHR 27-29 MaxHR 31-33
         RecHR 35-37 TimeMin 39-40 TimeSec 42-43
         Tolerance $ 45;
   if tolerance='D';
   TotalTime=(timemin*60)+timesec;
run;
proc print data=clinic.admittan;
run;
```

Reading Instream Data

```
libname clinic 'c:\bethesda\patients\admit';
data clinic.group1;
   input ID $ 1-4 Name $ 6-25 RestHR 27-29 MaxHR 31-33
         RecHR 35-37 TimeMin 39-40 TimeSec 42-43
         Tolerance $ 45;
   if tolerance='D';
   TotalTime=(timemin*60)+timesec;
datalines;
2458 Murray, W         72  185 128 12 38 D
```

```
2462 Almers, C          68  171 133 10  5 I
2501 Bonaventure, T     78  177 139 11 13 I
2523 Johnson, R         69  162 114  9 42 S
2539 LaMance, K         75  168 141 11 46 D
2544 Jones, M           79  187 136 12 26 N
2595 Warren, C          77  170 136 12 10 S
;
proc print data=clinic.group1;
run;
```

Reading Excel Data

```
libname sasuser 'c:\users\admit.xlsx' mixed=yes;
proc contents data=sasuser._all_;
run;
proc print data=sasuser.'worksheet1$'n;
run;
```

Creating an Excel Worksheet

```
libname clinic 'c:\Users\mylaptop\admitxl.xlsx' mixed=yes;
data clinic.admit;
   set work.admit;
run;
```

Points to Remember

- LIBNAME and FILENAME statements are global. Librefs and filerefs remain in effect until you change them, cancel them, or end your SAS session.

- For each field of raw data that you read into your SAS data set, you *must* specify the following in the INPUT statement: a valid SAS variable name, a type (character or numeric), a starting column, and if necessary, an ending column.

- When you use column input, you can read any or all fields from the raw data file, read the fields in any order, and specify only the starting column for variables whose values occupy only one column.

- Column input is appropriate only in some situations. When you use column input, your data *must* be standard character and numeric values, and these values *must* be in fixed fields. That is, values for a particular variable must be in the same location in all records.

Chapter Quiz

Select the best answer for each question. After completing the quiz, you can check your answers using the answer key in the appendix.

1. Which SAS statement associates the fileref Crime with the raw data file `C:\States\Data\crime.dat`?

   a. `filename crime 'c:\states\data\crime.dat';`

   b. `filename crime c:\states\data\crime.dat;`

   c. `fileref crime 'c:\states\data\crime.dat';`

d. `filename 'c:\states\data\crime' crime.dat;`

2. Filerefs remain in effect until . . .

   a. you change them.

   b. you cancel them.

   c. you end your SAS session.

   d. all of the above

3. Which statement identifies the name of a raw data file to be read with the fileref Products and specifies that the DATA step read-only records 1-15?

   a. `infile products obs 15;`

   b. `infile products obs=15;`

   c. `input products obs=15;`

   d. `input products 1-15;`

4. Which of the following programs correctly writes the observations from the data set below to a raw data file?

   Figure 5.32 *Data Set work.patients*

   SAS data set Work.Patients

ID	Sex	Age	Height	Weight	Pulse
2304	F	16	61	102	100
1128	M	43	71	218	76
4425	F	48	66	162	80
1387	F	57	64	142	70
9012	F	39	63	157	68
6312	M	52	72	240	77
5438	F	42	62	168	83
3788	M	38	73	234	71
9125	F	56	64	159	70
3438	M	15	66	140	67

   a.
   ```
   data _null_;
       set work.patients;
       infile 'c:\clinic\patients\referals.dat';
       input id $ 1-4 sex 6 $ age 8-9 height 11-12
           weight 14-16 pulse 18-20;
   run;
   ```

   b.
   ```
   data referals.dat;
       set work.patients;
       input id $ 1-4 sex $ 6 age 8-9 height 11-12
           weight 14-16 pulse 18-20;
   run;
   ```

   c.
   ```
   data _null_;
       set work.patients;
       file c:\clinic\patients\referals.dat;
       put id $ 1-4 sex 6 $ age 8-9 height 11-12
           weight 14-16 pulse 18-20;
   run;
   ```

d. ```
data _null_;
    set work.patients;
    file 'c:\clinic\patients\referals.dat';
    put id $ 1-4 sex 6 $ age 8-9 height 11-12
        weight 14-16 pulse 18-20;
run;
```

5. Which raw data file can be read using column input?

   a.

   **Figure 5.33** *Raw Data File A*

   ```
   1---+----10---+----20---+
   Henderson CA 26 ADM
   Josephs SC 33 SALES
   Williams MN 40 HRD
   Rogan NY RECRTN
   ```

   b.

   Figure 5.34 *Raw Data File B*

   ```
   1---+----10---+----20---+----30
   2803 Deborah Campos   173.97
   2912 Bill    Marin    205.14
   3015 Helen   Stinson  194.08
   3122 Nicole  Terry    187.65
   ```

   c.

   Figure 5.35 *Raw Data File C*

   ```
   1---+----10---+----20---+
   Avery John        $601.23
   Davison Sherrill  $723.15
   Holbrook Grace    $489.76
   Jansen Mike       $638.42
   ```

   d. all of the above.

6. Which program creates the output shown below?

   Figure 5.36 *Raw Data and SAS Output Data Set*

   ```
   1---+----10---+----20---+----30
   3427 Chen    Steve Raleigh
   1436 Davis   Lee   Atlanta
   2812 King    Vicky Memphis
   1653 Sanchez Jack  Atlanta
   ```

Obs	ID	LastName	FirstName	City
1	3427	Chen	Steve	Raleigh
2	1436	Davis	Lee	Atlanta
3	2812	King	Vicky	Memphis
4	1653	Sanchez	Jack	Atlanta

   a. ```
   data work.salesrep;
       infile empdata;
       input ID $ 1-4 LastName $ 6-12
             FirstName $ 14-18 City $ 20-29;
   ```

```
run;
proc print data=work.salesrep;
run;
```

b.
```
data work.salesrep;
    infile empdata;
    input ID $ 1-4 Name $ 6-12
          FirstName $ 14-18 City $ 20-29;
run;
proc print data=work.salesrep;
run;
```

c.
```
data work.salesrep;
    infile empdata;
    input ID $ 1-4 name1 $ 6-12
          name2 $ 14-18 City $ 20-29;
run;
proc print data=work.salesrep;
run;
```

d. all of the above.

7. Which statement correctly reads the fields in the following order: StockNumber, Price, Item, Finish, Style?

Field Name	Start Column	End Column	Data Type
StockNumber	1	3	character
Finish	5	9	character
Style	11	18	character
Item	20	24	character
Price	27	32	numeric

Figure 5.37 *Raw Data*

```
1---+----10---+----20---+----30---+
310 oak    pedestal table   329.99
311 maple pedestal table   369.99
312 brass floor    lamp     79.99
313 glass table    lamp     59.99
313 oak    rocking  chair   153.99
```

a.
```
input StockNumber $ 1-3 Finish $ 5-9 Style $ 11-18
      Item $ 20-24 Price 27-32;
```

b.
```
input StockNumber $ 1-3 Price 27-32
      Item $ 20-24 Finish $ 5-9 Style $ 11-18;
```

c.
```
input $ StockNumber 1-3 Price 27-32 $
      Item 20-24 $ Finish 5-9 $ Style 11-18;
```

d.
```
input StockNumber $ 1-3 Price $ 27-32
      Item $ 20-24 Finish $ 5-9 Style $ 11-18;
```

8. Which statement correctly re-defines the values of the variable Income as 100 percent higher?

   a. `income=income*1.00;`

   b. `income=income+(income*2.00);`

   c. `income=income*2;`

   d. `income=*2;`

9. Which program correctly reads instream data?

   a.
   ```
   data finance.newloan;
        input datalines;
        if country='JAPAN';
        MonthAvg=amount/12;
        1998 US      CARS    194324.12
        1998 US      TRUCKS  142290.30
        1998 CANADA CARS     10483.44
        1998 CANADA TRUCKS   93543.64
        1998 MEXICO CARS     22500.57
        1998 MEXICO TRUCKS   10098.88
        1998 JAPAN   CARS    15066.43
        1998 JAPAN   TRUCKS  40700.34
        ;
   ```

   b.
   ```
   data finance.newloan;
        input Year 1-4 Country $ 6-11
              Vehicle $ 13-18 Amount 20-28;
        if country='JAPAN';
        MonthAvg=amount/12;
        datalines;
   run;
   ```

   c.
   ```
   data finance.newloan;
        input Year 1-4 Country 6-11
              Vehicle 13-18 Amount 20-28;
        if country='JAPAN';
        MonthAvg=amount/12;
        datalines;
        1998 US      CARS    194324.12
        1998 US      TRUCKS  142290.30
        1998 CANADA CARS     10483.44
        1998 CANADA TRUCKS   93543.64
        1998 MEXICO CARS     22500.57
        1998 MEXICO TRUCKS   10098.88
        1998 JAPAN   CARS    15066.43
        1998 JAPAN   TRUCKS  40700.34
        ;
   ```

   d.
   ```
   data finance.newloan;
        input Year 1-4 Country $ 6-11
              Vehicle $ 13-18 Amount 20-28;
        if country='JAPAN';
        MonthAvg=amount/12;
        datalines;
        1998 US      CARS    194324.12
        1998 US      TRUCKS  142290.30
        1998 CANADA CARS     10483.44
   ```

```
1998 CANADA TRUCKS  93543.64
1998 MEXICO CARS    22500.57
1998 MEXICO TRUCKS  10098.88
1998 JAPAN  CARS    15066.43
1998 JAPAN  TRUCKS  40700.34
;
```

10. Which SAS statement subsets the raw data shown below so that only the observations in which Sex (in the second field) has a value of **F** are processed?

Figure 5.38 *Raw Data*

```
----+----10---+----20---+--
Alfred  M 14 69.0 112.5
Becka   F 13 65.3 98.0
Gail    F 14 64.3 90.0
Jeffrey M 13 62.5 84.0
John    M 12 59.0 99.5
Karen   F 12 56.3 77.0
Mary    F 15 66.5 112.0
Philip  M 16 72.0 150.0
Sandy   F 11 51.3 50.5
Tammy   F 14 62.8 102.5
William M 15 66.5 112.0
```

a. `if sex=f;`

b. `if sex=F;`

c. `if sex='F';`

d. a or b

Chapter 6
Understanding DATA Step Processing

Overview

Introduction

"Creating SAS Data Sets from External Files" on page 152 explained how to read data, perform basic modifications, and create a new SAS data set.

This chapter teaches you what happens *behind the scenes* when the DATA step reads raw data. You'll examine the program data vector, which is a logical framework that SAS uses when creating SAS data sets.

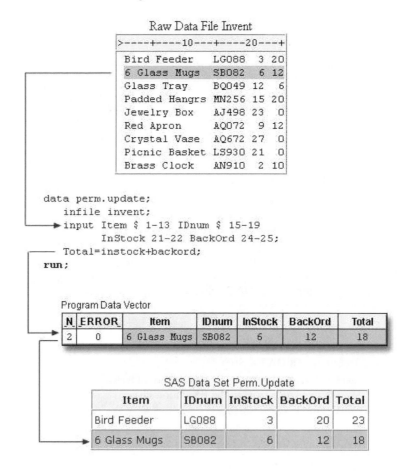

Understanding how the program operates can help you to anticipate how variables will be created and processed, to plan your modifications, and to interpret and debug program errors. It also gives you useful strategies for preventing and correcting common DATA step errors.

Objectives

In this chapter, you learn to

- identify the two phases that occur when a DATA step is processed
- interpret automatic variables
- identify the processing phase in which an error occurs
- debug SAS DATA steps
- validate and clean invalid data
- test programs by limiting the number of observations that are created
- flag errors in the SAS log.

Writing Basic DATA Steps

"Creating SAS Data Sets from External Files" on page 152 explained how to write a DATA step to create a permanent SAS data set from raw data that is stored in an external file.

```
data clinic.stress;
   infile tests;
   input ID 1-4 Name $ 6-25 RestHR 27-29
         MaxHR 31-33 RecHR 35-37
         TimeMin 39-40 TimeSec 42-43
         Tolerance $ 45;
run;
```

```
              Partial Raw Data File Tests
1---+----10---+----20---+----30---+----40---+-
2458 Murray, W          72    185 128 12 38 D
2462 Almers, C          68    171 133 10  5 I
2501 Bonaventure, T     78    177 139 11 13 I
2523 Johnson, R         69    162 114  9 42 S
2539 LaMance, K         75    168 141 11 46 D
2544 Jones, M           79    187 136 12 26 N
2552 Reberson, P        69    158 139 15 41 D
2555 King, E            70    167 122 13 13 I
```

You learned how to submit the DATA step and how to check the log to see whether the step ran successfully.

```
NOTE: The infile TESTS is:
      Filename=c:\rawdata\tests,
      RECFM=V,LRECL=256,File Size (bytes)=376,
      Last Modified=16May2011:11:23:09,
      Create Time=17May2011:07:42:32

NOTE: 8 records were read from the infile TESTS.
      The minimum record length was 45.
      The maximum record length was 45.
NOTE: The data set CLINIC.STRESS has 8 observations and 8 variables.
NOTE: DATA statement used (Total process time):
      real time            0.01 seconds
      cpu time             0.00 seconds
```

You also learned how to display the contents of the data set with the PRINT procedure.

```
proc print data=clinic.stress;
run;
```

Figure 6.1 *Output from the PRINT Procedure*

Obs	ID	Name	RestHr	MaxHR	RecHR	TimeMin	TimeSec	Tolerance
1	2458	Murray, W	72	185	128	12	38	D
2	2462	Almers, C	68	171	133	10	5	I
3	2501	Bonaventure, T	78	177	139	11	13	I
4	2523	Johnson, R	69	162	114	9	42	S
5	2539	LaMance, K	75	168	141	11	46	D
6	2544	Jones, M	79	187	136	12	26	N
7	2552	Reberson, P	69	158	139	15	41	D
8	2555	King, E	70	167	122	13	13	I

How SAS Processes Programs

When you submit a DATA step, SAS processes the DATA step and then creates a new SAS data set. Let's see exactly how that happens.

A SAS DATA step is processed in two phases:

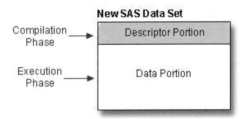

- During the compilation phase, each statement is scanned for syntax errors. Most syntax errors prevent further processing of the DATA step. When the compilation phase is complete, the descriptor portion of the new data set is created.

- If the DATA step compiles successfully, then the execution phase begins. During the execution phase, the DATA step reads and processes the input data. The DATA step executes once for each record in the input file, unless otherwise directed.

The diagram below shows the general flow of DATA step processing for reading raw data. We'll examine both the compilation phase and the execution phase in this chapter.

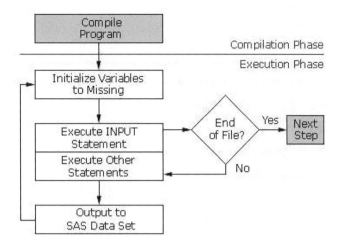

Let's start with the compilation phase.

Compilation Phase

Input Buffer

At the beginning of the compilation phase, the input buffer (an area of memory) is created to hold a record from the external file. The input buffer is created only when raw data is read, not when a SAS data set is read.

Input Buffer

1	2	3	4	5	6	7	8	9	10	11	12	13	14	15	16	17	18	19	20	21

Program Data Vector

After the input buffer is created, the program data vector is created. The program data vector is the area of memory where SAS holds one observation at a time.

The program data vector contains two automatic variables that can be used for processing but which are not written to the data set as part of an observation.

- _N_ counts the number of times that the DATA step begins to execute.

- _ERROR_ signals the occurrence of an error that is caused by the data during execution. The default value is 0, which means there is no error. When one or more errors occur, the value is set to 1.

Program Data Vector

_N_	_ERROR_	

Syntax Checking

During the compilation phase, SAS also scans each statement in the DATA step, looking for syntax errors. Syntax errors include

- missing or misspelled keywords

- invalid variable names

- missing or invalid punctuation

- invalid options.

Data Set Variables

As the INPUT statement is compiled, a slot is added to the program data vector for each variable in the new data set. Generally, variable attributes such as length and type are determined the first time a variable is encountered.

```
data perm.update;
   infile invent;
   input Item $ 1-13 IDnum $ 15-19
         InStock 21-22 BackOrd 24-25;
    Total=instock+backord;
run;
```

Program Data Vector

N	ERROR	Item	IDnum	InStock	BackOrd	

Any variables that are created with an assignment statement in the DATA step are also added to the program data vector. For example, the assignment statement below creates the variable Total. As the statement is compiled, the variable is added to the program data vector. The attributes of the variable are determined by the expression in the statement. Because the expression contains an arithmetic operator and produces a numeric value, Total is defined as a numeric variable and is assigned the default length of 8.

```
data perm.update;
   infile invent;
   input Item $ 1-13 IDnum $ 15-19
         InStock 21-22 BackOrd 24-25;
    Total=instock+backord;
run;
```

Program Data Vector

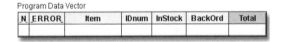

N	ERROR	Item	IDnum	InStock	BackOrd	Total

Descriptor Portion of the SAS Data Set

At the bottom of the DATA step (in this example, when the RUN statement is encountered), the compilation phase is complete, and the descriptor portion of the new SAS data set is created. The descriptor portion of the data set includes

- the name of the data set

- the number of variables

- the names and attributes of the variables.

```
                            The SAS System
                         The CONTENTS Procedure

    Data Set Name        PERM.UPDATE                 Observations         9
    Member Type          DATA                        Variables            5
    Engine               V9                          Indexes              0
    Created              Thursday, May 05, 2011 03:38:15 PM  Observation Length   48
    Last Modified        Thursday, May 05, 2011 03:38:15 PM  Deleted Observations 0
    Protection                                       Compressed           NO
    Data Set Type                                    Sorted               NO
    Label
    Data Representation  WINDOWS_32
    Encoding             wlatin1  Western (Windows)


                      Engine/Host Dependent Information

    Data Set Page Size       4096
    Number of Data Set Pages 1
    First Data Page          1
    Max Obs per Page         84
    Obs in First Data Page   9
    Number of Data Set Repairs  0
    Filename                 C:\Documents and Settings\user-name\sasuser\update.sas7bdat
    Release Created          9.0301M0
    Host Created             XP_PRO


                 Alphabetic List of Variables and Attributes

                      #    Variable    Type    Len

                      4    BackOrd     Num      8
                      2    IDnum       Char     5
                      3    InStock     Num      8
                      1    Item        Char    13
                      5    Total       Num      8
```

At this point, the data set contains the five variables that are defined in the input data set and in the assignment statement. Remember, _N_ and _ERROR_ are not written to the data set. There are no observations because the DATA step has not yet executed. During execution, each raw data record is processed and is then written to the data set as an observation.

For additional information about assigning attributes to variables, see the SAS documentation.

Summary of the Compilation Phase

Let's review the compilation phase.

```
data perm.update;
   infile invent;
   input Item $ 1-13 IDnum $ 15-19
         InStock 21-22 BackOrd 24-25;
   Total=instock+backord;
run;
```

During the compilation phase, the input buffer is created to hold a record from the external file.

Input Buffer

```
  1  2  3  4  5  6  7  8  9 10 11 12 13 14 15 16 17 18 19 20 21
+--+--+--+--+--+--+--+--+--+--+--+--+--+--+--+--+--+--+--+--+--+
|  |  |  |  |  |  |  |  |  |  |  |  |  |  |  |  |  |  |  |  |  |
+--+--+--+--+--+--+--+--+--+--+--+--+--+--+--+--+--+--+--+--+--+
```

The program data vector is created to hold the current observation.

Program Data Vector

N_	ERROR_	Item	IDnum	InStock	BackOrd	Total

The descriptor portion of the SAS data set is created.

Data Set Descriptor (Partial)

```
Data Set Name:        PERM.UPDATE
Member Type:          DATA
Engine:               V9
Created:              Tuesday, May 03, 2011  07:26:52 AM
Observations:         9
Variables:            5
Indexes:              0
Observation Length:   48
```

Execution Phase

Overview

After the DATA step is compiled, it is ready for execution. During the execution phase, the data portion of the data set is created. The data portion contains the data values.

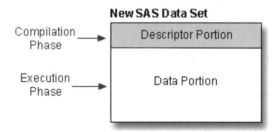

During execution, each record in the input raw data file is read into the input buffer, copied to the program data vector, and then written to the new data set as an observation. The DATA step executes once for each record in the input file, unless otherwise directed by additional statements.

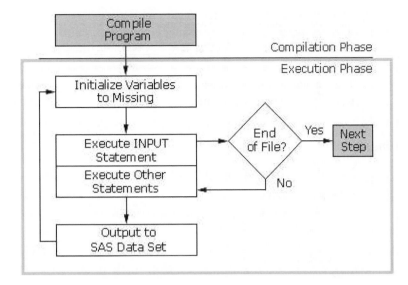

Example

The following DATA step reads values from the file Invent and executes nine times because there are nine records in the file.

```
data perm.update;
 infile invent;
 input Item $ 1-13 IDnum $ 15-19
   InStock 21-22 BackOrd 24-25;
 Total=instock+backord;
run;
```

Raw Data File Invent

```
1---+----10---+----20---+-
Bird Feeder    LG088  3 20
6 Glass Mugs   SB082  6 12
Glass Tray     BQ049 12  6
Padded Hangrs  MN256 15 20
Jewelry Box    AJ498 23  0
Red Apron      AQ072  9 12
Crystal Vase   AQ672 27  0
Picnic Basket  LS930 21  0
Brass Clock    AN910  2 10
```

Initializing Variables

At the beginning of the execution phase, the value of _N_ is **1**. Because there are no data errors, the value of _ERROR_ is **0**.

```
data perm.update;
   infile invent;
   input Item $ 1-13 IDnum $ 15-19
         InStock 21-22 BackOrd 24-25;
   Total=instock+backord;
run;
```

Program Data Vector

N	ERROR	Item	IDnum	InStock	BackOrd	Total
1	0			•	•	•

└─────────── Initialized to Missing ───────────┘

The remaining variables are initialized to missing. Missing numeric values are represented by periods, and missing character values are represented by blanks.

INFILE Statement

Next, the INFILE statement identifies the location of the raw data.

```
data perm.update;
   infile invent;
   input Item $ 1-13 IDnum $ 15-19
         InStock 21-22 BackOrd 24-25;
   Total=instock+backord;
run;
```

Raw Data File Invent

```
1---+----10---+----20---+-
Bird Feeder   LG088   3 20
6 Glass Mugs  SB082   6 12
Glass Tray    BQ049  12  6
Padded Hangrs MN256  15 20
Jewelry Box   AJ498  23  0
Red Apron     AQ072   9 12
Crystal Vase  AQ672  27  0
Picnic Basket LS930  21  0
Brass Clock   AN910   2 10
```

Program Data Vector

N	ERROR	Item	IDnum	InStock	BackOrd	Total
1	0			•	•	•

INPUT Statement

Next, the INPUT statement reads a record into the input buffer. Then, the raw data in columns 1-13 is read and is assigned to Item in the program data vector.

```
data perm.update;
   infile invent;
   input Item $ 1-13 IDnum $ 15-19
         InStock 21-22 BackOrd 24-25;
   Total=instock+backord;
run;
```

Raw Data File Invent

```
1---+----10---+----20---+-
Bird Feeder   LG088   3 20
6 Glass Mugs  SB082   6 12
Glass Tray    BQ049  12  6
Padded Hangrs MN256  15 20
Jewelry Box   AJ498  23  0
Red Apron     AQ072   9 12
Crystal Vase  AQ672  27  0
Picnic Basket LS930  21  0
Brass Clock   AN910   2 10
```

Program Data Vector

N	ERROR	Item	IDnum	InStock	BackOrd	Total
1	0	Bird Feeder		•	•	•

```
data perm.update;
   infile invent;
   input Item $ 1-13 IDnum $ 15-19
        InStock 21-22 BackOrd 24-25;
   Total=instock+backord;
run;
```

Raw Data File Invent

```
1---+----10--V+----20---+-
Bird Feeder   •LG088   3 20
6 Glass Mugs  SB082   6 12
Glass Tray    BQ049  12  6
Padded Hangrs MN256  15 20
Jewelry Box   AJ498  23  0
Red Apron     AQ072   9 12
Crystal Vase  AQ672  27  0
Picnic Basket LS930  21  0
Brass Clock   AN910   2 10
```

Program Data Vector

N	ERROR	Item	IDnum	InStock	BackOrd	Total
1	0	Bird Feeder		•	•	•

Next, the data in columns 15-19 is read and is assigned to IDnum in the program data vector.

```
data perm.update;
   infile invent;
   input Item $ 1-13 IDnum $ 15-19
        InStock 21-22 BackOrd 24-25;
   Total=instock+backord;
run;
```

Raw Data File Invent

```
1---+----10---+----V0---+-
Bird Feeder   LG088• 3 20
6 Glass Mugs  SB082  6 12
Glass Tray    BQ049 12  6
Padded Hangrs MN256 15 20
Jewelry Box   AJ498 23  0
Red Apron     AQ072  9 12
Crystal Vase  AQ672 27  0
Picnic Basket LS930 21  0
Brass Clock   AN910  2 10
```

Program Data Vector

N	ERROR	Item	IDnum	InStock	BackOrd	Total
1	0	Bird Feeder	LG088	•	•	•

Likewise, the INPUT statement reads the values for InStock from columns 21-22, and it reads the values for BackOrd from columns 24-25.

```
data perm.update;
   infile invent;
   input Item $ 1-13 IDnum $ 15-19
         InStock 21-22 BackOrd 24-25;
   Total=instock+backord;
run;
```

Raw Data File Invent

```
1---+----10---+----20---+V
Bird Feeder   LG088  3 20•
6 Glass Mugs  SB082  6 12
Glass Tray    BQ049 12  6
Padded Hangrs MN256 15 20
Jewelry Box   AJ498 23  0
Red Apron     AQ072  9 12
Crystal Vase  AQ672 27  0
Picnic Basket LS930 21  0
Brass Clock   AN910  2 10
```

Program Data Vector

N	ERROR	Item	IDnum	InStock	BackOrd	Total
1	0	Bird Feeder	LG088	3	20	•

Next, the assignment statement executes. The values for InStock and BackOrd are added to produce the values for Total.

```
data perm.update;
  infile invent;
  input Item $ 1-13 IDnum $ 15-19
        InStock 21-22 BackOrd 24-25;
  Total=instock+backord;
run;
```

Raw Data File Invent

```
1---+----10---+----20---+V
Bird Feeder   LG088  3 20•
6 Glass Mugs  SB082  6 12
Glass Tray    BQ049 12  6
Padded Hangrs MN256 15 20
Jewelry Box   AJ498 23  0
Red Apron     AQ072  9 12
Crystal Vase  AQ672 27  0
Picnic Basket LS930 21  0
Brass Clock   AN910  2 10
```

Program Data Vector

N	ERROR	Item	IDnum	InStock	BackOrd	Total
1	0	Bird Feeder	LG088	3	20	23

End of the DATA Step

At the end of the DATA step, several actions occur. First, the values in the program data vector are written to the output data set as the first observation.

```
data perm.update;
    infile invent;
    input Item $ 1-13 IDnum $ 15-19
          InStock 21-22 BackOrd 24-25;
    Total=instock+backord;
run;
```

Raw Data File Invent

```
1---+----10---+----20---+V
Bird Feeder   LG088  3 20•
6 Glass Mugs  SB082  6 12
Glass Tray    BQ049 12  6
Padded Hangrs MN256 15 20
Jewelry Box   AJ498 23  0
Red Apron     AQ072  9 12
Crystal Vase  AQ672 27  0
Picnic Basket LS930 21  0
Brass Clock   AN910  2 10
```

Program Data Vector

N	ERROR	Item	IDnum	InStock	BackOrd	Total
1	0	Bird Feeder	LG088	3	20	23

SAS Data Set Perm.Update

Item	IDnum	InStock	BackOrd	Total
Bird Feeder	LG088	3	20	23

Next, control returns to the top of the DATA step and the value of _N_ increments from 1 to **2**. Finally, the variable values in the program data vector are re-set to missing. Notice that the automatic variable _ERROR_ is reset to zero if necessary.

```
data perm.update;
    infile invent;
    input Item $ 1-13 IDnum $ 15-19
          InStock 21-22 BackOrd 24-25;
```

```
        Total=instock+backord;
run;
```

Raw Data File Invent

```
V---+----10---+----20---+-
Bird Feeder    LG088  3 20
6 Glass Mugs   SB082  6 12
Glass Tray     BQ049 12  6
Padded Hangrs  MN256 15 20
Jewelry Box    AJ498 23  0
Red Apron      AQ072  9 12
Crystal Vase   AQ672 27  0
Picnic Basket  LS930 21  0
Brass Clock    AN910  2 10
```

Program Data Vector

N	ERROR	Item	IDnum	InStock	BackOrd	Total
2	0			•	•	•

Set to Missing

SAS Data Set Perm.Update

Item	IDnum	InStock	BackOrd	Total
Bird Feeder	LG088	3	20	23

Iterations of the DATA Step

You can see that the DATA step works like a loop, repetitively executing statements to read data values and create observations one by one. Each loop (or cycle of execution) is called an iteration. At the beginning of the second iteration, the value of _N_ is 2, and _ERROR_ is still 0. Notice that the input pointer points to the second record.

```
data perm.update;
    infile invent;
    inpute Item $ 1-13 IDnum $ 15-19
            InStock 21-22 BackOrd 24-25;
    Total=instock+backord;
run;
```

Raw Data File Invent

```
V---+----10---+----20---+-
Bird Feeder   LG088  3 20
6 Glass Mugs  SB082  6 12
Glass Tray    BQ049 12  6
Padded Hangrs MN256 15 20
Jewelry Box   AJ498 23  0
Red Apron     AQ072  9 12
Crystal Vase  AQ672 27  0
Picnic Basket LS930 21  0
Brass Clock   AN910  2 10
```

Program Data Vector

N	ERROR	Item	IDnum	InStock	BackOrd	Total
2	0			•	•	•

SAS Data Set Perm.Update

Item	IDnum	InStock	BackOrd	Total
Bird Feeder	LG088	3	20	23

As the INPUT statement executes for the second time, the values from the second record are read into the input buffer and then into the program data vector.

Raw Data File Invent

```
1---+----10---+----20---+V
Bird Feeder   LG088  3 20
6 Glass Mugs  SB082  6 12•
Glass Tray    BQ049 12  6
Padded Hangrs MN256 15 20
Jewelry Box   AJ498 23  0
Red Apron     AQ072  9 12
Crystal Vase  AQ672 27  0
Picnic Basket LS930 21  0
Brass Clock   AN910  2 10
```

Program Data Vector

N	ERROR	Item	IDnum	InStock	BackOrd	Total
2	0	6 Glass Mugs	SB082	6	12	

SAS Data Set Perm.Update

Item	IDnum	InStock	BackOrd	Total
Bird Feeder	LG088	3	20	23

Next, the value for Total is calculated based on the current values for InStock and BackOrd. The RUN statement indicates the end of the DATA step loop.

```
data perm.update;
   infile invent;
   input Item $ 1-13 IDnum $ 15-19
         InStock 21-22 BackOrd 24-25;
   Total=instock+backord;
run;
```

Raw Data File Invent

```
1---+----10---+----20---+V
Bird Feeder   LG088  3 20
6 Glass Mugs  SB082  6 12•
Glass Tray    BQ049 12  6
Padded Hangrs MN256 15 20
Jewelry Box   AJ498 23  0
Red Apron     AQ072  9 12
Crystal Vase  AQ672 27  0
Picnic Basket LS930 21  0
Brass Clock   AN910  2 10
```

Program Data Vector

N	ERROR	Item	IDnum	InStock	BackOrd	Total
2	0	6 Glass Mugs	SB082	6	12	18

SAS Data Set Perm.Update

Item	IDnum	InStock	BackOrd	Total
Bird Feeder	LG088	3	20	23

At the bottom of the DATA step, the values in the program data vector are written to the data set as the second observation.

```
data perm.update;
   infile invent;
   input Item $ 1-13 IDnum $ 15-19
        InStock 21-22 BackOrd 24-25;
   Total=instock+backord;
run;
```

Next, the value of _N_ increments from 2 to **3**, control returns to the top of the DATA step, and the values for Item, IDnum, InStock, BackOrd, and Total are re-set to missing.

```
data perm.update;
   infile invent;
   input Item $ 1-13 IDnum $ 15-19
        InStock 21-22 BackOrd 24-25;
   Total=instock+backord;
run;
```

```
        Raw Data File Invent
    v---+----10---+----20---+-
    Bird Feeder    LG088   3 20
    6 Glass Mugs   SB082   6 12
    Glass Tray     BQ049  12  6
    Padded Hangrs  MN256  15 20
    Jewelry Box    AJ498  23  0
    Red Apron      AQ072   9 12
    Crystal Vase   AQ672  27  0
    Picnic Basket  LS930  21  0
    Brass Clock    AN910   2 10
```

Program Data Vector

N	ERROR	Item	IDnum	InStock	BackOrd	Total
3	0			•	•	•

———————————— Reset to Missing ————————————

SAS Data Set Perm.Update

Item	IDnum	InStock	BackOrd	Total
Bird Feeder	LG088	3	20	23
6 Glass Mugs	SB082	6	12	18

End-of-File Marker

The execution phase continues in this manner until the end-of-file marker is reached in the raw data file. When there are no more records in the raw data file to be read, the data portion of the new data set is complete and the DATA step stops.

```
        Raw Data File Invent
    1---+----10---+----20---+-
    Bird Feeder    LG088   3 20
    6 Glass Mugs   SB082   6 12
    Glass Tray     BQ049  12  6
    Padded Hangrs  MN256  15 20
    Jewelry Box    AJ498  23  0
    Red Apron      AQ072   9 12
    Crystal Vase   AQ672  27  0
    Picnic Basket  LS930  21  0
    Brass Clock    AN910   2 10
```

This is the output data set that SAS creates:

SAS Data Set Perm.Udpate

Item	IDnum	InStock	BackOrd	Total
Bird Feeder	LG088	3	20	23
6 Glass Mugs	SB082	6	12	18
Glass Tray	BQ049	12	6	18
Padded Hangrs	MN256	15	20	35
Jewelry Box	AJ498	23	0	23
Red Apron	AQ072	9	12	21
Crystal Vase	AQ672	27	0	27
Picnic Basket	LS930	21	0	21
Brass Clock	AN910	2	10	12

When reading variables from raw data, SAS sets the value of each variable in the DATA step to missing at the beginning of each cycle of execution, with these exceptions:

- variables that are named in a RETAIN statement

- variables that are created in a sum statement

- data elements in a _TEMPORARY_ array

- any variables that are created with options in the FILE or INFILE statements

- automatic variables.

In contrast, when reading variables from a SAS data set, SAS sets the values to missing only before the first cycle of execution of the DATA step. Therefore, the variables retain their values until new values become available (for example, through an assignment statement or through the next execution of a SET or MERGE statement). Variables that are created with options in the SET or MERGE statements also retain their values from one cycle of execution to the next. (You learn about reading SAS data sets, working with arrays, and using the SET and MERGE statements in later chapters.)

Summary of the Execution Phase

You've seen how the DATA step iteratively reads records in the raw data file. Now take a minute to review execution-phase processing.

During the execution phase

- variables in the program data vector are initialized to missing before each execution of the DATA step

- each statement is executed sequentially

- the INPUT statement reads into the input buffer the next record from the external file identified by the INFILE statement, and then reads the data into the program data vector

- other statements can then further modify the current observation

- the values in the program data vector are written to the SAS data set at the end of the DATA step

- program flow returns to the top of the DATA step
- the DATA step is executed until the end-of-file marker is reached in the external file.

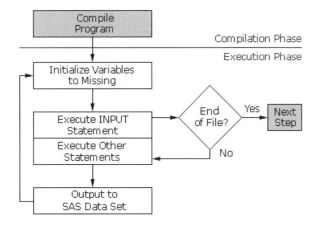

End of the Execution Phase

At the end of the execution phase, the SAS log confirms that the raw data file was read, and it displays the number of observations and variables in the data set.

SAS Log

```
NOTE: 9 records were read from the infile INVENT.
NOTE: The data set PERM.UPDATE has 9 observations
      and 5 variables.
```

You already know how to display the data set with the PRINT procedure.

```
proc print data=perm.update;
run;
```

Figure 6.2   *Output from the PRINT Procedure*

Obs	Item	IDnum	InStock	BackOrd	Total
1	Bird Feeder	LG088	3	20	23
2	6 Glass Mugs	SB082	6	12	18
3	Glass Tray	BQ049	12	6	18
4	Padded Hangrs	MN256	15	20	35
5	Jewelry Box	AJ498	23	0	23
6	Red Apron	AQ072	9	12	21
7	Crystal Vase	AQ672	27	0	27
8	Picnic Basket	LS930	21	0	21
9	Brass Clock	AN910	2	10	12

Debugging a DATA Step

Diagnosing Errors in the Compilation Phase

Now that you know how a DATA step is processed, you can use that knowledge to correct errors. Chapter 3, "Editing and Debugging SAS Programs," on page 75 provided examples of errors that are detected during the compilation phase, including

- misspelled keywords and data set names

- unbalanced quotation marks

- invalid options.

During the compilation phase, SAS can interpret some syntax errors (such as the keyword DATA misspelled as DAAT). If it cannot interpret the error, SAS

- prints the word ERROR followed by an error message in the log

- compiles but does not execute the step where the error occurred, and prints the following message to warn you:

  NOTE: The SAS System stopped processing this step because of errors.

Some errors are explained fully by the message that SAS prints; other error messages are not as easy to interpret. For example, because SAS statements are free-format, when you fail to end a SAS statement with a semicolon, SAS cannot detect the error.

Diagnosing Errors in the Execution Phase

As you have seen, errors can occur in the compilation phase, resulting in a DATA step that is compiled but not executed. Errors can also occur during the execution phase. When SAS detects an error in the execution phase, the following can occur, depending on the type of error:

- A note, warning, or error message is displayed in the log.

- The values that are stored in the program data vector are displayed in the log.

- The processing of the step either continues or stops.

Example

Suppose you misspelled the fileref in the INFILE statement below. This is not a syntax error, because SAS does not validate the file that you reference until the execution phase. During the compilation phase, the fileref Invnt is assumed to reference some external raw data file.

```
data perm.update;
   infile invnt;
   input Item $ 1-13 IDnum $ 15-19
        InStock 21-22 BackOrd 24-25;
   Total=instock+backord;
run;
```

This error is not detected until the execution phase begins. Because there is no external file that is referenced by the fileref Invnt, the DATA step stops processing.

```
3486    data perm.update;
3487        infile invnt;
3488        input Item $ 1-13 IDnum $ 15-19 InStock 21-22 BackOrd 24-25;
3489        Total=instock+backord;
3490
3491    run;

ERROR: No logical assign for filename INVNT.
NOTE: The SAS System stopped processing this step because of errors.
WARNING: The data set  PERM.UPDATE may be incomplete.  When this step was
         stopped there were 0 observations and 5 variables.
WARNING: Data set PERM.UPDATE was not replaced because this step was stopped.
         stopped.
NOTE: DATA statement used (Total process time):
      real time           0.01 seconds
      cpu time            0.01 seconds
```

Because Invent is misspelled, the statement in the DATA step that identifies the raw data is incorrect. Note, however, that the correct number of variables was defined in the descriptor portion of the data set.

Incorrectly identifying a variable's type is another common execution-time error. As you know, the values for IDnum are character values. Suppose you forget to place the dollar sign ($) after the variable's name in your INPUT statement. This is not a compile-time error, because SAS cannot verify IDnum's type until the data values for IDnum are read.

Raw Data File Invent

```
----+----10---+----20---+-
Bird Feeder   LG088  3 20
6 Glass Mugs  SB082  6 12
Glass Tray    BQ049 12  6
Padded Hangrs MN256 15 20
Jewelry Box   AJ498 23  0
Red Apron     AQ072  9 12
Crystal Vase  AQ672 27  0
Picnic Basket LS930 21  0
Brass Clock   AN910  2 10
```

```
data perm.update;
   infile invent;
   input Item $ 1-13 IDnum 15-19
         InStock 21-22 BackOrd 24-25;
   Total=instock+backord;
run;
```

In this case, the DATA step completes the execution phase, and the observations are written to the data set. However, several notes appear in the log.

SAS Log

```
NOTE: Invalid data for IDnum in line 7 15-19.
RULE: ----+----1----+----2----+----3----+----4--
07    Crystal Vase  AQ672 27 0
Item=Crystal Vase IDnum=. InStock=27 BackOrd=0
Total=27 _ERROR_=1 _N_=7
NOTE: Invalid data for IDnum in line 8 15-19.
08    Picnic Basket LS930 21 0
Item=Picnic Basket IDnum=. InStock=21 BackOrd=0
Total=21 _ERROR_=1 _N_=8
NOTE: Invalid data for IDnum in line 9 15-19.
09    Brass Clock   AN910 2 10
Item=Brass Clock IDnum=. InStock=2 BackOrd=10
Total=12 _ERROR_=1 _N_=9

NOTE: 9 records were read from the infile INVENT.
NOTE: The data set PERM.UPDATE has 9 observations
      and 5 variables.
```

Each note identifies the location of the invalid data for each observation. In this example, the invalid data is located in columns 15-19 for all observations.

The second line in each note (excluding the RULE line) displays the raw data record. Notice that the second field displays the values for IDnum, which are obviously character values.

SAS Log

```
NOTE: Invalid data for IDnum in line 7 15-19.
RULE: ----+----1----+----2----+----3----+----4--
07    Crystal Vase  AQ672 27 0
Item=Crystal Vase IDnum=. InStock=27 BackOrd=0
Total=27 _ERROR_=1 _N_=7
NOTE: Invalid data for IDnum in line 8 15-19.
08    Picnic Basket LS930 21 0
Item=Picnic Basket IDnum=. InStock=21 BackOrd=0
Total=21 _ERROR_=1 _N_=8
NOTE: Invalid data for IDnum in line 9 15-19.
09    Brass Clock   AN910 2 10
Item=Brass Clock IDnum=. InStock=2 BackOrd=10
Total=12 _ERROR_=1 _N_=9

NOTE: 9 records were read from the infile INVENT.
NOTE: The data set PERM.UPDATE has 9 observations
      and 5 variables.
```

The third and fourth lines display the values that are stored in the program data vector. Here, the values for IDnum are missing, although the other values have been correctly assigned to their respective variables. Notice that _ERROR_ has a value of **1**, indicating that a data error has occurred.

SAS Log

```
NOTE: Invalid data for IDnum in line 7 15-19.
RULE: ----+----1----+----2----+----3----+----4--
07     Crystal Vase  AQ672 27 0
Item=Crystal Vase IDnum=. InStock=27 BackOrd=0
Total=27 _ERROR_=1 _N_=7
NOTE: Invalid data for IDnum in line 8 15-19.
08     Picnic Basket LS930 21 0
Item=Picnic Basket IDnum=. InStock=21 BackOrd=0
Total=21 _ERROR_=1 _N_=8
NOTE: Invalid data for IDnum in line 9 15-19.
09     Brass Clock   AN910 2 10
Item=Brass Clock IDnum=. InStock=2 BackOrd=10
Total=12 _ERROR_=1 _N_=9


NOTE: 9 records were read from the infile INVENT.
NOTE: The data set PERM.UPDATE has 9 observations
      and 5 variables.
```

The PRINT procedure displays the data set, showing that the values for IDnum are missing. In this example, the periods indicate that IDnum is a numeric variable, although it should have been defined as a character variable.

```
proc print data=perm.update;
run;
```

Figure 6.3 *Output from the PRINT Procedure Showing Missing Values for IDnum*

Obs	Item	IDnum	InStock	BackOrd	Total
1	Bird Feeder	.	3	20	23
2	6 Glass Mugs	.	6	12	18
3	Glass Tray	.	12	6	18
4	Padded Hangrs	.	15	20	35
5	Jewelry Box	.	23	0	23
6	Red Apron	.	9	12	21
7	Crystal Vase	.	27	0	27
8	Picnic Basket	.	21	0	21
9	Brass Clock	.	2	10	12

When you read raw data with the DATA step, it's important to check the SAS log to verify that your data was read correctly. Here is a typical message:

SAS Log

```
WARNING: The data set PERM.UPDATE may be incomplete.
         When this step was stopped there were
         0 observations and 5 variables.
```

When no observations are written to the data set, you should check to see whether your DATA step was completely executed. Most likely, a syntax error or another error was detected at the beginning of the execution phase.

Validating and Cleaning Data

Data errors occur when data values are not appropriate for the SAS statements that are specified in a program. SAS detects data errors during program execution. When a data error is detected, SAS continues to execute the program.

In general, SAS procedures analyze data, produce output, or manage SAS files. In addition, SAS procedures can be used to detect invalid data. The following procedures can be used to detect invalid data:

- PROC PRINT
- PROC FREQ
- PROC MEANS

The PRINT procedure displays the data set, showing that the values for IDnum are missing. Periods indicate that IDnum is a numeric variable, although it should have been defined as a character variable.

```
proc print data=perm.update;
run;
```

Display 6.1   Output from the PRINT Procedure Showing Missing Values for IDnum

Obs	Item	IDnum	InStock	BackOrd	Total
1	Bird Feeder	.	3	20	23
2	6 Glass Mugs	.	6	12	18
3	Glass Tray	.	12	6	18
4	Padded Hangrs	.	15	20	35
5	Jewelry Box	.	23	0	23
6	Red Apron	.	9	12	21
7	Crystal Vase	.	27	0	27
8	Picnic Basket	.	21	0	21
9	Brass Clock	.	2	10	12

The FREQ procedure detects invalid character and numeric values by looking at distinct values. You can use PROC FREQ to identify any variables that were not given an expected value.

General form, FREQ procedure:

PROC FREQ DATA=*SAS-data-set <NLEVELS>*;

TABLES *variable(s);*

RUN;

where

- The TABLES statement specifies the frequency tables to produce.
- The NLEVELS option displays a table that provides the number of distinct values for each variable named in the TABLES statement.

In the following example, the data set contains invalid characters for the variables Gender and Age. PROC FREQ displays the distinct values of variables and is therefore useful for finding invalid values in data. You can use PROC FREQ with the TABLES statement to produce a frequency table for specific variables.

```
proc freq data=work.Patients;
     tables Gender Age;
run;
```

In the output you can see both the valid (**M** and **F**) and invalid (**G**) values for Gender, and the valid and invalid (242) values for age. In both the gender and age FREQ tables, one observation needs data cleaned as shown below:

The FREQ Procedure

Gender	Frequency	Percent	Cumulative Frequency	Cumulative Percent
F	2	50.00	2	50.00
G	1	25.00	3	75.00
M	1	25.00	4	100.00

Age	Frequency	Percent	Cumulative Frequency	Cumulative Percent
44	1	25.00	1	25.00
61	1	25.00	2	50.00
64	1	25.00	3	75.00
242	1	25.00	4	100.00

The MEANS procedure can also be used to validate data because it produces summary reports displaying descriptive statistics. For example, PROC MEANS can show whether the values for a particular variable are within their expected range.

General form, MEANS procedure:

PROC MEANS DATA=_SAS-data-set_ _<statistics>_;

VAR _variable(s)_;

RUN;

- The VAR statement specifies the analysis variables and their order in the results.
- The statistics to display can be specified in the PROC MEANS statement.

Using the same data set as in the previous example, we can submit PROC MEANS to determine if the Age of all test subjects is within a reasonable range. Notice the VAR statement is specified with that particular variable to get the statistical information, or range, of the data values.

```
proc means data=work.Patients;
     var Age;
run;
```

The output for the MEANS procedure displays a range of 44 to 242, which clearly indicates that there is invalid data somewhere in the Age column.

The MEANS Procedure

	Analysis Variable : Age			
N	Mean	Std Dev	Minimum	Maximum
4	102.7500000	93.2501117	44.0000000	242.0000000

Cleaning the Data

You can use an assignment statement or a conditional clause to programmatically clean invalid data once it is identified. For example, if your input data contains a field called Gender, and that field has an invalid value (a value other than M or F), then you can clean your data by changing the invalid value to a valid value for Gender. To avoid overwriting your original data set, you can use the DATA statement to create a new data set. The new data set will contain all of the data from your original data set, along with the correct values for invalid data.

The following example assumes that in your input data, Gender has an invalid value of G. This error might be the result of a data entry error. After examining your data, you determine that G should actually be F. You can fix the invalid data for Gender by using an assignment statement along with an IF-THEN statement:

```
data work.clean_data;
   set work.patients;
   gender=upcase(Gender);
   if Gender='G' then Gender='F';
run;

proc print data=work.clean_data;
run;
```

When you examine your input data further, you find that two observations contain invalid values for Age. These values exceed a maximum value of 110. You can uniquely identify each of the observations by the variable Empid. By checking the date of birth in each of the observations, you determine the correct value for Age. To change the data, you can use an IF-THEN-ELSE statement:

```
data work.clean_data;
   set work.patients;
   if empid=3294 then age=65;
   else if empid=7391 then age=75;
run;
```

```
proc print data=work.clean_data;
run;
```

Another way of ensuring that your output data set contains valid data is to programmatically identify invalid data and delete the associated observations from your output data set:

```
data work.clean_data;
   set work.patients;
   if Age>110 then delete;
run;


proc print data=work.clean_data;
run;
```

Testing Your Programs

Limiting Observations

Remember that you can use the OBS= option in the INFILE statement to limit the number of observations that are read or created during the execution of the DATA step.

```
data perm.update;
   infile invent obs=10;
   input Item $ 1-13 IDnum $ 15-19
         InStock 21-22 BackOrd 24-25;
   Total=instock+backord;
run;
```

When processed, this DATA step creates the perm.update data set with variables but with only ten observations.

PUT Statement

When the source of program errors is not apparent, you can use the PUT statement to examine variable values and to print your own message in the log. For diagnostic purposes, you can use IF-THEN/ELSE statements to conditionally check for values. You can learn about IF-THEN/ELSE statements in detail in "Creating and Managing Variables" on page 304.

```
data work.test;
   infile loan;
   input Code $ 1 Amount 3-10 Rate 12-16
         Account $ 18-25 Months 27-28;
   if code='1' then type='variable';
   else if code='2' then type='fixed';
   else type='unknown';
   if type=unknown then put 'MY NOTE: invalid value: '
   code=;
run;
```

In this example, if CODE does not have the expected values of **1** or **2**, the PUT statement writes a message to the log:

Display 6.2 *SAS Log*

```
MY NOTE: invalid value: Code=V
NOTE: 9 records were read from the infile LOAN.
      The minimum record length was 28.
      The maximum record length was 28.
NOTE: The data set WORK.TEST has 9 observations and 6 variables.
NOTE: DATA statement used (Total process time):
      real time            0.01 seconds
      cpu time             0.03 seconds
```

General form, simple PUT statement:

PUT *specification(s)*;

where *specification* specifies what is written, how it is written, and where it is written. This can include

- a character string

- one or more data set variables

- the automatic variables _N_ and _ERROR_

- the automatic variable _ALL_

and much more. The following pages show examples of PUT specifications.

Character Strings

You can use a PUT statement to specify a character string to identify your message in the log. The character string must be enclosed in quotation marks.

```
put 'MY NOTE: The condition was met.';
```

Display 6.3 *SAS Log*

```
MY NOTE: The condition was met.
NOTE: 9 records were read from the infile LOAN.
      The minimum record length was 28.
      The maximum record length was 28.
NOTE: The data set WORK.TEST has 9 observations and 6 variables.
NOTE: DATA statement used (Total process time):
      real time            0.01 seconds
      cpu time             0.01 seconds
```

Data Set Variables

You can use a PUT statement to specify one or more data set variables to be examined for that iteration of the DATA step:

```
put 'MY NOTE: invalid value: '
    code type;
```

Display 6.4 *SAS Log*

```
MY NOTE: invalid value: V unknown
NOTE: 9 records were read from the infile LOAN.
      The minimum record length was 28.
      The maximum record length was 28.
NOTE: The data set WORK.TEST has 9 observations and 6 variables.
NOTE: DATA statement used (Total process time):
      real time           0.00 seconds
      cpu time            0.00 seconds
```

Note that when you specify a variable in the PUT statement, only its value is written to the log. To write both the variable name and its value in the log, add an equal sign (=) to the variable name.

```
put 'MY NOTE: invalid value: '
    code= type=;
```

Display 6.5 *SAS Log*

```
MY NOTE: invalid value: Code=V type=unknown
NOTE: 9 records were read from the infile LOAN.
      The minimum record length was 28.
      The maximum record length was 28.
NOTE: The data set WORK.TEST has 9 observations and 6 variables.
NOTE: DATA statement used (Total process time):
      real time           0.01 seconds
      cpu time            0.00 seconds
```

Automatic Variables

You can use a PUT statement to display the values of the automatic variables _N_ and _ERROR_. In some cases, knowing the value of _N_ can help you locate an observation in the data set:

```
put 'MY NOTE: invalid value: '
    code= _n_=_error_=;
```

Display 6.6 *SAS Log*

```
MY NOTE: invalid value: Code=V _N_=3 _ERROR_=0
NOTE: 9 records were read from the infile LOAN.
      The minimum record length was 28.
      The maximum record length was 28.
NOTE: The data set WORK.TEST has 9 observations and 6 variables.
NOTE: DATA statement used (Total process time):
      real time           0.01 seconds
      cpu time            0.01 seconds
```

You can also use a PUT statement to write all variable names and variable values, including automatic variables, to the log. Use the _ALL_ specification:

```
put 'MY NOTE: invalid value: ' _all_ ;
```

Display 6.7 SAS Log

```
MY NOTE: invalid value: Code=V Amount=10000 Rate=10.5 Account=101-1289 Months=8
type=unknown _ERROR_=0 _N_=3
NOTE: 9 records were read from the infile LOAN.
      The minimum record length was 28.
      The maximum record length was 28.
NOTE: The data set WORK.TEST has 9 observations and 6 variables.
NOTE: DATA statement used (Total process time):
      real time           0.01 seconds
      cpu time            0.01 seconds
```

Conditional Processing

You can use a PUT statement with conditional processing (that is, with IF-THEN/ELSE statements) to flag program errors or data that is out of range. In the example below, the PUT statement is used to flag any missing or zero values for the variable Rate.

```
data finance.newcalc;
   infile newloans;
   input LoanID $ 1-4 Rate 5-8 Amount 9-19;
   if rate>0 then
      Interest=amount*(rate/12);
   else put 'DATA ERROR ' rate= _n_=;
run;
```

The PUT statement can accomplish a wide variety of tasks. This chapter shows a few ways to use the PUT statement to help you debug a program or examine variable values. For a complete description of the PUT statement, see the SAS documentation.

Chapter Summary

Text Summary

How SAS Processes Programs

A SAS DATA step is processed in two distinct phases. During the compilation phase, each statement is scanned for syntax errors. During the execution phase, the DATA step writes observations to the new data set.

Compilation Phase

At the beginning of the compilation phase, the input buffer and the program data vector are created. The program data vector is the area of memory where one observation at a time is processed. Two automatic variables are also created: _N_ counts the number of DATA step executions, and _ERROR_ signals the occurrence of a data error. DATA step statements are checked for syntax errors, such as invalid options or misspellings.

Execution Phase

During the execution phase, one record at a time from the input file is read into the input buffer, copied to the program data vector, and then written to the new data set as an observation. The DATA step executes once for each record in the input file, unless otherwise directed.

Diagnosing Errors in the Compilation Phase

Missing semicolons, misspelled keywords, and invalid options will cause syntax errors in the compilation phase. Detected errors are underlined and are identified with a number and message in the log. If SAS can interpret a syntax error, the DATA step compiles and executes; if SAS cannot interpret the error, the DATA step compiles but doesn't execute.

Diagnosing Errors in the Execution Phase

Illegal mathematical operations or processing a character variable as numeric will cause data errors in the execution phase. Depending on the type of error, the log might show a warning and might include invalid data from the input buffer or the program data vector, and the DATA step either stops or continues.

Testing Your Programs

To detect common errors and save development time, use the OBS= option in the INFILE statement to limit the number of observations that are read or created during the DATA step. You can also use the PUT statement to examine variable values and to generate your own message in the log. For information about Informats, see "Using Informats" on page 521.

Syntax

DATA < _NULL_ |*SAS-data-set*>;

    INFILE *file-specification* **OBS=***n*;

    INPUT *variable-1 informat-1*

    <. . .*variable-n informat-n*;>

    PUT *specification(s)*;

RUN;

Sample Programs

```
data perm.update;
   infile invent;
   input Item $ 1-13 IDnum $ 15-19
         InStock 21-22 BackOrd 24-25;
   Total=instock+backord;
run;

data work.test;
   infile loan;
   input Code $ 1 Amount 3-10;
   if code='1' then type='variable';
   else if code='2' then type='fixed';
   else put 'MY NOTE: invalid value: '
        code=;
run;
```

Points to Remember

- Making, diagnosing, and resolving errors is part of the process of writing programs. However, checking for common errors will save you time and trouble. Make sure that

- each SAS statement ends with a semicolon

- filenames are spelled correctly

- keywords are spelled correctly.

- In SAS output, missing numeric values are represented by periods, and missing character values are left blank.

- The order in which variables are defined in the DATA step determines the order in which the variables are stored in the data set.

- Standard character values can include numbers, but numeric values cannot include characters.

Chapter Quiz

Select the best answer for each question. After completing the quiz, you can check your answers using the answer key in the appendix.

1. Which of the following is *not* created during the compilation phase?

    a. the data set descriptor

    b. the first observation

    c. the program data vector

    d. the _N_ and _ERROR_ automatic variables

2. During the compilation phase, SAS scans each statement in the DATA step, looking for syntax errors. Which of the following is *not* considered a syntax error?

    a. incorrect values and formats

    b. invalid options or variable names

    c. missing or invalid punctuation

    d. missing or misspelled keywords

3. Unless otherwise directed, the DATA step executes...

    a. once for each compilation phase.

    b. once for each DATA step statement.

    c. once for each record in the input file.

    d. once for each variable in the input file.

4. At the beginning of the execution phase, the value of _N_ is 1, the value of _ERROR_ is 0, and the values of the remaining variables are set to:

    a. 0

    b. 1

    c. undefined

    d. missing

5. Suppose you run a program that causes three DATA step errors. What is the value of the automatic variable _ERROR_ when the observation that contains the third error is processed?

a. 0

b. 1

c. 2

d. 3

6. Which of the following actions occurs at the beginning of an iteration of the DATA step?

    a. The automatic variables _N_ and _ERROR_ are incremented by one.

    b. The DATA step stops execution.

    c. The descriptor portion of the data set is written.

    d. The values of variables created in programming statements are re-set to missing in the program data vector.

7. Look carefully at the DATA step shown below. Based on the INPUT statement, in what order will the variables be stored in the new data set?

```
data perm.update;
   infile invent;
   input IDnum $ Item $ 1-13 Instock 21-22
        BackOrd 24-25;
   Total=instock+backord;
run;
```

    a. IDnum Item InStock BackOrd Total

    b. Item IDnum InStock BackOrd Total

    c. Total IDnum Item InStock BackOrd

    d. Total Item IDnum InStock BackOrd

8. If SAS cannot interpret syntax errors, then...

    a. data set variables will contain missing values.

    b. the DATA step does not compile.

    c. the DATA step still compiles, but it does not execute.

    d. the DATA step still compiles and executes.

9. What is wrong with this program?

```
data perm.update;
   infile invent
   input Item $ 1-13 IDnum $ 15-19 Instock 21-22
        BackOrd 24-25;
   total=instock+backord;
run;
```

    a. missing semicolon on second line

    b. missing semicolon on third line

    c. incorrect order of variables

    d. incorrect variable type

10. Look carefully at this section of a SAS session log. Based on the note, what was the most likely problem with the DATA step?

```
NOTE: Invalid data for IDnum in line 7 15-19.
RULE: ----+----1----+----2----+----3----+----4
7         Bird Feeder LG088 3 20
Item=Bird Feeder IDnum=. InStock=3 BackOrd=20
Total=23 _ERROR_=1 _N_=1
```

a. A keyword was misspelled in the DATA step.

b. A semicolon was missing from the INFILE statement.

c. A variable was misspelled in the INPUT statement.

d. A dollar sign was missing in the INPUT statement.

Chapter 7

Creating and Applying User-Defined Formats

Overview

Introduction

If you read "Creating List Reports" on page 112, you learned to associate formats with variables either temporarily or permanently.

Figure 7.1   *Variable and Format*

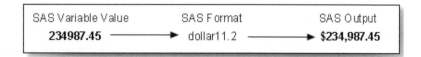

But sometimes you might want to create custom formats for displaying variable values. For example, you can format a product number so that it is displayed as descriptive text, as shown below.

Figure 7.2 *Variable and Custom Format*

Using the FORMAT procedure, you can define your own formats for variables. You can store your formats temporarily or permanently, and you can display a list of all your formats and descriptions of their values.

Objectives

In this chapter, you learn to

- create your own formats for displaying variable values
- permanently store the formats that you create
- associate your formats with variables.

Introduction to PROC FORMAT

Overview

Sometimes variable values are stored according to a code. For example, when the PRINT procedure displays the data set Perm.Employee, notice that the values for JobTitle are coded, and they are not easily interpreted.

Display 7.1 *Perm.Employee Data Set*

	FirstName	LastName	JobTitle	Salary
1	Donny	Evans	112	29996.63
2	Lisa	Helms	105	18567.23
3	John	Higgins	111	25309.24
4	Amy	Larson	113	32696.78
5	Mary	Moore	112	28945.89
6	Jason	Powell	103	35099.55
7	Judy	Riley	111	25309.37
8	Neal	Ryan	112	28180.85

You can display more descriptive values for these variables. Here is how a report that contains formatted values for the variable JobTitle might look. The predefined SAS formats cannot help here.

Display 7.2   Perm.Employee Data Set with Formatted Values for JobTitle

	FirstName	LastName	JobTitle	Salary
1	Donny	Evans	technical writer	29996.63
2	Lisa	Helms	text processor	18567.23
3	John	Higgins	assoc. technical writer	25309.24
4	Amy	Larson	senior technical writer	32696.78
5	Mary	Moore	technical writer	28945.89
6	Jason	Powell	manager	35099.55
7	Judy	Riley	assoc. technical writer	25309.37
8	Neal	Ryan	technical writer	28180.85

However, you can use the FORMAT procedure to define your own formats for displaying values of variables.

Invoking PROC FORMAT

To begin a PROC FORMAT step, you use a PROC FORMAT statement.

General form, PROC FORMAT statement:

PROC FORMAT *<options>*;

where *options* include

- **LIBRARY=***libref* specifies the libref for a SAS data library to contain a permanent catalog of user-defined formats

- **FMTLIB** displays a list of all of the formats in your catalog, along with descriptions of their values.

Any time you use PROC FORMAT to create a format, the format is stored in a format catalog. If the SAS data library does not already contain a format catalog, SAS automatically creates one. If you do not specify the LIBRARY= option, the formats are stored in a default format catalog named Work.Formats.

As the libref Work implies, any format that is stored in Work.Formats is a temporary format that exists only for the current SAS session. At the end of the current session, the catalog is erased.

Permanently Storing Your Formats

You can store your formats in a permanent format catalog named Formats when you specify the LIBRARY= option in the PROC FORMAT statement.

```
PROC FORMAT LIBRARY=libref;
```

But first, you need a LIBNAME statement that associates the libref with the permanent SAS data library in which the format catalog is to be stored.

```
libname library 'c:\sas\formats\lib';
```

When you associate a permanent format with a variable in a subsequent DATA or PROC step, you would then use the Library libref to reference the location of the format catalog.

We'll discuss the use of permanent user-defined formats later, after you learn how to create them.

Now, any format that you create in this PROC FORMAT step is stored in a permanent format catalog called Library.Formats.

```
libname library 'c:\sas\formats\lib';
proc format library=library;
   ... ;
run;
```

In the program above, the catalog Library.Formats is located in the SAS data library **C:\Sas\Formats\Lib**, which is referenced by the libref Library.

Notice that LIB= is an acceptable abbreviation for the LIBRARY= option.

```
proc format lib=library;
```

Also notice that you can specify a catalog name in the LIBRARY= option, and you can store formats in any catalog. The catalog name must conform to SAS naming conventions.

```
proc format lib=library.catalog;
```

Now that you know how to store your own formats, let's see how to create them.

Defining a Unique Format

Overview

You can use the VALUE statement to define a format for displaying one or more values.

General form, VALUE statement:

VALUE *format-name*

       range1='label1'

       range2='label2'

       ...;

where *format-name*

- must begin with a dollar sign ($) if the format applies to character data

- must be a valid SAS name

- cannot be the name of an existing SAS format

- cannot end in a number

- does not end in a period when specified in a VALUE statement.

In SAS 8.2, the format name must be a SAS name up to eight characters in length, not ending in a number.

Beginning with SAS®9, a numeric format name can be up to 32 characters in length. A character format name can be up to 31 characters in length.

Notice that the statement begins with the keyword VALUE and ends with a semicolon after all the labels have been defined. The following VALUE statement creates the JOBFMT format to specify descriptive labels that will later be assigned to the variable JobTitle:

```
proc format lib=library;
   value jobfmt
         103='manager'
```

```
            105='text processor'
            111='assoc. technical writer'
            112='technical writer'
            113='senior technical writer';
run;
```

The VALUE range specifies

- a single value, such as **24** or **'S'**.

- a range of numeric values, such as **0-1500**.

- a range of character values enclosed in quotation marks, such as **'A'**-**'M'**.

- a list of unique values separated by commas, such as **90,180,270** or **'B'**, **'D'**, **'F'**.
  These values can be character values or numeric values, but not a combination of
  character and numeric values (because formats themselves are either character or
  numeric).

When the specified values are character values, they must be enclosed in quotation
marks and must match the case of the variable's values. The format's name must also
start with a dollar sign ($). For example, the VALUE statement below defines the
$GRADE format, which displays the character values as text labels.

```
proc format lib=library;
    value $grade
          'A'='Good'
          'B'-'D'='Fair'
          'F'='Poor'
          'I','U'='See Instructor';
run;
```

When the specified values are numeric values, they are not enclosed in quotation marks,
and the format's name should not begin with a dollar sign ($). The VALUE statement
that defines the format JOBFMT assigns labels to numeric values.

Specifying Value Ranges

You can specify a non-inclusive range of numeric values by using the less than symbol
(<) to avoid any overlapping. In this example, the range of values from **0** to less than **13**
is labeled as child. The next range begins at **13**, so the label teenager would be assigned
to the values **13** to **19**.

```
proc format lib=library;
    value agefmt
          0-<13='child'
          13-<20='teenager'
          20-<65='adult'
          65-100='senior citizen';
run;
```

You can also use the keywords LOW and HIGH to specify the lower and upper limits of
a variable's value range. The keyword LOW does not include missing numeric values.
The keyword OTHER can be used to label missing values as well as any values that are
not specifically addressed in a range.

```
proc format lib=library;
    value agefmt
          low-<13='child'
          13-<20='teenager'
```

```
        20-<65='adult'
        65-high='senior citizen'
        other='unknown';
run;
```

If applied to a character format, the keyword LOW includes missing character values.

When specifying a label for displaying each range, remember to

- enclose the label in quotation marks

- limit the label to 256 characters

- use two single quotation marks if you want an apostrophe to appear in the label, as in this example:

```
000='employee''s jobtitle unknown';
```

To define several formats, you can use multiple VALUE statements in a single PROC FORMAT step. In this example, each VALUE statement defines a different format.

```
proc format lib=library;
   value jobfmt
         103='manager'
         105='text processor'
         111='assoc. technical writer'
         112='technical writer'
         113='senior technical writer';
   value $respnse
         'Y'='Yes'
         'N'='No'
         'U'='Undecided'
         'NOP'='No opinion';
run;
```

The SAS log prints notes informing you that the formats have been created.

Figure 7.3 *SAS Log*

```
                    SAS Log (Partial Listing)

01   proc format lib=library;
02      value jobfmt
03            103='manager'
04            105='text processor'
05            111='assoc. technical writer'
06            112='technical writer'
07            113='senior technical writer';
     NOTE: Format JOBFMT has been written to LIBRARY.FORMATS.
```

Because you have defined the JOBFMT format for displaying the values of JobTitle, the format can be used with PROC PRINT for more legible output.

Display 7.3 *Data Set with Unformatted Values for JobTitle*

	FirstName	LastName	JobTitle	Salary
1	Donny	Evans	112	29996.63
2	Lisa	Helms	105	18567.23
3	John	Higgins	111	25309.24
4	Amy	Larson	113	32696.78
5	Mary	Moore	112	28945.89
6	Jason	Powell	103	35099.55
7	Judy	Riley	111	25309.37
8	Neal	Ryan	112	28180.85

Display 7.4 *Data Set with Formatted Values for JobTitle*

	FirstName	LastName	JobTitle	Salary
1	Donny	Evans	technical writer	29996.63
2	Lisa	Helms	text processor	18567.23
3	John	Higgins	assoc. technical writer	25309.24
4	Amy	Larson	senior technical writer	32696.78
5	Mary	Moore	technical writer	28945.89
6	Jason	Powell	manager	35099.55
7	Judy	Riley	assoc. technical writer	25309.37
8	Neal	Ryan	technical writer	28180.85

The next section will show how to apply your formats to variables.

Associating User-Defined Formats with Variables

Referencing Your Formats

Remember that permanent, user-defined formats are stored in a format catalog. For example, the program below stores the format JOBFMT in a catalog named Library.Formats, which is located in the directory **C:\Sas\Formats\Lib** in the Windows environment.

```
libname library 'c:\sas\formats\lib';
proc format lib=library;
   value jobfmt
        103='manager'
        105='text processor'
        111='assoc. technical writer'
        112='technical writer'
        113='senior technical writer';
run;
```

To use the JOBFMT format in a subsequent SAS session, you must assign the libref Library again.

```
libname library 'c:\sas\formats\lib';
```

SAS searches for the format JOBFMT in two libraries, in this order:

- the temporary library referenced by the libref Work

- a permanent library referenced by the libref Library.

SAS uses the first instance of a specified format that it finds.

 You can delete formats using PROC CATALOG or the SAS Explorer window.

Assigning Your Formats to Variables

Just as with SAS formats, you associate a user-defined format with a variable in a FORMAT statement.

```
data perm.employee;
   infile empdata;
   input @9 FirstName $5. @1 LastName $7. +7 JobTitle 3.
         @19 Salary comma9.;
   format salary comma9.2 jobtitle jobfmt.;
run;
```

Don't worry about @ pointer controls in programs in this chapter, as in @9 FirstName—they don't affect the behavior of formats. To learn more about using @ pointer controls in SAS programs, see "Reading Raw Data in Fixed Fields" on page 514.

Remember, you can place the FORMAT statement in either a DATA step or a PROC step. By placing the FORMAT statement in a DATA step, you can permanently associate a format with a variable. Note that you do not have to specify a width value when using a user-defined format.

When you submit the PRINT procedure, the output for Perm.Employee now shows commas in the values for Salary, and it shows descriptive labels in place of the values for JobTitle.

```
proc print data=perm.employee;
run;
```

Display 7.5 *Data Set with Formatted Values for Salary and JobTitle*

	FirstName	LastName	JobTitle	Salary
1	Donny	Evans	technical writer	29,996.63
2	Lisa	Helms	text processor	18,567.23
3	John	Higgins	assoc. technical writer	25,309.24
4	Amy	Larson	senior technical writer	32,696.78
5	Mary	Moore	technical writer	28,945.89
6	Jason	Powell	manager	35,099.55
7	Judy	Riley	assoc. technical writer	25,309.37
8	Neal	Ryan	technical writer	28,180.85

When associating a format with a variable, remember to

- use the same format name in the FORMAT statement that you specified in the VALUE statement

- place a period at the end of the format name when it is used in the FORMAT statement.

Note: The following example shows you how to create a format and then use the format in a DATA step. If you have already created the JOBFMT format, you do not need to use PROC FORMAT to recreate it here.

```
libname library 'c:\sas\formats\lib';
proc format lib=library;
   value jobfmt
         103='manager'
```

```
                105='text processor'
                111='assoc. technical writer'
                112='technical writer'
                113='senior technical writer';
        run;
        libname perm 'c:\data\perm';
        filename empdata 'c:\data\temp\newhires.txt';
        data perm.employee;
            infile empdata;
            input @9 FirstName $5. @1 LastName $7. +7 JobTitle 3.
                @19 Salary comma9.;
            format salary comma9.2 jobtitle jobfmt.;
        run;
```

Notice that a period is *not* required at the end of the SAS format COMMA9.2 in the FORMAT statement. The period in this format occurs between the width specification and the decimal place specification. All formats contain periods, but only user-defined formats invariably require periods at the end of the name.

If you do not format all of a variable's values, then those that are not listed in the VALUE statement are printed as they appear in the SAS data set, as shown in the following example:

```
proc format lib=library;
    value jobfmt
            103='manager'
            105='text processor';
run;
proc print data=perm.employee;
run;
```

Display 7.6 *Perm.Employee Data Set*

	FirstName	LastName	JobTitle	Salary
1	Donny	Evans	112	29996.63
2	Lisa	Helms	text processor	18567.23
3	John	Higgins	111	25309.24
4	Amy	Larson	113	32696.78
5	Mary	Moore	112	28945.89
6	Jason	Powell	manager	35099.55
7	Judy	Riley	111	25309.37
8	Neal	Ryan	112	28180.85

When you build a large catalog of permanent formats, it can be easy to forget the exact spelling of a specific format name or its range of values. Adding the keyword FMTLIB to the PROC FORMAT statement displays a list of all the formats in your catalog, along with descriptions of their values.

```
libname library 'c:\sas\formats\lib';
proc format library=library fmtlib;
run;
```

When you submit this PROC step, a description of each format in your permanent catalog is displayed as output.

Display 7.7 *SAS Output: Format Description*

```
┌───────────────────────────────────────────────────────────────────────┐
│       FORMAT NAME: JOBFMT   LENGTH:   23   NUMBER OF VALUES:    5        │
│   MIN LENGTH:   1  MAX LENGTH:  40  DEFAULT LENGTH:  23  FUZZ: STD       │
├──────────────────────┬──────────────────┬──────────────────────────────┤
│ START                │ END              │ LABEL  (VER. V7|V8   13MAY2011:10:02:13 ) │
├──────────────────────┼──────────────────┼──────────────────────────────┤
│                  103 │              103 │ manager                      │
│                  105 │              105 │ text processor               │
│                  111 │              111 │ assoc. technical  writer     │
│                  112 │              112 │ technical  writer            │
│                  113 │              113 │ senior technical  writer     │
└──────────────────────┴──────────────────┴──────────────────────────────┘
```

In addition to the name, range, and label, the format description includes the

- length of the longest label

- number of values defined by this format

- version of SAS that was used to create the format

- date and time of creation.

Chapter Summary

Text Summary

Invoking PROC FORMAT

The FORMAT procedure enables you to substitute descriptive text for the values of variables. The LIBRARY= option stores the new formats in a specified format catalog; otherwise, they are stored in a default catalog named Work.Formats. The keyword FMTLIB displays the formats and values that are currently stored in the catalog. The VALUE statement defines a new format for the values of a variable.

Defining a Unique Format

Formats can be specified for a single value, a range of values, or a list of unique values. Unique values must be separated by commas. When character values are specified, the values must be enclosed in quotation marks, and the format name must begin with a dollar sign ($). You can specify non-inclusive numeric ranges by using the less than sign (<). The keywords HIGH, LOW, and OTHER can be used to label values that are not specifically addressed in a range.

Associating User-Defined Formats with Variables

To access the permanent, user-defined formats in a format catalog, you'll need to use a LIBNAME statement to reference the catalog library. To associate user-defined formats with variables in the FORMAT statement, use the same format names in both the FORMAT and VALUE statements, but place a period at the end of the format name when it is used in the FORMAT statement.

Syntax

LIBNAME *libref 'SAS-data-library'*;

```
PROC FORMAT LIBRARY=libref FMTLIB;
       VALUE format-name
           range1='label1'
           range2 'label2'
           ...;
RUN;
DATA SAS-data-set;
       INFILE data-file;
       INPUT pointer variable-name informat.;
       FORMAT variable(s) format-name.;
RUN;
```

Sample Programs

```
libname library 'c:\sas\formats\lib';
proc format library=library fmtlib;
    value jobfmt
          103='manager'
          105='text processor';
run;

data perm.empinfo;
    infile empdata;
    input @9 FirstName $5. @1 LastName $7.
          +7 JobTitle 3. @19 Salary comma9.;
    format salary comma9.2 jobtitle jobfmt.;
run;
```

Points to Remember

- Formats—even permanently associated ones—do not affect the values of variables in a SAS data set. Only the appearance of the values is altered.

- A user-defined format name must begin with a dollar sign ($) when it is assigned to character variables. A format name cannot end with a number.

- Use two single quotation marks when you want an apostrophe to appear in a label.

- Place a period at the end of the format name when you use the format name in the FORMAT statement.

Chapter Quiz

Select the best answer for each question. After completing the quiz, check your answers using the answer key in the appendix

1. If you don't specify the LIBRARY= option, your formats are stored in Work.Formats, and they exist ...

   a. only for the current procedure.

   b. only for the current DATA step.

    c. only for the current SAS session.

    d. permanently.

2. Which of the following statements will store your formats in a permanent catalog?

    a.
```
libname library 'c:\sas\formats\lib';
proc format lib=library
    ...;
```

    b.
```
libname library 'c:\sas\formats\lib';
format lib=library
    ...;
```

    c.
```
library='c:\sas\formats\lib';
proc format library
    ...;
```

    d.
```
library='c:\sas\formats\lib';
proc library
    ...;
```

3. When creating a format with the VALUE statement, the new format's name

    • cannot end with a number

    • cannot end with a period

    • cannot be the name of a SAS format, and...

    a. cannot be the name of a data set variable.

    b. must be at least two characters long.

    c. must be at least eight characters long.

    d. must begin with a dollar sign ($) if used with a character variable.

4. Which of the following FORMAT procedures is written correctly?

    a.
```
proc format lib=library
    value colorfmt;
        1='Red'
        2='Green'
        3='Blue'
run;
```

    b.
```
proc format lib=library;
    value colorfmt
        1='Red'
        2='Green'
        3='Blue';
run;
```

    c.
```
proc format lib=library;
    value colorfmt;
        1='Red'
        2='Green'
        3='Blue'
run;
```

    d.
```
proc format lib=library;
    value colorfmt
        1='Red';
        2='Green';
```

```
        3='Blue';
run;
```

5. Which of these is *false*? Ranges in the VALUE statement can specify...

   a. a single value, such as `24` or `'S'`.

   b. a range of numeric values, such as `0-1500`.

   c. a range of character values, such as `'A'-'M'`.

   d. a list of numeric and character values separated by commas, such as `90,'B', 180,'D',270`.

6. How many characters can be used in a label?

   a. 40

   b. 96

   c. 200

   d. 256

7. Which keyword can be used to label missing numeric values as well as any values that are not specified in a range?

   a. LOW

   b. MISS

   c. MISSING

   d. OTHER

8. You can place the FORMAT statement in either a DATA step or a PROC step. What happens when you place it in a DATA step?

   a. You temporarily associate the formats with variables.

   b. You permanently associate the formats with variables.

   c. You replace the original data with the format labels.

   d. You make the formats available to other data sets.

9. The format JOBFMT was created in a FORMAT procedure. Which FORMAT statement will apply it to the variable JobTitle in the program output?

   a. `format jobtitle jobfmt;`

   b. `format jobtitle jobfmt.;`

   c. `format jobtitle=jobfmt;`

   d. `format jobtitle='jobfmt';`

10. Which keyword, when added to the PROC FORMAT statement, will display all the formats in your catalog?

    a. CATALOG

    b. LISTFMT

    c. FMTCAT

    d. FMTLIB

Chapter 8
Producing Descriptive Statistics

Overview

Introduction

As you have seen, one of the many features of PROC REPORT is the ability to summarize large amounts of data by producing descriptive statistics. However, there are SAS procedures that are designed specifically to produce various types of descriptive statistics and to display them in meaningful reports. The type of descriptive statistics that you need and the SAS procedure that you should use depend on whether you need to summarize continuous data values or discrete data values.

If the data values that you want to describe are continuous numeric values (for example, people's ages), then you can use the MEANS procedure or the SUMMARY procedure to calculate statistics such as the mean, sum, minimum, and maximum.

Figure 8.1   *MEANS Procedure Output*

Variable	N	Mean	Std Dev	Minimum	Maximum
Age	20	47	13	15	63
Height	20	67	4	61	75
Weight	20	175	36	102	240
Pulse	20	75	8	65	100
FastGluc	20	299	126	152	568
PostGluc	20	355	126	206	625

If the data values that you want to describe are discrete (for example, the color of people's eyes), then you can use the FREQ procedure to show the distribution of these values, such as percentages and counts.

Figure 8.2   *FREQ Procedure Output*

Eye Color	Frequency	Percent	Cumulative Frequency	Cumulative Percent
Brown	92	58.60	92	58.60
Blue	65	41.40	157	100.00

This chapter will show you how to use the MEANS, SUMMARY, and FREQ procedures to describe your data.

Objectives

In this chapter, you learn to

- determine the *n*-count, mean, standard deviation, minimum, and maximum of numeric variables using the MEANS procedure

- control the number of decimal places used in PROC MEANS output

- specify the variables for which to produce statistics

- use the PROC SUMMARY procedure to produce the same results as the PROC MEANS procedure

- describe the difference between the SUMMARY and MEANS procedures

- create one-way frequency tables for categorical data using the FREQ procedure

- create two-way and -way crossed frequency tables

- control the layout and complexity of crossed frequency tables.

Computing Statistics Using PROC MEANS

Descriptive statistics such as the mean, minimum, and maximum provide useful information about numeric data. The MEANS procedure provides these and other data summarization tools, as well as helpful options for controlling your output.

Procedure Syntax

The MEANS procedure can include many statements and options for specifying needed statistics. For simplicity, let's consider the procedure in its basic form.

General form, basic MEANS procedure:

PROC MEANS <DATA=*SAS-data-set*>

  <*statistic-keyword(s)*> <*option(s)*>;

RUN;

where

- *SAS-data-set* is the name of the data set to be used

- *statistic-keyword(s)* specify the statistics to compute

- *option(s)* control the content, analysis, and appearance of output.

In its simplest form, PROC MEANS prints the *n*-count (number of nonmissing values), the mean, the standard deviation, and the minimum and maximum values of every numeric variable in a data set.

```
proc means data=perm.survey;
run;
```

Figure 8.3 *Default PROC MEANS Output*

The MEANS Procedure

Variable	N	Mean	Std Dev	Minimum	Maximum
Item1	4	3.7500000	1.2583057	2.0000000	5.0000000
Item2	4	3.0000000	1.6329932	1.0000000	5.0000000
Item3	4	4.2500000	0.5000000	4.0000000	5.0000000
Item4	4	3.5000000	1.2909944	2.0000000	5.0000000
Item5	4	3.0000000	1.6329932	1.0000000	5.0000000
Item6	4	3.7500000	1.2583057	2.0000000	5.0000000
Item7	4	3.0000000	1.8257419	1.0000000	5.0000000
Item8	4	2.7500000	1.5000000	1.0000000	4.0000000
Item9	4	3.0000000	1.4142136	2.0000000	5.0000000
Item10	4	3.2500000	1.2583057	2.0000000	5.0000000
Item11	4	3.0000000	1.8257419	1.0000000	5.0000000
Item12	4	2.7500000	0.5000000	2.0000000	3.0000000
Item13	4	2.7500000	1.5000000	1.0000000	4.0000000
Item14	4	3.0000000	1.4142136	2.0000000	5.0000000
Item15	4	3.0000000	1.6329932	1.0000000	5.0000000
Item16	4	2.5000000	1.9148542	1.0000000	5.0000000
Item17	4	3.0000000	1.1547005	2.0000000	4.0000000
Item18	4	3.2500000	1.2583057	2.0000000	5.0000000

Selecting Statistics

The default statistics that the MEANS procedure produces (*n*-count, mean, standard deviation, minimum, and maximum) are not always the ones that you need. You might prefer to limit output to the mean of the values. Or you might need to compute a different statistic, such as the median or range of the values.

To specify statistics, include statistic keywords as options in the PROC MEANS statement. When you specify a statistic in the PROC MEANS statement, default statistics are not produced. For example, to see the median and range of Perm.Survey numeric values, add the MEDIAN and RANGE keywords as options.

```
proc means data=perm.survey median range;
run;
```

Figure 8.4 *MEANS Procedure Output Displaying Median and Range*

MEANS Procedure Output Displaying Median and Range

The MEANS Procedure

Variable	Median	Range
Item1	4.0000000	3.0000000
Item2	3.0000000	4.0000000
Item3	4.0000000	1.0000000
Item4	3.5000000	3.0000000
Item5	3.0000000	4.0000000
Item6	4.0000000	3.0000000
Item7	3.0000000	4.0000000
Item8	3.0000000	3.0000000
Item9	2.5000000	3.0000000
Item10	3.0000000	3.0000000
Item11	3.0000000	4.0000000
Item12	3.0000000	1.0000000
Item13	3.0000000	3.0000000
Item14	2.5000000	3.0000000
Item15	3.0000000	4.0000000
Item16	2.0000000	4.0000000
Item17	3.0000000	2.0000000
Item18	3.0000000	3.0000000

The following keywords can be used with PROC MEANS to compute statistics:

Table 8.1 *Descriptive Statistics*

Keyword	Description
CLM	Two-sided confidence limit for the mean
CSS	Corrected sum of squares
CV	Coefficient of variation
KURTOSIS	Kurtosis
LCLM	One-sided confidence limit below the mean
MAX	Maximum value
MEAN	Average
MODE	Value that occurs most frequently (new in SAS 9.2)
MIN	Minimum value
N	Number of observations with nonmissing values
NMISS	Number of observations with missing values
RANGE	Range
SKEWNESS	Skewness

Keyword	Description
STDDEV / STD	Standard deviation
STDERR	Standard error of the mean
SUM	Sum
SUMWGT	Sum of the Weight variable values
UCLM	One-sided confidence limit above the mean
USS	Uncorrected sum of squares
VAR	Variance

Table 8.2 Quantile Statistics

Keyword	Description
MEDIAN / P50	Median or 50th percentile
P1	1st percentile
P5	5th percentile
P10	10th percentile
Q1 / P25	Lower quartile or 25th percentile
Q3 / P75	Upper quartile or 75th percentile
P90	90th percentile
P95	95th percentile
P99	99th percentile
QRANGE	Difference between upper and lower quartiles: Q3-Q1

Table 8.3 Hypothesis Testing

Keyword	Description
PROBT	Probability of a greater absolute value for the t value
T	Student's t for testing the hypothesis that the population mean is 0

Limiting Decimal Places

By default, PROC MEANS output automatically uses the BEST*w.* format to display numeric values in the report.

The BEST*w.* format is the default format that SAS uses for writing numeric values. When there is no format specification, SAS chooses the format that provides the most information about the value according to the available field width. At times, this can result in unnecessary decimal places, making your output hard to read.

```
proc means data=clinic.diabetes min max;
run;
```

Figure 8.5 *Variables Fromatted with BESTw. Format*

The MEANS Procedure

Variable	Minimum	Maximum
Age	15.0000000	63.0000000
Height	61.0000000	75.0000000
Weight	102.0000000	240.0000000
Pulse	65.0000000	100.0000000
FastGluc	152.0000000	568.0000000
PostGluc	206.0000000	625.0000000

To limit decimal places, use the MAXDEC= option in the PROC MEANS statement, and set it equal to the length that you prefer.

General form, PROC MEANS statement with MAXDEC= option:

PROC MEANS <DATA=*SAS-data-set*>

      <*statistic-keyword(s)*> **MAXDEC=***n***;**

where *n* specifies the maximum number of decimal places.

```
proc means data=clinic.diabetes min max maxdec=0;
run;
```

Figure 8.6 *Variables Fromatted Using the MAXDEC= Option*

The MEANS Procedure

Variable	Minimum	Maximum
Age	15	63
Height	61	75
Weight	102	240
Pulse	65	100
FastGluc	152	568
PostGluc	206	625

Specifying Variables in PROC MEANS

By default, the MEANS procedure generates statistics for every numeric variable in a data set. But you'll typically want to focus on just a few variables, particularly if the data set is large. It also makes sense to exclude certain types of variables. The values of ID, for example, are unlikely to yield useful statistics.

To specify the variables that PROC MEANS analyzes, add a VAR statement and list the variable names.

General form, VAR statement:

VAR *variable(s)*;

where *variable(s)* lists numeric variables for which to calculate statistics.

```
proc means data=clinic.diabetes min max maxdec=0;
   var age height weight;
run;
```

Figure 8.7 *Output with Selected Variables Age, Height, and Weight*

The MEANS Procedure

Variable	Minimum	Maximum
Age	15	63
Height	61	75
Weight	102	240

In addition to listing variables separately, you can use a numbered range of variables.

```
proc means data=perm.survey mean stderr maxdec=2;
   var item1-item5;
run;
```

Figure 8.8 *Output with a Range of Variables Selected*

The MEANS Procedure

Variable	Mean	Std Error
Item1	3.75	0.63
Item2	3.00	0.82
Item3	4.25	0.25
Item4	3.50	0.65
Item5	3.00	0.82

Group Processing Using the CLASS Statement

You will often want statistics for groups of observations, instead of over the entire data set. For example, census numbers are more useful when grouped by region than when viewed as a national total. To produce separate analyses of grouped observations, add a CLASS statement to the MEANS procedure.

General form, CLASS statement:

CLASS *variable(s)*;

where *variable(s)* specifies category variables for group processing.

CLASS variables are used to categorize data. CLASS variables can be either character or numeric, but they should contain a limited number of discrete values that represent meaningful groupings. If a CLASS statement is used, then the N Obs statistic is calculated. The N Obs statistic is based on the CLASS variables, as shown in the output below.

The output of the program shown below is categorized by values of the variables Survive and Sex. The order of the variables in the CLASS statement determines their order in the output table.

```
proc means data=clinic.heart maxdec=1;
   var arterial heart cardiac urinary;
   class survive sex;
run;
```

Figure 8.9 *Output Categorized by Values of the Variables Survive and Sex*

The MEANS Procedure

Survive	Sex	N Obs	Variable	N	Mean	Std Dev	Minimum	Maximum
DIED	1	4	Arterial	4	92.5	10.5	83.0	103.0
			Heart	4	111.0	53.4	54.0	183.0
			Cardiac	4	176.8	75.2	95.0	260.0
			Urinary	4	98.0	186.1	0.0	377.0
	2	6	Arterial	6	94.2	27.3	72.0	145.0
			Heart	6	103.7	16.7	81.0	130.0
			Cardiac	6	318.3	102.6	156.0	424.0
			Urinary	6	100.3	155.7	0.0	405.0
SURV	1	5	Arterial	5	77.2	12.2	61.0	88.0
			Heart	5	109.0	32.0	77.0	149.0
			Cardiac	5	298.0	139.8	66.0	410.0
			Urinary	5	100.8	60.2	44.0	200.0
	2	5	Arterial	5	78.8	6.8	72.0	87.0
			Heart	5	100.0	13.4	84.0	111.0
			Cardiac	5	330.2	87.0	256.0	471.0
			Urinary	5	111.2	152.4	12.0	377.0

Group Processing Using the BY Statement

Like the CLASS statement, the BY statement specifies variables to use for categorizing observations.

General form, BY statement:

BY *variable(s)*;

where *variable(s)* specifies category variables for group processing.

But BY and CLASS differ in two key ways:

1. Unlike CLASS processing, BY processing requires that your data already be sorted or indexed in the order of the BY variables. Unless data set observations are already sorted, you will need to run the SORT procedure before using PROC MEANS with any BY group.

   Be careful when sorting data sets to enable group processing. If you don't specify an output data set by using the OUT= option, PROC SORT will overwrite your initial data set with the newly sorted observations.

2. BY group results have a layout that is different from the layout of CLASS group results. Note that the BY statement in the program below creates four small tables; a CLASS statement would produce a single large table.

```
proc sort data=clinic.heart out=work.heartsort;
   by survive sex;
run;
proc means data=work.heartsort maxdec=1;
   var arterial heart cardiac urinary;
   by survive sex;
run;
```

Figure 8.10 *BY Groups Created by PROC MEANS*

The MEANS Procedure

Survive=DIED Sex=1

Variable	N	Mean	Std Dev	Minimum	Maximum
Arterial	4	92.5	10.5	83.0	103.0
Heart	4	111.0	53.4	54.0	183.0
Cardiac	4	176.8	75.2	95.0	260.0
Urinary	4	98.0	186.1	0.0	377.0

Survive=DIED Sex=2

Variable	N	Mean	Std Dev	Minimum	Maximum
Arterial	6	94.2	27.3	72.0	145.0
Heart	6	103.7	16.7	81.0	130.0
Cardiac	6	318.3	102.6	156.0	424.0
Urinary	6	100.3	155.7	0.0	405.0

Survive=SURV Sex=1

Variable	N	Mean	Std Dev	Minimum	Maximum
Arterial	5	77.2	12.2	61.0	88.0
Heart	5	109.0	32.0	77.0	149.0
Cardiac	5	298.0	139.8	66.0	410.0
Urinary	5	100.8	60.2	44.0	200.0

Survive=SURV Sex=2

Variable	N	Mean	Std Dev	Minimum	Maximum
Arterial	5	78.8	6.8	72.0	87.0
Heart	5	100.0	13.4	84.0	111.0
Cardiac	5	330.2	87.0	256.0	471.0
Urinary	5	111.2	152.4	12.0	377.0

Because it doesn't require a sorting step, the CLASS statement is easier to use than the BY statement. However, BY group processing can be more efficient when your categories might contain many levels.

Creating a Summarized Data Set Using PROC MEANS

Overview

You might want to create an output SAS data set. You can do this by using the OUTPUT statement in PROC MEANS.

General form, OUTPUT statement:

OUTPUT OUT=*SAS-data-set statistic=variable(s)***;**

where

- **OUT=** specifies the name of the output data set

- *statistic=* specifies the summary statistic written out

- *variable(s)* specifies the names of the variables to create. These variables represent the statistics for the analysis variables that are listed in the VAR statement.

When you use the OUTPUT statement, the summary statistics N, MEAN, STD, MIN, and MAX are produced for all of the numeric variables or for *all* of the variables that are listed in a VAR statement *by default*. To specify which statistics to produce, use the OUT= option in the OUTPUT statement.

Specifying Statistics Using PROC MEANS

You can specify which statistics to produce in the output data set by using the OUT= option in the OUTPUT statement.. To do so, you must specify the statistic and then list all of the variables. The variables must be listed in the same order as in the VAR statement. You can specify more than one statistic in the OUTPUT statement.

The following program creates a typical PROC MEANS report and also creates a summarized output data set.

```
proc means data=clinic.diabetes;
   var age height weight;
   class sex;
   output out=work.sum_gender
      mean=AvgAge AvgHeight AvgWeight
      min=MinAge MinHeight MinWeight;
run;
```

Figure 8.11   *Report Created by PROC MEANS*

The MEANS Procedure

Sex	N Obs	Variable	N	Mean	Std Dev	Minimum	Maximum
F	11	Age	11	48.9090909	13.3075508	16.0000000	63.0000000
		Height	11	63.9090909	2.1191765	61.0000000	68.0000000
		Weight	11	150.4545455	18.4464828	102.0000000	168.0000000
M	9	Age	9	44.0000000	12.3895117	15.0000000	54.0000000
		Height	9	70.6666667	2.6457513	66.0000000	75.0000000
		Weight	9	204.2222222	30.2893454	140.0000000	240.0000000

To see the contents of the output data set, submit the following PROC PRINT step.

```
proc print data=work.sum_gender;
run;
```

Figure 8.12 Data Set Created by PROC PRINT

Obs	Sex	_TYPE_	_FREQ_	AvgAge	AvgHeight	AvgWeight	MinAge	MinHeight	MinWeight
1		0	20	46.7000	66.9500	174.650	15	61	102
2	F	1	11	48.9091	63.9091	150.455	16	61	102
3	M	1	9	44.0000	70.6667	204.222	15	66	140

You can use the NOPRINT option in the PROC MEANS statement to suppress the default report. For example, the following program creates only the output data set:

```
proc means data=clinic.diabetes noprint;
   var age height weight;
   class sex;
   output out=work.sum_gender
      mean=AvgAge AvgHeight AvgWeight;
run;
```

In addition to the variables that you specify, the procedure adds the _TYPE_ and _FREQ_ variables to the output data set. When no statistic keywords are specified, PROC MEANS also adds the variable _STAT_. For more information about these variables, see the SAS documentation for the MEANS procedure.

Creating a Summarized Data Set Using PROC SUMMARY

You can also create a summarized output data set with PROC SUMMARY. When you use PROC SUMMARY, you use the same code to produce the output data set that you would use with PROC MEANS. The difference between the two procedures is that PROC MEANS produces a report by default (remember that you can use the NOPRINT option to suppress the default report). By contrast, to produce a report in PROC SUMMARY, you must include a PRINT option in the PROC SUMMARY statement.

Example

The following example creates an output data set but does not create a report:

```
proc summary data=clinic.diabetes;
   var age height weight;
   class sex;
   output out=work.sum_gender
      mean=AvgAge AvgHeight AvgWeight;
run;
```

If you placed a PRINT option in the PROC SUMMARY statement above, this program would produce the same report as if you replaced the word SUMMARY with MEANS.

```
proc summary data=clinic.diabetes print;
   var age height weight;
   class sex;
   output out=work.sum_gender
      mean=AvgAge AvgHeight AvgWeight;
run;
```

Figure 8.13 *Output Created by the SUMMARY Procedure with the PRINT Option Specified*

The SUMMARY Procedure

Sex	N Obs	Variable	N	Mean	Std Dev	Minimum	Maximum
F	11	Age	11	48.9090909	13.3075508	16.0000000	63.0000000
		Height	11	63.9090909	2.1191765	61.0000000	68.0000000
		Weight	11	150.4545455	18.4464828	102.0000000	168.0000000
M	9	Age	9	44.0000000	12.3895117	15.0000000	54.0000000
		Height	9	70.6666667	2.6457513	66.0000000	75.0000000
		Weight	9	204.2222222	30.2893454	140.0000000	240.0000000

Producing Frequency Tables Using PROC FREQ

The FREQ procedure is a descriptive procedure as well as a statistical procedure. It produces one-way and *n*-way frequency tables, and it concisely describes your data by reporting the distribution of variable values. You can use the FREQ procedure to create crosstabulation tables that summarize data for two or more categorical variables by showing the number of observations for each combination of variable values.

Procedure Syntax

The FREQ procedure can include many statements and options for controlling frequency output. For simplicity, let's consider the procedure in its basic form.

General form, basic FREQ procedure:

PROC FREQ <DATA=*SAS-data-set* >;
RUN;

where *SAS-data-set* is the name of the data set to be used.

By default, PROC FREQ creates a one-way table with the frequency, percent, cumulative frequency, and cumulative percent of every value of all variables in a data set. This can produce excessive or inappropriate output. It is recommended that you always use a TABLES statement with PROC FREQ.

Figure 8.14 *Default PROC FREQ Output*

Variable	Frequency	Percent	Cumulative Frequency	Cumulative Percent
Value	Number of observations with the value.	Frequency of the value divided by the total number of observations.	Sum of the frequency counts of the value and all other values listed above it in the table.	Cumulative frequency of the value divided by the total number of observations.

For example, the following FREQ procedure creates a frequency table for each variable in the data set Finance.Loans. All the unique values are shown for the variables Account and Amount.

```
proc freq data=finance.loans;
run;
```

Figure 8.15 *Frequency Table for Account and Amount*

The SAS System

The FREQ Procedure

Account	Frequency	Percent	Cumulative Frequency	Cumulative Percent
101-1092	1	11.11	1	11.11
101-1289	1	11.11	2	22.22
101-1731	1	11.11	3	33.33
101-3144	1	11.11	4	44.44
103-1135	1	11.11	5	55.56
103-1994	1	11.11	6	66.67
103-2335	1	11.11	7	77.78
103-3864	1	11.11	8	88.89
103-3891	1	11.11	9	100.00

Amount	Frequency	Percent	Cumulative Frequency	Cumulative Percent
$3,500	1	11.11	1	11.11
$5,000	1	11.11	2	22.22
$8,700	1	11.11	3	33.33
$10,000	1	11.11	4	44.44
$18,500	1	11.11	5	55.56
$22,000	1	11.11	6	66.67
$30,000	1	11.11	7	77.78
$87,500	1	11.11	8	88.89
$114,000	1	11.11	9	100.00

Specifying Variables in PROC FREQ

Overview

By default, the FREQ procedure creates frequency tables for every variable in your data set. But this isn't always what you want. A variable that has continuous numeric values —such as DateTime—can result in a lengthy and meaningless table. Likewise, a variable that has a unique value for each observation—such as FullName—is unsuitable for PROC FREQ processing. Frequency distributions work best with variables whose values can be described as categorical, and whose values are best summarized by counts rather than by averages.

To specify the variables to be processed by the FREQ procedure, include a TABLES statement.

General form, TABLES statement:

TABLES *variable(s)*;

where *variable(s)* lists the variables to include.

Example

The order in which the variables appear in the TABLES statement determines the order in which they are listed in the PROC FREQ report.

Consider the SAS data set Finance.Loans. The variables Rate and Months are best described as categorical variables, so they are the best choices for frequency tables.

Figure 8.16   *PROC PRINT Report of the Data Set Finance.Loans*

Account	Amount	Rate	Months	Payment
101-1092	$22,000	10.00%	60	$467.43
101-1731	$114,000	9.50%	360	$958.57
101-1289	$10,000	10.50%	36	$325.02
101-3144	$3,500	10.50%	12	$308.52
103-1135	$8,700	10.50%	24	$403.47
103-1994	$18,500	10.00%	60	$393.07
103-2335	$5,000	10.50%	48	$128.02
103-3864	$87,500	9.50%	360	$735.75
103-3891	$30,000	9.75%	360	$257.75

```
proc freq data=finance.loans;
   tables rate months;
run;
```

Figure 8.17 *Frequency Tables for Rate and Months*

The FREQ Procedure

Rate	Frequency	Percent	Cumulative Frequency	Cumulative Percent
9.50%	2	22.22	2	22.22
9.75%	1	11.11	3	33.33
10.00%	2	22.22	5	55.56
10.50%	4	44.44	9	100.00

Months	Frequency	Percent	Cumulative Frequency	Cumulative Percent
12	1	11.11	1	11.11
24	1	11.11	2	22.22
36	1	11.11	3	33.33
48	1	11.11	4	44.44
60	2	22.22	6	66.67
360	3	33.33	9	100.00

In addition to listing variables separately, you can use a numbered range of variables.

```
proc freq data=perm.survey;
   tables item1-item3;
run;
```

Figure 8.18 *Frequency Tables for Item1–Item3*

The FREQ Procedure

Item1	Frequency	Percent	Cumulative Frequency	Cumulative Percent
2	1	25.00	1	25.00
4	2	50.00	3	75.00
5	1	25.00	4	100.00

Item2	Frequency	Percent	Cumulative Frequency	Cumulative Percent
1	1	25.00	1	25.00
3	2	50.00	3	75.00
5	1	25.00	4	100.00

Item3	Frequency	Percent	Cumulative Frequency	Cumulative Percent
4	3	75.00	3	75.00
5	1	25.00	4	100.00

Adding the NOCUM option to your TABLES statement suppresses the display of cumulative frequencies and cumulative percentages in one-way frequency tables and in list output. The syntax for the NOCUM option is shown below.

```
TABLES variable(s) / NOCUM;
```

Creating Two-Way Tables

So far, you have used the FREQ procedure to create one-way frequency tables. However, it is often helpful to crosstabulate frequencies of two or more variables. For example, census data is typically crosstabulated with a variable that represents geographical regions.

The simplest crosstabulation is a two-way table. To create a two-way table, join two variables with an asterisk (*) in the TABLES statement of a PROC FREQ step.

General form, TABLES statement for crosstabulation:

TABLES *variable-1* *variable-2 <* ... *variable-n*>;

where (for two-way tables)

- *variable-1* specifies table rows
- *variable-2* specifies table columns.

When crosstabulations are specified, PROC FREQ produces tables with cells that contain

- cell frequency

- cell percentage of total frequency

- cell percentage of row frequency

- cell percentage of column frequency.

For example, the following program creates the two-way table shown below.

```
proc format;
    value wtfmt low-139='< 140'
                140-180='140-180'
                181-high='> 180';
    value htfmt low-64='< 5''5"'
                65-70='5''5-10"'
                71-high='> 5''10"';
run;
proc freq data=clinic.diabetes;
    tables weight*height;
    format weight wtfmt. height htfmt.;
run;
```

Figure 8.19 *Two-Way Table Created by PROC FREQ*

The FREQ Procedure

Frequency Percent Row Pct Col Pct	Table of Weight by Height			
		Height		
Weight	< 5'5"	5'5-10"	> 5'10"	Total
< 140	2	0	0	2
	10.00	0.00	0.00	10.00
	100.00	0.00	0.00	
	28.57	0.00	0.00	
140-180	5	5	0	10
	25.00	25.00	0.00	50.00
	50.00	50.00	0.00	
	71.43	62.50	0.00	
> 180	0	3	5	8
	0.00	15.00	25.00	40.00
	0.00	37.50	62.50	
	0.00	37.50	100.00	
Total	7	8	5	20
	35.00	40.00	25.00	100.00

Note that the first variable, Weight, forms the table rows, and the second variable, Height, forms the columns; reversing the order of the variables in the TABLES statement would reverse their positions in the table. Note also that the statistics are listed in the legend box.

Creating *N*-Way Tables

Overview

For a frequency analysis of more than two variables, use PROC FREQ to create *n*-way crosstabulations. A series of two-way tables is produced, with a table for each level of the other variables.

For example, suppose you want to add the variable Sex to your crosstabulation of Weight and Height in the data set Clinic.Diabetes. Add Sex to the TABLES statement, joined to the other variables with an asterisk (*).

```
tables sex*weight*height;
```

Determining the Table Layout

The order of the variables is important. In *n*-way tables, the last two variables of the TABLES statement become the two-way rows and columns. Variables that precede the last two variables in the TABLES statement stratify the crosstabulation tables.

```
        levels
        ↓
tables sex*weight*height;
          ↑        ↑
        rows + columns = two-way tables
```

Notice the structure of the output that is produced by the program shown below.

```
proc format;
   value wtfmt low-139='< 140'
               140-180='140-180'
               181-high='> 180';
   value htfmt low-64='< 5''5"'
               65-70='5''5-10"'
               71-high='> 5''10"';
run;
proc freq data=clinic.diabetes;
   tables sex*weight*height;
   format weight wtfmt. height htfmt.;
run;
```

Figure 8.20 *Creating N-Way Tables*

The FREQ Procedure

Frequency Percent Row Pct Col Pct	Table 1 of Weight by Height			
	Controlling for Sex=F			
	Height			
Weight	< 5'5"	5'5-10"	> 5'10"	Total
< 140	2 18.18 100.00 28.57	0 0.00 0.00 0.00	0 0.00 0.00 .	2 18.18
140-180	5 45.45 55.56 71.43	4 36.36 44.44 100.00	0 0.00 0.00 .	9 81.82
> 180	0 0.00 . 0.00	0 0.00 . 0.00	0 0.00 . .	0 0.00
Total	7 63.64	4 36.36	0 0.00	11 100.00

Frequency Percent Row Pct Col Pct	Table 2 of Weight by Height			
	Controlling for Sex=M			
	Height			
Weight	< 5'5"	5'5-10"	> 5'10"	Total
< 140	0 0.00 . .	0 0.00 . 0.00	0 0.00 . 0.00	0 0.00
140-180	0 0.00 0.00 .	1 11.11 100.00 25.00	0 0.00 0.00 0.00	1 11.11
> 180	0 0.00 0.00 .	3 33.33 37.50 75.00	5 55.56 62.50 100.00	8 88.89
Total	0 0.00	4 44.44	5 55.56	9 100.00

Creating Tables in List Format

Overview

When three or more variables are specified, the multiple levels of *n*-way tables can produce considerable output. Such bulky, often complex crosstabulations are often easier to read as a continuous list. Although this eliminates row and column frequencies and percents, the results are compact and clear.

The LIST option is not available when you also specify statistical options.

To generate list output for crosstabulations, add a slash (/) and the LIST option to the TABLES statement in your PROC FREQ step.

TABLES *variable-1* *variable-2 <* ... *variable-n*> / **LIST;**

Example

Adding the LIST option to our Clinic.Diabetes program puts its frequencies in a simple, short table.

```
proc format;
   value wtfmt low-139='< 140'
               140-180='140-180'
               181-high='> 180';
   value htfmt low-64='< 5''5"'
               65-70='5''5-10"'
               71-high='> 5''10"';
run;
proc freq data=clinic.diabetes;
   tables sex*weight*height / list;
   format weight wtfmt. height htfmt.;
run;
```

Figure 8.21 *Table Created by Using the LIST Option*

The FREQ Procedure

Sex	Weight	Height	Frequency	Percent	Cumulative Frequency	Cumulative Percent
F	< 140	< 5'5"	2	10.00	2	10.00
F	140-180	< 5'5"	5	25.00	7	35.00
F	140-180	5'5-10"	4	20.00	11	55.00
M	140-180	5'5-10"	1	5.00	12	60.00
M	> 180	5'5-10"	3	15.00	15	75.00
M	> 180	> 5'10"	5	25.00	20	100.00

Changing the Table Format

Beginning in SAS®9, adding the CROSSLIST option to your TABLES statement displays crosstabulation tables in ODS column format. This option creates a table that has a table definition that you can customize by using the TEMPLATE procedure.

Notice the structure of the output that is produced by the program shown below.

```
proc format;
   value wtfmt low-139='< 140'
               140-180='140-180'
               181-high='> 180';
   value htfmt low-64='< 5''5"'
               65-70='5''5-10"'
               71-high='> 5''10"';
run;
proc freq data=clinic.diabetes;
   tables sex*weight*height / crosslist;
   format weight wtfmt. height htfmt.;
run;
```

Figure 8.22   *Table Created by Using CROSSLIST Option: Sex=F*

The FREQ Procedure

Weight	Height	Frequency	Percent	Row Percent	Column Percent
Table of Weight by Height					
Controlling for Sex=F					
< 140	< 5'5"	2	18.18	100.00	28.57
	5'5-10"	0	0.00	0.00	0.00
	> 5'10"	0	0.00	0.00	.
	Total	2	18.18	100.00	
140-180	< 5'5"	5	45.45	55.56	71.43
	5'5-10"	4	36.36	44.44	100.00
	> 5'10"	0	0.00	0.00	.
	Total	9	81.82	100.00	
> 180	< 5'5"	0	0.00	.	0.00
	5'5-10"	0	0.00	.	0.00
	> 5'10"	0	0.00	.	.
	Total	0	0.00	.	
Total	< 5'5"	7	63.64		100.00
	5'5-10"	4	36.36		100.00
	> 5'10"	0	0.00		.
	Total	11	100.00		

Figure 8.23   *Table Created by Using CROSSLIST Option: Sex=M*

Weight	Height	Frequency	Percent	Row Percent	Column Percent
Table of Weight by Height					
Controlling for Sex=M					
< 140	< 5'5"	0	0.00	.	.
	5'5-10"	0	0.00	.	0.00
	> 5'10"	0	0.00	.	0.00
	Total	0	0.00	.	
140-180	< 5'5"	0	0.00	0.00	.
	5'5-10"	1	11.11	100.00	25.00
	> 5'10"	0	0.00	0.00	0.00
	Total	1	11.11	100.00	
> 180	< 5'5"	0	0.00	0.00	.
	5'5-10"	3	33.33	37.50	75.00
	> 5'10"	5	55.56	62.50	100.00
	Total	8	88.89	100.00	
Total	< 5'5"	0	0.00		.
	5'5-10"	4	44.44		100.00
	> 5'10"	5	55.56		100.00
	Total	9	100.00		

Suppressing Table Information

Another way to control the format of crosstabulations is to limit the output of the FREQ procedure to a few specific statistics. Remember that when crosstabulations are run, PROC FREQ produces tables with cells that contain:

- cell frequency
- cell percentage of total frequency

- cell percentage of row frequency

- cell percentage of column frequency.

You can use options to suppress any of these statistics. To control the depth of crosstabulation results, add any combination of the following options to the TABLES statement:

- NOFREQ suppresses cell frequencies

- NOPERCENT suppresses cell percentages

- NOROW suppresses row percentages

- NOCOL suppresses column percentages.

Example

Suppose you want to use only the percentages of Sex and Weight combinations in the data set Clinic.Diabetes. To block frequency counts and row and column percentages, add the NOFREQ, NOROW, and NOCOL options to your program's TABLES statement.

```
proc format;
   value wtfmt low-139='< 140'
               140-180='140-180'
               181-high='> 180';
run;
proc freq data=clinic.diabetes;
   tables sex*weight / nofreq norow nocol;
   format weight wtfmt.;
run;
```

Figure 8.24 *Suppressing Table Information*

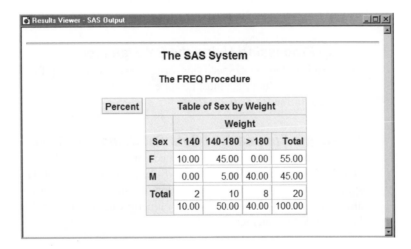

Notice that Percent is the only statistic that remains in the table's legend box.

Chapter Summary

Text Summary

Purpose of PROC MEANS

The MEANS procedure provides an easy way to compute descriptive statistics. Descriptive statistics such as the mean, minimum, and maximum provide useful information about numeric data.

Specifying Statistics

By default, PROC MEANS computes the *n*-count (the number of non-missing values), the mean, the standard deviation, and the minimum and maximum values for variables. To specify statistics, list their keywords in the PROC MEANS statement.

Limiting Decimal Places

Because PROC MEANS uses the BEST. format by default, procedure output can contain unnecessary decimal places. To limit decimal places, use the MAXDEC= option and set it equal to the length that you prefer.

Specifying Variables in PROC MEANS

By default, PROC MEANS computes statistics for all numeric variables. To specify the variables to include in PROC MEANS output, list them in a VAR statement.

Group Processing Using the CLASS Statement

Include a CLASS statement, specifying variable names, to group PROC MEANS output by variable values of classes.

Group Processing Using the BY Statement

Include a BY statement, specifying variable names, to group PROC MEANS output by variable values. Your data must be sorted according to those variables. Statistics are not computed for the BY variables.

Creating a Summarized Data Set Using PROC MEANS

You can create an output data set that contains summarized variables by using the OUTPUT statement in PROC MEANS. When you use the OUTPUT statement without specifying the statistic-keyword= option, the summary statistics N, MEAN, STD, MIN, and MAX are produced for all of the numeric variables or for all of the variables that are listed in a VAR statement.

Creating a Summarized Data Set Using PROC SUMMARY

You can also create a summarized output data set by using PROC SUMMARY. The PROC SUMMARY code for producing an output data set is exactly the same as the code for producing an output data set with PROC MEANS. The difference between the two procedures is that PROC MEANS produces a report by default, whereas PROC SUMMARY produces an output data set by default.

The FREQ Procedure

The FREQ Procedure is a descriptive procedure as well as a statistical procedure that produces one-way and n-way frequency tables. It describes your data by reporting the distribution of variable values.

Specifying Variables

By default, the FREQ procedure creates frequency tables for every variable in your data set. To specify the variables to analyze, include them in a TABLES statement.

Creating Two-Way Tables

When a TABLES statement contains two variables joined by an asterisk (*), PROC FREQ produces crosstabulations. The resulting table displays values for

- cell frequency

- cell percentage of total frequency

- cell percentage of row frequency

- cell percentage of column frequency.

Creating N-Way Tables

Crosstabulations can include more than two variables. When three or more variables are joined in a TABLES statement, the result is a series of two-way tables that are grouped by the values of the first variables listed.

Creating Tables in List Format

To reduce the bulk of n-way table output, add a slash (/) and the LIST option to the end of the TABLES statement. PROC FREQ then prints compact, multi-column lists instead of a series of tables. Beginning in SAS®9, you can use the CROSSLIST option to format your tables in ODS column format.

Suppressing Table Information

You can suppress the display of specific statistics by adding one or more options to the TABLES statement:

- NOFREQ suppresses cell frequencies

- NOPERCENT suppresses cell percentages

- NOROW suppresses row percentages

- NOCOL suppresses column percentages.

Syntax

PROC MEANS <DATA=*SAS-data-set*>
        <*statistic-keyword(s)*> <*option(s)*>;
        <VAR*variable(s)*>;
        <CLASS*variable(s)*>;
        <BY*variable(s)*>;
        <OUTPUT*out=SAS-data-set statistic=variable(s)*>;
RUN;

> **PROC SUMMARY** <DATA=*SAS-data-set*>
>   <*statistic-keyword(s)*> <*option(s)*>;
>   <VAR*variable(s)*>;
>   <CLASS*variable(s)*>;
>   <BY*variable(s)*>;
>   <OUTPUT*out=SAS-data-set*>;
> **RUN;**
> **PROC FREQ** <DATA=*SAS-data-set*>;
>   **TABLES** *variable-1 *variable-2 <* ... variable-n>*
>   / <NOFREQ|NOPERCENT|NOROW|NOCOL|CROSSLIST>;
> **RUN;**

Sample Programs

```
proc means data=clinic.heart min max maxdec=1;
   var arterial heart cardiac urinary;
   class survive sex;
run;

proc summary data=clinic.diabetes;
   var age height weight;
   class sex;
   output out=work.sum_gender
      mean=AvgAge AvgHeight AvgWeight;
run;

proc freq data=clinic.heart;
   tables sex*survive*shock / nopercent list;
run;
```

Points to Remember

- In PROC MEANS, use a VAR statement to limit output to relevant variables. Exclude statistics for variables such as dates.

- By default, PROC MEANS prints the full width of each numeric variable. Use the MAXDEC= option to limit decimal places and improve legibility.

- Data must be sorted for BY group processing. You might need to run PROC SORT before using PROC MEANS with a BY statement.

- PROC MEANS and PROC SUMMARY produce the same results; however, the default output is different. PROC MEANS produces a report, whereas PROC SUMMARY produces an output data set.

- If you do not include a TABLES statement, PROC FREQ produces statistics for every variable in the data set. Variables that have continuous numeric values can create a large amount of output. Use a TABLES statement to exclude such variables, or group their values by applying a FORMAT statement.

Chapter Quiz

Select the best answer for each question. After completing the quiz, check your answers using the answer key in the appendix.

1. The default statistics produced by the MEANS procedure are *n*-count, mean, minimum, maximum, and...

   a. median

   b. range

   c. standard deviation

   d. standard error of the mean.

2. Which statement will limit a PROC MEANS analysis to the variables Boarded, Transfer, and Deplane?

   a. `by boarded transfer deplane;`

   b. `class boarded transfer deplane;`

   c. `output boarded transfer deplane;`

   d. `var boarded transfer deplane;`

3. The data set Survey.Health includes the following variables. Which is a poor candidate for PROC MEANS analysis?

   a. `IDnum`

   b. `Age`

   c. `Height`

   d. `Weight`

4. Which of the following statements is true regarding BY group processing?

   a. BY variables must be either indexed or sorted.

   b. Summary statistics are computed for BY variables.

   c. BY group processing is preferred when you are categorizing data that contains few variables.

   d. BY group processing overwrites your data set with the newly grouped observations.

5. Which group processing statement produced the PROC MEANS output shown below?

Figure 8.25   *PROC MEANS Output*

The MEANS Procedure

Survive	Sex	N Obs	Variable	N	Mean	Std Dev	Minimum	Maximum
DIED	1	4	Arterial	4	92.5000000	10.4721854	83.0000000	103.0000000
			Heart	4	111.0000000	53.4103610	54.0000000	183.0000000
			Cardiac	4	176.7500000	75.2257713	95.0000000	260.0000000
			Urinary	4	98.0000000	186.1343601	0	377.0000000
	2	6	Arterial	6	94.1666667	27.3160514	72.0000000	145.0000000
			Heart	6	103.6666667	16.6573307	81.0000000	130.0000000
			Cardiac	6	318.3333333	102.6034437	156.0000000	424.0000000
			Urinary	6	100.3333333	155.7134120	0	405.0000000
SURV	1	5	Arterial	5	77.2000000	12.1942609	61.0000000	88.0000000
			Heart	5	109.0000000	31.9687347	77.0000000	149.0000000
			Cardiac	5	298.0000000	139.8499196	66.0000000	410.0000000
			Urinary	5	100.8000000	60.1722527	44.0000000	200.0000000
	2	5	Arterial	5	78.8000000	6.8337398	72.0000000	87.0000000
			Heart	5	100.0000000	13.3790882	84.0000000	111.0000000
			Cardiac	5	330.2000000	86.9839066	256.0000000	471.0000000
			Urinary	5	111.2000000	152.4096454	12.0000000	377.0000000

a. `class sex survive;`

b. `class survive sex;`

c. `by sex survive;`

d. `by survive sex;`

6. Which program can be used to create the following output?

Figure 8.26   *Output*

Sex	N Obs	Variable	N	Mean	Std Dev	Minimum	Maximum
F	11	Age	11	48.9090909	13.3075508	16.0000000	63.0000000
		Height	11	63.9090909	2.1191765	61.0000000	68.0000000
		Weight	11	150.4545455	18.4464828	102.0000000	168.0000000
M	9	Age	9	44.0000000	12.3895117	15.0000000	54.0000000
		Height	9	70.6666667	2.6457513	66.0000000	75.0000000
		Weight	9	204.2222222	30.2893454	140.0000000	240.0000000

a.
```
proc means data=clinic.diabetes;
    var age height weight;
    class sex;
    output out=work.sum_gender
        mean=AvgAge AvgHeight AvgWeight;
run;
```

b.
```
proc summary data=clinic.diabetes print;
    var age height weight; class sex;
    output out=work.sum_gender
        mean=AvgAge AvgHeight AvgWeight;
run;
```

c.
```
proc means data=clinic.diabetes noprint;
    var age height weight;
    class sex;
    output out=work.sum_gender
        mean=AvgAge AvgHeight AvgWeight;
run;
```

d. Both a and b.

7. By default, PROC FREQ creates a table of frequencies and percentages for which data set variables?

   a. character variables

   b. numeric variables

   c. both character and numeric variables

   d. none: variables must always be specified

8. Frequency distributions work best with variables that contain

   a. continuous values.

   b. numeric values.

   c. categorical values.

   d. unique values.

9. Which PROC FREQ step produced this two-way table?

   Figure 8.27 *Two-Way Table*

   The FREQ Procedure

Frequency Percent Row Pct Col Pct	Table of Weight by Height			
		Height		
Weight	< 5'5"	5'5-10"	> 5'10"	Total
< 140	2	0	0	2
	10.00	0.00	0.00	10.00
	100.00	0.00	0.00	
	28.57	0.00	0.00	
140-180	5	5	0	10
	25.00	25.00	0.00	50.00
	50.00	50.00	0.00	
	71.43	62.50	0.00	
> 180	0	3	5	8
	0.00	15.00	25.00	40.00
	0.00	37.50	62.50	
	0.00	37.50	100.00	
Total	7	8	5	20
	35.00	40.00	25.00	100.00

   a.
   ```
   proc freq data=clinic.diabetes;
       tables height weight;
       format height htfmt. weight wtfmt.;
   run;
   ```

   b.
   ```
   proc freq data=clinic.diabetes;
       tables weight height;
       format weight wtfmt. height htfmt.;
   run;
   ```

   c.
   ```
   proc freq data=clinic.diabetes;
       tables height*weight;
       format height htfmt. weight wtfmt.;
   run;
   ```

   d.
   ```
   proc freq data=clinic.diabetes;
       tables weight*height;
       format weight wtfmt. height htfmt.;
   run;
   ```

10. Which PROC FREQ step produced this table?

Figure 8.28 PROC FREQ Table

The FREQ Procedure

Percent	Table of Sex by Weight			
		Weight		
Sex	< 140	140-180	> 180	Total
F	10.00	45.00	0.00	55.00
M	0.00	5.00	40.00	45.00
Total	2	10	8	20
	10.00	50.00	40.00	100.00

a.
```
proc freq data=clinic.diabetes;
   tables sex weight / list;
   format weight wtfmt.;
run;
```

b.
```
proc freq data=clinic.diabetes;
   tables sex*weight / nocol;
   format weight wtfmt.;
run;
```

c.
```
proc freq data=clinic.diabetes;
   tables sex weight / norow nocol;
   format weight wtfmt.;
run;
```

d.
```
proc freq data=clinic.diabetes;
   tables sex*weight / nofreq norow nocol;
   format weight wtfmt.;
run;
```

Chapter 9
Producing HTML Output

Overview

Introduction

In previous chapters, you've seen both traditional SAS LISTING output and HTML output. HTML output is created by default in the SAS Windowing environment in the Windows and UNIX operating systems. If you are not using SAS in the SAS in the SAS Windowing environment in the Windows and UNIX operating systems, then you can set options to create HTML output. The HTML created by default is basic HTML 4.0 with default formatting.

You can use ODS to create many other types of output. You can create RTF, PDF, Excel, and many others using ODS statements. This chapter will discuss HTML output.

Using ODS, you can customize and manage HTML output by submitting programming statements. After you create HTML files, you can view them in the Results Viewer or Internet Explorer, Netscape Navigator, or any Web browser that fully supports HTML 4.0.

This chapter shows you how to create and view HTML output using ODS. You also learn how to apply styles to ODS output.

⚠ By default, all code that you submit to SAS with Enterprise Guide has ODS statements included to create HTML output.

Figure 9.1 *LISITING Output versus HTML Output*

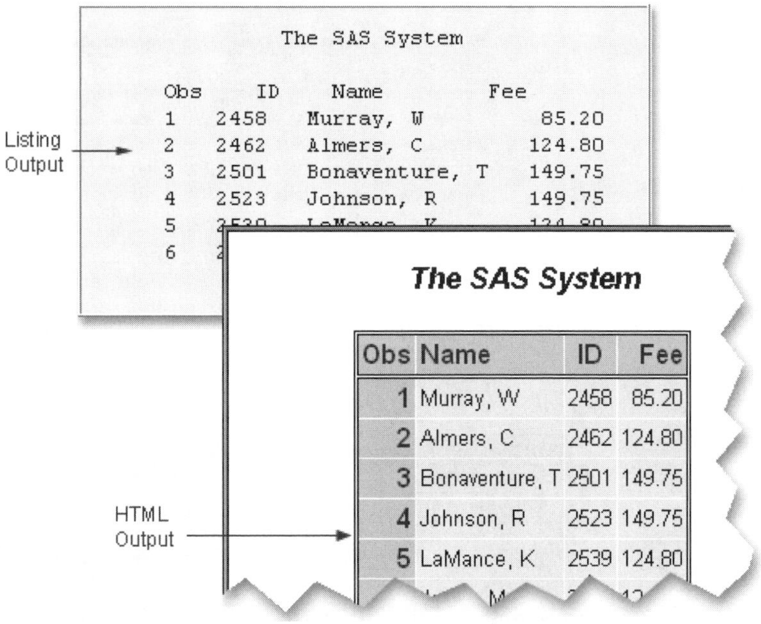

Objectives

In this chapter, you learn to

* open and close ODS destinations

* create a simple HTML file with the output of one or more procedures

* create HTML output with a linked table of contents in a frame

* use options to specify links and file paths

* view HTML output

* apply styles to HTML output.

The Output Delivery System

Before you learn to write ODS programming statements, it is helpful to understand a little about ODS.

Advantages of ODS

ODS gives you formatting options and makes procedure output much more flexible. With ODS, you can easily create output in a variety of formats, including

* HTML output

Figure 9.2 HTML Output

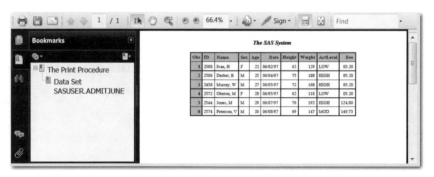

The SAS System

Obs	ID	Name	Sex	Age	Date	Height	Weight	ActLevel	Fee
1	2588	Ivan, H	F	22	06/02/97	63	139	LOW	85.20
2	2586	Derber, B	M	25	06/04/97	75	188	HIGH	85.20
3	2458	Murray, W	M	27	06/05/97	72	168	HIGH	85.20
4	2572	Oberon, M	F	28	06/05/97	62	118	LOW	85.20
5	2544	Jones, M	M	29	06/07/97	76	193	HIGH	124.80
6	2574	Peterson, V	M	30	06/08/97	69	147	MOD	149.75

- RTF output

Figure 9.3 RTF Output

The SAS System

Obs	ID	Name	Sex	Age	Date	Height	Weight	ActLevel	Fee
1	2588	Ivan, H	F	22	06/02/97	63	139	LOW	85.20
2	2586	Derber, B	M	25	06/04/97	75	188	HIGH	85.20
3	2458	Murray, W	M	27	06/05/97	72	168	HIGH	85.20
4	2572	Oberon, M	F	28	06/05/97	62	118	LOW	85.20
5	2544	Jones, M	M	29	06/07/97	76	193	HIGH	124.80
6	2574	Peterson, V	M	30	06/08/97	69	147	MOD	149.75

- PDF output

Figure 9.4 PDF Output

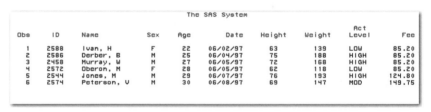

- traditional SAS LISTING output

Figure 9.5 LISTING Output

```
                                    The SAS System

                                                                         Act
    Obs      ID     Name          Sex    Age       Date     Height   Weight   Level      Fee

     1      2588    Ivan, H        F      22     06/02/97      63       139    LOW       85.20
     2      2586    Derber, B      M      25     06/04/97      75       188    HIGH      85.20
     3      2458    Murray, W      M      27     06/05/97      72       168    HIGH      85.20
     4      2572    Oberon, M      F      28     06/05/97      62       118    LOW       85.20
     5      2544    Jones, M       M      29     06/07/97      76       193    HIGH     124.80
     6      2574    Peterson, V    M      30     06/08/97      69       147    MOD      149.75
```

Also, ODS holds your output in its component parts (data and table definition) so that numerical data retains its full precision.

Let's see how ODS creates output.

How ODS Works

When you submit your ODS statements and the SAS program that creates your output, ODS does the following:

1. ODS creates your output in the form of output objects.

    Each output object contains the results of a procedure or DATA step (the data component) and can also contain information about how to render the results (the table definition).

    Figure 9.6   *ODS Processing: What Goes In*

    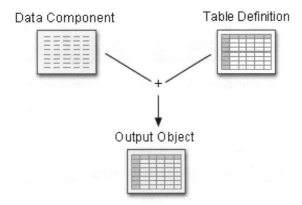

2. ODS sends the output object to the ODS destination(s) that you specify and creates formatted output as specified by the destination. For example, when the LISTING and HTML destinations are open, ODS creates LISTING and HTML output.

    Figure 9.7   *ODS Processing: What Comes Out*

    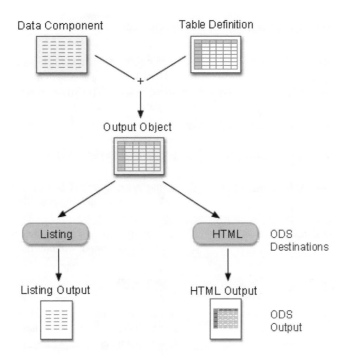

3. ODS creates a link to each output object in the Results window and identifies each output object by the appropriate icon.

Figure 9.8 *Results Window*

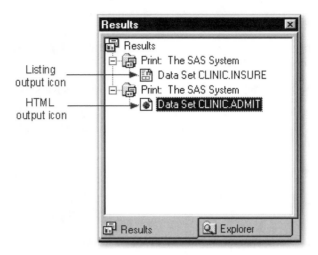

Listing output icon

HTML output icon

Let's look at how you open various ODS destinations.

Opening and Closing ODS Destinations

ODS Destinations

You use ODS statements to specify destinations for your output. Each destination creates a specific type of formatted output. The table below lists some of the ODS destinations that are currently supported.

This destination . . .	Produces . . .
HTML	output that is formatted in Hypertext Markup Language (HTML). This is the default destination. You do not have to specify the ODS HTML statement to produce basic HTML output.
LISTING	plain text output that is formatted like traditional SAS procedure (LISTING) output
Markup Languages Family	output that is formatted using markup languages such as Extensible Markup Language (XML)
Document	a hierarchy of output objects that enables you to render multiple ODS output without rerunning procedures
Output	SAS data sets
Printer Family	output that is formatted for a high-resolution printer such as PostScript (PS), Portable Document Format (PDF), or Printer Control Language (PCL) files
RTF	Rich Text Format output for use with Word

In this chapter, we'll primarily work with the LISTING destination and the HTML destination. For information about all ODS destinations, please see the SAS documentation for the Output Delivery System.

Using Statements to Open and Close ODS Destinations

For each type of formatted output that you want to create, you use an ODS statement to open the destination. At the end of your program, you use another ODS statement to close the destination so that you can access your output.

General form, ODS statement to open and close destinations:

ODS *open-destination*;

ODS *close-destination* **CLOSE;**

where

- *open-destination* is a keyword and any required options for the type of output that you want to create, such as

  - **HTML FILE=**'*html-file-pathname*'

  - **LISTING**

- *close-destination* is a keyword for the type of output.

You can issue ODS statements in any order, depending on whether you need to open or close the destination. Most ODS destinations are closed by default. You open them at the beginning of your program and close them at the end. The exception is the HTML destination, which is open by default.

Figure 9.9 *Default ODS Destination*

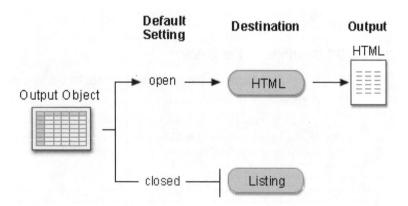

Example

The following program creates HTML output because the HTML destination is open by default. No other ODS destinations are open, so no other output objects are produced.

```
proc print data=sasuser.mydata;
run;
```

The following program produces HTML and LISTING output:

```
ods listing;
proc print data=sasuser.mydata;
run;
ods listing close;
```

This example is meant to demonstrate how you open and close ODS destinations. You learn the specifics of creating HTML output later in this chapter.

Closing the HTML Destination

As you have learned, the HTML destination is open by default. Because open destinations use system resources, it is a good idea to close the HTML destination at the beginning of your program if you don't want to produce HTML output. For example:

```
ods html close;
```

The HTML destination remains closed until you end your current SAS session or until you re-open the destination. It's a good programming practice to reset ODS to HTML output (the default setting) at the end of your programs. For example:

```
ods html;
```

Example

The following program produces only LISTING output:

```
ods html close;
ods listing;
proc print data=sasuser.mydata;
run;
ods listing close;
ods html;
```

Closing Multiple ODS Destinations at Once

You can produce output in multiple formats at once by opening each ODS destination at the beginning of the program.

When you have more than one open ODS destination, you can use the keyword _ALL_ in the ODS CLOSE statement to close all open destinations at once.

Example

The program below opens the RTF and PDF destinations before the PROC step and closes all ODS destinations at the end of the program:

```
ods rtf file='RTF-file-pathname';
ods pdf file='PDF-file-pathname';

proc print data=sasuser.admit;
run;

ods _all_ close;
ods html;
```

Notice that the last ODS statement reopens the HTML destination so that ODS returns to producing HTML output for subsequent DATA or PROC steps in the current session.

Now that you have learned how to open and close ODS destinations, let's look at creating HTML output with options.

Creating HTML Output with Options

ODS HTML Statement

To create simple HTML output, you do not have to specify the ODS HTML statement. However, to create HTML output with options specified, you open the HTML destination using the ODS HTML statement.

General form, ODS HTML statement:

ODS HTML BODY=*file-specification***;**

ODS HTML CLOSE;

where *file-specification* identifies the file that contains the HTML output. The specification can be

- a quoted string which contains the HTML filename (use only the filename to write the file to your current working directory location, such as **C:\Documents and Settings ***username***\My Documents\My SAS Files**) Example: **ODS HTML BODY=***"myreport.html"*;

- a quoted string which contains the complete directory path and HTML filename (include the complete pathname if you want to save the HTML file to a specific location other than your working directory) Example: **ODS HTML BODY=***"c:\reportdir\myreport.html"*;

- a fileref (unquoted file shortcut) that has been assigned to an HTML file using the FILENAME statement Example: **FILENAME MYHTML** *"c:\reportdir\myreport.html"*; **ODS HTML BODY=MYHTML;**

- a SAS catalog entry in the form *entry-name*.html. Note that the catalog name is specified in the PATH= option and the *entry-name*.html value for the BODY= option is unquoted. Example: **ODS HTML PATH=***work.mycat***BODY=***myentry* **BODY=.HTML;**

   FILE= can also be used to specify the file that contains the HTML output. FILE= is an alias for BODY=.

   You can also use the **PATH=** option to explicitly specify a directory path for your file. This option will be discussed in an upcoming topic.

Example

The program below creates PROC PRINT output in an HTML file. The BODY= option specifies the file **F:\admit.html** in the Windows operating environment as the file that contains the PROC PRINT results.

```
ods html body='f:\admit.html';
proc print data=clinic.admit label;
   var sex age height weight actlevel;
   label actlevel='Activity Level';
```

```
run;
ods html close;
ods html;
```

The HTML file admit.html contains the results of all procedure steps between the ODS HTML statement and ODS HTML CLOSE statement.

Figure 9.10 *HTML Output*

Obs	Sex	Age	Height	Weight	Activity Level
1	M	27	72	168	HIGH
2	F	34	66	152	HIGH
3	F	31	61	123	LOW
4	F	43	63	137	MOD
5	M	51	71	158	LOW
6	M	29	76	193	HIGH
7	F	32	67	151	MOD
8	M	35	70	173	MOD
9	M	34	73	154	LOW
10	F	49	64	172	LOW
11	F	44	66	140	HIGH
12	F	28	62	118	LOW
13	M	30	69	147	MOD
14	F	40	69	163	HIGH
15	M	47	72	173	MOD
16	M	60	71	191	LOW
17	F	43	65	123	MOD
18	M	25	75	188	HIGH
19	F	22	63	139	LOW
20	F	41	67	141	HIGH
21	M	54	71	183	MOD

Viewing Your HTML Output

If you're working in the Windows environment, when you submit the program, the body file will automatically appear in the Results Viewer (using the internal browser) or your preferred Web browser, as specified in the **Results** tab of the Preferences window.

In other operating environments, you can double-click the corresponding link in the Results window to view the HTML output in your Web browser. If you don't have a Web browser in your operating environment, you can transfer the HTML files to an operating environment where you can view them.

Creating HTML Output in Enterprise Guide

When you specify a path and filename for your HTML results, SAS Enterprise Guide will allow you to download and display the HTML file. In addition, you can use Windows Explorer to locate the file that you created and then double-click it to open it in your browser.

When you submit a program, two HTML results appear in the Project window. One is the HTML output that is created by SAS Enterprise Guide ODS statements. The other uses the ODS statements from the code that you submitted and creates a temporary file labeled with the path and filename that you designated. It is similar in style to the actual HTML file that is created in the location that you specify.

You can double-click the shortcut to open the HTML results in SAS Enterprise Guide, or you can right-click the shortcut and select **Open with default-browser** to open it in a browser. To see the actual file rather than the temporary file, you can use Windows Explorer to locate the HTML file that you created and then double-click it to open it in your browser.

Creating HTML Output from Multiple Procedures

You can also use the ODS HTML statement to direct the results from multiple procedures to the same HTML file.

The program below generates HTML output for the PRINT and TABULATE procedures. The results for both procedures are saved to the file **C:\Records \data.html** in the Windows environment. A link for each output object (one for each procedure) appears in the Results window.

```
ods html body='c:\records\data.html';
proc print data=clinic.admit label;
   var id sex age height weight actlevel;
   label actlevel='Activity Level';
run;
proc tabulate data=clinic.stress2;
   var resthr maxhr rechr;
   table min mean, resthr maxhr rechr;
run;
ods html close;
```

The following is a representation of the HTML file containing the results from the program above. The results from the TABULATE procedure are appended. The results also show up in the Results window, where you can select the PRINT procedure output or the TABULATE procedure output.

Figure 9.11 HTML Output

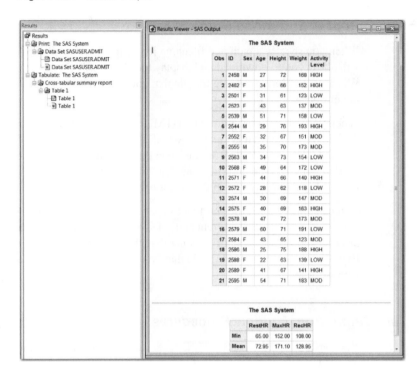

Creating HTML Output with a Table of Contents

Overview

So far in this chapter, you've used the BODY= specification to create an HTML file containing your procedure output. Suppose you want to create an HTML file that has a table of contents with links to the output of each specific procedure. You can do this by specifying additional files in the ODS HTML statement.

General form, ODS HTML statement to create a linked table of contents:

ODS HTML

    BODY=*body-file-specification*

    CONTENTS=*contents-file-specification*

    FRAME=*frame-file-specification*;

ODS HTML CLOSE;

where

- *body-file-specification* is the name of an HTML file that contains the procedure output.

- *contents-file-specification* is the name of an HTML file that contains a table of contents with links to the procedure output.

- *frame-file-specification* is the name of an HTML file that integrates the table of contents and the body file. If you specify FRAME=, you must also specify CONTENTS=.

To direct the HTML output to a specific storage location, specify the complete pathname of the HTML file in the *file-specification*.

Example

In the program below,

- the BODY= specification creates data.html in the **c:\records** directory. The body file contains the results of the two procedures.

- the CONTENTS= specification creates toc.html in the **c:\records** directory. The table of contents file has links to each procedure output in the body file.

- the FRAME= specification creates frame.html in the **c:\records** directory. The frame file integrates the table of contents and the body file.

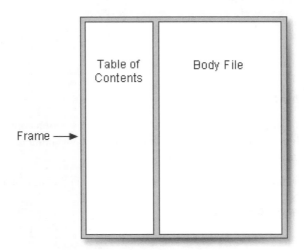

```
ods html body='c:\records\data.html'
         contents='c:\records\toc.html'
         frame='c:\records\frame.html';
proc print data=clinic.admit label;
   var id sex age height weight actlevel;
   label actlevel='Activity Level';
run;
proc print data=clinic.stress2;
   var id resthr maxhr rechr;
run;
ods html close;

ods html;
```

Viewing Frame Files

The Results window does not display links to frame files. In the Windows environment, only the body file will automatically appear in the internal browser or your preferred Web browser.

To view the frame file that integrates the body file and the table of contents, select **File** ⇨ **Open** from within the internal browser or your preferred Web browser. Then open the frame file that you specified using FRAME=. In the example above, this file is frame.html, which is stored in the Records directory in the Windows environment.

The frame file, frame.html, is shown below.

Figure 9.12   *Frame File, Frame.html*

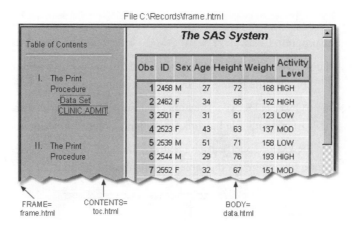

Using the Table of Contents

The table of contents created by the CONTENTS= option contains a numbered heading for each procedure that creates output. Below each heading is a link to the output for that procedure.

On some browsers, you can select a heading to contract or expand the table of contents.

Figure 9.13   *Table of Contents*

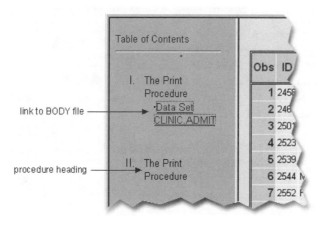

Using Options to Specify Links and Paths

Overview

When ODS generates HTML files for the body, contents, and frame, it also generates links between the files using the HTML filenames that you specify in the ODS HTML statement. If you specify complete pathnames, ODS uses those pathnames in the links that it generates.

The ODS statement below creates a frame file that has links to `c:\records\toc.html` and `c:\records\data.html`, and a contents file that has links to `c:\records\data.html`.

```
ods html body='c:\records\data.html'
         contents='c:\records\toc.html'
         frame='c:\records\frame.html';
```

A portion of the source code for the HTML file frame.html is shown below. Notice that the links have the complete pathnames specified in the file specifications for the contents and body files.

Figure 9.14 *Source Code for the HTML File Frame.html*

```
<FRAME MARGINWIDTH="4" MARGINHEIGHT="0" SRC="c:\records\toc.html"
     NAME="contents" SCROLLING=auto>
<FRAME MARGINWIDTH="9" MARGINHEIGHT="0" SRC="c:\records\data.html"
     NAME="body" SCROLLING=auto>
```

These links work when you are viewing the HTML files locally. If you want to place these files on a Web server so that other people could access them, then the link needs to include either the complete URL for an absolute link or the HTML filename for a relative link.

The URL= Suboption

By specifying the URL= suboption in the BODY= or CONTENTS= file specification, you can provide a URL that ODS uses in all the links that it creates to the file. You can use the URL= suboption in any ODS file specification except FRAME= (because no ODS file references the frame file).

General form, URL= suboption in a file specification:

(URL='*Uniform-Resource-Locator***')**

where *Uniform-Resource-Locator* is the name of an HTML file or the full URL of an HTML file. ODS uses this URL instead of the file specification in all the links and references that it creates that point to the file.

The URL= suboption is useful for building HTML files that might be moved from one location to another. If the links from the contents and page files are constructed with a simple URL (one name), they work as long as the contents, page, and body files are all in the same location.

Example: Relative URLs

In this ODS HTML statement, the URL= suboption specifies only the HTML filename. This is the most common style of linking between files because maintenance is easier. The files can be moved as long as they all remain in the same directory or storage location.

```
ods html body='c:\records\data.html' (url='data.html')
         contents='c:\records\toc.html' (url='toc.html')
         frame='c:\records\frame.html';
```

The source code for frame.html has only the HTML filename as specified on the URL= suboptions for the body and contents files.

Figure 9.15 *Source Code Frame.html*

```
<FRAME MARGINWIDTH="4" MARGINHEIGHT="0" SRC="toc.htm"
    NAME="contents" SCROLLING=auto>
<FRAME MARGINWIDTH="9" MARGINHEIGHT="0" SRC="data.htm"
    NAME="body" SCROLLING=auto>
```

Example: Absolute URLs

Alternatively, in this ODS HTML statement, the URL= suboptions specify complete URLs using Hypertext Transfer Protocol (HTTP). These files can be stored in the same or different locations.

```
ods html body='c:\records\data.html'
         (url='http://mysite.com/myreports/data.html')
         contents='c:\records\toc.html'
         (url='http://mysite.com/mycontents/toc.html')
         frame='c:\records\frame.html';
```

As you would expect, the source code for Frame.html has the entire HTTP addresses that you specified in the URL= suboptions for the body and contents files.

Figure 9.16 *Source Code Frame.html*

```
<FRAME MARGINWIDTH="4" MARGINHEIGHT="0" SRC="http://mysite.com/myreports/data.html"
    NAME="contents" SCROLLING=auto>
<FRAME MARGINWIDTH="9" MARGINHEIGHT="0" SRC="http://mysite.com/myreports/toc.html"
    NAME="body" SCROLLING=auto>
```

When you use the URL= suboption to specify a complete URL, you might need to move your files to that location before you can view them.

The PATH= Option

So far, you've learned to specify the complete pathname for HTML files in the BODY=, CONTENTS=, and FRAME= specifications when you want to save HTML files to specific locations. You can use the following options to streamline your ODS HTML statement:

PATH= option
  specifies the location where you want to store your HTML output

URL=NONE suboption
  prevents ODS from using the pathname in any links that it creates in your files

General form, PATH= option with the URL= suboption:

PATH=*file-location-specification* <(URL= NONE |*'Uniform-Resource-Locator'*)>

where

- *file-location-specification* identifies the location where you want HTML files to be saved. It can be one of the following:

  - the complete pathname to an aggregate storage location, such as a directory or partitioned data set

  - a fileref (file shortcut) that has been assigned to a storage location

  - a SAS catalog (*libname.catalog*).

- *Uniform-Resource-Locator* provides a URL for links in the HTML files that ODS generates. If you specify the keyword **NONE**, no information from the PATH= option appears in the links or references.

  If you do not use the URL= suboption, information from the PATH= option is added to links and references in the files that are created.

In the z/OS operating environment, if you store your HTML files as members in a partitioned data set, the PATH= value must be a PDSE, not a PDS. You can allocate a PDSE within SAS as shown in this example:

```
filename pdsehtml '.example.htm'

   dsntype=library dsorg=po

   disp=(new, catlg, delete);
```

Also, you should specify valid member names for the HTML files (without extensions).

Example: PATH= Option with URL=NONE

In the program below, the PATH= option directs the files data.html, toc.html, and frame.html to the **C:\Records** directory in the Windows operating environment. The links from the frame file to the body and contents files contain only the HTML filenames data.html and toc.html.

```
ods html path='c:\records' (url=none)
        body='data.html'
        contents='toc.html'
        frame='frame.html';
proc print data=clinic.admit;
run;
proc print data=clinic.stress2;
run;
ods html close;
ods html;
```

This program generates the same files and links as the previous example in which you learned to use the URL= suboption with the BODY= and CONTENTS= file specifications. However, it is a bit simpler to specify the path only once in the PATH= option and to specify URL=NONE.

If you plan to move your HTML files, you should specify URL=NONE with the PATH= option to prevent information from the PATH= option from creating URLs that are invalid or incorrect.

Example: PATH= Option without the URL= Suboption

In the program below, the PATH= option directs the files data.html, toc.html, and frame.html to the `C:\Records` directory in the Windows operating environment. The links from the frame file to the body and contents files contain the complete pathname, `c:\records\data.html` and `c:\records\toc.html`:

```
ods html path='c:\records'
         body='data.html'
         contents='toc.html'
         frame='frame.html';
proc print data=clinic.admit;
run;
proc print data=clinic.stress2;
run;
ods html close;
ods html;
```

Example: PATH= Option with a Specified URL

In the program below, the PATH= option directs the files data.html, toc.html, and frame.html to the `C:\Records` directory in the Windows operating environment. The links from the frame file to the body and contents files contain the specified URL, `http://mysite.com/myreports/data.html` and `http://mysite.com/myreports/toc.html`:

```
ods html path='c:\records'(url='http://mysite.com/myreports/')
         body='data.html'
         contents='toc.html'
         frame='frame.html';
proc print data=clinic.admit;
run;
proc print data=clinic.stress2;
run;
ods html close;
ods html;
```

Changing the Appearance of HTML Output

Style Definitions

You can change the appearance of your HTML output by specifying a style in the STYLE= option in the ODS HTML statement. Some of the style definitions that are currently shipped with SAS are

- Banker

- BarrettsBlue

- Default

- HTMLblue

- Minimal.

- Statistical

General form, STYLE= option:

STYLE=*style-name*

where *style-name* is the name of a valid SAS or user-defined style definition.

 Don't enclose *style-name* in quotation marks.

Example

In the program below, the STYLE= option applies the Banker style to the output for the PROC PRINT step:

```
ods html body='c:\records\data.html'
   style=banker;
proc print data=clinic.admit label;
    var sex age height weight actlevel;
run;
ods html close;
ods html;
```

The following example shows PROC PRINT output with the Banker style applied.

Figure 9.17   *PROC PRINT Output with the Banker Style Applied*

Obs	Sex	Age	Height	Weight	ActLevel
1	M	27	72	168	HIGH
2	F	34	66	152	HIGH
3	F	31	61	123	LOW
4	F	43	63	137	MOD
5	M	51	71	158	LOW
6	M	29	76	193	HIGH
7	F	32	67	151	MOD
8	M	35	70	173	MOD
9	M	34	73	154	LOW
10	F	49	64	172	LOW

To view a full list of the available style definitions, click the Results tab on the Explorer window. Then right-click the **Results icon** and select **Templates** from the pop-up menu. In the Templates window, open Sashelp.tmplmst. Then open the Styles folder.

A list of the available style definitions appears in the right panel of the Templates window.

 Your site might have its own, customized, style definitions.

Additional Features

Customizing HTML Output

You've seen that you can use the STYLE= option to apply predefined styles to your HTML output. However, you might want to further customize your results.

ODS provides ways for you to customize HTML output using definitions for tables, columns, headers, and so on. These definitions describe how to render the HTML output or part of the HTML output. You can create style definitions using the TEMPLATE procedure. See the online documentation for more information.

Chapter Summary

Text Summary

The Output Delivery System
The Output Delivery System (ODS) makes new report formatting options available in SAS. ODS separates your output into component parts so that the output can be sent to any ODS destination that you specify.

Opening and Closing ODS Destinations
Each ODS destination creates a different type of formatted output. By default, the HTML destination is open and SAS creates simple HTML output. Because an open destination uses system resources, it is a good idea to close the HTML destination if you don't need to create HTML output. Using ODS statements, you can create multiple output objects at the same time. When you have several ODS destinations open, you can close them all using the ODS _ALL_ CLOSE statement.

Creating HTML Output with Option
The HTML destination is open by default. However, you can use the ODS HTML statement to specify options. Use the BODY= or FILE= specification to create a custom named HTML body file containing procedure results. You can also use the ODS HTML statement to direct the HTML output from multiple procedures to the same HTML file.

Creating HTML Output with a Table of Contents
In order to manage multiple pieces of procedure output, you can use the CONTENTS= and FRAME = options with the ODS HTML statement to create a table of contents that links to your HTML output. The table of contents contains a heading for each procedure that creates output.

Specifying a Path for Output

You can also use the PATH= option to specify the directory where you want to store your HTML output. When you use the PATH= option, you don't need to specify the complete pathname for the body, contents, or frame files. By specifying the URL= suboption in the file specification, you can provide a URL that ODS uses in all the links that it creates to the file.

Changing the Appearance of HTML Output

You can change the appearance of your output using the STYLE= option in the ODS HTML statement. Several predefined styles are shipped with SAS.

Additional Features

ODS provides ways for you to customize HTML output using style definitions. Definitions are created using PROC TEMPLATE and describe how to render the HTML output or part of the HTML output.

Syntax

LIBNAME *libref 'SAS-data-library'*;

ODS LISTING CLOSE;

ODS HTML PATH=*file-specification*

      <(URL=*'Uniform-Resource-Locator'* | NONE)>

      BODY=*file-specification*

      CONTENTS=*<file-specification>*

      FRAME=*<file-specification>*

      STYLE=*<style-name>*;

PROC PRINT DATA=*SAS-data-set*;

RUN;

ODS HTML CLOSE;

ODS LISTING;

Sample Program

```
libname clinic 'c:\data98\patients';
ods html path='c:\records'(url=none)
         body='data.html'
         contents='toc.html'
         frame='frame.html'
         style=brick;
proc print data=clinic.admit label;
   var id sex age height weight actlevel;
   label actlevel='Activity Level';
run;
proc print data=clinic.stress2;
   var id resthr maxhr rechr;
run;
ods html close;
ods html;
```

Points to Remember

- An open destination uses system resources. Therefore, it is a good idea to close the HTML destination before you create other types of output and reopen the HTML destination after you close all destinations destination.

- You do not need the ODS HTML CLOSE statement to create simple HTML output. However, if you want to create HTML with options, then use the ODS HTML CLOSE statement to close the HTML destination. The ODS HTML CLOSE statement is added *after* the RUN statement for the procedure.

- If you use the CONTENTS= and FRAME= options, open the frame file from within your Web browser to view the procedure output *and* the table of contents.

Chapter Quiz

Select the best answer for each question. After completing the quiz, you can check your answers using the answer key in the appendix.

1. Using ODS statements, how many types of output can you generate at once?

   a. 1 (only LISTING output)

   b. 2

   c. 3

   d. as many as you want

2. If ODS is set to its default settings, what types of output are created by the code below?

   ```
   ods html file='c:\myhtml.htm';
   ods pdf file='c:\mypdf.pdf';
   ```

   a. HTML and PDF

   b. PDF only

   c. HTML, PDF, and LISTING

   d. No output is created because ODS is closed by default.

3. What is the purpose of closing the HTML destination in the code shown below?

   ```
   ods HTML close;
   ods pdf ... ;
   ```

   a. It conserves system resources.

   b. It simplifies your program.

   c. It makes your program compatible with other hardware platforms.

   d. It makes your program compatible with previous versions of SAS.

4. When the code shown below is run, what will the file **D:\Output\body.html** contain?

   ```
   ods html body='d:\output\body.html';
   proc print data=work.alpha;
   run;
   ```

```
proc print data=work.beta;
run;
ods html close;
```

    a. The PROC PRINT output for Work.Alpha.

    b. The PROC PRINT output for Work.Beta.

    c. The PROC PRINT output for both Work.Alpha and Work.Beta.

    d. Nothing. No output will be written to **D:\Output\body.html**.

5. When the code shown below is run, what file will be loaded by the links in **D:\Output\contents.html**?

```
ods html body='d:\output\body.html'
        contents='d:\output\contents.html'
        frame='d:\output\frame.html';
```

    a. **D:\Output\body.html**

    b. **D:\Output\contents.html**

    c. **D:\Output\frame.html**

    d. There are no links from the file **D:\Output\contents.html**.

6. The table of contents created by the CONTENTS= option contains a numbered heading for

    a. each procedure.

    b. each procedure that creates output.

    c. each procedure and DATA step.

    d. each HTML file created by your program.

7. When the code shown below is run, what will the file **D:\Output\frame.html** display?

```
ods html body='d:\output\body.html'
        contents='d:\output\contents.html'
        frame='d:\output\frame.html';
```

    a. The file **D:\Output\contents.html**.

    b. The file **D:\Output\frame.html**.

    c. The files **D:\Output\contents.html** and **D:\Output\body.html**.

    d. It displays no other files.

8. What is the purpose of the URL= suboptions shown below?

```
ods html body='d:\output\body.html' (url='body.html')
        contents='d:\output\contents.html'
        (url='contents.html')
        frame='d:\output\frame.html';
```

    a. To create absolute link addresses for loading the files from a server.

    b. To create relative link addresses for loading the files from a server.

    c. To allow HTML files to be loaded from a local drive.

    d. To send HTML output to two locations.

9. Which ODS HTML option was used in creating the following table?

Obs	Sex	Age	Height	Weight	ActLevel
1	M	27	72	168	HIGH
2	F	34	66	152	HIGH
3	F	31	61	123	LOW

a. `format=banker`

b. `format='bbanker'`

c. `style=banker`

d. `style='banker'`

10. What is the purpose of the PATH= option?

```
ods html path='d:\output' (url=none)
         body='body.html'
         contents='contents.html'
         frame='frame.html';
```

a. It creates absolute link addresses for loading HTML files from a server.

b. It creates relative link addresses for loading HTML files from a server.

c. It allows HTML files to be loaded from a local drive.

d. It specifies the location of HTML file output.

Chapter 10
Creating and Managing Variables

Overview

Introduction

You've learned how to create a SAS data set from raw data that is stored in an external file. You've also learned how to subset observations and how to assign values to variables.

This chapter shows you additional techniques for creating and managing variables. In this chapter, you learn how to retain and accumulate values, assign variable values conditionally, select variables, and assign permanent labels and formats to variables.

Figure 10.1 *Variables Displayed in a Data Set*

Obs	ID	Name	RestHR	MaxHR	RecHR	Tolerance	TotalTime	SumSec	TestLength
1	2458	Murray, W	72	185	128	D	758	6,158	Normal
2	2539	LaMance, K	75	168	141	D	706	6,864	Short
3	2572	Oberon, M	74	177	138	D	731	7,595	Short
4	2574	Peterson, V	80	164	137	D	849	8,444	Long
5	2584	Takahashi, Y	76	163	135	D	967	9,411	Long

Objectives

In this chapter, you learn how to

- create variables that accumulate variable values

- initialize retained variables

- assign values to variables conditionally

- specify an alternative action when a condition is false

- specify lengths for variables

- delete unwanted observations

- select variables

- assign permanent labels and formats.

Creating and Modifying Variables

Accumulating Totals

It is often useful to create a variable that accumulates the values of another variable.

Suppose you want to create the data set Clinic.Stress and to add a new variable, SumSec, to accumulate the total number of elapsed seconds in treadmill stress tests.

Figure 10.2   *Clinic.Stress Data Set*

SAS Data Set Clinic.Stress (Partial Listing)

ID	Name	RestHr	MaxHR	RecHR	TimeMin	TimeSec	Tolerance	TotalTime
2458	Murray, W	72	185	128	12	38	D	758
2462	Almers, C	68	171	133	10	5	I	605
2501	Bonaventure, T	78	177	139	11	13	I	673
2523	Johnson, R	69	162	114	9	42	S	582
2539	LaMance, K	75	168	141	11	46	D	706

To add the result of an expression to an accumulator variable, you can use a sum statement in your DATA step.

General form, sum statement:

variable+expression;

where

- *variable* specifies the name of the accumulator variable. This variable must be numeric. The variable is automatically set to 0 before the first observation is read. The variable's value is retained from one DATA step execution to the next.

- *expression* is any valid SAS expression.

 If the *expression* produces a missing value, the sum statement ignores it. (Remember, however, that assignment statements assign a missing value if the *expression* produces a missing value.)

The sum statement is one of the few SAS statements that doesn't begin with a keyword.

The sum statement adds the result of the expression that is on the right side of the plus sign (+) to the numeric variable that is on the left side of the plus sign. At the top of the DATA step, the value of the numeric variable is not set to missing as it usually is when reading raw data. Instead, the variable retains the new value in the program data vector for use in processing the next observation.

Example

To find the total number of elapsed seconds in treadmill stress tests, you need a variable (in this example, SumSec) whose value begins at 0 and increases by the amount of the total seconds in each observation. To calculate the total number of elapsed seconds in treadmill stress tests, you use the sum statement shown below:

```
data clinic.stress;
   infile tests;
   input ID $ 1-4 Name $ 6-25 RestHR 27-29 MaxHR 31-33
         RecHR 35-37 TimeMin 39-40 TimeSec 42-43
         Tolerance $ 45;
   TotalTime=(timemin*60)+timesec;
   SumSec+totaltime;
run;
```

The value of the variable on the left side of the plus sign (here, SumSec) begins at 0 and increases by the value of TotalTime with each observation.

SumSec	=	TotalTime	+	Previous total
0				
758	=	758	+	0
1363	=	605	+	758
2036	=	673	+	1363
2618	=	582	+	2036
3324	=	706	+	2618

Initializing Sum Variables

In a previous example, the sum variable SumSec was initialized to 0 by default before the first observation was read. But what if you want to initialize SumSec to a different number, such as the total seconds from previous treadmill stress tests?

You can use the RETAIN statement to assign an initial value other than the default value of 0 to a variable whose value is assigned by a sum statement.

The RETAIN statement

- assigns an initial value to a retained variable

- prevents variables from being initialized each time the DATA step executes.

General form, simple RETAIN statement for initializing sum variables:

RETAIN *variable <initial;-value>*;

where

- *variable* is a variable whose values you want to retain

- *initial-value* specifies an initial value (numeric or character) for the preceding variable.

 The RETAIN statement

- is a compile-time only statement that creates variables if they do not already exist.

- initializes the retained variable to missing before the first execution of the DATA step if you do not supply an initial value.

- has no effect on variables that are read with SET, MERGE, or UPDATE statements. (The SET and MERGE statements are discussed in later chapters.)

Example

Suppose you want to add 5400 seconds (the accumulated total seconds from a previous treadmill stress test) to the variable SumSec in the Clinic.Stress data set when you create the data set. To initialize SumSec with the value **5400**, you use the RETAIN statement shown below.

```
data clinic.stress;
   infile tests;
   input ID $ 1-4 Name $ 6-25 RestHR 27-29 MaxHR 31-33
         RecHR 35-37 TimeMin 39-40 TimeSec 42-43
         Tolerance $ 45;
   TotalTime=(timemin*60)+timesec;
   retain SumSec 5400;
   sumsec+totaltime;
run;
```

Now the value of SumSec begins at **5400** and increases by the value of TotalTime with each observation.

SumSec	=	TotalTime	+	Previous total
5400				
6158	=	758	+	5400
6763	=	605	+	6158
7436	=	673	+	6763
8018	=	582	+	7436
8724	=	706	+	8018

Assigning Values Conditionally

Overview

In the previous section, you created the variable SumSec by using a sum statement to add total seconds from treadmill stress. This time, let's create a variable that categorizes the length of time that a subject spends on the treadmill during a stress test. This new variable, TestLength, will be based on the value of the existing variable TotalTime. The value of TestLength will be assigned conditionally:

If TotalTime is . . .	then TestLength is . . .
greater than 800	Long
750 - 800	Normal
less than 750	Short

To perform an action conditionally, use an IF-THEN statement. The IF-THEN statement executes a SAS statement when the condition in the IF clause is true.

General form, IF-THEN statement:

IF *expression* **THEN** *statement*;

where

- *expression* is any valid SAS expression.
- *statement* is any executable SAS statement

Example

To assign the value **Long** to the variable TestLength when the value of TotalTime is greater than **800**, add the following IF-THEN statement to your DATA step:

```
data clinic.stress;
   infile tests;
   input ID $ 1-4 Name $ 6-25 RestHR 27-29 MaxHR 31-33
         RecHR 35-37 TimeMin 39-40 TimeSec 42-43
         Tolerance $ 45;
   TotalTime=(timemin*60)+timesec;
   retain SumSec 5400;
   sumsec+totaltime;
   if totaltime>800 then TestLength='Long';
run;
```

SAS executes the assignment statement only when the condition (TotalTime>800) is true. If the condition is false, the value of TestLength will be missing.

Comparison and Logical Operators

When writing IF-THEN statements, you can use any of the following comparison operators:

Operator	Comparison Operation
= or **eq**	equal to
^= or **ne**	not equal to
> or **gt**	greater than
< or **lt**	less than
>= or **ge**	greater than or equal to
<= or **le**	less than or equal to
in	equal to one of a list

Examples:

```
if test<85 and time<=20
   then Status='RETEST';
if region in ('NE','NW','SW')
   then Rate=fee-25;
if target gt 300 or sales ge 50000

   then Bonus=salary*.05;
```

You can also use these logical operators:

Operator	Logical Operation
&	and
|	or
^	not

Use the AND operator to execute the THEN statement if both expressions that are linked by AND are true.

```
if status='OK' and type=3
   then Count+1;
if (age^=agecheck | time^=3)
   & error=1 then Test=1;
```

Use the OR operator to execute the THEN statement if either expression that is linked by OR is true.

```
if (age^=agecheck | time^=3)
   & error=1 then Test=1;
if status='S' or cond='E'
   then Control='Stop';
```

Use the NOT operator with other operators to reverse the logic of a comparison.

```
if not(loghours<7500)
   then Schedule='Quarterly';
if region not in ('NE','SE')
   then Bonus=200;
```

Character values must be specified in the same case in which they appear in the data set and must be enclosed in quotation marks.

```
if status='OK' and type=3
   then Count+1;
if status='S' or cond='E'
   then Control='Stop';
if not(loghours<7500)
   then Schedule='Quarterly';
if region not in ('NE','SE')
   then Bonus=200;
```

Logical comparisons that are enclosed in parentheses are evaluated as true or false before they are compared to other expressions. In the example below, the OR comparison in parentheses is evaluated before the first expression and the AND operator are evaluated.

Figure 10.3 *Example of a Logical Comparison*

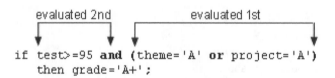

In SAS, any numeric value other than 0 or missing is true, and a value of 0 or missing is false. Therefore, a numeric variable or expression can stand alone in a condition. If its value is a number other than 0 or missing, the condition is true. If its value is 0 or missing, the condition is false.

- 0 = False

- . = False

- 1 = True

As a result, you need to be careful when using the OR operator with a series of comparisons. Remember that only one comparison in a series of OR comparisons must be true to make a condition true, and any nonzero, nonmissing constant is always evaluated as true. Therefore, the following subsetting IF statement is always true:

```
if x=1 or 2;
```

SAS first evaluates x=1, and the result can be either true or false. However, since the 2 is evaluated as nonzero and nonmissing (true), the entire expression is true. In this statement, however, the condition is not necessarily true because either comparison can evaluate as true or false:

```
if x=1 or x=2;
```

Providing an Alternative Action

Now suppose you want to assign a value to TestLength based on the other possible values of TotalTime. One way to do this is to add IF-THEN statements for the other two conditions, as shown below.

```
if totaltime>800 then TestLength='Long';
if 750<=totaltime<=800 then TestLength='Normal';
if totaltime<750 then TestLength='Short';
```

However, when the DATA step executes, each IF statement is evaluated in order, even if the first condition is true. This wastes system resources and slows the processing of your program.

Instead of using a series of IF-THEN statements, you can use the ELSE statement to specify an alternative action to be performed when the condition in an IF-THEN statement is false. As shown below, you can write multiple ELSE statements to specify a series of mutually exclusive conditions.

```
if totaltime>800 then TestLength='Long';
else if 750<=totaltime<=800 then TestLength='Normal';
else if totaltime<750 then TestLength='Short';
```

The ELSE statement must immediately follow the IF-THEN statement in your program. An ELSE statement executes only if the previous IF-THEN/ELSE statement is false.

General form, ELSE statement:

ELSE *statement*;

where *statement* is any executable SAS statement, including another IF-THEN statement.

So, to assign a value to TestLength when the condition in your IF-THEN statement is false, you can add the ELSE statement to your DATA step, as shown below:

```
data clinic.stress;
    infile tests;
    input ID $ 1-4 Name $ 6-25 RestHR 27-29 MaxHR 31-33
        RecHR 35-37 TimeMin 39-40 TimeSec 42-43
        Tolerance $ 45;
    TotalTime=(timemin*60)+timesec;
    retain SumSec 5400;
    sumsec+totaltime;
    if totaltime>800 then TestLeng h='Long';
    else if 750<=totaltime<=800 then TestLength='Normal';
    else if totaltime<750 then TestLength='Short';
run;
```

Using ELSE statements with IF-THEN statements can save resources:

- Using IF-THEN statements *without* the ELSE statement causes SAS to evaluate all IF-THEN statements.

- Using IF-THEN statements *with* the ELSE statement causes SAS to execute IF-THEN statements until it encounters the first true statement. Subsequent IF-THEN statements are not evaluated.

For greater efficiency, construct your IF-THEN/ELSE statements with conditions of decreasing probability.

Remember that you can use PUT statements to test your conditional logic.

```
data clinic.stress;
   infile tests;
   input ID $ 1-4 Name $ 6-25 RestHR 27-29 MaxHR 31-33
         RecHR 35-37 TimeMin 39-40 TimeSec 42-43
         Tolerance $ 45;
   TotalTime=(timemin*60)+timesec;
   retain SumSec 5400;
   sumsec+totaltime;
   if totaltime>800 then TestLength='Long';
   else if 750<=totaltime<=800 then TestLength='Normal';
   else put 'NOTE: Check this Length: ' totaltime=;
run;
```

Specifying Lengths for Variables

Overview

Previously, you added IF-THEN and ELSE statements to a DATA step in order to create the variable TestLength. Values for TestLength were assigned conditionally, based on the value for TotalTime.

```
data clinic.stress;
   infile tests;
   input ID $ 1-4 Name $ 6-25 RestHR 27-29 MaxHR 31-33
         RecHR 35-37 TimeMin 39-40 TimeSec 42-43
         Tolerance $ 45;
   TotalTime=(timemin*60)+timesec;
   retain SumSec 5400;
   sumsec+totaltime;
   if totaltime>800 then TestLength='Long';
   else if 750<=totaltime<=800 then TestLength='Normal';
   else if totaltime<750 then TestLength='Short';
run;
```

But look what happens when you submit this program.

During compilation, when creating a new character variable in an assignment statement, SAS allocates as many bytes of storage space as there are characters in the first value that it encounters for that variable. In this case, the first value for TestLength occurs in the IF-THEN statement, which specifies a four-character value (**Long**). So TestLength is assigned a length of 4, and any longer values (**Normal** and **Short**) are truncated.

Figure 10.4 *Truncated Variable Values*

Variable TestLength
(Partial Listing)

TestLength
Norm
Shor
Shor
Shor
Norm
Shor
Long
...

The example above assigns a character constant as the value of the new variable. You can also see other examples of the default type and length that SAS assigns when the type and length of a variable are not explicitly set.

You can use a LENGTH statement to specify a length (the number of bytes) for TestLength before the first value is referenced elsewhere in the DATA step.

General form, LENGTH statement:

LENGTH *variable(s)* <$> *length*;

where

- *variable(s)* names the variable(s) to be assigned a length
- $ is specified if the variable is a character variable
- *length* is an integer that specifies the length of the variable.

Examples:

```
length Type $ 8;
length Address1 Address2 Address3
$200;
length FirstName  $12 LastName
$16;
```

Within your program, you include a LENGTH statement to assign a length to accommodate the longest value of the variable TestLength. The longest value is **Normal**, which has six characters. Because TestLength is a character variable, you must follow the variable name with a dollar sign ($).

```
data clinic.stress;
   infile tests;
   input ID $ 1-4 Name $ 6-25 RestHR 27-29 MaxHR 31-33
         RecHR 35-37 TimeMin 39-40 TimeSec 42-43
         Tolerance $ 45;
   TotalTime=(timemin*60)+timesec;
   retain SumSec 5400;
```

```
    sumsec+totaltime;
    length TestLength $ 6;
    if totaltime>800 then testlength='Long';
    else if 750<=totaltime<=800 then testlength='Normal';
    else if totaltime<750 then TestLength='Short';
run;
```

Make sure the LENGTH statement appears before any other reference to the variable in the DATA step. If the variable has been created by another statement, then a later use of the LENGTH statement will not change its length.

Now that you have added the LENGTH statement to your program, the values of TestLength are no longer truncated.

Figure 10.5 *Variable Values That Are Not Truncated*

Variable TestLength
(Partial Listing)

TestLength
Normal
Short
Short
Short
Normal
Short
Long
...

Subsetting Data

Deleting Unwanted Observations

So far in this chapter, you've learned to use IF-THEN statements to execute assignment statements conditionally. But you can specify any executable SAS statement in an IF-THEN statement. For example, you can use an IF-THEN statement with a DELETE statement to determine which observations to omit as you read data.

- The IF-THEN statement executes a SAS statement when the condition in the IF clause is true.

- The DELETE statement stops the processing of the current observation.

General form, DELETE statement:

DELETE;

To conditionally execute a DELETE statement, you submit a statement in the following general form:

IF *expression* **THEN DELETE;**

If the expression is

- *true*, the DELETE statement executes, and control returns to the top of the DATA step (the observation is deleted).

- *false*, the DELETE statement does not execute, and processing continues with the next statement in the DATA step.

Example

The IF-THEN and DELETE statements below omit any observations whose values for RestHR are lower than **70**.

```
data clinic.stress;
   infile tests;
    input ID $ 1-4 Name $ 6-25 RestHR 27-29 MaxHR 31-33
          RecHR 35-37 TimeMin 39-40 TimeSec 42-43
          Tolerance $ 45;
   if resthr<70 then delete;
   TotalTime=(timemin*60)+timesec;
   retain SumSec 5400;
   sumsec+totaltime;
   length TestLength $ 6;
   if totaltime>800 then testlength='Long';
   else if 750<=totaltime<=800 then testlength='Normal';
   else if totaltime<750 then TestLength='Short';
run;
```

Selecting Variables

Sometimes you might need to read and process variables that you don't want to keep in your data set. In this case, you can use the DROP= and KEEP= data set options to specify the variables that you want to drop or keep.

Use the KEEP= option instead of the DROP= option if more variables are dropped than kept. You specify data set options in parentheses after a SAS data set name.

General form, DROP= and KEEP= data set options:

(DROP=*variable(s)***)**

(KEEP=*variable(s)***)**

where

- the **DROP=KEEP=** or option, in parentheses, follows the name of the data set that contains the variables to be dropped or kept

- *variable(s)* identifies the variables to drop or keep.

Example

Suppose you are interested in keeping only the new variable TotalTime and not the original variables TimeMin and TimeSec. You can drop TimeMin and TimeSec when you create the Stress data set.

```
data clinic.stress (drop=timemin timesec);
   infile tests;
   input ID $ 1-4 Name $ 6-25 RestHR 27-29 MaxHR 31-33
         RecHR 35-37 TimeMin 39-40 TimeSec 42-43
         Tolerance $ 45;
   if tolerance='D';
   TotalTime=(timemin*60)+timesec;
   retain SumSec 5400;
   sumsec+totaltime;
   length TestLength $ 6;
   if totaltime>800 then testlength='Long';
   else if 750<=totaltime<=800 then testlength='Normal';
   else if totaltime<750 then TestLength='Short';
run;
```

Figure 10.6 *Stress Data Set with Dropped Variables*

SAS Data Set Clinic.Stress (Partial Listing)

ID	Name	RestHR	MaxHR	RecHR	Tolerance	TotalTime	SumSec	TestLength
2458	Murray, W	72	185	128	D	758	6158	Normal
2539	LaMance, K	75	168	141	D	706	6864	Short
2552	Reberson, P	69	158	139	D	941	7805	Long
2572	Oberon, M	74	177	138	D	731	8536	Short
2574	Peterson, V	80	164	137	D	849	9385	Long
2584	Takahashi, Y	76	163	135	D	967	10352	Long

Another way to exclude variables from your data set is to use the DROP statement or the KEEP statement. Like the DROP= and KEEP= data set options, these statements drop or keep variables. However, the DROP and KEEP statements differ from the DROP= and KEEP data set options in the following ways:

- You cannot use the DROP and KEEP statements in SAS procedure steps.

- The DROP and KEEP statements apply to all output data sets that are named in the DATA statement. To exclude variables from some data sets but not from others, use the DROP= and KEEP data set options in the DATA statement.

The KEEP statement is similar to the DROP statement, except that the KEEP statement specifies a list of variables to write to output data sets. Use the KEEP statement instead of the DROP statement if the number of variables to keep is smaller than the number to drop.

General form, DROP and KEEP statements:

DROP *variable(s);*

KEEP *variable(s);*

where v*ariable(s)* identifies the variables to drop or keep.

Example

The two programs below produce the same results. The first example uses the DROP=
data set option. The second example uses the DROP statement.

```
data clinic.stress (drop=timemin timesec);
   infile tests;
   input ID $ 1-4 Name $ 6-25 RestHR 27-29 MaxHR 31-33
         RecHR 35-37 TimeMin 39-40 TimeSec 42-43
         Tolerance $ 45;
   if tolerance='D';
   TotalTime=(timemin*60)+timesec;
   retain SumSec 5400;
   sumsec+totaltime;
   length TestLength $ 6;
   if totaltime>800 then testlength='Long';
   else if 750<=totaltime<=800 then testlength='Normal';
   else if totaltime<750 then TestLength='Short';
run;

data clinic.stress;
   infile tests;
   input ID $ 1-4 Name $ 6-25 RestHR 27-29 MaxHR 31-33
         RecHR 35-37 TimeMin 39-40 TimeSec 42-43
         Tolerance $ 45;
   if tolerance='D';
   drop timemin timesec;
   TotalTime=(timemin*60)+timesec;
   retain SumSec 5400;
   sumsec+totaltime;
   length TestLength $ 6;
   if totaltime>800 then testlength='Long';
   else if 750<=totaltime<=800 then testlength='Normal';
   else if totaltime<750 then TestLength='Short';
run;
```

Assigning Permanent Labels and Formats

Overview

At this point, you've seen raw data read and manipulated to obtain the observations,
variables, and variable values that you want. Our final task in this chapter is to
permanently assign labels and formats to variables.

If you read "Creating List Reports" on page 112, you learned to temporarily assign
labels and formats within a PROC step. These temporary labels and formats are
applicable only for the duration of the step. To permanently assign labels and formats,
you use LABEL and FORMAT statements in DATA steps.

Remember that labels and formats do not affect how data is stored in the data set,
but only how it appears in output.

To do this...	Use this type of statement...
Reference a SAS data libraryReference an external file	`libname clinic 'c:\users\may\data';` `filename tests 'c:\users\tmill.dat';`

Name a SAS data setIdentify an external fileDescribe raw data	`data clinic.stress;`
	`    infile tests obs=10;`
	`    input ID $ 1-4 Name $ 6-25 ...;`

Subset data	`if resthr<70 then delete;` `if tolerance='D';`
Drop unwanted variables	`drop timemin timesec;`
Create or modify a variable	`TotalTime=(timemin*60)+timesec;`
Initialize and retain variable	`retain SumSec 5400;`
Accumulate values	`sumsec+totaltime;`
Specify a variable's length	`length TestLength $ 6;`

Execute statements conditionally	`if totaltime>800 then TestLength='Long';`
	`else if 750<=totaltime<=800`
	`    then TestLength='Normal';`
	`else if totaltime<750`
	`    then TestLength='Short';`

Label a variable	LABEL statement
Format a variable	FORMAT statement
Execute the DATA step	`run;`
List the data	`proc print data=clinic.stress label;`
Execute the final program step	`run;`

Example

To specify the label Cumulative Total Seconds (+5,400) and the format COMMA6. for the variable SumSec, you can submit the following program:

```
data clinic.stress;
   infile tests;
   input ID $ 1-4 Name $ 6-25 RestHR 27-29 MaxHR 31-33
         RecHR 35-37 TimeMin 39-40 TimeSec 42-43
         Tolerance $ 45;
   if resthr<70 then delete;
   if tolerance='D';
   drop timemin timesec;
   TotalTime=(timemin*60)+timesec;
   retain SumSec 5400;
   sumsec+totaltime;
```

```
      length TestLength $ 6;
      if totaltime>800 then testlength='Long';
      else if 750<=totaltime<=800 then testlength='Normal';
      else if totaltime<750 then TestLength='Short';
      label sumsec='Cumulative Total Seconds (+5,400)';
      format sumsec comma6.;
run;
```

You're done! When we print the new data set, SumSec is labeled and formatted as specified. (Don't forget the LABEL option in the PROC PRINT statement.)

```
proc print data=clinic.stress label;
run;
```

Figure 10.7   *Completed Clinic.Stress Data Set*

Specifying the Label Comulative Total Seconds

Obs	ID	Name	RestHR	MaxHR	RecHR	Tolerance	TotalTime	Cumulative Total Seconds (+5,400)	TestLength
1	2458	Murray, W	72	185	128	D	758	6,158	Normal
2	2539	LaMance, K	75	168	141	D	706	6,864	Short
3	2572	Oberon, M	74	177	138	D	731	7,595	Short
4	2574	Peterson, V	80	164	137	D	849	8,444	Long
5	2584	Takahashi, Y	76	163	135	D	967	9,411	Long

Remember that most SAS procedures automatically use permanent labels and formats in output, without requiring additional statements or options.

If you assign temporary labels or formats within a PROC step, they override any permanent labels or formats that were assigned during the DATA step.

Assigning Values Conditionally Using SELECT Groups

Overview

Earlier in this chapter, you learned to assign values conditionally using IF-THEN/ELSE statements. You can also use SELECT groups in DATA steps to perform conditional processing. A SELECT group contains these statements:

This statement...	Performs this action...
SELECT	begins a SELECT group.
WHEN	identifies SAS statements that are executed when a particular condition is true.

This statement...	Performs this action...
OTHERWISE (optional)	specifies a statement to be executed if no WHEN condition is met.
END	ends a SELECT group.

You can decide whether to use IF-THEN/ELSE statements or SELECT groups based on personal preference.

General form, SELECT group:

SELECT *<(select-expression)>*;

      WHEN-1 *(when-expression-1<..., when-expression-n>) statement*;

      WHEN-n *(when-expression-1 <..., when-expression-n>) statement*;

      <OTHERWISE*statement*;>

END;

where

- **SELECT** begins a SELECT group.

- the optional *select-expression* specifies any SAS expression that evaluates to a single value.

- **WHEN** identifies SAS statements that are executed when a particular condition is true.

- *when-expression* specifies any SAS expression, including a compound expression. You must specify at least one *when-expression*.

- *statement* is any executable SAS statement. You must specify the *statement* argument.

- the optional **OTHERWISE** statement specifies a statement to be executed if no WHEN condition is met.

- **END** ends a SELECT group.

Example: Basic SELECT Group

The following code is a simple example of a SELECT group. Notice that the variable a is specified in the SELECT statement, and various values to compare to a are specified in the WHEN statements. When the value of the variable a is

- **1**, x is multiplied by 10

- **3**, **4**, or **5**, x is multiplied by 100

- **2** or any other value, nothing happens.

```
select (a);
   when (1) x=x*10;
   when (3,4,5) x=x*100;
   otherwise;
end
```

Example: SELECT Group in a DATA Step

Now let's look at a SELECT group in context. In the DATA step below, the SELECT group assign values to the variable Group based on values of the variable JobCode. Most of the assignments are one-to-one correspondences, but ticket agents (the JobCode values **TA1**, **TA2**, and **TA3**) are grouped together, as are values in the Other category.

```
data emps(keep=salary group);
   set sasuser.payrollmaster;
   length Group $ 20;
   select(jobcode);
      when ("FA1") group="Flight Attendant I";
      when ("FA2") group="Flight Attendant II";
      when ("FA3") group="Flight Attendant III";
      when ("ME1") group="Mechanic I";
      when ("ME2") group="Mechanic II";
      when ("ME3") group="Mechanic III";
      when ("NA1") group="Navigator I";
      when ("NA2") group="Navigator II";
      when ("NA3") group="Navigator III";
      when ("NA1") group="Navigator I";
      when ("NA2") group="Navigator II";
      when ("NA3") group="Navigator III";
      when ("PT1") group="Pilot I";
      when ("PT2") group="Pilot II";
      when ("PT3") group="Pilot III";
      when ("TA1","TA2","TA3") group="Ticket Agents";
      otherwise group="Other";
   end;
run;
```

Notice that in this case the SELECT statement contains a select-expression. You are checking values of a single variable by using select(jobcode), which is more concise than eliminating the select-expression and repeating the variable in each when-expression, as in when(jobcode="FA1").

Notice that the LENGTH statement in the DATA step above specifies a length of 20 for Group. Remember that without the LENGTH statement, values for Group might be truncated, as the first value for Group (**Flight Attendant I**) is not the longest possible value.

When you are comparing values in the when-expression, be sure to express the values exactly as they are coded in the data. For example, the when-expression below would be evaluated as false because the values for JobCode in Sasuser.Payrollmaster are stored in uppercase letters.

```
when ('fa1') group="Flight Attendant I";
```

In this case, given the SELECT group above, Group would be assigned the value **Other**.

As you saw in the general form for SELECT groups, you can optionally specify a select-expression in the SELECT statement. The way SAS evaluates a when-expression depends on whether you specify a select-expression.

Specifying SELECT Statements with Expressions

If you *do* specify a select-expression in the SELECT statement, SAS compares the value of the select-expression with the value of each when-expression. That is, SAS evaluates the select-expression and when-expression, compares the two for equality, and returns a value of true or false.

- If the comparison is *true*, SAS executes the statement in the WHEN statement.

- If the comparison is *false*, SAS proceeds either to the next when-expression in the current WHEN statement, or to the next WHEN statement if no more expressions are present. If no WHEN statements remain, execution proceeds to the OTHERWISE statement, if one is present.

⚠ If the result of all SELECT-WHEN comparisons is false and no OTHERWISE statement is present, SAS issues an error message and stops executing the DATA step.

In the following SELECT group, SAS determines the value of toy and compares it to values in each WHEN statement in turn. If a WHEN statement is true compared to the toy value, SAS assigns the related price and continues processing the rest of the DATA step. If none of the comparisons is true, SAS executes the OTHERWISE statement and writes a debugging message to the SAS log.

```
select (toy);
   when ("Bear") price=35.00;
   when ("Violin") price=139.00;
   when ("Top","Whistle","Duck") price=7.99;
   otherwise put "Check unknown toy: " toy=;
end;
```

Specifying SELECT Statements without Expressions

If you *don't* specify a select-expression, SAS evaluates each when-expression to produce a result of true or false.

- If the result is *true*, SAS executes the statement in the WHEN statement.

- If the result is *false*, SAS proceeds either to the next when-expression in the current WHEN statement, or to the next WHEN statement if no more expressions are present, or to the OTHERWISE statement if one is present. (That is, SAS performs the action that is indicated in the first true WHEN statement.)

If more than one WHEN statement has a true when-expression, only the *first* WHEN statement is used. Once a when-expression is true, no other when-expressions are evaluated.

⚠ If the result of all when-expressions is false and no OTHERWISE statement is present, SAS issues an error message.

In the example below, the SELECT statement does not specify a select-expression. The WHEN statements are evaluated in order, and only one is used. For example, if the value of toy is **Bear** and the value of month is **FEB**, only the second WHEN statement is used, even though the condition in the third WHEN statement is also met. In this case, the variable price is assigned the value **25.00**:

```
select;
   when (toy="Bear" and month in ('OCT', 'NOV', 'DEC')) price=45.00;
   when (toy="Bear" and month in ('JAN', 'FEB')) price=25.00;
   when (toy="Bear") price=35.00;
   otherwise;
end;
```

Grouping Statements Using DO Groups

Overview

So far in this chapter, you've seen examples of conditional processing (IF-THEN/ELSE statements and SELECT groups) that execute only a single SAS statement when a condition is true. However, you can also execute a group of statements as a unit by using DO groups.

To construct a DO group, you use the DO and END statements along with other SAS statements.

General form, simple DO group:

DO;

   SAS statements

END;

where

- the **DO** statement begins DO group processing

- *SAS statements* between the DO and END statements are called a DO group and execute as a unit

- the **END** statement terminates DO group processing.

 You can nest DO statements within DO groups.

You can use DO groups in IF-THEN/ELSE statements and SELECT groups to execute many statements as part of the conditional action.

Examples

In this simple DO group, the statements between DO and END are performed only when TotalTime is greater than 800. If TotalTime is less than or equal to 800, statements in the DO group do not execute, and the program continues with the assignment statement that follows the appropriate ELSE statement.

```
data clinic.stress;
   infile tests;
   input ID $ 1-4 Name $ 6-25 RestHR 27-29 MaxHR 31-33
         RecHR 35-37 TimeMin 39-40 TimeSec 42-43
         Tolerance $ 45;
   TotalTime=(timemin*60)+timesec;
   retain SumSec 5400;
   sumsec+totaltime;
   length TestLength $ 6 Message $ 20;
   if totaltime>800 then
      do;
         testlength='Long';
         message='Run blood panel';
      end;
   else if 750<=totaltime<=800 then testlength='Normal';
```

```
         else if totaltime<750 then TestLength='Short';
run;
```

In the SELECT group below, the statements between DO and END are performed only when the value of Payclass is **hourly**. Notice that an IF-THEN statement appears in the DO group; the PUT statement executes only when Hours is greater than **40**. The second END statement in the program closes the SELECT group.

```
data payroll;
   set salaries;
   select(payclass);
   when ('monthly') amt=salary;
   when ('hourly')
     do;
        amt=hrlywage*min(hrs,40);
        if hrs>40 then put 'CHECK TIMECARD';
     end;
   otherwise put 'PROBLEM OBSERVATION';
   end;
run;
```

Indenting and Nesting DO Groups

You can nest DO groups to any level, just like you nest IF-THEN/ELSE statements. (The memory capabilities of your system may limit the number of nested DO statements that you can use. For details, see the SAS documentation about how many levels of nested DO statements your system's memory can support.)

The following is an example of nested DO groups:

```
do;
    statements;
      do;
             statements ;
               do;
                      statements;
               end;
      end;
end;
```

It is good practice to indent the statements in DO groups, as shown in the preceding statements, so that their position indicates the levels of nesting.

There are three other forms of the DO statement:

- The iterative DO statement executes statements between DO and END statements repetitively based on the value of an index variable. The iterative DO statement can contain a WHILE or UNTIL clause.

- The DO UNTIL statement executes statements in a DO loop repetitively until a condition is true, checking the condition after each iteration of the DO loop.

- The DO WHILE statement executes statements in a DO loop repetitively while a condition is true, checking the condition before each iteration of the DO loop.

You can learn about these forms of the DO statement in "Generating Data with DO Loops" on page 463 .

Chapter Summary

Text Summary

Accumulating Totals

Use a sum statement to add the result of an expression to an accumulator variable.

Initializing and Retaining Variables

You can use the RETAIN statement to assign an initial value to a variable whose value is assigned by a sum statement.

Assigning Values Conditionally

To perform an action conditionally, use an IF-THEN statement. The IF-THEN statement executes a SAS statement when the condition in the IF expression is true. You can include comparison and logical operators; logical comparisons that are enclosed in parentheses are evaluated as true or false before other expressions are evaluated. Use the ELSE statement to specify an alternative action when the condition in an IF-THEN statement is false.

Specifying Lengths for Variables

When creating a new character variable, SAS allocates as many bytes of storage space as there are characters in the first value that it encounters for that variable at compile time. This can result in truncated values. You can use the LENGTH statement to specify a length before the variable's first value is referenced in the DATA step.

Subsetting Data

To omit observations as you read data, include the DELETE statement in an IF-THEN statement. If you need to read and process variables that you don't want to keep in the data set, use the DROP= and KEEP= data set options or the DROP and KEEP statements.

Assigning Permanent Labels and Formats

You can use LABEL and FORMAT statements in DATA steps to permanently assign labels and formats. These do not affect how data is stored in the data set, only how it appears in output.

Assigning Values Conditionally Using SELECT Groups

As an alternative to IF-THEN/ELSE statements, you can use SELECT groups in DATA steps to perform conditional processing.

Grouping Statements Using DO Groups

You can execute a group of statements as a unit with DO groups in DATA steps. You can use DO groups in IF-THEN/ELSE statements and SELECT groups to perform many statements as part of the conditional action.

Syntax

LIBNAME *libref 'SAS-data-library'*;

FILENAME *fileref 'filename'*;

DATA *SAS-data-set*(**DROP**=*variable(s)*|**KEEP**=*variable(s)*);

    INFILE *file-specification* <OBS=*n*>;

    INPUT *variable* <$> *startcol-endcol*...;

    DROP *variable(s)*;

    KEEP *variable(s)*;

    RETAIN *variable initial-value*;

    variable+expression;

    LENGTH *variable(s)* <$> *length*;

    IF *expression***THEN** *statement*;

    ELSE *statement*;

    IF *expression***THEN DELETE**;

    LABEL *variable1='label1' variable2='label2'* ...;

    FORMAT *variable(s) format-name*;

    SELECT <*(select-expression)*>;

      WHEN-1 (*when-expression-1* <..., *when-expression-n*>) *statement*;

      WHEN-n (*when-expression-1* <..., *when-expression-n*>) *statement*;

      <OTHERWISE*statement*;>

    END;

RUN;

PROC PRINT DATA=*SAS-data set* **LABEL**;

RUN;

Sample Program

```
data clinic.stress;
   infile tests;
   input ID $ 1-4 Name $ 6-25 RestHR 27-29 MaxHR 31-33
         RecHR 35-37 TimeMin 39-40 TimeSec 42-43
         Tolerance $ 45;
   if tolerance='D'and resthr ge 70 then delete;
   drop timemin timesec;
   TotalTime=(timemin*60)+timesec;
   retain SumSec 5400;
   sumsec+totaltime;
   length TestLength $ 6;
   if totaltime>800 then testlength='Long';
   else if 750<=totaltime<=800 then testlength='Normal';
   else if totaltime<750 then TestLength='Short';
   label sumsec='Cumulative Total Seconds (+5,400)';
   format sumsec comma6.;
run;
```

Points to Remember

- Like the assignment statement, the sum statement does not contain a keyword.

- If the expression in a sum statement produces a missing value, the sum statement ignores it. (Remember, however, that assignment statements assign a missing value if the expression produces a missing value.)

- Using ELSE statements with IF-THEN statements can save resources. For greater efficiency, construct your IF-THEN/ELSE statements with conditions of decreasing probability.

- Make sure the LENGTH statement appears before any other reference to the variable in the DATA step. If the variable has been created by another statement, a later use of the LENGTH statement will not change its length.

- Labels and formats do not affect how data is stored in the data set, only how it appears in output. You assign labels and formats temporarily in PROC steps and permanently in DATA steps.

Chapter Quiz

Select the best answer for each question. After completing the quiz, you can check your answers using the answer key in the appendix.

1. Which program creates the output shown below?

Figure 10.8 Output 1

Raw Data File Furnture
```
1---+----10---+----20---+----30---+
310 oak    pedestal table 329.99
311 maple pedestal table 369.99
312 brass floor     lamp   79.99
313 glass table     lamp   59.99
314 oak    rocking   chair 153.99
```

Figure 10.9 Output 2

StockNum	Finish	Style	Item	TotalPrice
310	oak	pedestal	table	329.99
311	maple	pedestal	table	699.98
312	brass	floor	lamp	779.97
313	glass	table	lamp	839.96

a.
```
data test2;
    infile furnture;
    input StockNum $ 1-3 Finish $ 5-9 Style $ 11-18
        Item $ 20-24 Price 26-31;
    if finish='oak' then delete;
    retain TotPrice 100;
    totalprice+price;
    drop price;
```

```
      run;
      proc print data=test2 noobs;
      run;
```

b.
```
   data test2;
       infile furnture;
       input StockNum $ 1-3 Finish $ 5-9 Style $ 11-18
           Item $ 20-24 Price 26-31;
       if finish='oak' and price<200 then delete;
       TotalPrice+price;
   run;
   proc print data=test2 noobs;
   run;
```

c.
```
   data test2(drop=price);
       infile furnture;
       input StockNum $ 1-3 Finish $ 5-9 Style $ 11-18
           Item $ 20-24 Price 26-31;
       if finish='oak' and price<200 then delete;
       TotalPrice+price;
   run;
   proc print data=test2 noobs;
   run;
```

d.
```
   data test2;
       infile furnture;
       input StockNum $ 1-3 Finish $ 5-9 Style $ 11-18
           Item $ 20-24 Price 26-31;
       if finish=oak and price<200 then delete price;
       TotalPrice+price;
   run;
   proc print data=test2 noobs;
   run;
```

2. How is the variable Amount labeled and formatted in the PROC PRINT output?

```
data credit;
    infile creddata;
    input Account $ 1-5 Name $ 7-25 Type $ 27
            Transact $ 29-35 Amount 37-50;
    label amount='Amount of Loan';
    format amount dollar12.2;
run;
proc print data=credit label;
    label amount='Total Amount Loaned';
    format amount comma10.;
run;
```

   a. label Amount of Loan, format DOLLAR12.2

   b. label Total Amount Loaned, format COMMA10.

   c. label Amount, default format

   d. The PROC PRINT step does not execute because two labels and two formats are assigned to the same variable.

3. Consider the IF-THEN statement shown below. When the statement is executed, which expression is evaluated first?

```
if finlexam>=95
   and (research='A' or
        (project='A' and present='A'))
   then Grade='A+';
```

a. `finlexam>=95`

b. `research='A'`

c. `project='A' and present='A'`

d. `research='A' or`
   `(project='A' and present='A')`

4. Consider the small raw data file and program shown below. What is the value of Count after the fourth record is read?

```
data work.newnums;
   infile numbers;
   input Tens 2-3;
   Count+tens;
run;
```

```
1---+----10
  10
  20

  40
  50
```

a. missing

b. **0**

c. **30**

d. **70**

5. Now consider the revised program below. What is the value of Count after the third observation is read?

```
data work.newnums;
   infile numbers;
   input Tens 2-3;
   retain Count 100;
   count+tens;
run;
```

```
1---+----10
  10
  20

  40
  50
```

a. missing

b. **0**

c. **100**

d. **130**

6. For the observation shown below, what is the result of the IF-THEN statements?

Status	Type	Count	Action	Control
Ok	3	12	E	Go

```
if status='OK' and type=3
   then Count+1;
if status='S' or action='E'
   then Control='Stop';
```

a. `Count = 12     Control = Go`

b. `Count = 13     Control =Stop`

c. `Count = 12     Control =Stop`

d. `Count = 13     Control = Go`

7. Which of the following can determine the length of a new variable?

   a. the length of the variable's first reference in the DATA step

   b. the assignment statement

   c. the LENGTH statement

   d. all of the above

8. Which set of statements is equivalent to the code shown below?

```
if code='1' then Type='Fixed';
if code='2' then Type='Variable';
if code^='1' and code^='2' then Type='Unknown';
```

   a.
```
if code='1' then Type='Fixed';
   else if code='2' then Type='Variable';
   else Type='Unknown';
```

   b.
```
 if code='1' then Type='Fixed';
   if code='2' then Type='Variable';
   else Type='Unknown';
```

   c.
```
if code='1' then type='Fixed';
   else code='2' and type='Variable';
   else type='Unknown';
```

   d.
```
if code='1' and type='Fixed';
   then code='2' and type='Variable';
   else type='Unknown';
```

9. What is the length of the variable Type, as created in the DATA step below?

```
data finance.newloan;
   set finance.records;
   TotLoan+payment;
   if code='1' then Type='Fixed';
   else Type='Variable';
   length type $ 10;
run;
```

   a. 5

   b. 8

   c. 10

    d. it depends on the first value of Type

10. Which program contains an error?

    a.
```
data clinic.stress(drop=timemin timesec);
    infile tests;
    input ID $ 1-4 Name $ 6-25 RestHR 27-29 MaxHR 31-33
          RecHR 35-37 TimeMin 39-40 TimeSec 42-43
          Tolerance $ 45;
    TotalTime=(timemin*60)+timesec;
    SumSec+totaltime;
run;
```

    b.
```
proc print data=clinic.stress;
    label totaltime='Total Duration of Test';
    format timemin 5.2;
    drop sumsec;
run;
```

    c.
```
proc print data=clinic.stress(keep=totaltime timemin);
    label totaltime='Total Duration of Test';
    format timemin 5.2;
run;
```

    d.
```
data clinic.stress;
    infile tests;
    input ID $ 1-4 Name $ 6-25 RestHR 27-29 MaxHR 31-33
          RecHR 35-37 TimeMin 39-40 TimeSec 42-43
          Tolerance $ 45;
    TotalTime=(timemin*60)+timesec;
    keep id totaltime tolerance;
run;
```

Chapter 11
Reading SAS Data Sets

Overview

Introduction

You've learned about creating a SAS data set from raw data. However, you might often want to create a new data set from an existing SAS data set. To create the new data set, you can read a data set using the DATA step. As you read the data set, you can use all the programming features of the DATA step to manipulate your data.

Figure 11.1 *Data Set Diagram*

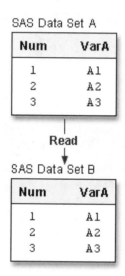

This chapter shows you how to use the DATA step to read an existing SAS data set. When you create your new data set, you can choose variables, select observations based on one or more conditions, and assign values conditionally. You can also assign variable attributes such as formats and labels.

You can also merge, concatenate, or interleave two or more data sets. For details, see Chapter 5, "Creating SAS Data Sets from External Files," on page 151 and "Combining SAS Data Sets" on page 360.

Objectives

In this chapter, you learn to

- create a new data set from an existing data set

- use BY groups to process observations

- read observations by observation number

- stop processing when necessary

- explicitly write observations to an output data set

- detect the last observation in a data set

- identify differences in DATA step processing for raw data and SAS data sets.

Reading a Single Data Set

The data set **sasuser.admit** contains health information about patients in a clinic, their activity level, height and weight. Suppose you want to create a small data set containing all the men in the group who are older than fifty.

To create the data set, you must first reference the library in which **admit** is stored and then the library in which you want to store the **males** data set. Then you write a DATA step to read your data and create a new data set.

General form, basic DATA step for reading a single data set:

DATA *SAS-data-set*;

      SET *SAS-data-set*;

    <more SAS statements>

RUN;

where

- *SAS-data-set* in the DATA statement is the name (**libref.filename**) of the SAS data set to be created

- *SAS-data-set* in the SET statement is the name (**libref.filename**) of the SAS data set to be read.

After you write a DATA step to name the SAS data set to be created, you specify the data set that will be read in the SET statement. The DATA step below reads all observations and variables from the existing data set **admit** into the new data set **males**. The DATA statement creates the permanent SAS data set **males**, which is stored in the SAS library **Men50** . The SET statement reads the permanent SAS data set **admit** and subsets the data using a WHERE statement. The new data set, **males**, contains all males in **sasuser.admit** who are older than 50.

```
libname sasuser "C:\Users\name\sasuser\";
libname Men50 "C:\Users\name\sasuser\Men50";
data Men50.males;
set sasuser.admit;


       where sex='M' and age>50;
run;
```

When you submit this DATA step, the following messages appear in the log, confirming that the new data set was created:

Figure 11.2 *SAS Log Output*

```
134  data Men50.males;
135    set sasuser.admit;
136    where sex="M" and age>50;
137  run;

NOTE: There were 3 observations read from the data set SASUSER.ADMIT.
      WHERE (sex='M') and (age>50);
NOTE: The data set MEN50.MALES has 3 observations and 9 variables.
NOTE: DATA statement used (Total process time):
      real time           0.00 seconds
      cpu time            0.00 seconds
```

You can add a PROC PRINT statement to this same example to see the output of `Men50.males`.

```
proc print data=Men50.males;
   title "Men Over 50";
run;
```

Figure 11.3 *PROC PRINT Output For Data Set males*

Men Over 50

Obs	ID	Name	Sex	Age	Date	Height	Weight	ActLevel	Fee
1	2539	LaMance, K	M	51	4	71	158	LOW	124.80
2	2579	Underwood, K	M	60	22	71	191	LOW	149.75
3	2595	Warren, C	M	54	7	71	183	MOD	149.75

Manipulating Data

Overview

In the previous example, you read in the data set **admit** and used the WHERE statement in the DATA step to subset the data. For any data set you read in, you can use any of the programming features of the DATA step to manipulate your data.

For example, you can use any of the statements and data set options that you learned in previous chapters.

Table 11.1 *Manipulating Data Using the DATA Step*

To do this...	Use this type of statement...
Subset data	`if resthr<70 then delete;` `if tolerance='D';`
Drop unwanted variables	`drop timemin timesec;`
Create or modify a variable	`TotalTime=(timemin*60)+timesec;`
Initialize and retain a variable Accumulate values	`retain SumSec 5400;` `sumsec+totaltime;`
Specify a variable's length	`length TestLength $ 6;`

Execute statements conditionally	`if totaltime>800 then TestLength='Long';`
	`else if 750<=totaltime<=800`
	`        then TestLength='Normal';`
	`else if totaltime<750`
	`        then TestLength='Short';`

To do this...	Use this type of statement...
Label a variableFormat a variable	`label sumsec='Cumulative Total Seconds';` `format sumsec comma6.;`

Example

The following DATA step reads the data set lab23.drug1h, selects observations and variables, and creates new variables.

```
data lab23.drug1h(drop=placebo uric);
   set research.cltrials(drop=triglyc);
      if sex='M' then delete;
      if placebo='YES';
      retain TestDate='22MAY1999'd;
      retain Days 30;
      days+1;
      length Retest $ 5;
      if cholesterol>190 then retest='YES';
        else if 150<=cholesterol<=190 then retest='CHECK';
        else if cholesterol<150 then retest='NO';
   label retest='Perform Cholesterol Test 2?';
   format enddate mmddyy10.;
run;
```

Where to Specify the DROP= and KEEP= Data Set Options

You've learned that you can specify the DROP= and KEEP= data set options anywhere you name a SAS data set. You can specify DROP= and KEEP= in either the DATA statement or the SET statement, depending on whether you want to drop variables onto output or input:

- If you never reference certain variables and you don't want them to appear in the new data set, use a DROP= option in the SET statement.

  In the DATA step shown below, the DROP= or KEEP= option in the SET statement prevents the variables Triglycerides and UricAcid from being read. These variables won't appear in the Lab23.Drug1H data set.

  ```
  data lab23.drug1h(drop=placebo);
     set research.cltrials(drop=triglycerides uricacid);
     if placebo='YES';
  run;
  ```

- If you do need to reference a variable in the original data set (in a subsetting IF statement, for example), you can specify the variable in the DROP= or KEEP= option in the DATA statement. Otherwise, the statement that references the variable uses a missing value for that variable.

  This DATA step uses the variable Placebo to select observations. To drop Placebo from the new data set, the DROP= option must appear in the DATA statement.

  ```
  data lab23.drug1h(drop=placebo);
     set research.cltrials(drop=triglycerides uricacid);
     if placebo='YES';
  run;
  ```

When used in the DATA statement, the DROP= option simply drops the variables from the new data set. However, they are still read from the original data set and are available within the DATA step.

Using BY-Group Processing

Finding the First and Last Observations in a Group

"Creating List Reports" on page 112 explained how to use a BY statement in PROC SORT to sort observations and in PROC PRINT to group observations for subtotals. You can also use the BY statement in the DATA step to group observations for processing.

```
data temp;
   set salary;
   by dept;
run;
```

When you use the BY statement with the SET statement,

- the data sets that are listed in the SET statement must be sorted by the values of the BY variable(s), or they must have an appropriate index.

- the DATA step creates two temporary variables for each BY variable. One is named FIRST.variable, where variable is the name of the BY variable, and the other is named LAST.variable. Their values are either 1 or 0. FIRST.variable and LAST.variable. identify the first and last observation in each BY group.

Table 11.2 *Finding the First and Last Observations in a Group*

This variable . . .	Equals . . .
FIRST.variable	1 for the first observation in a BY group
	0 for any other observation in a BY group
LAST.variable	1 for the last observation in a BY group
	0 for any other observation in a BY group

Example

To work with FIRST.variable and LAST.variable, let's look at a different set of data. The Company.USA data set contains payroll information for individual employees. Suppose you want to compute the annual payroll by department. Assume 2,000 work hours per year for hourly employees.

Before computing the annual payroll, you need to group observations by values of the variable Dept.

Figure 11.4 *Partial Listing of Data Set*

SAS Data Set Company.USA
(Partial Listing)

Dept	WageCat	WageRate
ADM20	S	3392.50
ADM30	S	5093.75
CAM10	S	1813.30
CAM10	S	1572.50
CAM10	H	13.48
ADM30	S	2192.25

The following program computes the annual payroll by department. Notice that the variable name Dept has been appended to FIRST. and LAST.

```
proc sort data=company.usa out=work.temp;
   by dept;
run;
data company.budget(keep=dept payroll);
   set work.temp;
   by dept;
   if wagecat='S' then Yearly=wagerate*12;
   else if wagecat='H' then Yearly=wagerate*2000;
   if first.dept then Payroll=0;
   payroll+yearly;
   if last.dept;
run;
```

If you could look behind the scenes at the program data vector (PDV) as the Company.Budget data set is created, you would see the following. Notice the values for FIRST.Dept and LAST.Dept.

Figure 11.5 *Program Data Vector*

Selected PDV Variables

_N_	Dept	Payroll	FIRST.Dept	LAST.Dept
1	ADM10	70929.0	1	0
2	ADM10	119479.2	0	0
3	ADM10	173245.2	0	0
4	ADM10	255516.0	0	0
5	ADM10	293472.0	0	1
1	ADM20	40710.0	1	0
2	ADM20	68010.0	0	0
3	ADM20	94980.0	0	0
4	ADM20	136020.0	0	0
5	ADM20	177330.0	0	1
1	ADM30	61125.0	1	0

When you print the new data set, you can now list and sum the annual payroll by department.

```
proc print data=company.budget noobs;
   sum payroll;
   format payroll dollar12.2;
run;
```

Figure 11.6 *Payroll Sum*

Dept	Payroll
ADM10	$293,472.00
ADM20	$177,330.00
ADM30	$173,388.00
CAM10	$130,709.60
CAM20	$156,731.20
	$931,630.80

Finding the First and Last Observations in Subgroups

When you specify multiple BY variables,

- FIRST.variable for each variable is set to 1 at the first occurrence of a new value for the primary variable

- a change in the value of a primary BY variable forces LAST.variable to equal *1* for the secondary BY variables.

Example

Suppose you now want to compute the annual payroll by job type for each manager. In your program, you specify two BY variables, Manager and JobType.

```
proc sort data=company.usa out=work.temp2;
   by manager jobtype;
   data company.budget2(keep=manager jobtype payroll);
   set work.temp2;
   by manager jobtype;
   if wagecat='S' then Yearly=wagerate*12;
   else if wagecat='H' then Yearly=wagerate*2000;
   if first.jobtype then Payroll=0;
   payroll+yearly;
   if last.jobtype;
run;
```

If you could look at the PDV now, you would see the following. Notice that the values for FIRST.JobType and LAST.JobType change according to values of FIRST.Manager and LAST.Manager.

Figure 11.7 PDV

Selected PDV Variables

_N_	Manager	JobType	Payroll	FIRST. Manager	LAST. Manager	FIRST. JobType	LAST. Jobtype
1	Coxe	3	40710.0	1	0	1	1
2	Coxe	50	41040.0	0	0	1	0
3	Coxe	50	82350.0	0	0	0	1
4	Coxe	240	27300.0	0	0	1	0
5	Coxe	240	54270.0	0	1	0	1
6	Delgado	240	35520.0	1	0	1	0
7	Delgado	240	63120.0	0	0	0	1
8	Delgado	420	18870.0	0	0	1	0
9	Delgado	420	45830.0	0	0	0	1
10	Delgado	440	21759.6	0	1	1	1
11	Overby	1	82270.8	1	0	1	1
12	Overby	5	48550.2	0	0	1	1
13	Overby	10	53766.0	0	0	1	1
14	Overby	20	70929.0	0	0	1	0
15	Overby	20	108885.0	0	1	0	1

Now you can sum the annual payroll by job type for each manager. Here, the payroll for only two managers (Coxe and Delgado) is listed.

```
proc print data=company.budget2 noobs;;
   by manager;
   var jobtype;
   sum payroll;
   where manager in ('Coxe','Delgado');
   format payroll dollar12.2;
run;
```

Figure 11.8 Payroll Sum by Job Type and Manager

Manager=Coxe

JobType	Payroll
3	$40,710.00
50	$123,390.00
240	$81,570.00
Manager	$245,670.00

Manager=Delgado

JobType	Payroll
240	$98,640.00
420	$64,700.00
440	$21,759.00
Manager	$185,099.00

Reading Observations Using Direct Access

So far we have read the observations in an input data set sequentially. That is, we have accessed observations in the order in which they appear in the physical file. However, you can also access observations directly, by going straight to an observation in a SAS data set without having to process each observation that precedes it.

To access observations directly by their observation number, you use the POINT= option in the SET statement.

General form, POINT= option:

POINT=_variable_;

where _variable_

- specifies a temporary numeric variable that contains the observation number of the observation to read

- must be given a value before the SET statement is executed.

Example

Let's suppose you want to read only the fifth observation from a data set. In the following DATA step, the value 5 is assigned to the variable ObsNum. The POINT= option reads the value of ObsNum to determine which observation to read from the data set Company.USA.

```
data work.getobs5;
   obsnum=5;
   set company.usa(keep=manager payroll) point=obsnum;
run;
```

But let's see what would happen if you submitted this program.

As you learned in a previous chapter, the DATA step continues to read observations until it reaches the end-of-file marker in the input data. However, because the POINT= option reads only specified observations, SAS cannot read an end-of-file indicator as it would if the file were being read sequentially. So submitting the following program would cause continuous looping.

```
data work.getobs5;
   obsnum=5;
   set company.usa(keep=manager payroll) point=obsnum;
run;
```

Preventing Continuous Looping with POINT=

Because there is no end-of-file condition when you use direct access to read data, you must use a STOP statement to prevent continuous looping. The STOP statement causes SAS to stop processing the current DATA step immediately and to resume processing statements after the end of the current DATA step.

General form, STOP statement:

STOP;

So, if you add a STOP statement, your program no longer loops continuously.

```
data work.getobs5;
   obsnum=5;
   set company.usa(keep=manager payroll) point=obsnum;
   stop;
run;
```

But it doesn't write any observations to output, either! Remember from "Understanding DATA Step Processing" on page 200 that the DATA step writes observations to output at the end of the DATA step. However, in this program, the STOP statement immediately stops processing before the end of the DATA step.

Let's see how you can write the observation to output before processing stops.

Writing Observations Explicitly

To override the default way in which the DATA step writes observations to output, you can use an OUTPUT statement in the DATA step. Placing an explicit OUTPUT statement in a DATA step overrides the automatic output, so that observations are added to a data set only when the explicit OUTPUT statement is executed.

General form, OUTPUT statement:

OUTPUT *<SAS-data-set(s)>*;

where *SAS-data-set(s)* names the data set(s) to which the observation is written. All data set names that are specified in the OUTPUT statement *must* also appear in the DATA statement.

Using an OUTPUT statement without a following data set name causes the current observation to be written to all data sets that are named in the DATA statement.

With an OUTPUT statement, your program now writes a single observation to output—observation 5.

```
data work.getobs5;
   obsnum=5;
   set company.usa(keep=manager payroll) point=obsnum;
   output;
   stop;
run;
proc print data=work.getobs5 noobs;
run;
```

Figure 11.9 *Single Observation*

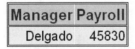

Manager	Payroll
Delgado	45830

Suppose your DATA statement contains two data set names, and you include an OUTPUT statement that references only one of the data sets. The DATA step will create both data sets, but only the data set that is named in the OUTPUT statement will contain output. For example, the program below creates two temporary data sets, Empty and Full. The result of this DATA step is that the data set Empty is created but contains no observations, and the data set Full contains all of the observations from Company.Usa.

```
data empty full;
   set company.usa;
   output full;
run;
```

More Complex Ways of Using Direct Access

To convey concepts clearly, the examples in this section have been as simple as possible. However, most uses of the POINT= option are more complex. For example, POINT= is commonly used to efficiently select a random sample from a data set.

You can see more complex examples of using POINT= in "Generating Data with DO Loops" on page 463 .

Detecting the End of a Data Set

Overview

Instead of reading specific observations, you might want to determine when the last observation in an input data set has been read, so that you can perform specific processing. For example, you might want to output only one observation that contains grand totals for numeric variables.

To create a temporary numeric variable whose value is used to detect the last observation, you can use the END= option in the SET statement.

General form, END= option:

END=*variable*

where *variable* creates and names a temporary variable that contains an end-of-file marker. The variable, which is initialized to 0, is set to 1 when the SET statement reads the last observation of the data set.

This variable is *not* added to the data set.

⚠ Do not specify END= with POINT=. POINT= reads only a specific observation, so the last observation in the data set is not encountered.

Example

Suppose you want to sum the number of seconds for treadmill stress tests. If you submit the following program, you produce a new data set that contains cumulative totals for each of the values of TotalTime.

```
data work.addtoend(drop=timemin timesec);
   set sasuser.stress2(keep=timemin timesec);
   TotalMin+timemin;
   TotalSec+timesec;
   TotalTime=totalmin*60+totalsec;
run;
proc print data=work.addtoend noobs;
run;
```

Figure 11.10 *Data Set with Cumulative Totals for Each of the Values of TotalTime*

TotalMin	TotalSec	TotalTime
12	38	758
22	43	1363
33	56	2036
42	98	2618
53	144	3324
65	170	4070
80	211	5011
93	224	5804
103	246	6426
119	295	7435
134	297	8337
146	308	9068
160	317	9917
171	343	10603
185	370	11470
198	389	12269
214	396	13236
231	431	14291
246	472	15232
260	529	16129
272	539	16859

But what if you want only the *final* total (the last observation) in the new data set? The following program uses the END= variable last to select only the last observation of the data set. You specify END= in the SET statement and last wherever you need it in processing (here, in the subsetting IF statement).

```
data work.addtoend(drop=timemin timesec);
   set sasuser.stress2(keep=timemin timesec) end=last;
   TotalMin+timemin;
   TotalSec+timesec;
   TotalTime=totalmin*60+totalsec;
   if last;
run;
proc print data=work.addtoend noobs;
run;
```

Now the new data set has one observation:

Figure 11.11 Data Set with One Observation

TotalMin	TotalSec	TotalTime
272	539	16859

Understanding How Data Sets Are Read

In a previous chapter, you learned about the compilation and execution phases of the DATA step as they pertain to reading raw data. DATA step processing for reading existing SAS data sets is very similar. The main difference is that while reading an existing data set with the SET statement, SAS retains the values of existing variables from one observation to the next.

Let's briefly look at the compilation and execution phases of DATA steps that use a SET statement. In this example, the DATA step reads the data set Finance.Loans, creates the variable Interest, and creates the new data set Finance.DueJan.

```
data finance.duejan;
   set finance.loans;
   Interest=amount*(rate/12);
run;
```

Figure 11.12 New Data Set Finance.DueJan

SAS Data Set Finance.Loans

Account	Amount	Rate	Months	Payment
101-1092	22000	0.100	60	467.43
101-1731	114000	0.095	360	958.57
101-1289	10000	0.105	36	325.02
101-3144	3500	0.105	12	308.52

Compilation Phase

1. The program data vector is created and contains the automatic variables _N_ and _ERROR_.

   Figure 11.13 program data vector with Automatic Variables

   Program Data Vector

_N_	_ERROR_	
1	0	

2. SAS also scans each statement in the DATA step, looking for syntax errors.

3. When the SET statement is compiled, a slot is added to the program data vector for each variable in the input data set. The input data set supplies the variable names, as well as attributes such as type and length.

Program Data Vector

_N_	_ERROR_	Account	Amount	Rate	Months	Payment

4. Any variables that are created in the DATA step are also added to the program data vector. The attributes of each of these variables are determined by the expression in the statement.

Program Data Vector

_N_	_ERROR_	Account	Amount	Rate	Months	Payment	Interest

5. At the bottom of the DATA step, the compilation phase is complete, and the descriptor portion of the new SAS data set is created. There are no observations because the DATA step has not yet executed.

When the compilation phase is complete, the execution phase begins.

Execution Phase

1. The DATA step executes once for each observation in the input data set. For example, this DATA step will execute four times because there are four observations in the input data set Finance.Loans.

2. At the beginning of the execution phase, the value of _N_ is **1**. Because there are no data errors, the value of _ERROR_ is **0**. The remaining variables are initialized to missing. Missing numeric values are represented by a period, and missing character values are represented by a blank.

```
data finance.duejan;
   set finance.loans;
   Interest=amount*(rate/12);
run;
```

SAS Data Set Finance.Loans

Account	Amount	Rate	Months	Payment
101-1092	22000	0.1000	60	467.43
101-1731	114000	0.0950	360	958.57
101-1289	10000	0.1050	36	325.02
101-3144	3500	0.1050	12	308.52

Program Data Vector

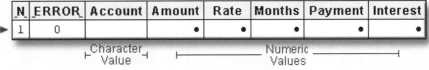

SAS Data Set Finance.Duejan

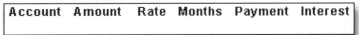

3. The SET statement reads the first observation from the input data set into the program data vector.

```
data finance.duejan;
   set finance.loans;
   Interest=amount*(rate/12);
run;
```

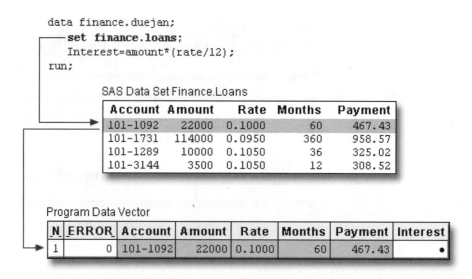

SAS Data Set Finance.Loans

Account	Amount	Rate	Months	Payment
101-1092	22000	0.1000	60	467.43
101-1731	114000	0.0950	360	958.57
101-1289	10000	0.1050	36	325.02
101-3144	3500	0.1050	12	308.52

Program Data Vector

N	ERROR	Account	Amount	Rate	Months	Payment	Interest
1	0	101-1092	22000	0.1000	60	467.43	•

SAS Data Set Finance.Duejan

Account	Amount	Rate	Months	Payment	Interest

4. Then, the assignment statement executes to compute the value for Interest.

```
data finance.duejan;
   set finance.loans;
   Interest=amount*(rate/12);
run;
```

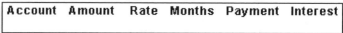

SAS Data Set Finance.Loans

Account	Amount	Rate	Months	Payment
101-1092	22000	0.1000	60	467.43
101-1731	114000	0.0950	360	958.57
101-1289	10000	0.1050	36	325.02
101-3144	3500	0.1050	12	308.52

Program Data Vector

N	ERROR	Account	Amount	Rate	Months	Payment	Interest
1	0	101-1092	22000	0.1000	60	467.43	183.333

SAS Data Set Finance.Duejan

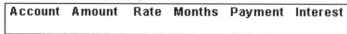

Account	Amount	Rate	Months	Payment	Interest

5. At the end of the first iteration of the DATA step, the values in the program data vector are written to the new data set as the first observation.

```
data finance.duejan;
   set finance.loans;
   Interest=amount*(rate/12);
run;
```

SAS Data Set Finance.Loans

Account	Amount	Rate	Months	Payment
101-1092	22000	0.1000	60	467.43
101-1731	114000	0.0950	360	958.57
101-1289	10000	0.1050	36	325.02
101-3144	3500	0.1050	12	308.52

Program Data Vector

N	ERROR	Account	Amount	Rate	Months	Payment	Interest
1	0	101-1092	22000	0.1000	60	467.43	183.333

SAS Data Set Finance.Duejan

Account	Amount	Rate	Months	Payment	Interest
101-1092	22000	0.1000	60	467.43	183.333

6. The value of _N_ increments from **1** to **2**, and control returns to the top of the DATA step. Remember, the automatic variable _N_ keeps track of how many times the DATA step has begun to execute.

```
data finance.duejan;
   set finance.loans;
   Interest=amount*(rate/12);
run;
```

SAS Data Set Finance.Loans

Account	Amount	Rate	Months	Payment
101-1092	22000	0.1000	60	467.43
101-1731	114000	0.0950	360	958.57
101-1289	10000	0.1050	36	325.02
101-3144	3500	0.1050	12	308.52

Program Data Vector

N	ERROR	Account	Amount	Rate	Months	Payment	Interest
2	0	101-1092	22000	0.1000	60	467.43	183.333

SAS Data Set Finance.Duejan

Account	Amount	Rate	Months	Payment	Interest
101-1092	22000	0.1000	60	467.43	183.333

7. SAS retains the values of variables that were read from a SAS data set with the SET statement, or that were created by a sum statement. All other variable values, such as the values of the variable Interest, are set to missing.

```
data finance.duejan;
    set finance.loans;
    Interest=amount*(rate/12);
run;
```

SAS Data Set Finance.Loans

Account	Amount	Rate	Months	Payment
101-1092	22000	0.1000	60	467.43
101-1731	114000	0.0950	360	958.57
101-1289	10000	0.1050	36	325.02
101-3144	3500	0.1050	12	308.52

Program Data Vector

N	ERROR	Account	Amount	Rate	Months	Payment	Interest
2	0	101-1092	22000	0.1000	60	467.43	•

Values Retained ————————— Reset to Missing

SAS Data Set Finance.Duejan

Account	Amount	Rate	Months	Payment	Interest
101-1092	22000	0.1000	60	467.43	183.333

When SAS reads raw data, the situation is different. In that case, SAS sets the value of each variable in the DATA step to missing at the beginning of each iteration, with these exceptions:

- variables named in a RETAIN statement
- variables created in a sum statement
- data elements in a _TEMPORARY_ array
- any variables created by using options in the FILE or INFILE statements
- automatic variables.

8. At the beginning of the second iteration, the value of _ERROR_ is reset to 0.

```
data finance.duejan;
   set finance.loans;
   Interest=amount*(rate/12);
run;
```

SAS Data Set Finance.Loans

Account	Amount	Rate	Months	Payment
101-1092	22000	0.1000	60	467.43
101-1731	114000	0.0950	360	958.57
101-1289	10000	0.1050	36	325.02
101-3144	3500	0.1050	12	308.52

Program Data Vector

N	ERROR	Account	Amount	Rate	Months	Payment	Interest
2	0	101-1092	22000	0.1000	60	467.43	•

SAS Data Set Finance.Duejan

Account	Amount	Rate	Months	Payment	Interest
101-1092	22000	0.1000	60	467.43	183.333

9. As the SET statement executes, the values from the second observation are read into the program data vector.

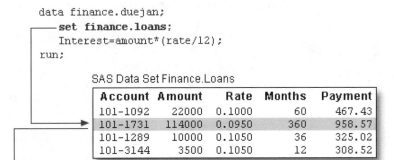

```
data finance.duejan;
   set finance.loans;
   Interest=amount*(rate/12);
run;
```

SAS Data Set Finance.Loans

Account	Amount	Rate	Months	Payment
101-1092	22000	0.1000	60	467.43
101-1731	114000	0.0950	360	958.57
101-1289	10000	0.1050	36	325.02
101-3144	3500	0.1050	12	308.52

Program Data Vector

N	ERROR	Account	Amount	Rate	Months	Payment	Interest
2	0	101-1731	114000	0.0950	360	958.57	•

SAS Data Set Finance.Duejan

Account	Amount	Rate	Months	Payment	Interest
101-1092	22000	0.1000	60	467.43	183.333

10. The assignment statement executes again to compute the value for Interest for the second observation.

```
data finance.duejan;
   set finance.loans;
   Interest=amount*(rate/12);
run;
```

SAS Data Set Finance.Loans

Account	Amount	Rate	Months	Payment
101-1092	22000	0.1000	60	467.43
101-1731	114000	0.0950	360	958.57
101-1289	10000	0.1050	36	325.02
101-3144	3500	0.1050	12	308.52

Program Data Vector

_N_	_ERROR_	Account	Amount	Rate	Months	Payment	Interest
2	0	101-1731	114000	0.0950	360	958.57	902.5

SAS Data Set Finance.Duejan

Account	Amount	Rate	Months	Payment	Interest
101-1092	22000	0.1000	60	467.43	183.333

11. At the bottom of the DATA step, the values in the program data vector are written to the data set as the second observation.

```
data finance.duejan;
   set finance.loans;
   Interest=amount*(rate/12);
run;
```

SAS Data Set Finance.Loans

Account	Amount	Rate	Months	Payment
101-1092	22000	0.1000	60	467.43
101-1731	114000	0.0950	360	958.57
101-1289	10000	0.1050	36	325.02
101-3144	3500	0.1050	12	308.52

Program Data Vector

_N_	_ERROR_	Account	Amount	Rate	Months	Payment	Interest
2	0	101-1731	114000	0.0950	360	958.57	902.5

SAS Data Set Finance.Duejan

Account	Amount	Rate	Months	Payment	Interest
101-1092	22000	0.1000	60	467.43	183.333
101-1731	114000	0.0950	360	958.57	902.5

12. The value of _N_ increments from 2 to 3, and control returns to the top of the DATA step. SAS retains the values of variables that were read from a SAS data set with the SET statement, or that were created by a sum statement. All other variable values, such as the values of the variable Interest, are set to missing.

```
data finance.duejan;
   set finance.loans;
   Interest=amount*(rate/12);
run;
```

SAS Data Set Finance.Loans

Account	Amount	Rate	Months	Payment
101-1092	22000	0.1000	60	467.43
101-1731	114000	0.0950	360	958.57
101-1289	10000	0.1050	36	325.02
101-3144	3500	0.1050	12	308.52

Program Data Vector

N	ERROR_	Account	Amount	Rate	Months	Payment	Interest
3	. 0	101-1731	114000	0.0950	360	958.57	.

Values Retained — Reset to Missing

SAS Data Set Finance.Duejan

Account	Amount	Rate	Months	Payment	Interest
101-1092	22000	0.1000	60	467.43	183.333
101-1731	114000	0.0950	360	958.57	902.5

This process continues until all of the observations are read.

Additional Features

The DATA step provides many other programming features for manipulating data sets. For example, you can

- use IF-THEN/ELSE logic with DO groups and DO loops to control processing based on one or more conditions
- specify additional data set options
- process variables in arrays
- use SAS functions.

You can also combine SAS data sets in several ways, including match merging, interleaving, one-to-one merging, and updating.

Chapter Summary

Text Summary

Setting Up

Before you can create a new data set, you must assign a libref to the SAS library that will store the data set.

Reading a Single Data Set

After you have referenced the library in which your data set is stored, you can write a DATA step to name the SAS data set to be created. You then specify the data set to be read in the SET statement.

Selecting Variables

You can select the variables that you want to drop or keep by using the DROP= or KEEP= data set options in parentheses after a SAS data set name. For convenience, use DROP= if more variables are kept than dropped.

BY-Group Processing

Use the BY statement in the DATA step to group observations for processing. When you use the BY statement with the SET statement, the DATA step automatically creates two temporary variables, FIRST. and LAST. When you specify multiple BY variables, a change in the value of a primary BY variable forces LAST.variable to equal **1** for the secondary BY variables.

Reading Observations Using Direct Access

In addition to reading input data sequentially, you can access observations directly by using the POINT= option to go directly to a specified observation. There is no end-of-file condition when you use direct access, so include an explicit OUTPUT statement and then the STOP statement to prevent continuous looping.

Detecting the End of a Data Set

To determine when the last observation in an input data set has been read, use the END= option in the SET statement. The specified variable is initialized to 0, then set to 1 when the SET statement reads the last observation of the data set.

Syntax

LIBNAME *libref 'SAS-data-library'*;

DATA *SAS-data-set*;

    SET *SAS-data-set* (**KEEP=** *variable-1 <...variable-n>*) | (**DROP=** *variable-1 <...variable-n>*)

    POINT=*variable* | **END=***variable*;

    OUTPUT *<SAS-data-set>*;

    STOP;

RUN;

Sample Program

```
proc sort data=company.usa out=work.temp2;
   by manager jobtype;
run;

data company.budget2(keep=manager jobtype payroll);
   set work.temp2;
   by manager jobtype;
   if wagecat='S' then Yearly=wagerate*12;
   else if wagecat='H' then Yearly=wagerate*2000;
   if first.jobtype then Payroll=0;
```

```
       payroll+yearly;
       if last.jobtype;
run;

data work.getobs5;
   obsnum=5;
   set company.usa(keep=manager payroll) point=obsnum;
   output;
   stop;
run;

data work.addtoend(drop=timemin timesec);
   set sasuser.stress2(keep=timemin timesec) end=last;
   TotalMin+timemin;
   TotalSec+timesec;
   TotalTime=totalmin*60+totalsec;
   if last;
run;
```

Points to Remember

- When you perform BY-group processing, the data sets listed in the SET statement must either be sorted by the values of the BY variable(s), or they must have an appropriate index.

- When using direct access to read data, you *must* prevent continuous looping. Add a STOP statement to the DATA step.

- Do not specify the END= option with the POINT= option in a SET statement.

Chapter Quiz

Select the best answer for each question. After completing the quiz, you can check your answers using the answer key in the appendix.

1. If you submit the following program, which variables appear in the new data set?

   ```
   data work.cardiac(drop=age group);
      set clinic.fitness(keep=age weight group);
      if group=2 and age>40;
   run;
   ```

   a. none

   b. Weight

   c. Age, Group

   d. Age, Weight, Group

2. Which of the following programs correctly reads the data set Orders and creates the data set FastOrdr?

   a. ```
       data catalog.fastordr(drop=ordrtime);
          set july.orders(keep=product units price);
          if ordrtime<4;
       ```
       ```

```
        Total=units*price;
    run;
```

b.
```
    data catalog.orders(drop=ordrtime);
        set july.fastordr(keep=product units price);
        if ordrtime<4;
        Total=units*price;
    run;
```

c.
```
    data catalog.fastordr(drop=ordrtime);
        set july.orders(keep=product units price
                            ordrtime);
        if ordrtime<4;
        Total=units*price;
    run;
```

d. none of the above

3. Which of the following statements is *false* about BY-group processing?

   When you use the BY statement with the SET statement:

   a. The data sets listed in the SET statement must be indexed or sorted by the values of the BY variable(s).

   b. The DATA step automatically creates two variables, FIRST. and LAST., for each variable in the BY statement.

   c. FIRST. and LAST. identify the first and last observation in each BY group, respectively.

   d. FIRST. and LAST. are stored in the data set.

4. There are 500 observations in the data set Usa. What is the result of submitting the following program?

```
data work.getobs5;
    obsnum=5;
    set company.usa(keep=manager payroll) point=obsnum;
    stop;
run;
```

   a. an error

   b. an empty data set

   c. continuous loop

   d. a data set that contains one observation

5. There is no end-of-file condition when you use direct access to read data, so how can your program prevent a continuous loop?

   a. Do not use a POINT= variable.

   b. Check for an invalid value of the POINT= variable.

   c. Do not use an END= variable.

   d. Include an OUTPUT statement.

6. Assuming that the data set Company.USA has five or more observations, what is the result of submitting the following program?

```
data work.getobs5;
    obsnum=5;
    set company.usa(keep=manager payroll) point=obsnum;
```

```
        output;
        stop;
    run;
```

    a. an error

    b. an empty data set

    c. a continuous loop

    d. a data set that contains one observation

7. Which of the following statements is *true* regarding direct access of data sets?

    a. You cannot specify END= with POINT=.

    b. You cannot specify OUTPUT with POINT=.

    c. You cannot specify STOP with END=.

    d. You cannot specify FIRST. with LAST.

8. What is the result of submitting the following program?

```
data work.addtoend;
    set clinic.stress2 end=last;
    if last;
run;
```

    a. an error

    b. an empty data set

    c. a continuous loop

    d. a data set that contains one observation

9. At the start of DATA step processing, during the compilation phase, variables are created in the program data vector (PDV), and observations are set to:

    a. blank

    b. missing

    c. 0

    d. there are no observations.

10. The DATA step executes:

    a. continuously if you use the POINT= option and the STOP statement.

    b. once for each variable in the output data set.

    c. once for each observation in the input data set.

    d. until it encounters an OUTPUT statement.

Chapter 12
Combining SAS Data Sets

Overview

Introduction

In SAS programming, a common task is to combine observations from two or more data sets into a new data set. Using the DATA step, you can combine data sets in several ways, including the following:

Method of Merging (Combining)	Illustration
One-to-one merging Creates observations that contain all of the variables from each contributing data set. Combines observations based on their relative position in each data set. Statement: SET	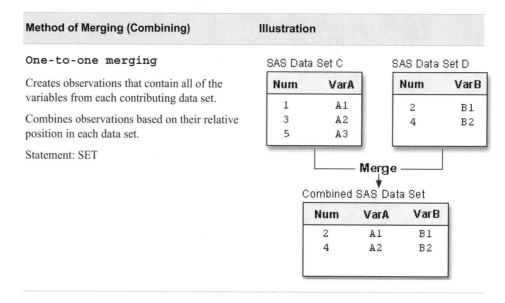

Method of Merging (Combining)	Illustration

Concatenating

Appends the observations from one data set to another.

Statement: SET

SAS Data Set A

Num	VarA
1	A1
2	A2
3	A3

SAS Data Set C

Num	VarB
1	B1
2	B2
4	B3

└─Concatenate─┘
↓

Combined SAS Data Set

Num	VarA	VarB
1	A1	
2	A2	
3	A3	
1		B1
2		B2
4		B3

Appending

Appending adds the observations in the second data set directly to the end of the original data set.

Procedure: APPEND

BASE= data set A

Num	Sex
1	M
2	F
3	F
4	M

DATA= data set B

Num	Sex
3	F
4	F
5	F
6	M

 Append
↓

BASE= data set A

Num	Sex
1	M
2	F
3	F
4	M
3	F
4	F
5	F
6	M

with data set B observations appended

Interleaving

Intersperses observations from two or more data sets, based on one or more common variables.

Statements: SET, BY

SAS Data Set C

Num	Var
1	C1
2	C2
2	C3
3	C4

SAS Data Set D

Num	Var
2	D1
3	D2
3	D3

└─Interleave─┘
↓

Combined SAS Data Set

Num	Var
1	C1
2	C2
2	C3
2	D1
3	C4
3	D2
3	D3

Method of Merging (Combining)	Illustration
Match-merging Matches observations from two or more data sets into a single observation in a new data set according to the values of a common variable. Statements: MERGE, BY	

SAS Data Set A

Num	VarA
1	A1
2	A2
3	A3

SAS Data Set B

Num	VarB
1	B1
2	B2
4	B3

—Match-Merge—

Combined SAS Data Set

Num	VarA	VarB
1	A1	B1
2	A2	B2
3	A3	
4		B3

You can also use PROC SQL to join data sets according to common values. PROC SQL enables you to perform many other types of data set joins. See the *SQL Processing with SAS* e-learning course for additional training.

This chapter shows you how to combine SAS data sets using one-to-one combing or merging, concatenating, appending, interleaving, and match-merging. When you use the DATA step to combine data sets, you have a high degree of control in creating and manipulating data sets.

Objectives

In this chapter, you learn to

- perform one-to-one merging (combining) of data sets
- concatenate data sets
- append data sets
- interleave data sets
- match-merge data sets
- rename any like-named variables to avoid overwriting values
- select only matched observations, if desired
- predict the results of match-merging.

One-to-One Merging

Overview

In "Reading SAS Data Sets" on page 334, you learned how to use the SET statement to read an existing SAS data set. You can also use multiple SET statements in a DATA step to combine data sets. This is called one-to-one merging (or combining). In one-to-one merging you can read different data sets, or you can read the same data set more than once, as if you were reading from separate data sets.

General form, basic DATA step for one-to-one reading:

DATA *output-SAS-data-set*;

    SET *SAS-data-set-1*;

    SET *SAS-data-set-2*;

RUN;

where

- *output-SAS-data-set* names the data set to be created

- *SAS-data-set-1* and *SAS-data-set-2* specify the data sets to be read.

How One-to-One Merging Selects Data

When you perform one-to-one merging,

- the new data set contains all the variables from all the input data sets. If the data sets contain variables that have the same names, the values that are read from the last data set overwrite the values that were read from earlier data sets.

- the number of observations in the new data set is the number of observations in the smallest original data set. Observations are combined based on their relative position in each data set. That is, the first observation in one data set is joined with the first observation in the other, and so on. The DATA step stops after it has read the last observation from the smallest data set.

```
data one2one;
   set a;
   setb;
run;
```

Figure 12.1   *One-to-One Merging*

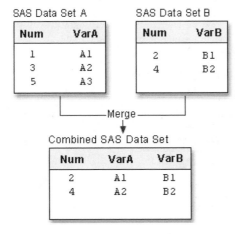

How One-to-One Reading Works

Let's look at a simple case of one-to-one merging.

```
data one2one;
   set a;
```

```
    set b;
run;
```

Figure 12.2 *How One-to-One Reading Works*

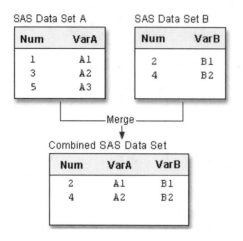

1. The first SET statement reads the first observation from data set A into the program data vector.

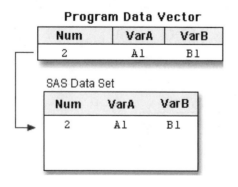

2. The second SET statement reads the first observation from data set B into the program data vector and SAS writes the contents of the program data vector to the new data set. The value for Num from data set B overwrites the value for Num from data set A.

3. The first SET statement reads the second observation from data set A into the program data vector.

Program Data Vector

Num	VarA	VarB
3	A2	

4. The second SET statement reads the second observation from data set B, and SAS writes the contents of the program data vector to the new data set. The value for Num from data set B overwrites the value for Num from data set A.

Program Data Vector

Num	VarA	VarB
4	A2	B2

SAS Data Set

Num	VarA	VarB
2	A1	B1
4	A2	B2

5. The first SET statement reads the third observation from data set A into the program data vector.

Program Data Vector

Num	VarA	VarB
5	A3	

6. The second SET statement reads the end of file in data set B, which stops the DATA step processing with no further output written to the data set. The last observation in data set A is read into the program data vector, but it is not written to the output data set.

Figure 12.3   *Final Combined SAS Data Set*

Num	VarA	VarB
2	A1	B1
4	A2	B2

Here is an example of how you might use one-to-one merging.

Example: One-to-One Merging

Suppose you have basic patient data (ID, sex, and age) in the data set Clinic.Patients and want to combine it with other patient data (height and weight) for patients under age 60. The height and weight data is stored in the data set Clinic.Measure. Both data sets are sorted by the variable ID.

Notice that Clinic.Patients contains 7 observations in which the patient age is less than 60, and Clinic.Measure contains 6 observations.

Figure 12.4   *Example: One-to-One Merging*

SAS Data Set Clinic.Patients

Obs	ID	Sex	Age
1	1129	F	48
2	1387	F	57
3	2304	F	16
4	2486	F	63
5	4759	F	60
6	5438	F	42
7	6488	F	59
8	9012	F	39
9	9125	F	56

SAS Data Set Clinic.Measure

Obs	ID	Height	Weight
1	1129	61	137
2	1387	64	142
3	2304	61	102
4	5438	62	168
5	6488	64	154
6	9012	63	157

To subset observations from the first data set and combine them with observations from the second data set, you can submit the following program:

```
data clinic.one2one;
   set clinic.patients;
   if age<60;
   set clinic.measure;
run;
```

The resulting data set, Clinic.One2one, contains 6 observations (the number of observations read from the smallest data set, which is Clinic.Measure). The last observation in Clinic.Patients is not written to the data set because the second SET statement reaches an end-of-file, which stops the DATA step processing.

Figure 12.5   *The Resulting Data Set for One-to-One Reading Example*

SAS Data Set Clinic.One2one

Obs	ID	Sex	Age	Height	Weight
1	1129	F	48	61	137
2	1387	F	57	64	142
3	2304	F	16	61	102
4	5438	F	42	62	168
5	6488	F	59	64	154
6	9012	F	39	63	157

Concatenating

Overview

Another way to combine SAS data sets with the SET statement is concatenating, which appends the observations from one data set to another data set. To concatenate SAS data sets, you specify a list of data set names in the SET statement.

General form, basic DATA step for concatenating:

DATA *output-SAS-data-set*;

      SET *SAS-data-set-1 SAS-data-set-2*;

RUN;

where

- *output-SAS-data-set* names the data set to be created
- *SAS-data-set-1* and *SAS-data-set-2* specify the data sets to be read.

How Concatenating Selects Data

When a program concatenates data sets, all of the observations are read from the first data set listed in the SET statement. Then all of the observations are read from the second data set listed, and so on, until all of the listed data sets have been read. The new data set contains all of the variables and observations from all of the input data sets.

```
data concat;
   set a c;
run;
```

Figure 12.6 *How Concatenating Selects*

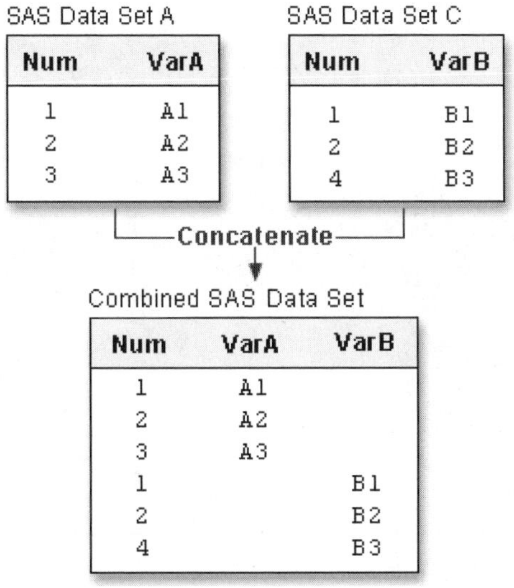

Notice that A and C contain a common variable named Num:

- Both instances of Num (or any common variable) must have the same type attribute, or SAS stops processing the DATA step and issues an error message stating that the variables are incompatible.

- However, if the length attribute is different, SAS takes the length from the first data set that contains the variable. In this case, the length of Num in A determines the length of Num in Concat.

- The same is true for the label, format, and informat attributes: If any of these attributes are different, SAS takes the attribute from the first data set that contains the variable with that attribute.

Example

The following DATA step creates Clinic.Concat by concatenating Clinic.Therapy1999 and Clinic.Therapy2000.

```
data clinic.concat;
   set clinic.therapy1999 clinic.therapy2000;
run;
```

Below is the listing of Clinic.Concat. The first 12 observations were read from Clinic.Therapy1999, and the last 12 observations were read from Clinic.Therapy2000.

Figure 12.7 *Example: Concatenating*

Obs	Month	Year	AerClass	WalkJogRun	Swim
1	01	1999	26	78	14
2	02	1999	32	109	19
3	03	1999	15	106	22
4	04	1999	47	115	24
5	05	1999	95	121	31
6	06	1999	61	114	67
7	07	1999	67	102	72
8	08	1999	24	76	77
9	09	1999	78	77	54
10	10	1999	81	62	47
11	11	1999	84	31	52
12	12	1999	92	44	55
13	01	2000	37	91	83
14	02	2000	41	102	27
15	03	2000	52	98	19
16	04	2000	61	118	22
17	05	2000	49	88	29
18	06	2000	24	101	54
19	07	2000	45	91	69
20	08	2000	63	65	53
21	09	2000	60	49	68
22	10	2000	78	70	41
23	11	2000	82	44	58
24	12	2000	93	57	47

Appending

Overview

Another way to combine SAS data sets is to append one data set to another using the APPEND procedure. Although appending and concatenating are similar, there are some important differences between the two methods. Whereas the DATA step creates an entirely new data set when concatenating, PROC APPEND simply adds the observations of one data set to the end of a "master" (or BASE) data set. SAS does not create a new data set nor does it read the base data set when executing the APPEND procedure.

To append SAS data sets, you specify a BASE= data set, which is the data set to which observations are added and then specify a DATA= data set, which is the data set containing the observations that are added to the base data set. The data set specified with DATA= is the only one of the two data sets that SAS actually reads.

General form of the APPEND procedure:

PROC APPEND *BASE=SAS-data-set*

       DATA=SAS-data-set;

RUN;

where

- *BASE=* names the data set to which observations are added

- *DATA=* names the data set containing observations that are added to the base data set

For example, the following PROC APPEND statement appends the observations in data set B to the end of data set A:

```
proc append base=A
            data=B;
run;
```

Figure 12.8 *How PROC APPEND Appends Data*

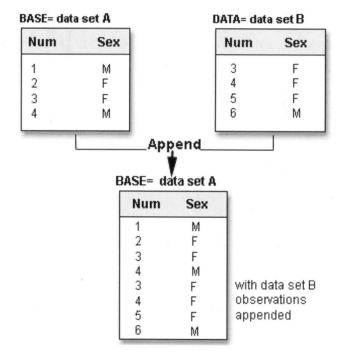

Requirements for the APPEND Procedure

The requirements for appending one data set to another are as follows:

- Only two data sets can be used at a time in one step.

- The observations in the base data set are not read.

- The variable information in the descriptor portion of the base data set cannot change.

Notice that the final data set is the original data set with appended observations and that no new data set was created.

Example

The following PROC APPEND statement appends the data set totals2011 to the base data set totals2005. The two data sets are like-structured data sets: that is, both data sets have the same variable information.

```
proc append base=totals2005 data=totals2011;
run;
```

Below is the listing of totals2005. The first 4 observations already existed in totals2005, and the last 4 observations were read from totals2011 and added to totals2005.

Figure 12.9   *Example: PROC APPEND*

Obs	Month	Therapy	NewAdmit	Treadmill
1	01	220	27	11
2	02	209	63	43
3	03	189	11	2
4	04	211	52	36
5	01	210	27	13
6	02	219	63	23
7	03	199	11	21
8	04	251	52	30

Using the FORCE Option with Unlike-Structured Data Sets

In order to use PROC APPEND with data sets that have unmatching variable definitions, you can use the FORCE option in the PROC APPEND statement.

General form of the APPEND procedure with the FORCE option:

PROC APPEND *BASE=SAS-data-set*

   DATA=SAS-data-set FORCE;

RUN;

The FORCE option is needed when the DATA= data set contains variables that meet any one of the following criteria:

- They are not in the BASE= data set.

- They are variables of a different type (for example, character or numeric).

- They are longer than the variables in the BASE= data set.

If the length of a variable is longer in the DATA= data set than in the BASE= data set, SAS truncates values from the DATA= data set to fit them into the length that is specified in the BASE= data set.

If the type of a variable in the DATA= data set is different than in the BASE= data set, SAS replaces all values for the variable in the DATA= data set with missing values and keeps the variable type of the variable specified in the BASE= data set.

If the BASE= data set contains a variable that is not in the DATA= data set, the observations are appended, but the observations from the DATA= data set have a missing value for the variable that was not present in the DATA= data set. If the DATA= data set contains a variable that is not in the BASE= data set, the variable is dropped from the output.

Example

For example:

```
proc append  base=sasuser.patients
             data=clinic.append force;
run;
```

Figure 12.10 *Example: PROC APPEND with FORCE Option*

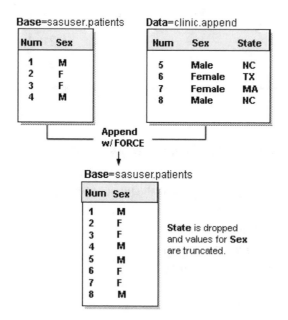

Interleaving

Overview

If you use a BY statement when you concatenate data sets, the result is interleaving. Interleaving intersperses observations from two or more data sets, based on one or more common variables.

To interleave SAS data sets, specify a list of data set names in the SET statement, and specify one or more BY variables in the BY statement.

General form, basic DATA step for interleaving:

DATA *output-SAS-data-set*;

       SET *SAS-data-set-1 SAS-data-set-2*;

       BY *variable(s)*;

RUN;

where

- *output-SAS-data-set* names the data set to be created

- *SAS-data-set-1* and *SAS-data-set-2* specify the data sets to be read.

- *variable(s)* specifies one or more variables that are used to interleave observations.

You can specify any number of data sets in the SET statement. Each input data set *must* be sorted or indexed in ascending order based on the BY variable(s).

How Interleaving Selects Data

When SAS interleaves data sets, observations in each BY group in each data set in the SET statement are read sequentially, in the order in which the data sets and BY variables are listed, until all observations have been processed. The new data set includes all the variables from all the input data sets, and it contains the total number of observations from all input data sets.

```
data interlv;
    set c d;
    by num;
run;
```

Figure 12.11 *How Interleaving Selects Data*

Example

The following DATA step creates Clinic.Interlv by interleaving Clinic.Therapy1999 and Clinic.Therapy2000.

```
data clinic.interlv;
   set clinic.therapy1999 clinic.therapy2000;
   by month;
run;
```

Below is the listing of Clinic.Interlv. Notice that, unlike the previous example, observations are interleaved by month instead of being concatenated.

Figure 12.12 *Example: How Interleaving Selects Data*

SAS Data Set Clinic.Interlv

Obs	Month	Year	AerClass	WalkJogRun	Swim
1	01	1999	26	78	14
2	01	2000	37	91	83
3	02	1999	32	109	19
4	02	2000	41	102	27
5	03	1999	15	106	22
6	03	2000	52	98	19
7	04	1999	47	115	24
8	04	2000	61	118	22
9	05	1999	95	121	31
10	05	2000	49	88	29
11	06	1999	61	114	67
12	06	2000	24	101	54
13	07	1999	67	102	72
14	07	2000	45	91	69
15	08	1999	24	76	77
16	08	2000	63	65	53
17	09	1999	78	77	54
18	09	2000	60	49	68
19	10	1999	81	62	47
20	10	2000	78	70	41
21	11	1999	84	31	52
22	11	2000	82	44	58
23	12	1999	92	44	55
24	12	2000	93	57	47

Match-Merging

Overview

So far in this chapter, we've combined data sets based on the order of the observations in the input data sets. But sometimes you need to combine observations from two or more data sets into a single observation in a new data set according to the values of a common variable. This is called match-merging.

When you match-merge, you use a MERGE statement rather than a SET statement to combine data sets.

General form, basic DATA step for match-merging:

DATA *output-SAS-data-set*;

      MERGE *SAS-data-set-1 SAS-data-set-2*;

      BY <DESCENDING> *variable(s)*;

RUN;

where

- *output-SAS-data-set* names the data set to be created.

- *SAS-data-set-1* and *SAS-data-set-2* specify the data sets to be read.

- *variable(s)* in the **BY** statement specifies one or more variables whose values are used to match observations.

- DESCENDING indicates that the input data sets are sorted in descending order (largest to smallest numerically, or reverse alphabetical for character variables) by the variable that is specified. If you have more that one variable in the BY statement, DESCENDING applies only to the variable that immediately follows it.

Each input data set in the MERGE statement *must* be sorted in order of the values of the BY variable(s), or it must have an appropriate index. Each BY variable must have the same type in all data sets to be merged.

You cannot use the DESCENDING option with indexed data sets because indexes are always stored in ascending order.

How Match-Merging Selects Data

Generally speaking, during match-merging, SAS sequentially checks each observation of each data set to see whether the BY values match, and then writes the combined observation to the new data set.

```
data merged;
   merge a b;
   by num;
run;
```

Figure 12.13 *How Match-Merging Selects Data*

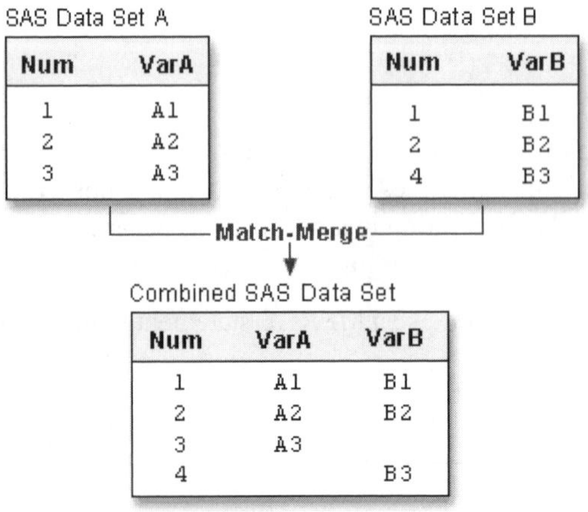

Basic DATA step match-merging produces an output data set that contains values from all observations in all input data sets. You can add statements and options to select only matching observations.

If an input data set doesn't have any observations for a particular value of the by-variable, then the observation in the output data set contains missing values for the variables that are unique to that input data set.

Figure 12.14 *Match-Merging with Missing Data*

Table 1		+	Table 2		=	All		
Year	Var_X		Year	Var_Y		Year	Var_X	Var_Y
1991	X1		1991	Y1		1991	X1	Y1
1992	X2		1991	Y2		1991	X1	Y2
1993	X3		1993	Y3		1992	X2	.
1994	X4		1994	Y4		1993	X3	Y3
1995	X5		1995	Y5		1994	X4	Y4
						1995	X5	Y5

In match-merging, often one data set contains unique values for the BY-variable and other data sets contain multiple values for the BY-variable.

Example

Suppose you have sorted the data sets Clinic.Demog and Clinic.Visit as follows:

```
proc sort data=clinic.demog;
   by id;
run;
proc print data=clinic.demog;
run;
```

Figure 12.15 *Example: Sorting clinic.demog*

Obs	ID	Age	Sex	Date
1	A001	21	m	05/22/75
2	A002	32	m	06/15/63
3	A003	24	f	08/17/72
4	A004	.		03/27/69
5	A005	44	f	02/24/52
6	A007	39	m	11/11/57

```
proc sort data=clinic.visit;
   by id;
run;
proc print data=clinic.visit;
run;
```

Figure 12.16 *Example: Sorting clinic.visit*

Obs	ID	Visit	SysBP	DiasBP	Weight	Date
1	A001	1	140	85	195	11/05/98
2	A001	2	138	90	198	10/13/98
3	A001	3	145	95	200	07/04/98
4	A002	1	121	75	168	04/14/98
5	A003	1	118	68	125	08/12/98
6	A003	2	112	65	123	08/21/98
7	A004	1	143	86	204	03/30/98
8	A005	1	132	76	174	02/27/98
9	A005	2	132	78	175	07/11/98
10	A005	3	134	78	176	04/16/98
11	A008	1	126	80	182	05/22/98

You can then submit this DATA step to create Clinic.Merged by merging Clinic.Demog and Clinic.Visit according to values of the variable ID.

```
data clinic.merged;
   merge clinic.demog clinic.visit;
   by id;
run;
proc print data=clinic.merged;
run;
```

Notice that all observations, including unmatched observations and observations that have missing data, are written to the output data set.

Figure 12.17 *Match-merging Output*

Obs	ID	Age	Sex	Date	Visit	SysBP	DiasBP	Weight
1	A001	21	m	11/05/98	1	140	85	195
2	A001	21	m	10/13/98	2	138	90	198
3	A001	21	m	07/04/98	3	145	95	200
4	A002	32	m	04/14/98	1	121	75	168
5	A003	24	f	08/12/98	1	118	68	125
6	A003	24	f	08/21/98	2	112	65	123
7	A004	.		03/30/98	1	143	86	204
8	A005	44	f	02/27/98	1	132	76	174
9	A005	44	f	07/11/98	2	132	78	175
10	A005	44	f	04/16/98	3	134	78	176
11	A007	39	m	11/11/57		.	.	.
12	A008	.		05/22/98	1	126	80	182

Example: Merge in Descending Order

The example above illustrates merging two data sets that are sorted in ascending order of the BY variable ID. To sort the data sets in descending order and then merge them, you can submit the following program.

```
proc sort data=clinic.demog;
   by descending id;
run;
proc sort data=clinic.visit;
   by descending id;
run;
data clinic.merged;
   merge clinic.demog clinic.visit;
   by descending id;
run;
proc print data=clinic.merged;
run;
```

Notice that you specify the DESCENDING option in the BY statements in both the PROC SORT steps and the DATA step. If you omit the DESCENDING option in the DATA step, you generate error messages about improperly sorted BY variables.

Now the data sets are merged in descending order of the BY variable ID.

Figure 12.18   *Example: Merge in Descending Order*

Obs	ID	Age	Sex	Date	Visit	SysBP	DiasBP	Weight
1	A008	.		05/22/98	1	126	80	182
2	A007	39	m	11/11/57		.	.	.
3	A005	44	f	02/27/98	1	132	76	174
4	A005	44	f	07/11/98	2	132	78	175
5	A005	44	f	04/16/98	3	134	78	176
6	A004	.		03/30/98	1	143	86	204
7	A003	24	f	08/12/98	1	118	68	125
8	A003	24	f	08/21/98	2	112	65	123
9	A002	32	m	04/14/98	1	121	75	168
10	A001	21	m	11/05/98	1	140	85	195
11	A001	21	m	10/13/98	2	138	90	198
12	A001	21	m	07/04/98	3	145	95	200

Match-Merge Processing

Overview

The match-merging examples in this chapter are straightforward. However, match-merging can be more complex, depending on your data and on the output data set that you want to create. To predict the results of match-merges correctly, you need to understand how the DATA step performs match-merges.

When you submit a DATA step, it is processed in two phases:

- the compilation phase, in which SAS checks the syntax of the SAS statements and compiles them (translates them into machine code). During this phase, SAS also sets up descriptor information for the output data set and creates the program data vector (PDV), an area of memory where SAS holds one observation at a time.

Figure 12.19   *Match-Merge Processing: Compilation Phase*

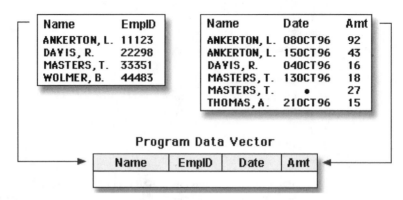

- the execution phase, in which the DATA step reads data and executes any subsequent programming statements. When the DATA step executes, data values are read into the appropriate variables in the program data vector. From here, the variables are written to the output data set as a single observation.

Figure 12.20   *Match-Merge Processing: Execution Phase*

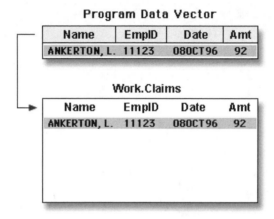

The following pages cover DATA step processing in greater detail. In those pages, you learn

- how the DATA step sets up the new output data set

- what happens when variables in different data sets have the same name

- how the DATA step matches observations in input data sets

- what happens when observations don't match

- how missing values are handled.

The Compilation Phase: Setting Up the New Data Set

To prepare to merge data sets, SAS

- reads the descriptor portions of the data sets that are listed in the MERGE statement

- reads the rest of the DATA step program

- creates the program data vector (PDV) for the merged data set

- assigns a tracking pointer to each data set that is listed in the MERGE statement.

If variables that have the same name appear in more than one data set, the variable from the first data set that contains the variable (in the order listed in the MERGE statement) determines the length of the variable.

Figure 12.21 *The Compilation Phase: Setting Up the New Data Set*

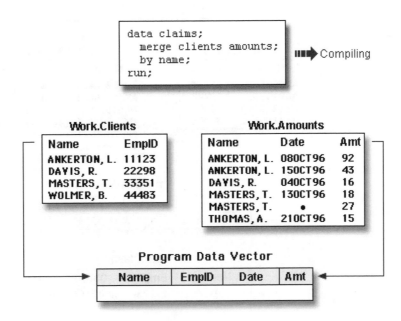

The illustration above shows match-merging during the compilation phase. After reading the descriptor portions of the data sets Clients and Amounts, SAS

1. creates a program data vector for the new Claims data set. The program data vector contains all variables from the two data sets. Note that although Name appears in both input data sets, it appears in the program data vector only once.

2. assigns tracking pointers to Clients and Amounts.

The Execution Phase: Match-Merging Observations

After compiling the DATA step, SAS sequentially match-merges observations by moving the pointers down each observation of each data set and checking to see whether the BY values match.

- If *Yes*, the observations are read into the PDV in the order in which the data sets appear in the MERGE statement. Values of any same-named variable are overwritten by values of the same-named variable in subsequent data sets. SAS writes the combined observation to the new data set and retains the values in the PDV until the BY value changes in all the data sets.

Figure 12.22 *The Execution Phase: Match-Merging Observations*

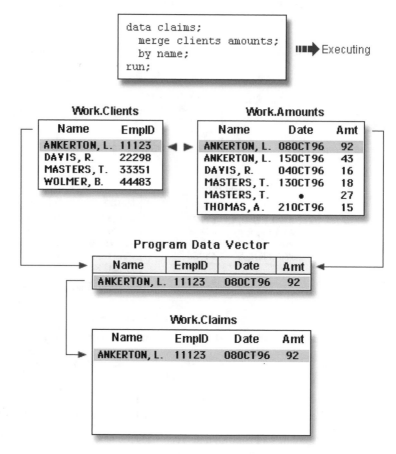

- If *No*, SAS determines which BY value comes first and reads the observation that contains this value into the PDV. Then the contents of the PDV are written

Figure 12.23   *The Execution Phase: PDV*

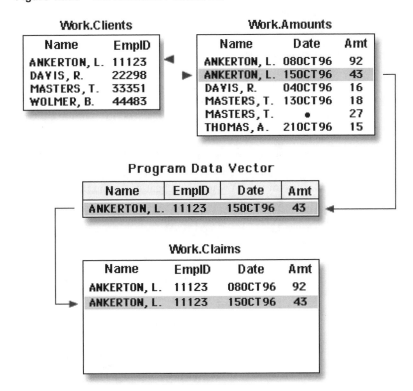

When the BY value changes in all the input data sets, the PDV is initialized to missing.

Figure 12.24   *Initializing the PDV to Missing*

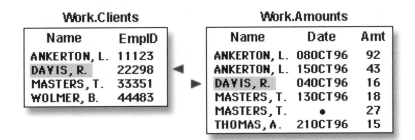

The DATA step merge continues to process every observation in each data set until it has processed all observations in all data sets.

Handling Unmatched Observations and Missing Values

By default, all observations that are read into the PDV, including observations that have missing data and no matching BY values, are written to the output data set. (If you specify a subsetting IF statement to select observations, then only those that meet the IF condition are written.)

- If an observation contains missing values for a variable, then the observation in the output data set contains the missing values as well. Observations that have missing values for the BY variable appear at the top of the output data set because missing values sort first in ascending order.

Figure 12.25 *Handling Unmatched Observations and Missing Values*

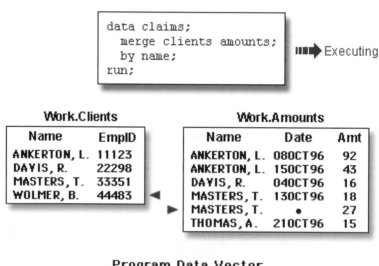

- If an input data set doesn't have a matching BY value, then the observation in the output data set contains missing values for the variables that are unique to that input data set.

Figure 12.26 *Handling Unmatched Observations and Missing Values: Unique Missing Values*

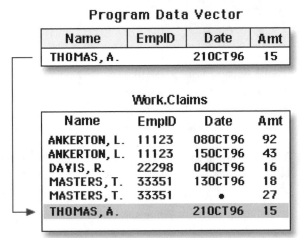

The last observation for Wolmer would be added after the Thomas observation.

Renaming Variables

Sometimes you might have same-named variables in more than one input data set. In this case, DATA step match-merging overwrites values of the like-named variable in the first data set in which it appears with values of the like-named variable in subsequent data sets.

For example, Clinic.Demog contains the variable Date (date of birth), and Clinic.Visit also contains Date (date of the clinic visit in 1998). The DATA step below overwrites the date of birth with the date of the clinic visit.

```
data clinic.merged;
   merge clinic.demog clinic.visit;
   by id;
run;
proc print data=clinic.merged;
run;
```

The following output shows the effects of overwriting the values of a variable in the Clinic.Merged data set. In most observations, the date is now the date of the clinic visit. In observation 11, the date is still the birth date because Clinic.Visit did not contain a matching ID value and did not contribute to the observation.

Figure 12.27 *Renaming Variables*

Obs	ID	Age	Sex	Date	Visit	SysBP	DiasBP	Weight
1	A001	21	m	11/05/98	1	140	85	195
2	A001	21	m	10/13/98	2	138	90	198
3	A001	21	m	07/04/98	3	145	95	200
4	A002	32	m	04/14/98	1	121	75	168
5	A003	24	f	08/12/98	1	118	68	125
6	A003	24	f	08/21/98	2	112	65	123
7	A004	.		03/30/98	1	143	86	204
8	A005	44	f	02/27/98	1	132	76	174
9	A005	44	f	07/11/98	2	132	78	175
10	A005	44	f	04/16/98	3	134	78	176
11	A007	39	m	11/11/57		.	.	.
12	A008	.		05/22/98	1	126	80	182

You now have a data set with values for Date that mean two different things: date of birth and date of clinic visit. Let's see how to prevent this problem.

To prevent overwriting, you can rename variables by using the RENAME= data set option in the MERGE statement.

General form, RENAME= data set option:

(RENAME=(old-variable-name=new-variable-name**))**

where

- the **RENAME=** option, in parentheses, follows the name of each data set that contains one or more variables to be renamed
- *old-variable-name* specifies the variable to be renamed
- *new-variable-name* specifies the new name for the variable.

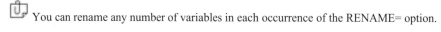 You can rename any number of variables in each occurrence of the RENAME= option.

You can also use RENAME= to rename variables in the SET statement or in the output data set that is specified in the DATA statement.

In the following example, the RENAME= option renames the variable Date in Clinic.Demog to BirthDate, and it renames the variable Date in Clinic.Visit to VisitDate.

```
data clinic.merged;
   merge clinic.demog (rename=(date=BirthDate))
         clinic.visit (rename=(date=VisitDate));
   by id;
run;
proc print data=clinic.merged;
run;
```

The following output shows the effect of the RENAME= option.

Figure 12.28   *Output for RENAME= Option*

Obs	ID	Age	Sex	BirthDate	Visit	SysBP	DiasBP	Weight	VisitDate
1	A001	21	m	05/22/75	1	140	85	195	11/05/98
2	A001	21	m	05/22/75	2	138	90	198	10/13/98
3	A001	21	m	05/22/75	3	145	95	200	07/04/98
4	A002	32	m	06/15/63	1	121	75	168	04/14/98
5	A003	24	f	08/17/72	1	118	68	125	08/12/98
6	A003	24	f	08/17/72	2	112	65	123	08/21/98
7	A004	.		03/27/69	1	143	86	204	03/30/98
8	A005	44	f	02/24/52	1	132	76	174	02/27/98
9	A005	44	f	02/24/52	2	132	78	175	07/11/98
10	A005	44	f	02/24/52	3	134	78	176	04/16/98
11	A007	39	m	11/11/57		.	.	.	.
12	A008	.		.	1	126	80	182	05/22/98

Excluding Unmatched Observations

Overview

By default, DATA step match-merging combines all observations in all input data sets. However, you may want to select only observations that match for two or more specific input data sets.

Figure 12.29   *Excluding Unmatched Observations*

Work.Clients

Name	EmpID
ANKERTON, L.	11123
DAVIS, R.	22298
MASTERS, T.	33351
WOLMER, B.	44483

Work.Amounts

Name	Date	Amt
ANKERTON, L.	08OCT96	92
ANKERTON, L.	15OCT96	43
DAVIS, R.	04OCT96	16
MASTERS, T.	13OCT96	18
MASTERS, T.	•	27
THOMAS, A.	21OCT96	15

To exclude unmatched observations from your output data set, you can use the IN= data set option and the subsetting IF statement in your DATA step. In this case, you use

- the IN= data set option to create and name a variable that indicates whether the data set contributed data to the current observation

- the subsetting IF statement to check the IN= values and write to the merged data set only matching observations.

Creating Temporary IN= Variables

Suppose you want to match-merge the data sets Clinic.Demog and Clinic.Visit and select only observations that appear in both data sets.

First, you use IN= to create two temporary variables, indemog and invisit. The IN= variable is a temporary variable that is available to program statements during the DATA step, but it is not included in the SAS data set that is being created.

General form, IN= data set option:

(IN= *variable***)**

where

- the **IN=** option, in parentheses, follows the data set name
- variable names the variable to be created.

Within the DATA step, the value of the variable is **1** if the data set contributed data to the current observation. Otherwise, its value is **0**.

The DATA step that contains the IN= options appears below. The first IN= creates the temporary variable indemog, which is set to **1** when an observation from Clinic.Demog contributes to the current observation. Otherwise, it is set to **0**. Likewise, the value of invisit depends on whether Clinic.Visit contributes to an observation or not.

```
data clinic.merged;
    merge clinic.demog (in=indemog)
          clinic.visit (in=invisit
                        rename=(date=BirthDate));
    by id;
run;
```

When you specify multiple data set options for a given data set, enclose them in a single set of parentheses.

Selecting Matching Observations

Next, to select only observations that appear in both Clinic.Demog and Clinic.Visit, you specify a subsetting IF statement in the DATA step.

In the DATA step below, the subsetting IF statement checks the values of indemog and invisit and continues processing only those observations that meet the condition of the expression. Here the condition is that both Clinic.Demog and Clinic.Visit contribute to the observation. If the condition is met, the new observation is written to Clinic.Merged. Otherwise, the observation is deleted.

```
data clinic.merged;
    merge clinic.demog (in=indemog
                        rename=(date=BirthDate))
          clinic.visit (in=invisit
                        rename=(date=VisitDate));
    by id;
    if indemog=1 and invisit=1;
run;
proc print data=clinic.merged;
run;
```

In previous examples, Clinic.Merged contained 12 observations. In the output below, notice that only 10 observations met the condition in the IF expression.

Figure 12.30 *Selecting Matching Observations*

Obs	ID	Age	Sex	BirthDate	Visit	SysBP	DiasBP	Weight	VisitDate
1	A001	21	m	05/22/75	1	140	85	195	11/05/98
2	A001	21	m	05/22/75	2	138	90	198	10/13/98
3	A001	21	m	05/22/75	3	145	95	200	07/04/98
4	A002	32	m	06/15/63	1	121	75	168	04/14/98
5	A003	24	f	08/17/72	1	118	68	125	08/12/98
6	A003	24	f	08/17/72	2	112	65	123	08/21/98
7	A004	.		03/27/69	1	143	86	204	03/30/98
8	A005	44	f	02/24/52	1	132	76	174	02/27/98
9	A005	44	f	02/24/52	2	132	78	175	07/11/98
10	A005	44	f	02/24/52	3	134	78	176	04/16/98

SAS evaluates the expression within an IF statement to produce a result that is either nonzero, zero, or missing. A nonzero and nonmissing result causes the expression to be true; a zero or missing result causes the expression to be false.

Thus, you can specify the subsetting IF statement from the previous example in either of the following ways. The first IF statement checks specifically for a value of **1**. The second IF statement checks for a value that is neither missing nor **0** (which for IN= variables is always **1**).

```
if indemog=1 and invisit=1;
```

```
if indemog and invisit;
```

Selecting Variables

Overview

As with reading raw data or reading SAS data sets, you can specify the variables you want to drop or keep by using the DROP= and KEEP= data set options.

For example, the DATA step below reads all variables from Clinic.Demog and all variables except Weight from Clinic.Visit. It then excludes the variable ID from Clinic.Merged after the merge processing is complete.

```
data clinic.merged (drop=id);
   merge clinic.demog(in=indemog
                  rename=(date=BirthDate))
         clinic.visit(drop=weight in=invisit
                  rename=(date=VisitDate));
   by id;
   if indemog and invisit;
run;
proc print data=clinic.merged;
run;
```

Figure 12.31 *Selecting Variables*

Obs	Age	Sex	BirthDate	Visit	SysBP	DiasBP	VisitDate
1	21	m	05/22/75	1	140	85	11/05/98
2	21	m	05/22/75	2	138	90	10/13/98
3	21	m	05/22/75	3	145	95	07/04/98
4	32	m	06/15/63	1	121	75	04/14/98
5	24	f	08/17/72	1	118	68	08/12/98
6	24	f	08/17/72	2	112	65	08/21/98
7	.		03/27/69	1	143	86	03/30/98
8	44	f	02/24/52	1	132	76	02/27/98
9	44	f	02/24/52	2	132	78	07/11/98
10	44	f	02/24/52	3	134	78	04/16/98

Where to Specify DROP= and KEEP=

As you've seen in previous chapters, you can specify the DROP= and KEEP= options wherever you specify a SAS data set. When match-merging, you can specify these options in either the DATA statement or the MERGE statement, depending on whether you want to reference the variables in that DATA step:

- If you *don't* reference certain variables and you don't want them to appear in the new data set, specify them in the DROP= option in the MERGE statement.

```
merge clinic.demog(in=indemog
                   rename=(date=BirthDate))
      clinic.visit(drop=weight in=invisit
                   rename=(date=VisitDate));
```

- If you *do* need to reference a variable in the original data set (in a subsetting IF statement, for example), then you must specify the variable in the DROP= option in the DATA statement. Otherwise, you may get unexpected results and your variable will be uninitialized.

```
data clinic.merged (drop=id);
```

When used in the DATA statement, the DROP= option simply drops the variables from the new data set. However, the variables are still read from the original data set and are available for processing within the DATA step.

Additional Features

The DATA step provides a large number of other programming features for manipulating data when you combine data sets. For example, you can

- use IF-THEN/ELSE logic to control processing based on one or more conditions

- specify additional data set options

- perform calculations

- create new variables

- process variables in arrays

- use SAS functions

- use special variables such as FIRST. and LAST. to control processing.

You can also combine SAS data sets in other ways:

- You can perform one-to-one merging, which creates a data set that contains all of the variables and observations from each contributing data set. Observations are combined based on their relative position in each data set.

  One-to-one merging is the same as one-to-one reading, with two exceptions:

  - You use the MERGE statement instead of multiple SET statements.

  - The DATA step reads all observations from all data sets.

  ```
  data work.onemerge;
     merge clinic.demog clinic.visit;
  run;
  ```

- You can perform a conditional merge, using DO loops or other conditional statements:

  ```
  data work.combine;
     set sales.pounds;
     do while(not(begin le date le last));
        set sales.rate;
     end;
     Dollars=(sales*1000)*rate;
  run;
  ```

  You can learn about DO loops in "Generating Data with DO Loops" on page 463 .

- You can read the same data set in more than one SET statement:

  ```
  data work.combine(drop=totpay);
     if _n_=1 then do until(last);
        set sales.budget(keep=payroll) end=last;
        totpay+payroll;
        end;
     set sales.budget;
     Percent=payroll/totpay;
  run;
  ```

Chapter Summary

Text Summary

One-to-One Merging

You can combine data sets with one-to-one merging (combining) by including multiple SET statements in a DATA step. When you perform one-to-one merging, the new data set contains all the variables from all the input data sets. If the data sets contain same-named variables, the values that are read in from the last data set replace those that were read in from earlier ones. The number of observations in the new data set is the number of observations in the smallest original data set.

```
data one2one;
   set a;
   set b;
run;
```

Concatenating

To append the observations from one data set to another data set, you concatenate them by specifying the data set names in the SET statement. When SAS concatenates, data sets in the SET statement are read sequentially, in the order in which they are listed. The new data set contains all the variables and the total number of observations from all input data sets.

```
data concat;
   set a b;
run;
```

Appending

Another way to combine SAS data sets is to append one data set to another using the APPEND procedure. Although appending and concatenating are similar, there are some important differences between the two methods. The DATA step creates a new data set when concatenating. PROC APPEND adds the observations of one data set to the end of a "master" (or BASE) data set. SAS does not create a new data set nor does it read the base data set when executing the APPEND procedure.

Interleaving

If you use a BY statement when you concatenate data sets, the result is interleaving. Interleaving intersperses observations from two or more data sets, based on one or more common variables. Each input data set must be sorted or indexed based on the BY variable(s). Observations in each BY group in each data set in the SET statement are read sequentially, in the order in which the data sets and BY variables are listed, until all observations have been processed. The new data set contains all the variables and the total number of observations from all input data sets.

```
data interlv;
   set a b;
   by num;
run;
```

Match-Merging

Sometimes you need to combine observations from two or more data sets into a single observation in a new data set according to the values of a BY variable. This is match-merging, which uses a MERGE statement rather than a SET statement to combine data sets. Each input data set must be sorted or indexed in ascending order based on the BY variable(s). During match-merging, SAS sequentially checks each observation of each data set to see whether the BY values match, and then writes the combined observation to the new data set.

```
data merged;
   merge a b;
   by num;
run;
```

Match-Merge Processing

To predict the results of match-merging correctly, you need to understand how the DATA step processes data in match-merges.

Compiling

To prepare to merge data sets, SAS

1. reads the descriptor portions of the data sets that are listed in the MERGE statement

2. reads and compiles the rest of the DATA step program

3. creates the program data vector (PDV), an area of memory where SAS holds one observation at a time

4. assigns a tracking pointer to each data set that is listed in the MERGE statement.

If variables with the same name appear in more than one data set, the variable from the first data set that contains the variable (in the order listed in the MERGE statement) determines the length of the variable.

Executing

After compiling the DATA step, SAS sequentially match-merges observations by moving the pointers down each observation of each data set and checking to see whether the BY values match.

- If *Yes*, the observations are read into the PDV in the order in which the data sets appear in the MERGE statement. Values of any same-named variable are overwritten by values of the same-named variable in subsequent data sets. SAS writes the combined observation to the new data set and retains the values in the PDV until the BY value changes in all the data sets.

- If *No*, SAS determines which BY value comes first and reads the observation that contains this value into the PDV. Then the observation is written to the new data set.

When the BY value changes in all the input data sets, the PDV is initialized to missing. The DATA step merge continues to process every observation in each data set until it has processed all observations in all data sets.

Handling Unmatched Observations and Missing Values

All observations that are written to the PDV, including observations that have missing data and no matching BY values, are written to the output data set.

- If an observation contains missing values for a variable, then the observation in the output data set contains the missing values as well. Observations that have missing values for the BY variable appear at the top of the output data set.

- If an input data set doesn't have a matching BY value, then the observation in the output data set contains missing values for the variables that are unique to that input data set.

Renaming Variables

Sometimes you might have same-named variables in more than one input data set. In this case, match-merging overwrites values of the same-named variable in the first data set with values of the same-named variable in subsequent data sets. To prevent overwriting, use the RENAME= data set option in the MERGE statement to rename variables.

Excluding Unmatched Observations

By default, match-merging combines all observations in all input data sets. However, you might want to select only observations that match for two or more input data sets. To exclude unmatched observations, use the IN= data set option and the subsetting IF statement in your DATA step. The IN= data set option creates a variable to indicate whether the data set contributed data to the current observation. The subsetting IF statement then checks the IN= values and writes to the merged data set only matching observations.

Selecting Variables

You can specify the variables you want to drop or keep by using the DROP= and KEEP= data set options. When match-merging, you can specify these options in either the DATA statement or the MERGE statement, depending on whether or not you want to reference values of the variables in that DATA step. When used in the DATA statement, the DROP= option simply drops the variables from the new data set. However, they are still read from the original data set and are available within the DATA step.

Syntax

One-to-One Merging

LIBNAME *libref 'SAS-data-library'*;
DATA *output-SAS-data-set*;
    SET *SAS-data-set-1*;
    SET *SAS-data-set-2*;
RUN;

Concatenating

DATA *output-SAS-data-set*;
    SET *SAS-data-set-1 SAS-data-set-2*;
RUN;

Interleaving

PROC SORT DATA=*SAS-data-set* **OUT=***SAS-data-set*;
    BY *variable(s)*;
RUN;
DATA *output-SAS-data-set*;
    SET *SAS-data-set-1 SAS-data-set-2*;
    BY *variable(s)*;
RUN;

Match-Merging

PROC SORT DATA=*SAS-data-set* **OUT=***SAS-data-set*;
    BY *variable(s)*;
RUN;

DATA *output-SAS-data-set* (**DROP**=*variable(s)* | **KEEP**=*variable(s)*);
    MERGE *SAS-data-set-1 SAS-data-set-2*
    (**RENAME**=(*old-variable-name=new-variable-name*)
    IN= *variable* **DROP**=*variable(s)* | **KEEP**=*variable(s)*);
    BY *variable(s)*;
    IF *expression*;
RUN;

Sample Programs

One-to-One Reading

```
data clinic.one2one;
   set clinic.patients;
   if age<60;
   set clinic.measure;
run;
```

Concatenating

```
data clinic.concat;
   set clinic.therapy1999 clinic.therapy2000;
run;
```

Interleaving

```
data clinic.intrleav;
   set clinic.therapy1999 clinic.therapy2000;
   by month;
run;
```

Match-Merging

```
data clinic.merged(drop=id);
   merge clinic.demog(in=indemog
                     rename=(date=BirthDate))
         clinic.visit(drop=weight in=invisit
                     rename=(date=VisitDate));
   by id;
   if indemog and invisit;
run;
```

Points to Remember

- You can rename any number of variables in each occurrence of the RENAME= option.

- In match-merging, the IN= data set option can apply to any data set in the MERGE statement. The RENAME=, DROP=, and KEEP= options can apply to any data set in the DATA or MERGE statements.

- Use the KEEP= option instead of the DROP= option if more variables are dropped than kept.

- When you specify multiple data set options for a particular data set, enclose them in a single set of parentheses.

Chapter Quiz

Select the best answer for each question. After completing the quiz, check your answers using the answer key in the appendix.

1. Which program will combine Brothers.One and Brothers.Two to produce Brothers.Three?

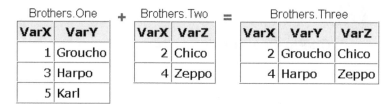

   a. ```
      data brothers.three;
          set brothers.one;
          set brothers.two;
      run;
      ```

   b. ```
      data brothers.three;
          set brothers.one brothers.two;
      run;
      ```

   c. ```
      data brothers.three;
          set brothers.one brothers.two;
          by varx;
      run;
      ```

   d. ```
      data brothers.three;
          merge brothers.one brothers.two;
          by varx;
      run;
      ```

2. Which program will combine Actors.Props1 and Actors.Props2 to produce Actors.Props3?

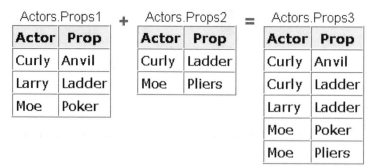

   a. ```
      data actors.props3;
          set actors.props1;
          set actors.props2;
      run;
      ```

   b. ```
      data actors.props3;
          set actors.props1 actors.props2;
      run;
      ```

c. ```
data actors.props3;
   set actors.props1 actors.props2;
   by actor;
run;
```

d. ```
data actors.props3;
   merge actors.props1 actors.props2;
   by actor;
run;
```

3. If you submit the following program, which new data set is created?

Work.Dataone

Career	Supervis	Finance
72	26	9
63	76	7
96	31	7
96	98	6
84	94	6

Work.Datatwo

Variety	Feedback	Autonomy
10	11	70
85	22	93
83	63	73
82	75	97
36	77	97

```
data work.jobsatis;
   set work.dataone work.datatwo;
run;
```

a.

Career	Supervis	Finance	Variety	Feedback	Autonomy
72	26	9	.	.	.
63	76	7	.	.	.
96	31	7	.	.	.
96	98	6	.	.	.
84	94	6	.	.	.
.	.	.	10	11	70
.	.	.	85	22	93
.	.	.	83	63	73
.	.	.	82	75	97
.	.	.	36	77	97

b.

Career	Supervis	Finance	Variety	Feedback	Autonomy
72	26	9	10	11	70
63	76	7	85	22	93
96	31	7	83	63	73
96	98	6	82	75	97
84	94	6	36	77	97

c.

Career	Supervis	Finance
72	26	9
63	76	7
96	31	7
96	98	6
84	94	6
10	11	70
85	22	93
83	63	73
82	75	97
36	77	97

    d. none of the above

4. If you concatenate the data sets below in the order shown, what is the value of Sale in observation 2 of the new data set?

Sales.Reps

ID	Name
1	Nay Rong
2	Kelly Windsor
3	Julio Meraz
4	Richard Krabill

Sales.Close

ID	Sale
1	$28,000
2	$30,000
2	$40,000
3	$15,000
3	$20,000
3	$25,000
4	$35,000

Sales.Bonus

ID	Bonus
1	$2,000
2	$4,000
3	$3,000
4	$2,500

    a. missing

    b. `$30,000`

    c. `$40,000`

    d. you cannot concatenate these data sets

5. What happens if you merge the following data sets by the variable SSN?

1st

SSN	Age
029-46-9261	39
074-53-9892	34
228-88-9649	32
442-21-8075	12
446-93-2122	36
776-84-5391	28
929-75-0218	27

2nd

SSN	Age	Date
029-46-9261	37	02/15/95
074-53-9892	32	05/22/97
228-88-9649	30	03/04/96
442-21-8075	10	11/22/95
446-93-2122	34	07/08/96
776-84-5391	26	12/15/96
929-75-0218	25	04/30/97

    a. The values of Age in the 1st data set overwrite the values of Age in the 2nd data set.

    b. The values of Age in the 2nd data set overwrite the values of Age in the 1st data set.

c. The DATA step fails because the two data sets contain same-named variables that have different values.

d. The values of Age in the 2nd data set are set to missing.

6. Suppose you merge data sets Health.Set1 and Health.Set2 below:

Health.Set1

ID	Sex	Age
1129	F	48
1274	F	50
1387	F	57
2304	F	16
2486	F	63
4425	F	48
4759	F	60
5438	F	42
6488	F	59
9012	F	39
9125	F	56

Health.Set2

ID	Height	Weight
1129	61	137
1387	64	142
2304	61	102
5438	62	168
6488	64	154
9012	63	157
9125	64	159

Which output does the following program create?

```
data work.merged;
   merge health.set1(in=in1) health.set2(in=in2);
   by id;
   if in1 and in2;
run;
proc print data=work.merged;
run;
```

a.

Obs	ID	Sex	Age	Height	Weight
1	1129	F	48	61	137
2	1274	F	50	.	.
3	1387	F	57	64	142
4	2304	F	16	61	102
5	2486	F	63	.	.
6	4425	F	48	.	.
7	4759	F	60	.	.
8	5438	F	42	62	168
9	6488	F	59	64	154
10	9012	F	39	63	157
11	9125	F	56	64	159

b.

Obs	ID	Sex	Age	Height	Weight
1	1129	F	48	61	137
2	1387	F	50	64	142
3	2304	F	57	61	102
4	5438	F	16	62	168
5	6488	F	63	64	154
6	9012	F	48	63	157
7	9125	F	60	64	159
8	5438	F	42	.	.
9	6488	F	59	.	.
10	9012	F	39	.	.
11	9125	F	56	.	.

c.

Obs	ID	Sex	Age	Height	Weight
1	1129	F	48	61	137
2	1387	F	57	64	142
3	2304	F	16	61	102
4	5438	F	42	62	168
5	6488	F	59	64	154
6	9012	F	39	63	157
7	9125	F	56	64	159

d. none of the above

7. The data sets Ensemble.Spring and Ensemble.Sum both contain a variable named Blue. How do you prevent the values of the variable Blue from being overwritten when you merge the two data sets?

a.
```
data ensemble.merged;
    merge ensemble.spring(in=blue)
        ensemble.summer;
    by fabric;
run;
```

b.
```
data ensemble.merged;
    merge ensemble.spring(out=blue)
        ensemble.summer;
    by fabric;
run;
```

c.
```
data ensemble.merged;
    merge ensemble.spring(blue=navy)
        ensemble.summer;
    by fabric;
run;
```

d.
```
data ensemble.merged;
    merge ensemble.spring(rename=(blue=navy))
        ensemble.summer;
    by fabric;
run;
```

8. What happens if you submit the following program to merge Blood.Donors1 and Blood.Donors2, shown below?

```
data work.merged;
   merge blood.donors1 blood.donors2;
   by id;
run
```

Blood.Donors1		
ID	**Type**	**Units**
2304	O	16
1129	A	48
1129	A	50
1129	A	57
2486	B	63

Blood.Donors2		
ID	**Code**	**Units**
6488	65	27
1129	63	32
5438	62	39
2304	61	45
1387	64	67

   a. The Merged data set contains some missing values because not all observations have matching observations in the other data set.

   b. The Merged data set contains eight observations.

   c. The DATA step produces errors.

   d. Values for Units in Blood.Donors2 overwrite values of Units in Blood.Donors1.

9. If you merge Company.Staff1 and Company.Staff2 below by ID, how many observations does the new data set contain?

Company.Staff1			
ID	**Name**	**Dept**	**Project**
000	Miguel	A12	Document
111	Fred	B45	Survey
222	Diana	B45	Document
888	Monique	A12	Document
999	Vien	D03	Survey

Company.Staff2		
ID	**Name**	**Hours**
111	Fred	35
222	Diana	40
777	Steve	0
888	Monique	37

   a. 4

   b. 5

   c. 6

   d. 9

10. If you merge data sets Sales.Reps, Sales.Close, and Sales.Bonus by ID, what is the value of Bonus in the third observation in the new data set?

Sales.Reps	
ID	**Name**
1	Nay Rong
2	Kelly Windsor
3	Julio Meraz
4	Richard Krabill

Sales.Close	
ID	**Sale**
1	$28,000
2	$30,000
2	$40,000
3	$15,000
3	$20,000
3	$25,000
4	$35,000

Sales.Bonus	
ID	**Bonus**
1	$2,000
2	$4,000
3	$3,000
4	$2,500

a. `$4,000`

b. `$3,000`

c. missing

d. can't tell from the information given

Chapter 13
Transforming Data with SAS Functions

Overview

Introduction

When planning modifications to SAS data sets, be sure to examine the many SAS
functions that are available. SAS functions are pre-written routines that provide
programming shortcuts for many calculations and manipulations of data.

This chapter teaches you how to use a variety of functions, such as those shown in the
table below. You learn to convert data from one data type to another, to work with SAS
date and time values, and to manipulate the values of character variables.

Table 13.1 *Selected SAS Functions and Their Descriptions*

Function	Description	Form	Sample Value
YEAR	Extracts the year value from a SAS date value.	YEAR*(date)*	2002
QTR	Extracts the quarter value from a SAS date value.	QTR*(date)*	1

Function	Description	Form	Sample Value
MONTH	Extracts the month value from a SAS date value.	MONTH*(date)*	12
DAY	Extracts the day value from a SAS date value.	DAY*(date)*	5

Objectives

In this chapter, you learn to

- convert character data to numeric data

- convert numeric data to character data

- create SAS date values

- extract the month, year, and interval from a SAS date value

- perform calculations with date and datetime values and time intervals

- extract, edit, and search the values of character variables

- replace or remove all occurrences of a particular word within a character string.

Understanding SAS Functions

Overview

SAS functions are built-in routines that enable you to complete many types of data manipulations quickly and easily. Generally speaking, functions provide programming shortcuts. There are many types of SAS functions: arithmetic functions, financial functions, character functions, probability functions, and many more.

Table 13.2   *SAS Function Catagories*

Categories of SAS Functions	
Array	Probability
Bitwise Logical Operations	Quantile
Character	Random Number
Character String Matching	Sample Statistics
Date and Time	SAS File I/O
Dynamic Link Library	Special

Categories of SAS Functions	
External Files	State and ZIP Code
Financial	Trigonometric
Hyperbolic	Truncation
Macro	Variable Control
Mathematical	Variable Information
MultiByte Character Set	

SAS Functions That Compute Sample Statistics

Function	Syntax	Calculates...
SUM	sum(*argument, argument,...*)	sum of values
MEAN	mean(*argument, argument,...*)	average of nonmissing values
MIN	min(*argument, argument,...*)	minimum value
MAX	max(*argument, argument,...*)	maximum value
VAR	var(*argument, argument,...*)	variance of the values
STD	std(*argument, argument,...*)	standard deviation of the values

Uses of SAS Functions

Using SAS functions, you can

- calculate sample statistics
- create SAS date values
- convert U.S. ZIP codes to state postal codes
- round values
- generate random numbers
- extract a portion of a character value
- convert data from one data type to another.

This chapter concentrates on functions that

- convert data
- manipulate SAS date values
- modify values of character variables.

However, be sure to explore the many other SAS functions, which are described in the SAS documentation.

Example of a SAS Function

SAS functions can be used in DATA step programming statements. A SAS function can be specified anywhere that you would use a SAS expression, as long as the function is part of a SAS statement.

Let's look at a simple example of a SAS function. The assignment statement below uses the MEAN function to calculate the average of three exam scores that are stored in the variables Exam1, Exam2, and Exam3.

```
AvgScore=mean(exam1,exam2,exam3);
```

When you reference a SAS function, the function returns a value that is based on the function arguments. The MEAN function above contains three arguments: the variables Exam1, Exam2, and Exam3. The function calculates the mean of the three variables that are listed as arguments.

Some functions require a specific number of arguments, whereas other functions can contain any number of arguments. Some functions require no arguments.

General Form of SAS Functions

Arguments, Variable Lists, Arrays

To use a SAS function, specify the function name followed by the function arguments, which are enclosed in parentheses.

General form, SAS function:

function-name*(argument-1<,argument-n>)***;**

where *arguments* can be

- variables, mean(*x,y,z*)
- constants, mean(*456,502,612,498*)
- expressions, mean(37*2,192/5,*mean(22,34,56)*)

⚠ Even if the function does not require arguments, the function name must still be followed by parentheses (for example, *function-name()*).

When a function contains more than one argument, the arguments are usually separated by commas.

```
function-name(argument-1,argument-2,argument-n)
```

However, for some functions, variable lists and arrays can also be used as arguments, as long as the list or the array is preceded by the word OF. (You can learn about arrays in "Processing Variables with Arrays" on page 484 .)

Example

Here is an example of a function that contains multiple arguments. Notice that the arguments are separated by commas.

```
mean(x1,x2,x3)
```

The arguments for this function can also be written as a variable list.

```
mean(of x1-x3)
```

Or, the variables can be referenced by an array.

```
mean(of newarray{*})
```

When specifying function arguments with a variable list or an array, be sure to precede the list or the array with the word OF. If you omit the word OF, the function arguments might not be interpreted as you expect. For example, the function below returns X1 minus X3, not the average of the variables X1, X2, and X3.

```
mean(x1-x3)
```

Target Variables

Now that you are familiar with the purpose and general form of SAS functions, let's think about target variables. A target variable is the variable to which the result of a function is assigned. For example, in the statement below, the variable AvgScore is the target variable.

```
AvgScore=mean(exam1,exam2,exam3);
```

Unless the length of the target variable has been previously defined, a default length is assigned. The default length depends on the function; the default for character functions can be as long as 200.

Default lengths could cause character variables to use more space than necessary in your data set. So, when using SAS functions, consider the appropriate length for any character target variables. If necessary, add a LENGTH statement to specify a length for the character target variable before the statement that creates the values of that variable.

Converting Data with Functions

Introduction to Converting Data

Suppose you are asked to complete a number of modifications to the data set Hrd.Temp. The first modification is to create a new variable that contains the salary of temporary employees. Examining the data set, you realize that one of the variables needed to calculate salaries is the character variable PayRate. To complete the calculation, you need to convert PayRate from character to numeric.

SAS Data Set Hrd.Temp

City	State	Zip	Phone	StartDate	EndDate	PayRate	Days	Hours
CARY	NC	27513	6224549	14567	14621	10	11	88
CARY	NC	27513	6223251	14524	14565	8	25	200
CHAPEL HILL	NC	27514	9974749	14570	14608	40	26	208

```
data hrd.newtemp;
   set hrd.temp;
   Salary=payrate*hours;
run;
```

In such cases, you should use the INPUT function before attempting the calculation. The INPUT function converts character data values to numeric values. The PUT function converts numeric data values to character values. Both functions are discussed in this section.

Potential Problems of Omitting INPUT or PUT

What happens if you skip the INPUT function or the PUT function when converting data?

SAS will detect the mismatched variables and will try an automatic character-to-numeric or numeric-to-character conversion. However, this process doesn't always work. Suppose each value of PayRate begins with a dollar sign ($). When SAS tries to automatically convert the values of PayRate to numeric values, the dollar sign blocks the process. The values cannot be converted to numeric values. Similar problems can occur with automatic numeric-to-character conversion.

Therefore, it is *always* best to include INPUT and PUT functions in your programs to avoid data type mismatches and circumvent automatic conversion.

Automatic Character-to-Numeric Conversion

Let's begin with the automatic conversion of character values to numeric values.

By default, if you reference a character variable in a numeric context such as an arithmetic operation, SAS tries to convert the variable values to numeric. For example, in the DATA step below, the character variable PayRate appears in a numeric context. It is multiplied by the numeric variable Hours to create a new variable named Salary.

```
data hrd.newtemp;
   set hrd.temp;
   Salary=payrate*hours;
run;
```

When this step executes, SAS automatically attempts to convert the character values of PayRate to numeric values so that the calculation can occur. This conversion is completed by creating a temporary numeric value for each character value of PayRate. This temporary value is used in the calculation. The character values of PayRate are *not* replaced by numeric values.

Whenever data is automatically converted, a message is written to the SAS log stating that the conversion has occurred.

SAS Log

```
4        data hrd.newtemp;
5           set hrd.temp;
6           Salary=payrate*hours;
7        run;

NOTE: Character values have been converted
      to numeric values at the places given
      by: (Line):(Column).

      6:11

NOTE: The data set HRD.NEWTEMP has 40 observations
      and 19 variables.
NOTE: The data statement used 0.78 seconds.
```

When Automatic Conversion Occurs

Automatic character-to-numeric conversion occurs when a character value is

- assigned to a previously defined numeric variable, such as the numeric variable Rate

  ```
  Rate=payrate;
  ```

- used in an arithmetic operation

  ```
  Salary=payrate*hours;
  ```

- compared to a numeric value, using a comparison operator

  ```
  if payrate>=rate;
  ```

- specified in a function that requires numeric arguments.

  ```
  NewRate=sum(payrate,raise);
  ```

The automatic conversion

- uses the *w.* informat, where *w* is the width of the character value that is being converted

- produces a numeric missing value from any character value that does not conform to standard numeric notation (digits with an optional decimal point, leading sign or scientific notation).

Table 13.3 Automatic Conversion of Character Variables

Character Value	automatic conversion	Numeric Value
12.47	→	12.47
-8.96	→	-8.96
1.243E1	→	12.43
1,742.64	→	.

Restriction for WHERE Expressions

The WHERE statement does *not* perform automatic conversions in comparisons. The simple program below demonstrates what happens when a WHERE expression encounters the wrong data type. The variable Number contains a numeric value, and the variable Character contains a character value, but the two WHERE statements specify the wrong data type.

```
data work.convtest;
   Number=4;
   Character='4';
run;
proc print data=work.convtest;
   where character=4;
run;
proc print data=work.convtest;
   where number='4';
run;
```

This mismatch of character and numeric variables and values prevents the program from processing the WHERE statements. Automatic conversion is not performed. Instead, the program stops, and error messages are written to the SAS log.

SAS Log

```
1       data work.convtest;
2          Number=4;
3          Character='4';
4       run;
NOTE: The data set Work.ConvTest has 1 observations and 2
      variables.

5       proc print data=work.convtest;
6          where character=4;
7       run;
ERROR: Where clause operator requires compatible variables.

NOTE: The SAS System stopped processing this step because
      of errors.

8       proc print data=work.convtest;
9          where number='4';
10      run;
ERROR: Where clause operator requires compatible variables.

NOTE: The SAS System stopped processing this step because
      of errors.
```

Explicit Character-to-Numeric Conversion

Using the INPUT Function

In order to avoid the problems we have seen, use the INPUT function to convert character data values to numeric values. To learn how to use this function, let's examine one of the data set modifications needed for Hrd.Temp. As mentioned earlier, you need to calculate employee salaries by multiplying the character variable PayRate by the numeric variable Hours.

SAS Data Set Hrd.Temp

City	State	Zip	Phone	StartDate	EndDate	PayRate	Days	Hours
CARY	NC	27513	6224549	14567	14621	10	11	88
CARY	NC	27513	6223251	14524	14565	8	25	200
CHAPEL HILL	NC	27514	9974749	14570	14608	40	26	208
RALEIGH	NC	27612	6970450	14516	14527	15	10	80

To calculate salaries, you write the following DATA step. It creates a new data set, Hrd.Newtemp, to contain the original data plus the new variable Salary.

```
data hrd.newtemp;
   set hrd.temp;
   Salary=payrate*hours;
run;
```

However, you know that submitting this DATA step would cause an automatic character-to-numeric conversion, because the character variable PayRate is used in a numeric context. You can explicitly convert the character values of PayRate to numeric values by using the INPUT function.

General form, INPUT function:

INPUT(*source,format*)

where

- *source* indicates the character variable, constant, or expression to be converted to a numeric value

- a numeric *informat* must also be specified, as in this example:

```
input(payrate,2.)
```

When choosing the informat, be sure to select a numeric informat that can read the form of the values.

Table 13.4 Character Values and Associated Informats

Character Value	Informat
2115233	7.
2,115,233	COMMA9.

Example

Here's an example of the INPUT function:

```
Test=input(saletest,comma9.);
```

The function uses the numeric informat COMMA9. to read the values of the character variable SaleTest. Then the resulting numeric values are stored in the variable Test.

Now let's use the INPUT function to convert the character values of PayRate to numeric values.

You begin the function by specifying PayRate as the source.

Because PayRate has a length of 2, you choose the numeric informat 2. to read the values of the variable.

```
input(payrate,2.)
```

Finally, you add the function to the assignment statement in your DATA step.

```
data hrd.newtemp;
   set hrd.temp;
   Salary=input(payrate,2.)*hours;
run;
```

After the DATA step is executed, the new data set (which contains the variable Salary) is created.

SAS Data Set Hrd.Newtemp

City	State	Zip	Phone	StartDate	EndDate	PayRate	Days	Hours	BirthDate	Salary
CARY	NC	27513	6224549	14567	14621	10	11	88	7054	880
CARY	NC	27513	6223251	14524	14565	8	25	200	5757	1600

Notice that no conversion messages appear in the SAS log when you use the INPUT function.

SAS Log

```
13      data hrd.newtemp;
14         set hrd.temp;
15            Salary=input(payrate,2.)*hours;
16      run;


NOTE: The data set Hrd.Newtemp has 40 observations
      and 19 variables.
NOTE: The DATA statement used 0.55 seconds.
```

The form of the INPUT function is very similar to the form of the PUT function (which performs numeric-to-character conversions).

INPUT(*source,informat*)

PUT(*source,format*))

However, note that the INPUT function requires an informat, whereas the PUT function requires a format. To remember which function requires a format versus an informat, note that the INPUT function requires the informat.

Automatic Numeric-to-Character Conversion

The automatic conversion of numeric data to character data is very similar to character-to-numeric conversion. Numeric data values are converted to character values whenever they are used in a character context.

For example, the numeric values of the variable Site are converted to character values if you

- assign the numeric value to a previously defined character variable, such as the character variable SiteCode: SiteCode=site;

- use the numeric value with an operator that requires a character value, such as the concatenation operator: SiteCode=site‖dept;

- specify the numeric value in a function that requires character arguments, such as the SUBSTR function: Region=substr(site,1,4);

Specifically, SAS writes the numeric value with the BEST12. format, and the resulting character value is right-aligned. This conversion occurs before the value is assigned or used with any operator or function. Automatic numeric-to-character conversion can cause unexpected results. For example, suppose the original numeric value has fewer than 12 digits. The resulting character value will have leading blanks, which might cause problems when you perform an operation or function.

Automatic numeric-to-character conversion also causes a message to be written to the SAS log indicating that the conversion has occurred.

SAS Log

```
9     data hrd.newtemp;
10       set hrd.temp;
11        SiteCode=site;
12    run;


NOTE: Numeric values have been converted
      to character values at the
      places given by: (Line):(Column).

      11:13

NOTE: The data set HRD.NEWTEMP has 40 observations
      and 19 variables.
NOTE: The data statement used 1.06 seconds.
```

As we saw with the INPUT statement, it is best not to rely on automatic conversion. When you know that numeric data must be converted to character data, perform an explicit conversion by including a PUT function in your SAS program. We look at the PUT function in the next section.

Explicit Numeric-to-Character Conversion

You can use the PUT function to explicitly convert numeric data values to character data values.

Let's use this function to complete one of the modifications needed for the data set Hrd.Temp. Suppose you are asked to create a new character variable named Assignment that concatenates the values of the numeric variable Site and the character variable Dept. The new variable values must contain the value of Site followed by a slash (/) and then the value of Dept, for example, **26/DP**. The following figure shows selected variables from the Hrd.Temp data set.

SAS Data Set Hrd.Temp

Overtime	Job	Contact	Dept	Site
4	Word processing	Word Processor	DP	26
.	Filing, administrative duties	Admin. Asst.	PURH	57
.	Organizational dev. specialist	Consultant	PERS	34
.	Bookkeeping, word processing	Bookkeeper Asst.	BK	57

You write an assignment statement that contains the concatenation operator (‖) to indicate that Site should be concatenated with Dept, using a slash as a separator. Note that the slash is enclosed in quotation marks. All character constants must be enclosed in quotation marks.

```
data hrd.newtemp;
   set hrd.temp;
   Assignment=site||'/'||dept;
run;
```

You know that submitting this DATA step will cause SAS to automatically convert the numeric values of Site to character values, because Site is used in a character context. The variable Site appears with the concatenation operator, which requires character values. To explicitly convert the numeric values of Site to character values, you must add the PUT function to your assignment statement.

General form, PUT function:

PUT(*source,format*)

where

- *source* indicates the numeric variable, constant, or expression to be converted to a character value

- a *format* matching the data type of the source must also be specified, as in this example:

    put(site,2.)

- The PUT function always returns a character string.

    - The PUT function returns the *source* written with a *format*.

    - The *format* must agree with the *source* in type.

    - Numeric formats right-align the result; character formats left-align the result.

    - If you use the PUT function to create a variable that has not been previously identified, it creates a character variable whose length is equal to the format width.

Because you are listing a numeric variable as the source, you must specify a numeric format.

Now that you know the general form of the PUT function, you can rewrite the assignment statement in your DATA step to explicitly convert the numeric values of Site to character values.

To perform this conversion, write the PUT function, specifying Site as the source. Because Site has a length of 2, choose 2. as the numeric format. After you add this PUT function to the assignment statement, the DATA step creates the new data set that contains Assignment.

```
data hrd.newtemp;
   set hrd.temp;
```

```
Assignment=put(site,2.)||'/'||dept;
run;
```

SAS Data Set Hrd.Newtemp

Overtime	Job	Contact	Dept	Site	BirthDate	Assignment
4	Word processing	Word Processor	DP	26	7054	**26/DP**
.	Filing, administrative duties	Admin. Asst.	PURH	57	5757	**57/PURH**

Notice that no conversion messages appear in the SAS log when you use the PUT function.

SAS Log

```
13     data hrd.newtemp;
14        set hrd.temp;
15        Assignment=put(site,2.)||'/'||dept;
16     run;

NOTE: The data set Hrd.Newtemp has 40 observations
      and 19 variables.
NOTE: The DATA statement used 0.71 seconds.
```

Matching the Data Type

Remember that the format specified in the PUT function must match the data type of the source.

PUT(source,format)

So, to do an explicit numeric-to-character data conversion, you specify a numeric source and a numeric format. The form of the PUT function is very similar to the form of the INPUT function.

PUT(source,format)

INPUT(source,informat)

Note that the PUT function requires a format, whereas the INPUT function requires an informat. To remember which function requires a format versus an informat, note that the INPUT function requires the informat.

Manipulating SAS Date Values with Functions

SAS Date and Time Values

SAS includes a variety of functions that enable you to work with SAS date values. SAS stores a date value as the number of days from January 1, 1960, to a given date. For example:

Jan. 1, 1959	Jan. 1, 1960	Jan. 1, 1961
← ⁻365	0	366 →

A SAS time value is stored as the number of seconds since midnight. For example:

(12:00 am) midnight	12:15 pm	17:00 (or 5:00 pm)
0	44100	61200 →

Consequently, a SAS datetime value is stored as the number of seconds between midnight on January 1, 1960, and a given date and time. For example:

July 4, 1776 11:30:23	Jan. 1, 1960 midnight	July 4, 1994 16:10:45
← ⁻5790400177	0	1088957445 →

SAS stores date values as numbers so that you can easily sort the values or perform arithmetic computations. You can use SAS date values as you use any other numeric values.

```
data test;
   set hrd.temp;
   TotDay=enddate-startdate;
run;
```

When you execute the program, TotDay has a value of 54 based on the StartDate and EndDate values in the Hrd.Temp data set.

SAS Data Set Hrd.Temp

City	State	Zip	Phone	StartDate	EndDate	PayRate	Days	Hours
CARY	NC	27513	6224549	14567	14621	10	11	88
CARY	NC	27513	6223251	14524	14565	8	25	200

You can display SAS date values in a variety of forms by associating a SAS format with the values. The format affects *only* the display of the dates, not the date values in the data set. For example, the FORMAT statement below associates the DATE9. format with the variables StartDate and EndDate. A portion of the output created by this PROC PRINT step appears below.

```
proc print data=hrd.temp;
   format startdate enddate date9.;
run;
```

City	State	Zip	Phone	StartDate	EndDate	Pay Rate	Days	Hours
CARY	NC	27513	6224549	19NOV1999	12JAN2000	10	11	88
CARY	NC	27513	6223251	07OCT1999	17NOV1999	8	25	200
CHAPEL HILL	NC	27514	9974749	22NOV1999	30DEC1999	40	26	208
RALEIGH	NC	27612	6970450	29SEP1999	10OCT1999	15	10	80

SAS date values are valid for dates that are based on the Gregorian calendar from A.D. 1582 through A.D. 20,000.

Use caution when working with historical dates. The Gregorian calendar was used throughout most of Europe from 1582, but Great Britain and the American colonies did not adopt the calendar until 1752.

SAS Date Functions

SAS stores dates, times, and datetimes as numeric values. You can use several functions to create these values.

Table 13.5   Typical Use of SAS Date Functions

Function	Typical Use	Result
MDY	`date=mdy(mon,day,yr);`	SAS date
TODAYDATE	`now=today();` `now=date();`	today's date as a SAS date
TIME	`curtime=time();`	current time as a SAS time

You use other functions to extract months, quarters, days, and years from SAS date values.

Table 13.6   Selected Functions to Use with SAS Date Values

Function	Typical Use	Result
DAY	`day=day(date);`	day of month (1-31)
QTR	`quarter=qtr(date);`	quarter (1-4)
WEEKDAY	`wkday=weekday(date);`	day of week (1-7)
MONTH	`month=month(date);`	month (1-12)

Function	Typical Use	Result
YEAR	`yr=year(date);`	year (4 digits)
INTCK	`x=intck('day',d1,d2);`	days from D1 to D2
	`x=intck('week',d1,d2);`	weeks from D1 to D2
	`x=intck('month',d1,d2);`	months from D1 to D2
	`x=intck('qtr',d1,d2);`	quarters from D1 to D2
	`x=intck('year',d1,d2);`	years from D1 to D2
INTNX	`x=intnx('interval',start` `-from,increment);`	date, time, or datetime value
DATDIF YRDIF	`x=datdif(date1,date2,'AC` `T/ACT');`	days between date1 and date2
	`x=yrdif(date1,date2,'ACT` `/ACT');`	years between date1 and date2

In the following pages, you will see several SAS date functions, showing how they are used to both create and extract date values.

YEAR, QTR, MONTH, and DAY Functions

Overview of YEAR, QTR, MONTH, and DAY Functions

Every SAS date value can be queried for the values of its year, quarter, month, and day. You extract these values by using the functions YEAR, QTR, MONTH, and DAY. They each work the same way, so we'll discuss them as a group.

General form, YEAR, QTR, MONTH, and DAY functions:

YEAR(*date*)

QTR(*date*)

MONTH(*date*)

DAY(*date*)

where *date* is a SAS date value that is specified either as a variable or as a SAS date constant. For more information about SAS date constants, see the SAS documentation.

The YEAR function returns a four-digit numeric value that represents the year (for example, 2002). The QTR function returns a value of **1**, **2**, **3**, or **4** from a SAS date value to indicate the quarter of the year in which a date value falls. The MONTH function returns a numeric value that ranges from **1** to **12**, representing the month of the year. The value **1** represents January, **2** represents February, and so on. The DAY function returns a numeric value from **1** to **31**, representing the day of the month. As you can see, these functions are very similar in purpose and form.

Table 13.7 *Selected Date Functions and Their Uses*

Function	Description	Form	Sample Value
YEAR	Extracts the year value from a SAS date value.	YEAR*(date)*	2002
QTR	Extracts the quarter value from a SAS date value	QTR*(date)*	1
MONTH	Extracts the month value from a SAS date value.	MONTH*(date)*	12
DAY	Extracts the day value from a SAS date value	DAY*(date)*	5

Example: Finding the Year and Month

Let's use the YEAR and MONTH functions to complete a simple task.

Suppose you need to create a subset of the data set Hrd.Temp that contains information about all temporary employees who were hired in November 1999. Hrd.Temp contains the beginning and ending dates for staff employment, but there are no month or year variables in the data set.

SAS Data Set Hrd.Temp

City	State	Zip	Phone	StartDate	EndDate	PayRate	Days	Hours
CARY	NC	27513	6224549	14567	14621	10	11	88
CARY	NC	27513	6223251	14524	14565	8	25	200
CHAPEL HILL	NC	27514	9974749	14570	14608	40	26	208
RALEIGH	NC	27612	6970450	14516	14527	15	10	80

To determine the year in which employees were hired, you can apply the YEAR function to the variable that contains the employee start date, StartDate. You write the YEAR function as

```
year(startdate)
```

Likewise, to determine the month in which employees were hired, you apply the MONTH function to StartDate.

```
month(startdate)
```

To create the new data set, you include these functions in a subsetting IF statement within a DATA step. The subsetting IF statement specifies that the new data set includes only observations in which the YEAR function extracts a value of **1999** and the MONTH function extracts a value of **11** (for November).

```
data hrd.nov99;
   set hrd.temp;
   if year(startdate)=1999 and month(startdate)=11;
run;
```

Finally, you add a PROC PRINT step to the program so that you can view the new data set. Notice that the PROC PRINT step includes a FORMAT statement to display the variables StartDate and EndDate with the DATE9. format.

```
data hrd.nov99;
   set hrd.temp;
   if year(startdate)=1999 and month(startdate)=11;
proc print data=hrd.tempnov;
   format startdate enddate date9.;
run;
```

Here is a portion of the PROC PRINT output that is created by your program. Notice that the new data set contains information about only those employees who were hired in November 1999.

City	State	Zip	Phone	StartDate	EndDate	PayRate	Days	Hours
CARY	NC	27513	6224549	19NOV1999	12JAN2000	10	11	88
CHAPEL HILL	NC	27514	9974749	22NOV1999	30DEC1999	40	26	208
DURHAM	NC	27713	3633618	02NOV1999	13NOV1999	12	9	72
CARRBORO	NC	27510	9976732	16NOV1999	04JAN2000	15	7	64

Example: Finding the Year

Now let's use the YEAR function to complete a task.

Suppose you need to create a subset of the data set Hrd.Temp that contains information about all temporary employees who were hired during a specific year, such as 1998. Hrd.Temp contains the dates on which employees began work with the company and their ending dates, but there is no year variable.

SAS Data Set Hrd.Temp

City	State	Zip	Phone	StartDate	EndDate	PayRate	Days	Hours
CARY	NC	27513	6224549	14567	14621	10	11	88
CARY	NC	27513	6223251	14524	14565	8	25	200
CHAPEL HILL	NC	27514	9974749	14570	14608	40	26	208
RALEIGH	NC	27612	6970450	14516	14527	15	10	80

To determine the year in which employees were hired, you can apply the YEAR function to the variable that contains the employee start date, StartDate. You write the YEAR function as

```
year(startdate)
```

Then, to create the new data set, you include this function in a subsetting IF statement within a DATA step. This subsetting IF statement specifies that only observations in which the YEAR function extracts a value of **1998** are placed in the new data set.

```
data hrd.temp98;
   set hrd.temp;
   if year(startdate)=1998;
run;
```

Finally, you add a PROC PRINT step to the program so that you can view the new data set. Notice that the PROC PRINT step includes a FORMAT statement to display the variables StartDate and EndDate with the DATE9. format.

```
data hrd.temp98;
   set hrd.temp;
   if year(startdate)=1998;
```

```
run;
proc print data=hrd.temp98;
   format startdate enddate date9.;
run;
```

Here is a portion of the PROC PRINT output that is created by your program. Notice that the new data set contains information for only those employees who were hired in 1998.

City	State	Zip	Phone	StartDate	EndDate	Pay Rate	Days	Hours
CHAPEL HILL	NC	27514	9972070	02AUG1998	17AUG1998	12	12	96
DURHAM	NC	27713	3633020	06OCT1998	10OCT1998	10	5	40

WEEKDAY Function

The WEEKDAY function enables you to extract the day of the week from a SAS date value.

General form, WEEKDAY function:

WEEKDAY(*date***)**

where *date* is a SAS date value that is specified either as a variable or as a SAS date constant. For more information about SAS date constants, see the SAS documentation.

The WEEKDAY function returns a numeric value from **1** to **7**. The values represent the days of the week.

Table 13.8   Values for the WEEKDAY Function

Value	equals	Day of the Week
1	=	Sunday
2	=	Monday
3	=	Tuesday
4	=	Wednesday
5	=	Thursday
6	=	Friday
7	=	Saturday

For example, suppose the data set Radio.Sch contains a broadcast schedule. The variable AirDate contains SAS date values. To create a data set that contains only weekend broadcasts, you use the WEEKDAY function in a subsetting IF statement. You include only observations in which the value of AirDate corresponds to a Saturday or Sunday.

```
data radio.schwkend;
   set radio.sch;
   if weekday(airdate)in(1,7);
run;
```

Note: In the example above, the statement `if weekday(airdate) in (1,7);` is
the same as `if weekday(airdate)=7 or weekday(airdate)=1;`

MDY Function

Overview of MDY Function

The MDY function creates a SAS date value from numeric values that represent the
month, day, and year. For example, suppose the data set Hrd.Temp contains the
employee start date in three numeric variables, Month, Day, and Year.

SAS Data Set Hrd.Temp

City	State	Zip	Phone	Month	Day	Year	PayRate	Days	Hours
CARY	NC	27513	6224549	1	12	2000	10	11	88
CARY	NC	27513	6223251	11	17	1999	8	25	200
CHAPEL HILL	NC	27514	9974749	12	30	1999	40	26	208
RALEIGH	NC	27612	6970450	10	10	1999	15	10	80

Having the start date in three variables makes it difficult to perform calculations that are
based on the length of employment. You can convert these numeric values to useful SAS
date values by applying the MDY function.

General form, MDY function:

MDY(*month,day,year***)**

where

- *month* can be a variable that represents the month, or a number from 1-12

- *day* can be a variable that represents the day, or a number from 1-31

- *year* can be a variable that represents the year, or a number that has 2 or 4 digits.

In the data set Hrd.Temp, the values for month, day, and year are stored in the numeric
variables Month, Day, and Year. You write the following MDY function to create the
SAS date values:

```
mdy(month,day,year)
```

Then place this function in an assignment statement to create a new variable to contain
the SAS date values.

```
data hrd.newtemp(drop=month day year);
   set hrd.temp;
   Date=mdy(month,day,year);
run;
```

Here is the new data set that contains the variable Date.

SAS Data Set Hrd.Newtemp

City	State	Zip	Phone	PayRate	Days	Hours	Date
CARY	NC	27513	6224549	10	11	88	**14621**
CARY	NC	27513	6223251	8	25	200	**14565**
CHAPEL HILL	NC	27514	9974749	40	26	208	**14608**
RALEIGH	NC	27612	6970450	15	10	80	**14527**

Remember, to display SAS date values in a more readable form, you can associate a SAS format with the values. For example, the FORMAT statement below associates the DATE9. format with the variable Date. A portion of the output that is created by this PROC PRINT step appears below.

```
proc print data=hrd.newtemp;
   format date date9.;
run;
```

City	State	Zip	Phone	PayRate	Days	Hours	Date
CARY	NC	27513	6224549	10	11	88	12JAN2000
CARY	NC	27513	6223251	8	25	200	17NOV1999
CHAPEL HILL	NC	27514	9974749	40	26	208	30DEC1999
RALEIGH	NC	27612	6970450	15	10	80	10OCT1999

The MDY function can also add the same SAS date to every observation. This may be useful if you want to compare a fixed beginning date with differing end dates. Just use numbers instead of data set variables when providing values to the MDY function.

```
data hrd.newtemp;
   set hrd.temp;
   DateCons=mdy(6,17,2002);
proc print data=hrd.newtemp;
   format datecons date9.;
run;
```

City	State	Zip	Phone	PayRate	Days	Hours	DateCons
CARY	NC	27513	6224549	10	11	88	17JUN2002
CARY	NC	27513	6223251	8	25	200	17JUN2002
CHAPEL HILL	NC	27514	9974749	40	26	208	17JUN2002
RALEIGH	NC	27612	6970450	15	10	80	17JUN2002

Be careful when entering and formatting year values. The MDY function accepts two-digit values for the year, but SAS interprets two-digit values according to the 100-year span that is set by the YEARCUTOFF= system option. The default value of YEARCUTOFF= is 1920. For details, see "Reading Date and Time Values" on page 571 .

The use of four-digit year values in the MDY function is recommended:

• **MDY(5,10,20)** = May 10, 1920

• **MDY(5,10,2020)** = May 10, 2020

To display the years clearly, format SAS dates with the DATE9. format. This forces the year to appear with four digits, as shown above in the Date and DateCons variables of your Hrd.Newtemp output.

Example: Finding the Date

Let's look at another example of the MDY function. The data set Dec.Review contains a variable named Day. This variable contains the day of the month for each employee's performance appraisal. The appraisals were all completed in December of 2010.

SAS Data Set Dec.Review

Site	Day	Rate	Name
Westin	12	A2	Mitchell, K
Stockton	4	A5	Worton, M
Center City	17	B1	Smith, A

The following DATA step uses the MDY function to create a new variable named ReviewDate. This variable contains the SAS date value for the date of each performance appraisal.

```
data dec.review2010;
   set dec.review;
   ReviewDate=mdy(12,day,2010);
run;
```

SAS Data Set Dec.Review2010

Site	Day	Rate	Name	ReviewDate
Westin	12	A2	Mitchell, K	18608
Stockton	4	A5	Worton, M	18600
Center City	17	B1	Smith, A	18613

If you specify an invalid date in the MDY function, SAS assigns a missing value to the target variable.

```
data dec.review2010;
   set dec.review;
   ReviewDate=mdy(15,day,2010);
run;
```

SAS Data Set Dec.Review2010

Site	Day	Rate	Name	ReviewDate
Westin	12	A2	Mitchell, K	.
Stockton	4	A5	Worton, M	.
Center City	17	B1	Smith, A	.

DATE and TODAY Functions

The DATE and TODAY functions return the current date from the system clock as a SAS date value. The DATE and TODAY functions have the same form and can be used interchangeably.

General form, DATE and TODAY functions:

DATE()

TODAY()

These functions require no arguments, but they must still be followed by parentheses.

Let's add a new variable, which contains the current date, to the data set Hrd.Temp. To create this variable, write an assignment statement such as the following:

```
EditDate=date();
```

After this statement is added to a DATA step and the step is submitted, the data set that contains EditDate is created.

```
data hrd.newtemp;
   set hrd.temp;
   EditDate=date();
run;
```

Note: For this example, the SAS date values shown below were created by submitting this program on January 15, 2000.

SAS Data Set Hrd.Newtemp

EndDate	EditDate
14621	**14624**
14565	**14624**
14608	**14624**

Remember, to display these SAS date values in a different form, you can associate a SAS format with the values. For example, the FORMAT statement below associates the DATE9. format with the variable EditDate. A portion of the output that is created by this PROC PRINT step appears below.

```
proc print data=hrd.newtemp;
   format editdate date9.;
run;
```

EndDate	EditDate
14621	15JAN2000
14565	15JAN2000
14608	15JAN2000

INTCK Function

Overview of INTCK Function

The INTCK function returns the number of time intervals that occur in a given time span. You can use it to count the passage of days, weeks, months, and so on.

General form, INTCK function:

INTCK('*interval*',*from*,*to***)**

where

- '*interval*' specifies a character constant or variable. The value can be one of the following:
  - DAY
  - WEEKDAY
  - WEEK
  - TENDAY
  - SEMIMONTH
  - MONTH
  - QTR
  - SEMIYEAR
  - YEAR
- *from* specifies a SAS date, time, or datetime value that identifies the beginning of the time span.
- *to* specifies a SAS date, time, or datetime value that identifies the end of the time span.

 The type of interval (date, time, or datetime) must match the type of value in *from*.

The INTCK function counts intervals from fixed interval beginnings, not in multiples of an interval unit *from* the from value. Partial intervals are not counted. For example, WEEK intervals are counted by Sundays rather than seven-day multiples from the *from* argument. MONTH intervals are counted by day 1 of each month, and YEAR intervals are counted from 01JAN, not in 365-day multiples.

Consider the results in the following table. The values that are assigned to the variables Weeks, Months, and Years are based on consecutive days.

Table 13.9 Examples of SAS Statements and Their Values

SAS Statement	Value
Weeks=intck('week','31dec2000'd,'01jan2001'd);	0
Months=intck('month','31dec2000'd,'01jan2001'd);	1
Years=intck('year','31dec2000'd,'01jan2001'd);	1

Because December 31, 2000, is a Sunday, no WEEK interval is crossed between that day and January 1, 2001. However, both MONTH and YEAR intervals are crossed.

Examples

The following statement creates the variable Years and assigns it a value of **2**. The INTCK function determines that 2 years have elapsed between June 15, 1999, and June 15, 2001.

```
Years=intck('year','15jun1999'd,'15jun2001'd);
```

As shown here, the *from* and *to* dates are often specified as date constants.

Likewise, the following statement assigns the value **24** to the variable Months.

```
Months=intck('month','15jun1999'd,'15jun2001'd);
```

However, the following statement assigns **0** to the variable Years, even though 364 days have elapsed. In this case the YEAR boundary (01JAN) is not crossed.

```
Years=intck('year','01jan2002'd,'31dec2002'd);
```

Example: The INTCK Function

A common use of the INTCK function is to identify periodic events such as due dates and anniversaries.

The following program identifies mechanics whose 20th year of employment occurs in the current month. It uses the INTCK function to compare the value of the variable Hired to the date on which the program is run.

```
data work.anniv20;
        set flights.mechanics(keep=id lastname firstname hired);
    Years=intck('year',hired,today());
    if years=20 and month(hired)=month(today());
run;
proc print data=work.anniv20;
    title '20-Year Anniversaries This Month';
run;
```

The following output is created when the program is run in December of 1999.

20-Year Anniversaries This Month

Obs	ID	LastName	FirstName	Hired	Years
1	1403	BOWDEN	EARL	24DEC79	20
2	1121	HERNANDEZ	MICHAEL	10DEC79	20
3	1412	MURPHEY	JOHN	08DEC79	20

INTNX Function

The INTNX function is similar to the INTCK function. The INTNX function applies multiples of a given interval to a date, time, or datetime value and returns the resulting value. You can use the INTNX function to identify past or future days, weeks, months, and so on.

General form, INTNX function:

INTNX('*interval*',*start-from,increment*<,'*alignment*'>**)**

where

- '*interval*' specifies a character constant or variable.
- *start-from* specifies a starting SAS date, time, or datetime value
- *increment* specifies a negative or positive integer that represents time intervals toward the past or future
- '*alignment*' (optional) forces the alignment of the returned date to the beginning, middle, or end of the interval.

The type of interval (date, time, or datetime) must match the type of value in *start-from* and *increment*.

When you specify date intervals, the value of the character constant or variable that is used in *interval* can be one of the following:

- DAY
- WEEKDAY
- WEEK
- TENDAY
- SEMIMONTH
- MONTH
- QTR
- SEMIYEAR
- YEAR
- DATETIME
- TIME

For example, the following statement creates the variable TargetYear and assigns it a SAS date value of **13515**, which corresponds to January 1, 1997.

```
TargetYear=intnx('year','05feb94'd,3);
```

Likewise, the following statement assigns the value for the date July 1, 2001, to the variable TargetMonth.

```
TargetMonth=intnx('semiyear','01jan2001'd,1);
```

As you know, SAS date values are based on the number of days since January 1, 1960. Yet the INTNX function can use intervals of weeks, months, years, and so on. What day should be returned when these larger intervals are used?

That's the purpose of the optional **alignment** value: it lets you specify whether the returned value should be at the beginning, middle, or end of the interval. When specifying date alignment in the INTNX function, use the following values or their corresponding aliases:

- BEGINNING Alias: B
- MIDDLE Alias:M
- END Alias:E

• SAME Alias: SAMEDAY or S

The best way to understand the alignment option is to see its effect on identical statements. The following table shows the results of three INTNX statements that differ only in the value of `alignment`.

Table 13.10   Alignment Values for the INTNX Function

SAS Statement	Date Value
`MonthX=intnx('month','01jan1995'd,5,'b');`	**12935** (June 1, 1995)
`MonthX=intnx('month','01jan1995'd,5,'m');`	**12949** (June 15, 1995)
`MonthX=intnx('month','01jan1995'd,5,'e');`	**12964** (June 30, 1995)

These INTNX statements count five months from January, but the returned value depends on whether `alignment` specifies the beginning, middle, or end day of the resulting month. If `alignment` is not specified, the beginning day is returned by default.

DATDIF and YRDIF Functions

The DATDIF and YRDIF functions calculate the difference in days and years between two SAS dates, respectively. Both functions accept start dates and end dates that are specified as SAS date values. Also, both functions use a basis argument that describes how SAS calculates the date difference.

General form, DATDIF and YRDIF functions:

DATDIF(*start_date,end_date,basis)*)

YRDIF(*start_date,end_date,basis)*)

where

• *start_date* specifies the starting date as a SAS date value

• *end_date* specifies the ending date as a SAS date value

• *basis* specifies a character constant or variable that describes how SAS calculates the date difference.

There are two character strings that are valid for basis in the DATDIF function and four character strings that are valid for basis in the YRDIF function. These character strings and their meanings are listed in the table below.

Table 13.11   Character Strings in the DATDIF Function

Character String	Meaning	Valid in DATDIF	Valid in YRDIF
`'30/360'`	specifies a 30 day month and a 360 day year	yes	yes
`'ACT/ACT'`	uses the actual number of days or years between dates	yes	yes

Character String	Meaning	Valid in DATDIF	Valid in YRDIF
'ACT/360'	uses the actual number of days between dates in calculating the number of years (calculated by the number of days divided by 360)	no	yes
'ACT/365'	uses the actual number of days between dates in calculating the number of years (calculated by the number of days divided by 365)	no	yes

The best way to understand the different options for the basis argument is to see the different effects that they have on the value that the function returns. The table below lists four YRDIF functions that use the same start_date and end_date. Each function uses one of the possible values for basis, and each one returns a different value.

Table 13.12 *Examples of the YRDIF Function*

YRDIF Statement	Returned Value
`yrdif('16oct1998'd,'16feb2003'd,'30/360')`	4.333333333
`yrdif('16oct1998'd,'16feb2003'd,'ACT/ACT')`	4.3369863014
`yrdif('16oct1998'd,'16feb2003'd,'ACT/360')`	4.4
`yrdif('16oct1998'd,'16feb2003'd,'ACT/365')`	4.3397260274

Modifying Character Values with Functions

Introduction to Modifying Character Values

This section teaches you how to use SAS functions to manipulate the values of character variables. After completing this section, you will be able to

- replace the contents of a character value

- trim trailing blanks from a character value

- search a character value and extract a portion of the value

- convert a character value to uppercase or lowercase.

To begin, let's look at some of the modifications that need to be made to the character variables in Hrd.Temp. These modifications include

- separating the values of one variable into multiple variables

SAS Data

Name	LastName	FirstName	MiddleName
CICHOCK, ELIZABETH MARIE ▸	CICHOCK	ELIZABETH	MARIE

- replacing a portion of a character variable's values

SAS Data

Phone	Phone
6224549 ▸	**433**4549

- searching for a specific string within a variable's values.

SAS Data

Job
filing, administrative duties

Character Functions

The character functions listed below can help you complete these tasks. You will learn about these functions as you continue this chapter.

Table 13.13 *Selected Character Functions*

Function	Purpose
SCAN	returns a specified word from a character value.
SUBSTR	extracts a substring or replaces character values.
TRIM	trims trailing blanks from character values.
CATX	concatenates character strings, removes leading and trailing blanks, and inserts separators.
INDEX	searches a character expression for a string of characters, and returns the position of the string's first character for the first occurrence of the string.
FIND	searches for a specific substring of characters within a character string.
UPPERCASE	converts all letters in a value to uppercase.
LOWCASE	converts all letters in a value to lowercase.
PROPCASE	converts all letters in a value to proper case.
TRANWRD	replaces or removes all occurrences of a pattern of characters within a character string.

SCAN Function

The SCAN function enables you to separate a character value into words and to return a specified word. Let's look at the following example to see how the SCAN function works.

The data set Hrd.Temp stores the names of temporary employees in the variable Name. The Name variable contains the employees' first, middle, and last names.

SAS Data Set Hrd.Temp

Agency	ID	Name
Administrative Support, Inc.	F274	CICHOCK, ELIZABETH MARIE
Administrative Support, Inc.	F101	BENINCASA, HANNAH LEE
OD Consulting, Inc.	F054	SHERE, BRIAN THOMAS

However, suppose you want to separate the value of Name into three variables: one variable to store the first name, one to store the middle name, and one to store the last name. You can use the SCAN function to create these new variables.

SAS Data Set Hrd.Newtemp

Agency	ID	LastName	FirstName	MiddleName
Administrative Support, Inc.	F274	CICHOCK	ELIZABETH	MARIE
Administrative Support, Inc.	F101	BENINCASA	HANNAH	LEE
OD Consulting, Inc.	F054	SHERE	BRIAN	THOMAS

Specifying Delimiters

The SCAN function uses delimiters, which are characters that are specified as word separators, to separate a character string into words. For example, if you are working with the character string below and you specify the comma as a delimiter, the SCAN function separates the string into three words.

Then the function returns whichever word you specify. In this example, if you specify the third word, the SCAN function returns the word HIGH.

Here's another example. Once again, let's use the comma as a delimiter, and specify that the third word be returned.

```
209 RADCLIFFE ROAD, CENTER CITY, NY, 92716
```

In this example, if you specify the third word, the word returned by the SCAN function is NY (NY contains a leading blank).

Specifying Multiple Delimiters

When using the SCAN function, you can specify as many delimiters as needed to correctly separate the character expression. When you specify multiple delimiters, SAS uses any of the delimiters, singly or in any combination, as word separators. For example, if you specify both the slash and the hyphen as delimiters, the SCAN function separates the following text string into three words:

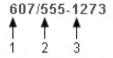

The SCAN function treats two or more contiguous delimiters, such as the parenthesis and slash below, as one delimiter. Also, leading delimiters have no effect.

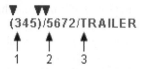

Default Delimiters

If you do not specify delimiters when using the SCAN function, default delimiters are used. The default delimiters are

```
blank . < ( + | & ! $ * ) ; ^ - / , %
```

SCAN Function Syntax

Now that you are familiar with how the SCAN function works, let's examine the syntax of the function.

General form, SCAN function:

SCAN(*argument,n*<,delimiters>**)**

where

- *argument* specifies the character variable or expression to scan.

- *n* specifies which word to return.

- *delimiters* are special characters that must be enclosed in single quotation marks (' '). If you do not specify *delimiters*, default delimiters are used.

Use the SCAN function to create your new name variables for Hrd.Temp. To begin, examine the values of the existing Name variable to determine which characters separate the names in the values. Notice that blanks and commas appear between the names and that the employee's last name appears first, then the first name, and then the middle name.

SAS Data Set Hrd.Temp

Agency	ID	Name
Administrative Support, Inc.	F274	**CICHOCK, ELIZABETH MARIE**
Administrative Support, Inc.	F101	**BENINCASA, HANNAH LEE**
OD Consulting, Inc.	F054	**SHERE, BRIAN THOMAS**
New Time Temps Agency	F077	**HODNOFF, RICHARD LEE**

To create the LastName variable to store the employee's last name, you write an assignment statement that contains the following SCAN function:

```
LastName=scan(name,1,' ,');
```

Note that a blank and a comma are specified as delimiters. You can also write the function without listing delimiters, because the blank and comma are default delimiters.

```
LastName=scan(name,1);
```

The complete DATA step that is needed to create LastName, FirstName, and MiddleName appears below. Notice that the original Name variable is dropped from the new data set.

```
data hrd.newtemp(drop=name);
   set hrd.temp;
   LastName=scan(name,1);
   FirstName=scan(name,2);
   MiddleName=scan(name,3);
run;
```

Specifying Variable Length

Note that the SCAN function assigns a length of 200 to each target variable. (Remember, a target variable is the variable that receives the result of the function.) So, if you submit the DATA step above, the LastName, FirstName, and MiddleName variables are each assigned a length of 200. This length is longer than necessary for these variables.

To save storage space, add a LENGTH statement to your DATA step, and specify an appropriate length for all three variables. Because SAS sets the length of a new character variable the first time it is encountered in the DATA step, be sure to place the LENGTH statement *before* the assignment statements that contain the SCAN function.

```
data hrd.newtemp(drop=name);
   set hrd.temp;
   length LastName FirstName MiddleName $ 10;
   lastname=scan(name,1);
   firstname=scan(name,2);
   middlename=scan(name,3);
run;
```

SCAN versus SUBSTR

The SUBSTR function is similar to the SCAN function. Let's briefly compare the two. Both the SCAN and SUBSTR functions can extract a substring from a character value:

- SCAN extracts words within a value that is marked by delimiters.

- SUBSTR extracts a portion of a value by starting at a specified location.

The SUBSTR function is best used when you know the exact position of the string that you want to extract from the character value. The string does not need to be marked by delimiters. For example, the first two characters of the variable ID identify the class level of college students. The position of these characters does not vary within the values of ID.

SAS Data

Name	ID
Trentonson, Matthew Robert	**SO**45467
Truell, Marcia Elizabeth	**SR**32881

The SUBSTR function is the best choice to extract class level information from ID. By contrast, the SCAN function is best used when

- you know the order of the words in the character value

- the starting position of the words varies

- the words are marked by some delimiter.

SUBSTR Function

The SUBSTR function can be used to

- extract a portion of a character value

- replace the contents of a character value.

Let's begin with the task of extracting a portion of a value. In the data set Hrd.Newtemp, the names of temporary employees are stored in three name variables: LastName, FirstName, and MiddleName.

SAS Data Set Hrd.Newtemp

Agency	ID	LastName	FirstName	MiddleName
Administrative Support, Inc.	F274	CICHOCK	ELIZABETH	**MARIE**
Administrative Support, Inc.	F101	BENINCASA	HANNAH	**LEE**
OD Consulting, Inc.	F054	SHERE	BRIAN	**THOMAS**
New Time Temps Agency	F077	HODNOFF	RICHARD	**LEE**

However, suppose you want to modify the data set to store only the middle initial instead of the full middle name. To do so, you must extract the first letter of the middle name values and assign these values to the new variable MiddleInitial.

SAS Data Set Work.Newtemp

Agency	ID	LastName	FirstName	MiddleInitial
Administrative Support, Inc.	F274	CICHOCK	ELIZABETH	**M**
Administrative Support, Inc.	F101	BENINCASA	HANNAH	**L**
OD Consulting, Inc.	F054	SHERE	BRIAN	**T**
New Time Temps Agency	F077	HODNOFF	RICHARD	**L**

The SUBSTR function enables you to extract any number of characters from a character string, starting at a specified position in the string.

General form, SUBSTR function:

SUBSTR(*argument,position* <,n>)

where

- *argument* specifies the character variable or expression to scan.

- *position* is the character position to start from.

- *n* specifies the number of characters to extract. If *n* is omitted, all remaining characters are included in the substring.

Using the SUBSTR function, you can extract the first letter of the MiddleName value to create the new variable MiddleInitial.

SAS Data Set Hrd.Newtemp

Agency	ID	LastName	FirstName	MiddleName
Administrative Support, Inc.	F274	CICHOCK	ELIZABETH	**MARIE**
Administrative Support, Inc.	F101	BENINCASA	HANNAH	**LEE**
OD Consulting, Inc.	F054	SHERE	BRIAN	**THOMAS**
New Time Temps Agency	F077	HODNOFF	RICHARD	**LEE**

You write the SUBSTR function as:

```
substr(middlename,1,1)
```

This function extracts a character string from the value of MiddleName. The string to be extracted begins in position 1 and contains one character. Then, you place this function in an assignment statement in your DATA step.

```
data work.newtemp(drop=middlename);
   set hrd.newtemp;
   length MiddleInitial $ 1;
   MiddleInitial=substr(middlename,1,1);
run;
```

The new MiddleInitial variable is given the same length as MiddleName. The MiddleName variable is then dropped from the new data set.

SAS Data Set Work.Newtemp

Agency	ID	LastName	FirstName	MiddleInitial
Administrative Support, Inc.	F274	CICHOCK	ELIZABETH	**M**
Administrative Support, Inc.	F101	BENINCASA	HANNAH	**L**
OD Consulting, Inc.	F054	SHERE	BRIAN	**T**
New Time Temps Agency	F077	HODNOFF	RICHARD	**L**

You can use the SUBSTR function to extract a substring from any character value if you know the position of the value.

Replacing Text Using SUBSTR

There is a second use for the SUBSTR function. This function can also be used to replace the contents of a character variable. For example, suppose the local phone exchange 622 was replaced by the exchange 433. You need to update the character variable Phone in Hrd.Temp to reflect this change.

SAS Data Set Hrd.Newtemp

City	State	Zip	Phone	StartDate	EndDate	PayRate	Days	Hours	BirthDate	Salary
CARY	NC	27513	6224549	14567	14621	10	11	88	7054	880
CARY	NC	27513	6223251	14524	14565	8	25	200	5757	1600

You can use the SUBSTR function to complete this modification. The syntax of the SUBSTR function, when used to replace a variable's values, is identical to the syntax for extracting a substring.

```
SUBSTR(argument,position,n)
```

However, in this case,

- the first argument specifies the character variable whose values are to be modified.

- the second argument specifies the position at which the replacement is to begin.

- the third argument specifies the number of characters to replace. If n is omitted, all remaining characters are replaced.

Positioning the SUBSTR Function

SAS uses the SUBSTR function to extract a substring or to modify a variable's values, depending on the position of the function in the assignment statement.

When the function is on the *right side* of an assignment statement, the function returns the requested string.

```
MiddleInitial=substr(middlename,1,1);
```

But if you place the SUBSTR function on the *left side* of an assignment statement, the function is used to modify variable values.

```
substr(region,1,3)='NNW';
```

When the SUBSTR function modifies variable values, the right side of the assignment statement must specify the value to place into the variable. For example, to replace the fourth and fifth characters of a variable named Test with the value **92**, you write the following assignment statement:

```
substr(test,4,2)='92';
```

```
Test                 Test


S7381K2    →    S7392K2
S7381K7    →    S7392K7
```

Now let's use the SUBSTR function to replace the 622 exchange in the variable Phone. You begin by writing this assignment statement:

```
data hrd.temp2;
   set hrd.temp;
   substr(phone,1,3)='433';
run;
```

} ijnu8

This statement specifies that the new exchange 433 should be placed in the variable Phone, starting at character position 1 and replacing three characters.

SAS Data Set Hrd.Temp

City	State	Zip	Phone	StartDate	EndDate	PayRate	Days	Hours
CARY	NC	27513	**622**4549	14567	14621	10	11	88
CARY	NC	27513	**622**3251	14524	14565	8	25	200
CHAPEL HILL	NC	27514	9974749	14570	14608	40	26	208

But executing this DATA step places the value **433** into all values of Phone. You need to replace only the values of Phone that contain the 622 exchange. So, you add an assignment statement to the DATA step to extract the exchange from Phone. Notice that the SUBSTR function is used on the right side of the assignment statement.

```
data hrd.temp2(drop=exchange);
   set hrd.temp;
   Exchange=substr(phone,1,3);
   substr(phone,1,3)='433';
run;
```

Now the DATA step needs an IF-THEN statement to verify the value of the variable Exchange. If the exchange is 622, the assignment statement executes to replace the value of Phone.

```
data hrd.temp2(drop=exchange);
   set hrd.temp;
   Exchange=substr(phone,1,3);
   if exchange='622' then substr(phone,1,3)='433';
run;
```

After the DATA step is executed, the appropriate values of Phone contain the new exchange.

SAS Data Set Hrd.Temp2

City	State	Zip	Phone	StartDate	EndDate	PayRate	Days	Hours
CARY	NC	27513	**433**4549	14567	14621	10	11	88
CARY	NC	27513	**433**3251	14524	14565	8	25	200
CHAPEL HILL	NC	27514	9974749	14570	14608	40	26	208

Once again, remember the rules for using the SUBSTR function. If the SUBSTR function is on the right side of an assignment statement, the function extracts a substring.

```
MiddleInitial=substr(middlename,1,1);
```

If the SUBSTR function is on the left side of an assignment statement, the function replaces the contents of a character variable.

```
substr(region,1,3)='NNW';
```

TRIM Function

The TRIM function enables you to remove trailing blanks from character values. To learn about the TRIM function, let's modify the data set Hrd.Temp.

The data set Hrd.Temp contains four address variables: Address, City, State, and Zip.

SAS Data Set Hrd.Temp

Agency	ID	Name	Address	City	State	Zip	Phone	StartDate	EndDate	PayRate	Days	Hours
Administrative Support, Inc.	F274	CICHOCK, ELIZABETH MARIE	65 ELM DR	CARY	NC	27513	6224549	14567	14621	10	11	88
Administrative Support, Inc.	F101	BENINCASA, HANNAH LEE	11 SUN DR	CARY	NC	27513	6223251	14524	14565	8	25	200

You need to create one address variable that contains the values of the three variables Address, City, and Zip. (Because all temporary employees are hired locally, the value of State does not need to be included in the new variable.)

SAS Data Set Hrd.NewTemp

Agency	ID	Name	NewAddress	Phone	StartDate	EndDate	PayRate	Days	Hours
Administrative Support, Inc.	F274	CICHOCK, ELIZABETH MARIE	65 ELM DR, CARY, 27513	6224549	14567	14621	10	11	88
Administrative Support, Inc.	F101	BENINCASA, HANNAH LEE	11 SUN DR, CARY, 27513	6223251	14524	14565	8	25	200

Writing a DATA step to create this new variable is easy. You include an assignment statement that contains the concatenation operator (||), as shown below.

```
data hrd.newtemp(drop=address city state zip);
   set hrd.temp;
   NewAddress=address||', '||city||', '||zip;
run;
```

The concatenation operator (||) enables you to concatenate character values. In this assignment statement, the character values of Address, City, and Zip are concatenated with two character constants that consist of a comma and a blank. The commas and blanks are needed to separate the street, city, and ZIP code values. The length of NewAddress is the sum of the length of each variable and constant that is used to create the new variable. Notice that this DATA step drops the original address variables from the new data set.

When the DATA step executes, you notice that the values of NewAddress do not appear as expected. The values of the new variable contain embedded blanks.

SAS Data Set Hrd.NewTemp

NewAddress
65 ELM DRIVE   , CARY   , 27513
11 SUN DRIVE   , CARY   , 27513
712 HARDWICK STREET   , CHAPEL HILL   , 27514
5372 WHITEBUD ROAD   , RALEIGH   , 27612

These blanks appear in the values of NewAddress because the values of the original address variables contained trailing blanks. Whenever the value of a character variable does not match the length of the variable, SAS pads the value with trailing blanks.

Address length=32	City length=15	Zip length=5
65 ELM DRIVE ·	RALEIGH · · · · · · · ·	27612
11 SUN DRIVE ·	DURHAM · · · · · · · · ·	27612
712 HARTWICK STREET · · · · · · · · · · · · · ·	CHAPEL HILL · · · ·	27514

So, when the original address values are concatenated to create NewAddress, the trailing blanks in the original values are included in the values of the new variable. The variable Zip is the only one that does not contain trailing blanks.

NewAddress length=56
65 ELM DRIVE · , RALEIGH · · · · · · · · , 27612
11 SUN DRIVE · , DURHAM · · · · · · · · · , 27612
712 HARTWICK STREET · · · · · · · · · · · · · , CHAPEL HILL · · · · · , 27514

The TRIM function enables you to remove trailing blanks from character values.

General form, TRIM function:

TRIM(*argument*)

where *argument* can be any character expression, such as

- a character variable: trim(address)

- another character function: trim(left(id)).

To remove the blanks from the variable NewAddress, include the TRIM function in your assignment statement. Trim the values of Address and City.

```
data hrd.newtemp(drop=address city state zip);
   set hrd.temp;
   NewAddress=trim(address)||', '||trim(city)||', '||zip;
run;
```

The revised DATA step creates the values that you expect for **NewAddress**.

SAS Data Set Hrd.Newtemp

NewAddress
65 ELM DRIVE, CARY, 27513
11 SUN DRIVE, CARY, 27513
712 HARDWICK STREET, CHAPEL HILL, 27514
5372 WHITEBUD ROAD, RALEIGH, 27612

Points to Remember

Keep in mind that the TRIM function does not affect how a variable is stored. Suppose you trim the values of a variable and then assign these values to a new variable. The trimmed values are padded with trailing blanks again if the values are shorter than the length of the new variable.

Here's an example. In the DATA step below, the trimmed value of Address is assigned to the new variable Street. When the trimmed value is assigned to Street, trailing blanks are added to the value to match the length of 20.

```
data temp;
   set hrd.temp;
   length Street $ 20;
```

```
     Street=trim(address);
run;
```

Address length=32	**Street** length=20
65 ELM DRIVE · · · · · · · · · · · · · · · · · ·	65 ELM DRIVE · · · · · · ·
11 SUN DRIVE · · · · · · · · · · · · · · · · · ·	11 SUN DRIVE · · · · · · ·
712 HARTWICK STREET · · · · · · · · · · · ·	712 HARTWICK STREET ·

CATX Function

Beginning in SAS®9, the CATX function enables you to concatenate character strings, remove leading and trailing blanks, and insert separators. The CATX function returns a value to a variable, or returns a value to a temporary buffer. The results of the CATX function are usually equivalent to those that are produced by a combination of the concatenation operator and the TRIM and LEFT functions.

In most cases, if the CATX function returns a value to a variable that has not previously been assigned a length, then that variable is given a length of 200 bytes. To save storage space, you can add a LENGTH statement to your DATA step, and specify an appropriate length for your variable. Because SAS sets the length of a new character variable the first time it is encountered in the DATA step, be sure to place the LENGTH statement before the assignment statements that contain the CATX function.

If the concatenation operator (||) returns a value to a variable that has not previously been assigned a length, then that variable is given a length that is the sum of the lengths of the values which are being concatenated.

Remember that you learned to use the TRIM function along with the concatenation operator to create one address variable that contains the values of the three variables Address, City, and Zip, and to remove extra blanks from the new values. You used the DATA step shown below.

```
data hrd.newtemp(drop=address city state zip);
   set hrd.temp;
   NewAddress=trim(address)||', '||trim(city)||', '||zip;
run;
```

You can accomplish the same concatenation using only the CATX function.

General form, CATX function:

CATX(*separator,string-1 <,...string-n>***)**

where

- *separator* specifies the character string that is used as a separator between concatenated strings

- *string* specifies a SAS character string.

You want to create the new variable NewAddress by concatenating the values of the Address, City, and Zip variables from the data set Hrd.Temp. You want to strip excess blanks from the old variables' values and separate the variable values with a comma and a space. The DATA step below uses the CATX function to create NewAddress.

```
data hrd.newtemp(drop=address city state zip);
   set hrd.temp;
   NewAddress=catx(', ',address,city,zip);
run;
```

The revised DATA step creates the values that you expect for NewAddress.

SAS Data Set Hrd.Newtemp

NewAddress
65 ELM DRIVE, CARY, 27513
11 SUN DRIVE, CARY, 27513
712 HARDWICK STREET, CHAPEL HILL, 27514
5372 WHITEBUD ROAD, RALEIGH, 27612

INDEX Function

The INDEX function enables you to search a character value for a specified string. The INDEX function searches values from left to right, looking for the first occurrence of the string. It returns the position of the string's first character; if the string is not found, it returns a value of **0**.

Suppose you need to search the values of the variable Job, which lists job skills. You want to create a data set that contains the names of all temporary employees who have word processing experience. The following figure shows a partial list of observations in the Hrd.Temp data set.

SAS Data Set Hrd.Temp

Job	Contact	Dept	Site
word processing	WORD PROCESSOR	DP	26
filing, administrative duties	ADMIN. ASST.	PURH	57
organizational dev. specialist	CONSULTANT	PERS	34
bookkeeping, word processing	BOOKKEEPER ASST.	BK	57

The INDEX function can complete this search.

General form, INDEX function:

INDEX(*source,excerpt*)

where

- *source* specifies the character variable or expression to search
- *excerpt* specifies a character string that is enclosed in quotation marks (")

To search for the occurrences of word processing in the values of the variable Job, you write the INDEX function as shown below. Note that the character string is enclosed in quotation marks.

```
index(job,'word processing')
```

Then, to create the new data set, include the INDEX function in a subsetting IF statement. Only those observations in which the function locates the string and returns a value greater than **0** are written to the data set.

```
data hrd.datapool;
   set hrd.temp;
   if index(job,'word processing') > 0;
run;
```

Here's your data set that shows the temporary employees who have word processing experience. The program processed all of the observations in the Hrd.Temp data set.

SAS Data Set Hrd.Datapool

Job	Contact	Dept	Site
word processing	WORD PROCESSOR	DP	26
bookkeeping, word processing	BOOKKEEPER AST	BK	57
word processing, sec. work	WORD PROCESSOR	DP	95
bookkeeping, word processing	BOOKKEEPER AST	BK	44
word processing	WORD PROCESSOR	DP	59
word processing, sec. work	WORD PROCESSOR	PUB	38
word processing	WORD PROCESSOR	DP	44
word processing	WORD PROCESSOR	DP	90

Note that the INDEX function is case sensitive, so the character string that you search for must be specified exactly as it is recorded in the data set. For example, the INDEX function shown below would not locate any employees who have word processing experience.

```
index(job,'WORD PROCESSING')
```

SAS Data Set Hrd.Temp

Job	Contact	Dept	Site
word processing	WORD PROCESSOR	DP	26
filing, administrative duties	ADMIN. ASST.	PURH	57
organizational dev. specialist	CONSULTANT	PERS	34
bookkeeping, word processing	BOOKKEEPER ASST.	BK	57

Finding a String Regardless of Case

To ensure that all occurrences of a character string are found, you can use the UPCASE or LOWCASE function with the INDEX function. The UPCASE and LOWCASE functions enable you to convert variable values to uppercase or lowercase letters. You can then specify the character string in the INDEX function accordingly.

```
index(upcase(job),'WORD PROCESSING')
```

```
index(lowcase(job),'word processing')
```

The UPCASE and LOWCASE functions are presented later in this section.

FIND Function

Overview

Beginning in SAS®9,the FIND function enables you to search for a specific substring of characters within a character string that you specify. The FIND function searches a string for the first occurrence of the substring, and returns the position of that substring. If the substring is not found in the string, FIND returns a value of **0**.

The FIND function is similar to the INDEX function. Remember that you used the INDEX function to search the values of the variable Job in Hrd.Temp in order to create a data set that contains the names of all temporary employees who have word processing experience.

SAS Data Set Hrd.Temp

Job	Contact	Dept	Site
word processing	WORD PROCESSOR	DP	26
filing, administrative duties	ADMIN. ASST.	PURH	57
organizational dev. specialist	CONSULTANT	PERS	34

You can also use the FIND function to complete this search.

General form, FIND function:

FIND(*string,substring<,modifiers><,startpos>*)

where

- *string* specifies a character constant, variable, or expression that will be searched for substrings
- *substring* is a character constant, variable, or expression that specifies the substring of characters to search for in *string*
- *modifiers* is a character constant, variable, or expression that specifies one or more modifiers
- *startpos* is an integer that specifies the position at which the search should start and the direction of the search. The default value for *startpos* is 1.

If *string* or *substring* is a character literal, you must enclose it in quotation marks.

The modifiers argument enables you to specify one or more modifiers for the function, as listed below.

- The modifier i causes the FIND function to ignore character case during the search. If this modifier is not specified, FIND searches for character substrings with the same case as the characters in substring.
- The modifier t trims trailing blanks from string and substring.

If the modifier is a constant, enclose it in quotation marks. Specify multiple constants in a single set of quotation marks. Modifier values are not case sensitive.

If startpos *is not* specified, FIND starts the search at the beginning of the string and searches the string from left to right. If startpos *is* specified, the absolute value of startpos determines the position at which to start the search. The sign of startpos determines the direction of the search. If startpos is *positive*, FIND searches from startpos to the right; and if startpos is *negative*, FIND searches from startpos to the left.

Example

The values of the variable Job are all lowercase. Therefore, to search for the occurrence of word processing in the values of the variable Job, you write the FIND function as shown below. Note that the character substring is enclosed in quotation marks.

```
find(job,'word processing')
```

Then, to create the new data set, include the FIND function in a subsetting IF statement. Only those observations in which the function locates the string and returns a value greater than **0** are written to the data set.

```
data hrd.datapool;
   set hrd.temp;
   if find(job,'word processing') > 0;
run;
```

UPCASE Function

The UPCASE function converts all letters in a character expression to uppercase.

General form, UPCASE function:

UPCASE(*argument*)

where *argument* can be any SAS character expression, such as a character variable or constant.

Let's use the UPCASE function to convert the values of a character variable in Hrd.Temp. The values of the variable Job appear in lowercase letters.

SAS Data Set Hrd.Temp

Job	Contact	Dept	Site
word processing	WORD PROCESSOR	DP	26
filing, administrative duties	ADMIN. ASST.	PURH	57
organizational dev. specialist	CONSULTANT	PERS	34
bookkeeping, word processing	BOOKKEEPER ASST.	BK	57

To convert the values of Job to uppercase, you write the UPCASE function as follows:

```
upcase(job)
```

Then place the function in an assignment statement in a DATA step. You can change the values of the variable Job in place.

```
data hrd.newtemp;
   set hrd.temp;
   Job=upcase(job);
run;
```

Here's the new data set that contains the converted values of Job.

SAS Data Set Hrd.Newtemp

Job	Contact	Dept	Site
WORD PROCESSING	WORD PROCESSOR	DP	26
FILING, ADMINISTRATIVE DUTIES	ADMIN. ASST.	PURH	57
ORGANIZATIONAL DEV. SPECIALIST	CONSULTANT	PERS	34
BOOKKEEPING, WORD PROCESSING	BOOKKEEPER ASST.	BK	57

LOWCASE Function

The LOWCASE function converts all letters in a character expression to lowercase.

General form, LOWCASE function:

LOWCASE(argument**)**

where *argument* can be any SAS character expression, such as a character variable or constant.

Here's an example of the LOWCASE function. In this example, the function converts the values of a variable named Title to lowercase letters.

```
lowcase(title)
```

Another example of the LOWCASE function is shown below. The assignment statement in this DATA step uses the LOWCASE function to convert the values of the variable Contact to lowercase.

SAS Data Set Hrd.Temp

Job	Contact	Dept	Site
word processing	**WORD PROCESSOR**	DP	26
filing, administrative duties	**ADMIN. ASST.**	PURH	57
organizational dev. specialist	**CONSULTANT**	PERS	34

```
data hrd.newtemp;
   set hrd.temp;
   Contact=lowcase(contact);
run;
```

After this DATA step executes, the new data set is created. Notice the converted values of the variable Contact.

SAS Data Set Hrd.Newtemp

Job	Contact	Dept	Site
word processing	**word processor**	DP	26
filing, administrative duties	**admin. asst.**	PURH	57
organizational dev. specialist	**consultant**	PERS	34

PROPCASE Function

Beginning in SAS®9, the PROPCASE function converts all words in an argument to proper case (so that the first letter in each word is capitalized).

General form, PROPCASE function:

PROPCASE(*argument<,delimiter(s)>***)**

where

- *argument* can be any SAS expression, such as a character variable or constant
- *delimiter(s)* specifies one or more delimiters that are enclosed in quotation marks. The default delimiters are blank, forward slash, hyphen, open parenthesis, period, and tab.

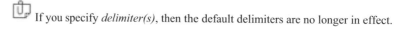 If you specify *delimiter(s)*, then the default delimiters are no longer in effect.

The PROPCASE function copies a character argument and converts all uppercase letters to lowercase letters. It then converts to uppercase the first character of a word that is preceded by a delimiter. PROPCASE uses the default delimiters unless you use the delimiter(s) argument.

Here's an example of the PROPCASE function. In this example, the function converts the values of a variable named Title to proper case and uses the default delimiters.

```
propcase(title)
```

Another example of the PROPCASE function is shown below. The assignment statement in this DATA step uses the PROPCASE function to convert the values for the variable Contact to proper case.

SAS Data Set Hrd.Temp

Job	Contact	Dept	Site
word processing	**WORD PROCESSOR**	DP	26
filing, administrative duties	**ADMIN. ASST.**	PURH	57
organizational dev. specialist	**CONSULTANT**	PERS	34

```
data hrd.newtemp;
   set hrd.temp;
   Contact=propcase(contact);
run;
```

After this DATA step executes, the new data set is created. Notice the converted values of the variable Contact.

SAS Data Set Hrd.Newtemp

Job	Contact	Dept	Site
word processing	**Word Processor**	DP	26
filing, administrative duties	**Admin. Asst.**	PURH	57
organizational dev. specialist	**Consultant**	PERS	34

TRANWRD Function

The TRANWRD function replaces or removes all occurrences of a pattern of characters within a character string. The translated characters can be located anywhere in the string.

In a DATA step, if the TRANWRD function returns a value to a variable that has not previously been assigned a length, then that variable is given a length of 200 bytes. To save storage space, you can add a LENGTH statement to your DATA step, and specify an appropriate length for your variable. Because SAS sets the length of a new character variable the first time it is encountered in the DATA step, be sure to place the LENGTH statement before the assignment statements that contain the TRANWRD function.

General form, TRANWRD function:

TRANWRD(*source,target,replacement*)

where

- *source* specifies the source string that you want to translate
- *target* specifies the string that SAS searches for in *source*
- *replacement* specifies the string that replaces *target*.

target and *replacement* can be specified as variables or as character strings. If you specify character strings, be sure to enclose the strings in quotation marks (' ' or " ").

You can use TRANWRD function to update variables in place. In this example, the function updates the values of Name by changing every occurrence of the string Monroe to Manson.

```
name=tranwrd(name,'Monroe','Manson')
```

Another example of the TRANWRD function is shown below. In this case, two assignment statements use the TRANWRD function to change all occurrences of Miss or Mrs. to Ms.

```
data work.after;
   set work.before;
   name=tranwrd(name,'Miss','Ms.');
   name=tranwrd(name,'Mrs.','Ms.');
run;
```

SAS Data Set Work.Before

Name
Mrs. Millicent Garrett Fawcett
Miss Charlotte Despard
Mrs. Emmeline Pankhurst
Miss Sylvia Pankhurst

After this DATA step executes, the new data set is created. Notice the changed strings within the variable Name.

SAS Data Set Work.After

Name
Ms. Millicent Garrett Fawcett
Ms. Charlotte Despard
Ms. Emmeline Pankhurst
Ms. Sylvia Pankhurst

Modifying Numeric Values with Functions

Introduction to Modifying Numeric Values

You've seen how SAS functions can be used to

- convert between character and numeric variable values

- manipulate SAS date values

- modify values of character variables.

SAS provides additional functions to create or modify numeric values. These include arithmetic, financial, and probability functions. There are far too many of these functions to explore them all in detail, but let's look at two examples.

INT Function

To return the integer portion of a numeric value, use the INT function. Any decimal portion of the INT function argument is discarded.

General form, INT function:

INT(*argument*)

where *argument* is a numeric variable, constant, or expression.

The two data sets shown below give before-and-after views of values that are truncated by the INT function.

SAS Data Set
Work.Before

Examples
326.54
98.20
-32.66
1401.75

```
data work.after;
   set work.before;
```

```
Examples=int(examples);
run;
```

SAS Data Set
Work.After

Examples
326
98
-32
1401

ROUND Function

To round ... ues to the nearest specified unit, use the ROUND function.

... ROUND function:

...ment,round-off-unit)

... s a numeric variable, constant, or expression.

...nit is numeric and nonnegative.

... unit is not provided, a default value of 1 is used, and the argument is ... e nearest integer. The two data sets shown below give before-and-after ... es that are modified by the ROUND function.

```
...ter;
...before;
...round(examples,.2);
```

SAS Data Set
Work.After

Examples
326.6
98.2
-32.6
1401.8

To learn more about SAS functions that modify numeric values, see the SAS documentation.

Nesting SAS Functions

Throughout this chapter, you've seen examples of individual functions. For example, in this assignment statement the SCAN function selects the middle name (third word) from the variable Name:

```
MiddleName=scan(name,3);
```

Then this assignment statement uses the SUBSTR function to select the first letter from the variable MiddleName:

```
MiddleInitial=substr(MiddleName,1,1);
```

To write more efficient programs, however, you can nest functions as appropriate. For example, you can nest the SCAN function within the SUBSTR function in an assignment statement to compute the value for MiddleInitial:

```
MiddleInitial=substr(scan(name,3),1,1);
```

This example of nested numeric functions determines the number of years between June 15, 1999, and today:

```
Years=intck('year','15jun1999'd,today());
```

You can nest any functions as long as the function that is used as the argument meets the requirements for the argument.

Chapter Summary

Text Summary

Using SAS Functions
SAS functions can be used to convert data and to manipulate the values of character variables. Functions are written by specifying the function name, then its arguments in parentheses. Arguments can include variables, constants, expressions, or other functions. Although arguments are typically separated by commas, they can also be specified as variable lists or arrays.

Automatic Character-to-Numeric Conversion
When character variables are used in a numeric context, SAS tries to convert the character values to numeric values. Numeric context includes arithmetic operations, comparisons with numeric values, and assignment to previously defined numeric variables. The original character values are not changed. The conversion creates temporary numeric values and places a note in the SAS log.

Explicit Character-to-Numeric Conversion

The INPUT function provides direct, controlled conversion of character values to numeric values. When a character variable is read with a numeric informat, the INPUT function generates numeric values without placing a note in the SAS log.

Automatic Numeric-to-Character Conversion

When numeric variables are used in a character context, SAS tries to convert the numeric values to character values. Character context includes concatenation operations, use in functions that require character arguments, and assignment to previously defined character variables. The original numeric values are not changed; the conversion creates temporary character values and places a note in the SAS log.

Explicit Numeric-to-Character Conversion

The PUT function provides direct, controlled conversion of numeric values to character values. The format specified in a PUT function must match the source, so use an appropriate numeric format to create the new character values. No note will appear in the SAS log.

SAS Date and Time Values

SAS date values are stored as the number of days from January 1, 1960; time values are stored as the number of seconds since midnight. These values can be displayed in a variety of forms by associating them with SAS formats.

YEAR, QTR, MONTH, and DAY Functions

To extract the year, quarter, month, or day value from a SAS date value, specify the YEAR, QTR, MONTH, or DAY function followed by the SAS date value in parentheses. The YEAR function returns a four-digit number; QTR returns a value of `1`, `2`, `3`, or `4`; MONTH returns a number from `1` to `12`; and DAY returns `1` to `31`.

WEEKDAY Function

To extract the day of the week from a SAS date value, specify the function WEEKDAY followed by the SAS date value in parentheses. The function returns a numeric value from `1` to `7`, representing the day of the week.

MDY Function

To create a SAS date value from a month, day, and year, specify the MDY function followed by the component values. The result can be displayed in several ways by applying a SAS date format. SAS interprets two-digit values according to the 100-year span that is set by the YEARCUTOFF= system option. The default value of YEARCUTOFF= is `1920`.

DATE and TODAY Functions

To convert the current date to a SAS date value, specify the DATE or TODAY function without arguments. The DATE and TODAY functions can be used interchangeably.

INTCK Function

To count the number of time intervals that occur in a time span, use the INTCK function and specify the interval constant or variable, the beginning date value, and the ending date value. The INTCK function counts intervals from fixed interval beginnings, not in multiples of an interval unit. Partial intervals are not counted.

INTNX Function

To apply multiples of an interval to a date value, use the INTNX function and specify the interval constant or variable, the start-from date value, and the increment. Include the alignment option to specify whether the date returned should be at the beginning, middle, or end of the interval, or on the same day.

DATDIF and YRDIF Functions

To calculate the difference between dates as a number of days or as a number of years, use the DATDIF or YRDIF function. These functions accept SAS date values and return a difference between the date values calculated according to the basis that you specify in the function.

SCAN Function

The SCAN function enables you to separate a character value into words and to return a word based on its position. It defines words according to delimiters, which are characters that are used as word separators. The name of the function is followed, in parentheses, by the name of the character variable, the desired word number, and the specified delimiters enclosed in quotation marks.

SUBSTR Function

The SUBSTR function can be used to extract or replace any portion of a character string. To extract values, place the function on the right side of an assignment statement and specify, in parentheses, the name of the character variable, the starting character position, and the number of characters to extract. To replace values, place the function on the left side of an assignment statement and specify, in parentheses, the name of the variable being modified, the starting character position, and the number of characters to replace.

SCAN versus SUBSTR

Both the SCAN and SUBSTR functions can extract a substring from a character value. SCAN relies on delimiters, whereas SUBSTR reads values from specified locations. Use SCAN when you know the delimiter and the order of words. Use SUBSTR when the positions of the characters don't vary.

TRIM Function

Because SAS pads character values with trailing blanks, unwanted spaces can sometimes appear after strings are concatenated. To remove trailing blanks from character values, specify the TRIM function with the name of a character variable. Remember that trimmed values will be padded with blanks again if they are shorter than the length of the new variable.

CATX Function

Beginning in SAS®9, you can concatenate character strings, remove leading and trailing blanks, and insert separators in one statement by using the CATX function. The results of the CATX function are usually equivalent to those that are produced by a combination of the concatenation operator and the TRIM and LEFT functions.

INDEX Function

To test character values for the presence of a string, use the INDEX function and specify, in parentheses, the name of the variable and the string enclosed in quotation marks. The INDEX function can be used with a subsetting IF statement when you create a data set. Only those observations in which the function finds the string and returns a value greater than 0 are written to the new data set.

FIND Function

Beginning in SAS®9, you can also use the FIND function to search for a specific substring of characters within a character string that you specify. The FIND function is similar to the INDEX function, but the FIND function enables you to ignore character case in your search and to trim trailing blanks. The FIND function can also begin the search at any position that you specify in the string.

UPCASE Function

Lowercase letters in character values can be converted to uppercase by using the UPCASE function. Include the function in an assignment statement, and specify the variable name in parentheses.

LOWCASE Function

Uppercase letters in character values can be converted to lowercase by using the LOWCASE function. Include the function in an assignment statement, and specify the variable name in parentheses.

PROPCASE Function

Beginning in SAS®9, character values can be converted to proper case by using the PROPCASE function. Include the function in an assignment statement, and specify the variable name in parentheses. Remember that you can specify delimiters or use the default delimiters.

TRANWRD Function

You can replace or remove patterns of characters in the values of character variables by using the TRANWRD function. Use the function in an assignment statement, and specify the source, target, and replacement strings or variables in parentheses.

INT Function

To return the integer portion of a numeric value, use the INT function. Any decimal portion of the INT function argument is discarded.

ROUND Function

To round values to the nearest specified unit, use the ROUND function. If a round-off unit is not provided, the argument is rounded to the nearest integer.

Nesting SAS Functions

To write more efficient programs, you can nest functions as appropriate. You can nest any functions as long as the function that is used as the argument meets the requirements for the argument.

Syntax

INPUT(*source,informat*)

PUT(*source,format*)

YEAR(*date*)

QTR(*date*)

MONTH(*date*)

DAY(*date*)

WEEKDAY(*date*)

MDY(*month,day,year*)

DATE()

TODAY()

INTCK(*'interval',from,to*)

INTNX(*'interval',start-from,increment<,'alignment'>*)

DATDIF(*start_date,end_date,basis*)

YRDIF(*start_date,end_date,basis*)

SCAN(*argument,n,<delimiters>*)

SUBSTR(*argument,position,<n>*)

TRIM(*argument*)

CATX(*separator,string-1<,...string-n>*)

INDEX(*source,excerpt*)

FIND(*string,substring<,modifiers><,startpos>*)

UPCASE(*argument*)

LOWCASE(*argument*)

PROPCASE(*(argument<,delimiter(s)>*)

TRANWRD(*source,target,replacement*)

INT(*argument*)

ROUND(*argument,round-off-unit*)

Sample Programs

```
data hrd.newtemp;
   set hrd.temp;
   Salary=input(payrate,2.)*hours;
run;

data hrd.newtemp;
   set hrd.temp;
   Assignment=put(site,2.)||'/'||dept;
run;

data hrd.tempnov;
   set hrd.temp;
   if month(startdate)=11;
run;

data hrd.temp2010;
   set hrd.temp;
   if year(startdate)=2010;
run;

data radio.schwkend;
   set radio.sch;
   if weekday(airdate) in(1,7);
run;

data hrd.newtemp(drop=month day year);
   set hrd.temp;
```

```
      Date=mdy(month,day,year);
run;

data hrd.newtemp;
   set hrd.temp;
   EditDate=today();
run;

data work.anniv20;
   set flights.mechanics(keep=id lastname firstname hired);
   Years=intck('year',hired,today());
   if years=20 and month(hired)=month(today());
proc print data=work.anniv20;
   title '20-Year Anniversaries this Month';
run;

data work.after;
   set work.before;
   TargetYear=intnx('year','05feb94'd,3);
run;

data hrd.newtemp(drop=name);
   set hrd.newtemp;
   length LastName FirstName MiddleName $ 10;
   lastname=scan(name,1,' ,');
   firstname=scan(name,2,' ,');
   middlename=scan(name,3,' ,');
run;

data hrd.temp2(drop=exchange);
   set hrd.temp;
   exchange=substr(phone,1,3);
   if exchange='622' then substr(phone,1,3)='433';
run;

data hrd.newtemp(drop=address city state zip);
   set hrd.temp;
   NewAddress=trim(address)||', '||trim(city)||', '||zip;
run;

data hrd.datapool;
   set hrd.temp;
   if index(job,'word processing') > 0;
run;

data hrd.newtemp;
   set hrd.temp;
   Job=upcase(job);
run;

data hrd.newtemp;
   set hrd.temp;
   Contact=lowcase(contact);
run;

data hrd.newtemp;
   set hrd.temp;
   location=propcase(city);
run;
```

```
data work.after;
    set work.before;
    name=tranwrd(name,'Miss','Ms.');
    name=tranwrd(name,'Mrs.','Ms.');
run;

data hrd.newtemp(drop=address city state zip);
    set hrd.temp;
    NewAddress=catx(', ',address,city,zip);
run;

data work.after;
    set work.before;
    Examples=int(examples);
run;

data work.after;
    set work.before;
    Examples=round(examples,.2);
run;

data hrd.datapool;
    set hrd.temp;
    if find(job,'word processing','t')>0;
run;
```

Points to Remember

- Even if a function doesn't require arguments, the function name must still be followed by parentheses.

- When specifying a variable list or an array as a function argument, be sure to precede the list or the array with the word OF.

- To remember which function requires a format versus an informat, note that the INPUT function requires the informat.

- If you specify an invalid date in the MDY function, a missing value is assigned to the target variable.

- The SCAN function treats contiguous delimiters as one delimiter; leading delimiters have no effect.

- When using the SCAN function, you can save storage space by adding a LENGTH statement to your DATA step to set an appropriate length for your new variable(s). Place the LENGTH statement before the assignment statements that contain the SCAN function.

- When the SUBSTR function is on the left side of an assignment statement, it replaces variable values. When SUBSTR is on the right side of an assignment statement, it extracts variable values. The syntax of the function is the same; only the placement of the function changes.

- The INDEX function is case sensitive. To ensure that all forms of a character string are found, use the UPCASE or LOWCASE function with the INDEX function. You can also use the FIND function with the i modifier.

Chapter Quiz

Select the best answer for each question. After completing the quiz, you can check your answers using the answer key in the appendix.

1. Which function calculates the average of the variables Var1, Var2, Var3, and Var4?

   a. `mean(var1,var4)`

   b. `mean(var1-var4)`

   c. `mean(of var1,var4)`

   d. `mean(of var1-var4)`

2. Within the data set Hrd.Temp, PayRate is a character variable and Hours is a numeric variable. What happens when the following program is run?

   ```
   data work.temp;
      set hrd.temp;
      Salary=payrate*hours;
   run;
   ```

   a. SAS converts the values of PayRate to numeric values. No message is written to the log.

   b. SAS converts the values of PayRate to numeric values. A message is written to the log.

   c. SAS converts the values of Hours to character values. No message is written to the log.

   d. SAS converts the values of Hours to character values. A message is written to the log.

3. A typical value for the character variable Target is `123,456`. Which statement correctly converts the values of Target to numeric values when creating the variable TargetNo?

   a. `TargetNo=input(target,comma6.);`

   b. `TargetNo=input(target,comma7.);`

   c. `TargetNo=put(target,comma6.);`

   d. `TargetNo=put(target,comma7.)`

4. A typical value for the numeric variable SiteNum is `12.3`. Which statement correctly converts the values of SiteNum to character values when creating the variable Location?

   a. `Location=dept||'/'||input(sitenum,3.1);`

   b. `Location=dept||'/'||input(sitenum,4.1);`

   c. `Location=dept||'/'||put(sitenum,3.1);`

   d. `Location=dept||'/'||put(sitenum,4.1);`

5. Suppose the YEARCUTOFF= system option is set to 1920. Which MDY function creates the date value for January 3, 2020?

   a. `MDY(1,3,20)`

b. `MDY(3,1,20)`

c. `MDY(1,3,2020)`

d. `MDY(3,1,2020)`

6. The variable Address2 contains values such as **Piscataway, NJ**. How do you assign the two-letter state abbreviations to a new variable named State?

a. `State=scan(address2,2);`

b. `State=scan(address2,13,2);`

c. `State=substr(address2,2);`

d. `State=substr(address2,13,2);`

7. The variable IDCode contains values such as **123FA** and **321MB**. The fourth character identifies sex. How do you assign these character codes to a new variable named Sex?

a. `Sex=scan(idcode,4);`

b. `Sex=scan(idcode,4,1);`

c. `Sex=substr(idcode,4);`

d. `Sex=substr(idcode,4,1);`

8. Due to growth within the 919 area code, the telephone exchange 555 is being reassigned to the 920 area code. The data set Clients.Piedmont includes the variable Phone, which contains telephone numbers in the form 919-555-1234. Which of the following programs will correctly change the values of Phone?

a.
```
data work.piedmont(drop=areacode exchange);
   set clients.piedmont;
   Areacode=substr(phone,1,3);
   Exchange=substr(phone,5,3);
   if areacode='919' and exchange='555'
      then scan(phone,1,3)='920';
run;
```

b.
```
data work.piedmont(drop=areacode exchange);
   set clients.piedmont;
   Areacode=substr(phone,1,3);
   Exchange=substr(phone,5,3);
   if areacode='919' and exchange='555'
      then phone=scan('920',1,3);
run;
```

c.
```
data work.piedmont(drop=areacode exchange);
   set clients.piedmont;
   Areacode=substr(phone,1,3);
   Exchange=substr(phone,5,3);
   if areacode='919' and exchange='555'
      then substr(phone,1,3)='920';
run;
```

d.
```
data work.piedmont(drop=areacode exchange);
   set clients.piedmont;
   Areacode=substr(phone,1,3);
   Exchange=substr(phone,5,3);
   if areacode='919' and exchange='555'
```

```
          then phone=substr('920',1,3);
     run;
```

9. Suppose you need to create the variable FullName by concatenating the values of
   FirstName, which contains first names, and LastName, which contains last names.
   What's the best way to remove extra blanks between first names and last names?

   a.
   ```
   data work.maillist;
        set retail.maillist;
        length FullName $ 40;
        fullname=trim firstname||' '||lastname;
   run;
   ```

   b.
   ```
   data work.maillist;
        set retail.maillist;
        length FullName $ 40;
        fullname=trim(firstname)||' '||lastname;
   run;
   ```

   c.
   ```
   data work.maillist;
        set retail.maillist;
        length FullName $ 40;
        fullname=trim(firstname)||' '||trim(lastname);
   run;
   ```

   d.
   ```
   data work.maillist;
        set retail.maillist;
        length FullName $ 40;
        fullname=trim(firstname||' '||lastname);
   run;
   ```

10. Within the data set Furnitur.Bookcase, the variable Finish contains values such as
    ash/cherry/teak/matte-black. Which of the following creates a subset of
    the data in which the values of Finish contain the string **walnut**? Make the search
    for the string case-insensitive.

    a.
    ```
    data work.bookcase;
         set furnitur.bookcase;
         if index(finish,walnut) = 0;
    run;
    ```

    b.
    ```
    data work.bookcase;
         set furnitur.bookcase;
         if index(finish,'walnut') > 0;
    run;
    ```

    c.
    ```
    data work.bookcase;
         set furnitur.bookcase;
         if index(lowcase(finish),walnut) = 0;
    run;
    ```

    d.
    ```
    data work.bookcase;
         set furnitur.bookcase;
         if index(lowcase(finish),'walnut') > 0;
    run;
    ```

Chapter 14

Generating Data with DO Loops

Overview

Introduction

You can execute SAS statements repeatedly by placing them in a DO loop. DO loops
can execute any number of times in a single iteration of the DATA step. Using DO loops
lets you write concise DATA steps that are easier to change and debug.

For example, the DO loop in this program eliminates the need for 12 separate
programming statements to calculate annual earnings:

```
data finance.earnings;
   set finance.master;
   Earned=0;
   do month=1 to 12;
      earned+(amount+earned)*(rate/12);
   end;
run;
```

You can also use DO loops to

- generate data

- conditionally execute statements

- read data.

This chapter shows you how to construct DO loops and how to include them in your programs.

Objectives

In this chapter, you learn to

- construct a DO loop to perform repetitive calculations

- control the execution of a DO loop

- generate multiple observations in one iteration of the DATA step

- construct nested DO loops.

Constructing DO Loops

Overview

DO loops process a group of statements repeatedly rather than once. This can greatly reduce the number of statements required for a repetitive calculation. For example, these twelve sum statements compute a company's annual earnings from investments. Notice that all 12 statements are identical.

```
data finance.earnings;
   set finance.master;
   Earned=0;
   earned+(amount+earned)*(rate/12);
   earned+(amount+earned)*(rate/12);
   earned+(amount+earned)*(rate/12);
   earned+(amount+earned)*(rate/12);
   earned+(amount+earned)*(rate/12);
   earned+(amount+earned)*(rate/12);
   earned+(amount+earned)*(rate/12);
   earned+(amount+earned)*(rate/12);
   earned+(amount+earned)*(rate/12);
   earned+(amount+earned)*(rate/12);
   earned+(amount+earned)*(rate/12);
   earned+(amount+earned)*(rate/12);
run;
```

Each sum statement accumulates the calculated interest earned for an investment for one month. The variable Earned is created in the DATA step to store the earned interest. The investment is compounded monthly, meaning that the value of the earned interest is cumulative.

A DO loop enables you to achieve the same results with fewer statements. In this case, the sum statement executes twelve times within the DO loop during each iteration of the DATA step.

```
data finance.earnings;
   set finance.master;
   Earned=0;
   do month=1 to 12;
      earned+(amount+earned)*(rate/12);
   end;
run;
```

To construct a DO loop, you use the DO and END statements along with other SAS statements.

General form, simple iterative DO loop:

DO *index-variable=start* **TO** *stop* **BY** *increment*;

        SAS statements

END;

where the *start*, *stop*, and *increment* values

- are set upon entry into the DO loop

- cannot be changed during the processing of the DO loop

- can be numbers, variables, or SAS expressions.

The **END** statement terminates the loop.

 The value of the index variable can be changed within the loop.

When creating a DO loop with the iterative DO statement, you must specify an index variable. The index variable stores the value of the current iteration of the DO loop. You may use any valid SAS name.

```
DO index-variable=start TO stop BY increment;
   SAS statements
END;
```

Next, specify the conditions that execute the DO loop. A simple specification contains a start value, a stop value, and an increment value for the DO loop.

```
Do index-variable=start TO stop BY increment;
   SAS statements
END;
```

The start value specifies the initial value of the index variable.

```
DO index-variable=start TO stop BY increment;
   SAS statements
END;
```

The TO clause specifies the stop value. The stop value is the last index value that executes the DO loop.

```
DO index-variable=start TO stop BY increment;
   SAS statements
END;
```

The optional BY clause specifies an increment value for the index variable. Typically, you want the DO loop to increment by 1 for each iteration. If you do not specify a BY clause, the default increment value is **1**.

```
DO index-variable=start TO stop BY increment;
   SAS statements
END;
```

For example, the specification below increments the index variable by 1, resulting in quiz values of **1**, **2**, **3**, **4**, and **5**:

```
do quiz=1 to 5;
```

By contrast, the following specification increments the index variable by 2, resulting in rows values of **2**, **4**, **6**, **8**, **10**, and **12**:

```
do rows=2 to 12 by 2;
```

DO Loop Execution

Using the form of the DO loop that was just presented, let's see how the DO loop executes in the DATA step. This example sums the interest that was earned each month for a one-year investment.

```
data finance.earnings;
   Amount=1000;
   Rate=.075/12;
   do month=1 to 12;
       Earned+(amount+earned)*rate;
   end;
run;
```

This DATA step does not read data from an external source. When submitted, it compiles and then executes only once to generate data. During compilation, the program data vector is created for the Finance.Earnings data set.

Program Data Vector

N_	Amount	Rate	month	Earned
.	.	.	.	.

When the DATA step executes, the values of Amount and Rate are assigned.

Program Data Vector

N_	Amount	Rate	month	Earned
1	1000	0.00625	.	.

Next, the DO loop executes. During each execution of the DO loop, the value of Earned is calculated and is added to its previous value; then the value of month is incremented. On the twelfth execution of the DO loop, the value of month is incremented to **12** and the value of Earned is **77.6326**.

Program Data Vector

N	Amount	Rate	month	Earned
1	1000	0.00625	12	77.6326

After the twelfth execution of the DO loop, the value of month is incremented to **13**. Because **13** exceeds the stop value of the iterative DO statement, the DO loop stops executing, and processing continues to the next DATA step statement. The end of the DATA step is reached, the values are written to the Finance.Earnings data set, and in this example, the DATA step ends. Only one observation is written to the data set.

SAS Data Set Finance.Earnings

Amount	Rate	month	Earned
1000	0.00625	13	77.6326

Notice that the index variable month is also stored in the data set. In most cases, the index variable is needed only for processing the DO loop and can be dropped from the data set.

Counting Iterations of DO Loops

In some cases, it is useful to create an index variable to count and store the number of iterations in the DO loop. Then you can drop the index variable from the data set.

```
data work.earn (drop=counter);
   Value=2000;
   do counter=1 to 20;
      Interest=value*.075;
      value+interest;
         Year+1;
   end;
run;
```

SAS Data Set Work.Earn

Value	Interest	Year
8495.70	592.723	20

The sum statement Year+1 accumulates the number of iterations of the DO loop and stores the total in the new variable Year. The final value of Year is then stored in the data set, whereas the index variable counter is dropped. The data set has one observation.

Explicit OUTPUT Statements

To create an observation for each iteration of the DO loop, place an OUTPUT statement inside the loop. By default, every DATA step contains an implicit OUTPUT statement at the end of the step. But placing an explicit OUTPUT statement in a DATA step overrides automatic output, causing SAS to add an observation to the data set only when the explicit OUTPUT statement is executed.

The previous example created one observation because it used automatic output at the end of the DATA step. In the following example, the OUTPUT statement overrides automatic output, so the DATA step writes 20 observations.

```
data work.earn;
   Value=2000;
   do Year=1 to 20;
      Interest=value*.075;
      value+interest;
      output;
   end;
run;
```

SAS Data Set Work.Earn
(Partial Listing)

Value	Year	Interest
2150.00	1	150.000
2311.25	2	161.250
2484.59	3	173.344
2670.94	4	186.345
2871.26	5	200.320
3086.60	6	215.344
3318.10	7	231.495
3566.96	8	248.857
...	...	...
8495.70	20	592.723

Decrementing DO Loops

You can decrement a DO loop's index variable by specifying a negative value for the BY clause. For example, the specification in this iterative DO statement decreases the index variable by 1, resulting in values of **5**, **4**, **3**, **2**, and **1**.

```
DO index-variable=5 to 1 by -1;
   SAS statements
END;
```

When you use a negative BY clause value, the start value must always be greater than the stop value in order to decrease the index variable during each iteration.

```
DO index-variable=5 to 1 by -1;
   SAS statements
END;
```

Specifying a Series of Items

You can also specify how many times a DO loop executes by listing items in a series.

General form, DO loop with a variable list:

DO *index-variable=value1, value2, value3... ;*

       SAS statements

END;

where *values* can be character or numeric.

When the DO loop executes, it executes once for each item in the series. The index variable equals the value of the current item. You must use commas to separate items in the series.

To list items in a series, you must specify either

- all numeric values

```
DO index-variable=2,5,9,13,27;
   SAS statements
END;
```

- all character values, with each value enclosed in quotation marks

```
DO index-variable='MON','TUE','WED','THR','FRI';
   SAS statements
END;
```

- all variable names—the index variable takes on the values of the specified variables.

```
DO index-variable=win,place,show;
   SAS statements
END;
```

Variable names must represent either all numeric or all character values. Do *not* enclose variable names in quotation marks.

Nesting DO Loops

Iterative DO statements can be executed within a DO loop. Putting a DO loop within a DO loop is called nesting.

```
do i=1 to 20;
   SAS statements
   do j=1 to 10;
      SAS statements
   end;
   SAS statements
end;
```

The DATA step below computes the value of a one-year investment that earns 7.5% annual interest, compounded monthly.

```
data work.earn;
   Capital=2000;
   do month=1 to 12;
      Interest=capital*(.075/12);
      capital+interest;
   end;
run;
```

Let's assume the same amount of capital is to be added to the investment each year for 20 years. The new program must perform the calculation for each month during each of the 20 years. To do this, you can include the monthly calculations within another DO loop that executes 20 times.

```
data work.earn;
   do year=1 to 20;
      Capital+2000;
      do month=1 to 12;
```

```
          Interest=capital*(.075/12);
          capital+interest;
       end;
    end;
 run;
```

During each iteration of the outside DO loop, an additional 2,000 is added to the capital, and the nested DO loop executes twelve times.

```
data work.earn;
   do year=1 to 20;
      Capital+2000;
      do month=1 to 12;
            Interest=capital*(.075/12);
            capital+interest;
      end;
   end;
run;
```

Remember, in order for nested DO loops to execute correctly, you must

- assign a unique index-variable name in each iterative DO statement.

```
data work.earn;
   do year=1 to 20;
      Capital+2000;
      do month=1 to 12;
         Interest=capital*(.075/12);
         capital+interest;
      end;
   end;
run;
```

- end each DO loop with an END statement.

```
data work.earn;
   do year=1 to 20;
      Capital+2000;
      do month=1 to 12;
         Interest=capital*(.075/12);
         capital+interest;
      end;
   end;
run;
```

It is easier to manage nested DO loops if you indent the statements in each DO loop as shown above.

Iteratively Processing Observations from a Data Set

So far, you have seen examples of DATA steps that use DO loops to generate one or more observations from one iteration of the DATA step. Now, let's look at a DATA step that reads a data set to compute the value of a new variable.

The SAS data set Finance.CDRates, shown below, contains interest rates for certificates of deposit (CDs) that are available from several institutions.

SAS Data Set Finance.CDRates

Institution	Rate	Years
MBNA America	0.0817	5
Metropolitan Bank	0.0814	3
Standard Pacific	0.0806	4

Suppose you want to compare how much each CD will earn at maturity with an investment of $5,000. The DATA step below creates a new data set, Work.Compare, that contains the added variable, Investment.

```
data work.compare(drop=i);
   set finance.cdrates;
   Investment=5000;
   do i=1 to years;
      investment+rate*investment;
   end;
run;
```

SAS Data Set Work.Compare

Institution	Rate	Years	Investment

The index variable is used only to execute the DO loop, so it is dropped from the new data set. Notice that the data set variable Years is used as the stop value in the iterative DO statement. As a result, the DO loop executes the number of times specified by the current value of Years.

During each iteration of the DATA step,

- an observation is read from Finance.CDRates

- the value **5000** is assigned to the variable Investment

- the DO loop executes, based on the current value of Years

- the value of Investment is incremented (each time that the DO loop executes), using the current value of Rate.

At the bottom of the DATA step, the first observation is written to the Work.Compare data set. Control returns to the top of the DATA step, and the next observation is read from Finance.CDRates. These steps are repeated for each observation in Finance.CDRates. The resulting data set contains the computed values of Investment for all observations that have been read from Finance.CDRates.

SAS Data Set Work.Compare

Institution	Rate	Years	Investment
MBNA America	0.0817	5	7404.64
Metropolitan Bank	0.0814	3	6323.09
Standard Pacific	0.0806	4	6817.57

Conditionally Executing DO Loops

Overview

The iterative DO statement specifies a fixed number of iterations for the DO loop. However, there are times when you want to execute a DO loop until a condition is reached or while a condition exists, but you don't know how many iterations are needed.

Suppose you want to calculate the number of years required for an investment to reach $50,000. In the DATA step below, using an iterative DO statement is inappropriate because you are trying to determine the number of iterations required for Capital to reach $50,000.

```
data work.invest;
   do year=1 to ? ;
      Capital+2000;
      capital+capital*.10;
   end;
run;
```

The DO WHILE and DO UNTIL statements enable you to execute DO loops based on whether a condition is true or false.

Using the DO UNTIL Statement

The DO UNTIL statement executes a DO loop until the expression becomes true.

General form, DO UNTIL statement:

DO UNTIL(*expression*);

    SAS statements

END;

where *expression* is a valid SAS expression enclosed in parentheses.

The expression is not evaluated until the bottom of the loop, so a DO UNTIL loop always executes at least once. When the expression is evaluated as true, the DO loop stops.

Assume you want to know how many years it will take to earn $50,000 if you deposit $2,000 each year into an account that earns 10% interest. The DATA step below uses a DO UNTIL statement to perform the calculation until $50,000 is reached. Each iteration of the DO loop represents one year.

```
data work.invest;
   do until(Capital>=50000);
      capital+2000;
      capital+capital*.10;
      Year+1;
   end;
run;
```

During each iteration of the DO loop,

- 2000 is added to the value of Capital to reflect the annual deposit of $2,000

- 10% interest is added to Capital

- the value of Year is incremented by 1.

Because there is no index variable in the DO UNTIL statement, the variable Year is created in a sum statement to count the number of iterations of the DO loop. This program produces a data set that contains the single observation shown below. To accumulate more than $50,000 in capital requires 13 years (and 13 iterations of the DO loop).

SAS Data Set Work.Invest

Capital	Year
53949.97	13

Using the DO WHILE Statement

Like the DO UNTIL statement, the DO WHILE statement executes DO loops conditionally. You can use the DO WHILE statement to execute a DO loop `while` the expression is true.

General form, DO WHILE statement:

DO WHILE(*expression*)**;**

      SAS statements

END;

where *expression* is a valid SAS expression enclosed in parentheses.

An important difference between the DO UNTIL and DO WHILE statements is that the DO WHILE expression is evaluated at the top of the DO loop. If the expression is false the first time it is evaluated, the DO loop never executes. For example, in the following program, because the value of Capital is initially zero, which is less than **50,000**, the DO loop does not execute.

```
data work.invest;
   do while(Capital>=50000);
      capital+2000;
      capital+capital*.10;
      Year+1;
   end;
run;
```

Using Conditional Clauses with the Iterative DO Statement

You have seen how the DO WHILE and DO UNTIL statements enable you to execute statements conditionally and how the iterative DO statement enables you to execute statements a set number of times, unconditionally.

```
DO WHILE(expression);
DO UNTIL(expression);
DO index-variable=start TO stop BY increment;
```

Now let's look at a form of the iterative DO statement that combines features of both conditional and unconditional execution of DO loops.

In this DATA step, the DO UNTIL statement determines how many years it takes (13) for an investment to reach $50,000.

```
data work.invest;
   do until(Capital>=50000);
      Year+1;
      capital+2000;
      capital+capital*.10;
   end;
run;
```

SAS Data Set Work.Invest

Capital	Year
53949.97	13

Suppose you also want to limit the number of years you invest your capital to 10 years. You can add the UNTIL or WHILE expression to an iterative DO statement to further control the number of iterations. This iterative DO statement enables you to execute the DO loop until Capital is greater than or equal to **50000** or until the DO loop executes ten times, whichever occurs first.

```
data work.invest;
   do year=1 to 10 until(Capital>=50000);
      capital+2000;
      capital+capital*.10;
   end;
   if year=11 then year=10;
run;
```

SAS Data Set Work.Invest

Capital	Year
35062.33	10

In this case, the DO loop stops executing after ten iterations, and the value of Capital never reaches **50000**. If you increase the amount added to Capital each year to **4000**, the DO loop stops executing after the eighth iteration when the value of Capital exceeds **50000**.

```
data work.invest;
   do year=1 to 10 until(Capital>=50000);
      capital+4000;
      capital+capital*.10;
   end;
   if year=11 then year=10;
run;
```

SAS Data Set Work.Invest

Capital	Year
50317.91	8

The UNTIL and WHILE specifications in an iterative DO statement function similarly to the DO UNTIL and DO WHILE statements. Both statements require a valid SAS expression enclosed in parentheses.

```
DO index-variable=start TO stop BY increment UNTIL(expression);
```

```
DO index-variable=start TO stop BY increment WHILE(expression);
```

The UNTIL expression is evaluated at the bottom of the DO loop, so the DO loop always executes at least once. The WHILE expression is evaluated before the execution of the DO loop. So, if the condition is initially false, the DO loop never executes.

Creating Samples

Because it performs iterative processing, a DO loop provides an easy way to draw sample observations from a data set. For example, suppose you would like to sample every tenth observation of the 5,000 observations in Factory.Widgets. Start with a simple DATA step:

```
data work.subset;
   set factory.widgets;
run;
```

You can create the sample data set by enclosing the SET statement in a DO loop. Use the start, stop, and increment values to select every tenth observation of the 5,000. Add the POINT= option to the SET statement, setting the POINT= option equal to the index variable that is used in the DO loop. (You learned about the POINT= option in "Reading SAS Data Sets" on page 334.)

```
data work.subset;
   do sample=10 to 5000 by 10;
      set factory.widgets point=sample;
   end;
run;
```

Remember that, in order to prevent continuous DATA step looping, you need to add a STOP statement when using the POINT= option. You also need to add an OUTPUT statement. Place the statement inside the DO loop to output each observation selected. (If the OUTPUT statement were placed after the DO loop, only the last observation would be written.)

```
data work.subset;
   do sample=10 to 5000 by 10;
      set factory.widgets point=sample;
      output;
   end;
   stop;
run;
```

When the program runs, the DATA step reads the observations in Factory.Widgets identified by the POINT= option. The values of the POINT= option are provided by the

DO loop, which starts at `10` and goes to `5,000` in increments of 10. The data set Work.Subset contains 500 observations.

Chapter Summary

Text Summary

Purpose of DO Loops
DO loops process groups of SAS statements repeatedly, reducing the number of statements required in repetitive calculations.

Syntax of Iterative DO Loops
To construct an iterative DO loop, specify an index variable and the conditions that will execute the loop. These conditions include a start value for the index variable, a stop value, and an increment value. Start, stop, and increment values can be any number, numeric variable, or SAS expression that results in a number.

DO Loop Execution
During each iteration of a DO loop, statements within the loop execute. Once the loop's index value exceeds the stop value, the DO loop stops, and processing continues with the following DATA step statement.

Generating Observations with DO Loops
Include an OUTPUT statement within the DO loop to write an observation for each iteration. This overrides the automatic generation of output at the end of the DATA step.

Decrementing DO Loops
You can decrement a DO loop by specifying a negative value for the BY clause. The start value must be greater than the stop value.

Specifying a Series of Items
You can specify how many times a DO loop executes by listing items in a series; the DO loop will execute once for each item, with the index variable equal to the value of each item. A series can consist of all numeric values, all character values (enclosed in quotation marks), or all variable names (without quotation marks).

Nesting DO Loops
DO loops can run within DO loops, as long as you assign a unique index variable to each loop and terminate each DO loop with its own END statement.

Iteratively Processing Observations from a Data Set
You can use a DO loop to read a data set and compute the value of a new variable. DO loop start and stop values, for example, can be read from a data set.

Conditionally Executing DO Loops
The DO UNTIL statement executes a DO loop until a condition becomes true. Because the expression is not evaluated until the bottom of the loop, a DO UNTIL loop will execute at least once. The DO WHILE statement is used to execute a DO loop while a

condition is true. Because the DO WHILE statement is evaluated at the top of the DO loop, if the expression is initially false, the DO loop never executes.

Using Conditional Clauses with the Iterative DO Statement

DO WHILE and DO UNTIL statements can be used within iterative DO loops to combine conditional and unconditional execution.

Creating Samples

DO loops provide an efficient way to create samples from data sets. Enclose the SET statement in a DO loop, using the start, stop, and increment values to select the observations. Add the POINT= option to the SET statement, setting it equal to the index variable of the DO loop. Then add a STOP statement to prevent DATA step looping, and add an OUTPUT statement to write observations.

Syntax

DO *index-variable=start* **TO** *stop* **BY** *increment*;

   SAS statements

END;

DO UNTIL(*expression***);**

DO WHILE(*expression***);**

DO *index-variable=start* **TO** *stop* **BY** *increment* **UNTIL(***expression***);**

DO *index-variable=start* **TO** *stop* **BY** *increment* **WHILE(***expression)***;**

Sample Programs

Simple DO Loop

```
data work.earn;
   Value=2000;
   do year=1 to 20;
      Interest=value*.075;
      value+interest;
   end;
run;
```

Nested DO Loops

```
data work.earn;
   do year=1 to 20;
      Capital+2000;
      do month=1 to 12;
         Interest=capital*(.075/12);
         capital+interest;
      end;
   end;
run;
```

Conditional Clause

```
data work.invest;
   do year=1 to 10 until(Capital>=50000);
      capital+2000;
```

```
            capital+capital*.10;
         end;
         if year=11 then year=10;
   run;
```

Creating Samples

```
   data work.subset;
      do sample=10 to 5000 by 10;
         set factory.widgets point=sample;
         output;
      end;
        stop;
     run;
```

Points to Remember

- If you do not specify a BY clause, the increment value for DO loops is **1**.

- In most cases, the index variable is needed only for processing the DO loop and can be dropped from the data set.

- The index variable is always incremented by one value beyond the stop value unless you terminate the DO loop in some other manner.

- It's easier to manage nested DO loops if you indent the statements in each loop.

Chapter Quiz

Select the best answer for each question. After completing the quiz, check your answers using the answer key in the appendix.

1. Which statement is *false* regarding the use of DO loops?

   a. They can contain conditional clauses.

   b. They can generate multiple observations.

   c. They can be used to combine DATA and PROC steps.

   d. They can be used to read data.

2. During each execution of the following DO loop, the value of Earned is calculated and is added to its previous value. How many times does this DO loop execute?

```
   data finance.earnings;
      Amount=1000;
      Rate=.075/12;
      do month=1 to 12;
         Earned+(amount+earned)*rate;
      end;
   run;
```

   a. 0

   b. 1

   c. 12

   d. 13

3. On January 1 of each year, $5000 is invested in an account. Complete the DATA step below to determine the value of the account after 15 years if a constant interest rate of ten percent is expected.

```
data work.invest;
   ...
      Capital+5000;
      capital+(capital*.10);
   end;
run;
```

   a. `do count=1 to 15;`

   b. `do count=1 to 15 by 10%;`

   c. `do count=1 to capital;`

   d. `do count=capital to (capital*.10);`

4. In the data set Work.Invest, what would be the stored value for Year?

```
data work.invest;
   do year=1990 to 2004;
      Capital+5000;
      capital+(capital*.10);
   end;
run;
```

   a. missing

   b. **1990**

   c. **2004**

   d. **2005**

5. Which of the following statements is *false* regarding the program shown below?

```
data work.invest;
   do year=1990 to 2004;
      Capital+5000;
      capital+(capital*.10);
      output;
   end;
run;
```

   a. The OUTPUT statement writes current values to the data set immediately.

   b. The last value for Year in the new data set is **2005**.

   c. The OUTPUT statement overrides the automatic output at the end of the DATA step.

   d. The DO loop performs 15 iterations.

6. How many observations will the data set Work.Earn contain?

```
data work.earn;
   Value=2000;
   do year=1 to 20;
      Interest=value*.075;
      value+interest;
      output;
   end;
run;
```

    a.   0

    b.   1

    c.   19

    d.   20

7. Which of the following would you use to compare the result of investing $4,000 a year for five years in three different banks that compound interest monthly? Assume a fixed rate for the five-year period.

    a.   DO WHILE statement

    b.   nested DO loops

    c.   DO UNTIL statement

    d.   a DO group

8. Which statement is *false* regarding DO UNTIL statements?

    a.   The condition is evaluated at the top of the loop, before the enclosed statements are executed.

    b.   The enclosed statements are always executed at least once.

    c.   SAS statements in the DO loop are executed until the specified condition is true.

    d.   The DO loop must have a closing END statement.

9. Select the DO WHILE statement that would generate the same result as the program below.

```
data work.invest;
capital=100000;
   do until(Capital gt 500000);
      Year+1;
      capital+(capital*.10);
   end;
run;
```

    a.   `do while(Capital ge 500000);`

    b.   `do while(Capital=500000);`

    c.   `do while(Capital le 500000);`

    d.   `do while(Capital>500000);`

10. In the following program, complete the statement so that the program stops generating observations when Distance reaches 250 miles or when 10 gallons of fuel have been used.

```
data work.go250;
   set perm.cars;
   do gallons=1 to 10 ... ;
      Distance=gallons*mpg;
      output;
   end;
run;
```

    a.   `while(Distance<250)`

    b.   `when(Distance>250)`

    c.   `over(Distance le 250)`

d. `until(Distance=250)`

Chapter 15
Processing Variables with Arrays

Overview

Introduction

In DATA step programming, you often need to perform the same action on more than one variable. Although you can process variables individually, it is easier to handle them as a group. You can do this by using array processing.

For example, using an array and DO loop, the program below eliminates the need for 365 separate programming statements to convert the daily temperature from Fahrenheit to Celsius for the year.

```
data work.report(drop=i);
   set master.temps;
   array daytemp{365} day1-day365;
   do i=1 to 365;
      daytemp{i}=5*(daytemp{i}-32)/9;
   end;
run;
```

You can use arrays to simplify the code needed to

- perform repetitive calculations

- create many variables that have the same attributes

- read data

- rotate SAS data sets by changing variables to observations or observations to variables

- compare variables

- perform a table lookup.

This chapter teaches you how to define an array and how to reference elements of the array in the DATA step.

Objectives

In this chapter, you learn to

- group variables into one- and two-dimensional arrays

- perform an action on array elements

- create new variables with an ARRAY statement

- assign initial values to array elements

- create temporary array elements with an ARRAY statement.

Creating One-Dimensional Arrays

Understanding SAS Arrays

A SAS array is a temporary grouping of SAS variables under a single name. An array exists only for the duration of the DATA step.

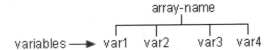

One reason for using an array is to reduce the number of statements that are required for processing variables. For example, in the DATA step below, the values of seven data set variables are converted from Fahrenheit to Celsius temperatures.

```
data work.report;
   set master.temps;
   mon=5*(mon-32)/9;
   tue=5*(tue-32)/9;
   wed=5*(wed-32)/9;
   thr=5*(thr-32)/9;
   fri=5*(fri-32)/9;
   sat=5*(sat-32)/9;
   sun=5*(sun-32)/9;
run;
```

As you can see, the assignment statements perform the same calculation on each variable in this series of statements. Only the name of the variable changes in each statement.

By grouping the variables into a one-dimensional array, you can process the variables in a DO loop. You use fewer statements, and the DATA step program is more easily modified or corrected.

```
data work.report(drop=day);
   set master.temps;
   array wkday{7} mon tue wed thr fri sat sun;
   do day=1 to 7;
      wkday{day}=5*(wkday{day}-32)/9;
   end;
run;
```

You will learn other uses for arrays as you continue through this chapter.

Defining an Array

To group data set variables into an array, use an ARRAY statement.

General form, ARRAY statement:

ARRAY *array-name{dimension} <elements>*;

where

- *array-name* specifies the name of the array.

- *dimension* describes the number and arrangement of array elements.

- *elements* lists the variables to include in the array. Array elements must be either all numeric or all character. If no elements are listed, new variables will be created with default names.

⚠ Do not give an array the same name as a variable in the same DATA step. Also, avoid using the name of a SAS function; the array will be correct, but you won't be able to use the function in the same DATA step, and a warning message will appear in the SAS log.

⚠ You cannot use array names in LABEL, FORMAT, DROP, KEEP, or LENGTH statements. Arrays exist only for the duration of the DATA step.

For example, in the data set Finance.Sales91, you might want to process the variables Qtr1, Qtr2, Qtr3, and Qtr4 in the same way.

Description of Finance.Sales91

Variable	Type	Length
SalesRep	char	8
Qtr1	num	8
Qtr2	num	8
Qtr3	num	8
Qtr4	num	8

Specifying the Array Name

To group the variables in the array, first give the array a *name*. In this example, make the array name **sales**.

```
array sales{4} qtr1 qtr2 qtr3 qtr4;
```

Specifying the Dimension

Following the array name, you must specify the dimension of the array. The dimension describes the number and arrangement of elements in the array. There are several ways to specify the dimension.

- In a one-dimensional array, you can simply specify the number of array elements. The array elements are the variables that you want to reference and process elsewhere in the DATA step.

```
array sales{4} qtr1 qtr2 qtr3 qtr4;
```

- The dimension of an array doesn't have to be the number of array elements. You can specify a range of values for the dimension when you define the array. For example, you can define the array sales as follows:

```
array sales{96:99} totals96 totals97 totals98 totals99;
```

- You can also indicate the dimension of a one-dimensional array by using an asterisk (*). This way, SAS determines the dimension of the array by counting the number of elements.

```
array sales{*} qtr1 qtr2 qtr3 qtr4;
```

- Enclose the dimension in either parentheses, braces, or brackets.

```
               ( )
array sales{4} qtr1 qtr2 qtr3 qtr4;
               [ ]
```

Specifying Array Elements

When specifying the elements of an array, you can list each variable name that you want to include in the array. When listing elements, separate each element with a space. As with all SAS statements, you end the ARRAY statement with a semicolon (;).

```
array sales{4} qtr1 qtr2 qtr3 qtr4;
```

You can also specify array elements as a variable list. Here is an example of an ARRAY statement that groups the variables Qtr1 through Qtr4 into a one-dimensional array, using a variable list.

```
array sales{4} qtr1-qtr4;
```

Let's look more closely at array elements that are specified as variable lists.

Variable Lists as Array Elements

You can specify variable lists in the forms shown below. Each type of variable list is explained in more detail following the table.

Table 15.1 *Variables and Their Forms*

Variables	Form
a numbered range of variables	Var1-Var*n*
all numeric variables	_NUMERIC_
all character variables	_CHARACTER_
all character or all numeric variables	_ALL_

A Numbered Range of Variables

```
Qtr1 Qtr2 Qtr3 Qtr4  →  Qtr1-Qtr4
```

- The variables must have the same name except for the last character or characters.
- The last character of each variable must be numeric.
- The variables must be numbered consecutively.

```
array sales{4} qtr1-qtr4;
```

In the preceding example, you would use sales(4) to reference Qtr4. However, the index of an array doesn't have to range from one to the number of array elements. You can specify a range of values for the index when you define the array. For example, you can define the array sales as follows:

```
array sales{96:99} totals96-totals99;
```

All Numeric Variables

```
Amount Rate Term  →  _NUMERIC_
```

NUMERIC specifies all numeric variables that have already been defined in the current DATA step.

```
array sales{*} _numeric_;
```

All Character Variables

```
FrstName LastName Address  →  _CHARACTER_
```

CHARACTER specifies all character variables that have already been defined in the current DATA step.

```
array sales{*} _character_;
```

All Variables

```
FrstName LastName Address  →  _ALL_

Amount Rate Term  →  _ALL_
```

ALL specifies all variables that have already been defined in the current DATA step. The variables must all be of the same type: all character or all numeric.

```
array sales{*} _all_;
```

Referencing Elements of an Array

Now let's look at some ways you can use arrays to process variables in the DATA step.

```
data work.report(drop=day);
   set master.temps;
   array wkday{7} mon tue wed thr fri sat sun;
   do day=1 to 7;
      if wkday{day}>95 then output;
   end;
run;

data work.weights(drop=day);
   set master.class;
   array wt{6} w1-w6;
   do day=1 to 6;
      wt{day}=wt{day}*2.2;
   end;
run;
```

```
data work.new(drop=day);
   set master.synyms;
   array term{9} also1-also9;
   do day=1 to 9;
      if term{day} ne ' ' then output;
   end;
run;
```

The ability to reference the elements of an array by an index value is what gives arrays their power. Typically, arrays are used with DO loops to process multiple variables and to perform repetitive calculations.

```
array quarter{4} jan apr jul oct;
do qtr=1 to 4;
   YearGoal=quarter{qtr}*1.2;
end;
```

When you define an array in a DATA step, a subscript value is assigned to each array element. The subscript values are assigned in the order of the array elements.

```
                     1   2   3   4
array quarter{4} jan apr jul oct;
do qtr=1 to 4;
   YearGoal=quarter{qtr}*1.2;
end;
```

You use an array reference to perform an action on an array element during execution. To reference an array element in the DATA step, specify the name of the array, followed by a subscript value enclosed in parentheses.

General form, ARRAY reference:

array-name(subscript)

where *subscript*

- is enclosed in parentheses, braces, or brackets

- specifies a variable, a SAS expression, or an integer

- is within the lower and upper bounds of the dimension of the array.

When used in a DO loop, the index variable of the iterative DO statement can reference each element of the array.

```
array quarter{4} jan apr jul oct;
do i=1 to 4;
   YearGoal=quarter{i}*1.2;
end;
```

For example, the DO loop above increments the index variable i from the lower bound of the quarter array, 1, to the upper bound, 4. The following sequence illustrates this process:

```
                 1
array quarter{4} jan apr jul oct;
do i=1 to 4;
   YearGoal=quarter{1}*1.2;
end;
```

```
               2
   array quarter{4} jan apr jul oct;
```

```
   do i=1 to 4;
      YearGoal=quarter{2}*1.2;
   end;
```

                                                 3
```
      array quarter{4} jan apr jul oct;
      do i=1 to 4;
         YearGoal=quarter{3}*1.2;
      end;
```

                                                          4
```
         array quarter{4} jan apr jul oct;
         do i=1 to 4;
            YearGoal=quarter{4}*1.2;
          end;
```

During each iteration of the DO loop, quarter{*i*} refers to an element of the array quarter in the order listed.

Let's look at another example of a DATA step that contains an array with a DO loop.

The Health Center of a company conducts a fitness class for its employees. Each week, participants are weighed so that they can monitor their progress. The weight data, currently stored in kilograms, needs to be converted to pounds.

SAS Data Set Hrd.Fitclass

Name	Weight1	Weight2	Weight3	Weight4	Weight5	Weight6
Alicia	69.6	68.9	68.8	67.4	66.0	66.2
Betsy	52.6	52.6	51.7	50.4	49.8	49.1
Brenda	68.6	67.6	67.0	66.4	65.8	65.2
Carl	67.6	66.6	66.0	65.4	64.8	64.2
Carmela	63.6	62.5	61.9	61.4	60.8	58.2
David	70.6	69.8	69.2	68.4	67.8	67.0

You can use a DO loop to update the variables Weight1 through Weight6 for each observation in the Hrd.Fitclass data set.

```
data hrd.convert(drop=i);
   set hrd.fitclass;
   array wt{6} weight1-weight6;
   do i=1 to 6;
      wt{i}=wt{i}*2.2046;
   end;
run;
```

Compilation and Execution

To understand how the DO loop processes the array elements, let's examine the compilation and execution phases of this DATA step.

During compilation, the program data vector is created.

Program Data Vector

N	Name	Weight1	Weight2	Weight3	Weight4	Weight5	Weight6	i

The DATA step is scanned for syntax errors. If there are any syntax errors in the ARRAY statement, they are detected at this time.

The index values of the array elements are assigned. Note that the array name is not included in the program data vector. The array exists only for the duration of the DATA step.

During the first iteration of the DATA step, the first observation in Hrd.Fitclass is read into the program data vector.

```
data hrd.convert;
   set hrd.fitclass;
   array wt{6} weight1-weight6;
   do i=1 to 6;
      wt{i}=wt{i}*2.2046;
   end;
run;
```

Program Data Vector

		wt{1}	wt{2}	wt{3}	wt{4}	wt{5}	wt{6}	
N	Name	Weight1	Weight2	Weight3	Weight4	Weight5	Weight6	i
1	Alicia	69.6	68.9	68.8	67.4	66.0	66.2	

Because the ARRAY statement is a compile-time only statement, it is ignored during execution. The DO statement is executed next.

During the first iteration of the DO loop, the index variable i is set to 1. As a result, the array reference wt{i} becomes wt{1}. Because wt{1} refers to the first array element, Weight1, the value of Weight1 is converted from kilograms to pounds.

```
data hrd.convert;
   set hrd.fitclass;
   array wt{6} weight1-weight6;
   do i=1 to 6;
      wt{1}=wt{1}*2.2046;
   end;
run;
```

Program Data Vector

		wt{1}	wt{2}	wt{3}	wt{4}	wt{5}	wt{6}	
N	Name	Weight1	Weight2	Weight3	Weight4	Weight5	Weight6	i
1	Alicia	153.4	68.9	68.8	67.4	66.0	66.2	1

Graphical Display of Array Processing

As the DATA step continues its DO loop iterations, the index variable i is changed from 1 to 2, 3, 4, 5, and 6, causing Weight1 through Weight6 to receive new values in the program data vector.

```
data hrd.convert;
   set hrd.fitclass;
   array wt{6} weight1-weight6;
   do i=1 to 6;
      wt{i}=wt{i}*2.2046;
   end;
run;
```

Using the DIM Function in an Iterative DO Statement

When using DO loops to process arrays, you can also use the DIM function to specify the TO clause of the iterative DO statement. For a one-dimensional array, specify the array name as the argument for the DIM function. The function returns the number of elements in the array.

General form, DIM function:

DIM(*array-name*)

where *array-name* specifies the array.

In this example, dim(wt) returns a value of **6**.

```
data hrd.convert;
   set hrd.fitclass;
   array wt{*} weight1-weight6;
   do i=1 to dim(wt);
      wt{i}=wt{i}*2.2046;
   end;
run;
```

When you use the DIM function, you do not have to re-specify the stop value of an iterative DO statement if you change the dimension of the array.

```
data hrd.convert;
   set hrd.fitclass;
   array wt{*} weight1-weight6;
   do i=1 to dim(wt);
      wt{i}=wt{i}*2.2046;
   end;
run;
```

```
data hrd.convert;
   set hrd.fitclass;
   array wt{*} weight1-weight10;
   do i=1 to dim(wt);
      wt{i}=wt{i}*2.2046;
   end;
run;
```

Expanding Your Use of Arrays

Creating Variables in an ARRAY Statement

So far, you have learned several ways to reference existing variables in an ARRAY statement. You can also create variables in an ARRAY statement by omitting the array elements from the statement. If you do not reference existing variables, SAS automatically creates new variables for you and assigns default names to them. The default name is the array-name followed by consecutive numbers 1 to the dimension of the array.

General form, ARRAY statement to create new variables:

ARRAY *array-name{dimension}*;

where

- *array-name* specifies the name of the array.

- *dimension* describes the number and arrangement of array elements.

For example, suppose you need to calculate the weight gain or loss from week to week for each member of a fitness class, shown below.

SAS Data Set Hrd.Convert

Name	Weight1	Weight2	Weight3	Weight4	Weight5	Weight6
Alicia	153.4	151.9	151.7	148.6	145.5	145.9
Betsy	116.0	116.0	114.0	111.1	109.8	108.2
Brenda	151.2	149.0	147.7	146.4	145.1	143.7
Carl	149.0	146.8	145.5	144.2	142.9	141.5
Carmela	140.2	137.8	136.5	135.4	134.0	128.3

You'd like to create variables that contain this weekly difference. To perform the calculation, you first group the variables Weight1 through Weight6 into an array.

```
data hrd.diff;
   set hrd.convert;
   array wt{6} weight1-weight6;
```

Next, you want to create the new variables to store the differences between the six recorded weights. You can use an additional ARRAY statement without elements to create the new variables, WgtDiff1 to WgtDiff5.

```
data hrd.diff;
   set hrd.convert;
   array wt{6} weight1-weight6;
   array WgtDiff{5};
```

SAS Data Set Hrd.Convert

Name	Weight1	Weight2	Weight3	Weight4	Weight5	Weight6
Alicia	153.4	<-1->151.9	<-2->151.7	<-3->148.6	<-4->145.5	<-5->145.9
Betsy	116.0	116.0	114.0	111.1	109.8	108.2
Brenda	151.2	149.0	147.7	146.4	145.1	143.7
Carl	149.0	146.8	145.5	144.2	142.9	141.5
Carmela	140.2	137.8	136.5	135.4	134.0	128.3

Remember, when creating variables in an ARRAY statement, you do not need to specify array elements as long as you specify how many elements will be in the array.

```
array WgtDiff{5};
```

Default Variable Names

The default variable names are created by concatenating the array name and the numbers 1, 2, 3, and so on, up to the array dimension.

```
array WgtDiff{5};
```

```
WgtDiff1   WgtDiff2   WgtDiff3   WgtDiff4   WgtDiff5
```

If you prefer, you can specify individual variable names. To specify variable names, you list each name as an element of the array. The following ARRAY statement creates the numeric variables Oct12, Oct19, Oct26, Nov02, and Nov09.

```
array WgtDiff{5} Oct12 Oct19 Oct26 Nov02 Nov09;
                array WgtDiff{5};
```

```
Oct12   Oct19   Oct26   Nov02   Nov09
```

Arrays of Character Variables

To create an array of character variables, add a dollar sign ($) after the array dimension.

```
array firstname{5} $;
```

By default, all character variables that are created in an ARRAY statement are assigned a length of 8. You can assign your own length by specifying the length after the dollar sign.

```
array firstname{5} $ 24;
```

The length that you specify is automatically assigned to all variables that are created by the ARRAY statement.

During the compilation of the DATA step, the variables that this ARRAY statement creates are added to the program data vector and are stored in the resulting data set.

```
data hrd.diff;
   set hrd.convert;
   array wt{6} Weight1-Weight6;
   array WgtDiff{5};
```

Program Data Vector

N	Name	Weight1	Weight2	Weight3	Weight4	Weight5	Weight6

WgtDiff1	WgtDiff2	WgtDiff3	WgtDiff4	WgtDiff5

When referencing the array elements, be careful not to confuse the array references WgtDiff{1} through WgtDiff{5} (note the braces) with the variable names WgtDiff1 through WgtDiff5. The program data vector below shows the relationship between the array references and the corresponding variable names.

Program Data Vector

N	Name	Weight1	Weight2	Weight3	Weight4	Weight5	Weight6

WgtDiff{1} WgtDiff{2} WgtDiff{3} WgtDiff{4} WgtDiff{5}

WgtDiff1	WgtDiff2	WgtDiff3	WgtDiff4	WgtDiff5

Now you can use a DO loop to calculate the differences between each of the recorded weights. Notice that each value of WgtDiff{i} is calculated by subtracting wt{i} from wt{$i+1$}. By manipulating the index variable, you can easily reference any array element.

```
data hrd.diff;
   set hrd.convert;
   array wt{6} weight1-weight6;
   array WgtDiff{5};
   do i=1 to 5;
      wgtdiff{i}=wt{i+1}-wt{i};
   end;
run;
```

A portion of the resulting data set is shown below.

SAS Data Set Hrd.Diff

Name	WgtDiff1	WgtDiff2	WgtDiff3	WgtDiff4	WgtDiff5
Alicia	-1.54322	-0.22046	-3.08644	-3.08644	0.44092
Betsy	0.00000	-1.98414	-2.86598	-1.32276	-1.54322
Brenda	-2.20460	-1.32276	-1.32276	-1.32276	-1.32276

Assigning Initial Values to Arrays

Sometimes it is useful to assign initial values to elements of an array when you define the array.

```
array goal{4} g1 g2 g3 g4 (initial values);
```

To assign initial values in an ARRAY statement:

1. Place the values after the array elements.

```
array goal{4} g1 g2 g3 g4 (9000 9300 9600 9900);
```

2. Specify one initial value for each corresponding array element.

```
                          ↓    ↓    ↓    ↓
array goal{4} g1 g2 g3 g4 (9000 9300 9600 9900);
                           ↑    ↑    ↑    ↑
```

3. Separate each value with a comma or blank.

```
                           ↓         ↓         ↓
array goal{4} g1 g2 g3 g4 (9000 9300 9600 9900);
```

4. Enclose the initial values in parentheses.

```
                          ↓                   ↓
array goal{4} g1 g2 g3 g4 (9000 9300 9600 9900);
```

5. Enclose each character value in quotation marks.

```
                        ↓   ↓ ↓     ↓ ↓      ↓
array col{3} $ color1-color3 ('red','green','blue');
```

It's also possible to assign initial values to an array without specifying each array element. The following statement creates the variables Var1, Var2, Var3, and Var4, and assigns them initial values of **1**, **2**, **3**, and **4**:

```
array Var{4} (1 2 3 4);
```

For this example, assume that you have the task of comparing the actual sales figures in the Finance.Qsales data set to the sales goals for each sales representative at the beginning of the year. The sales goals are not recorded in Finance.Qsales.

Description of Finance.Qsales

Variable	Type	Length
SalesRep	char	8
Sales1	num	8
Sales2	num	8
Sales3	num	8
Sales4	num	8

The DATA step below reads the Finance.Qsales data set to create the Finance.Report data set. The ARRAY statement creates an array to process sales data for each quarter.

```
data finance.report;
   set finance.qsales;
   array sale{4} sales1-sales4;
```

To compare the actual sales to the sales goals, you must create the variables for the sales goals and assign values to them.

```
data finance.report;
   set finance.qsales;
   array sale{4} sales1-sales4;
   array Goal{4} (9000 9300 9600 9900);
```

A third ARRAY statement creates the variables Achieved1 through Achieved4 to store the comparison of actual sales versus sales goals.

```
data finance.report;
   set finance.qsales;
   array sale{4} sales1-sales4;
   array Goal{4} (9000 9300 9600 9900);
   array Achieved{4};
   do i=1 to 4;
      achieved{i}=100*sale{i}/goal{i};
   end;
run;
```

A DO loop executes four times to calculate the value of each element of the achieved array (expressed as a percentage).

```
data finance.report;
   set finance.qsales;
   array sale{4} sales1-sales4;
   array Goal{4} (9000 9300 9600 9900);
   array Achieved{4};
   do i=1 to 4;
      achieved{i}=100*sale{i}/goal{i};
   end;
run;
```

Before submitting this DATA step, you can drop the index variable from the new data set by adding a DROP= option to the DATA statement.

```
data finance.report (drop=i);
   set finance.qsales;
   array sale{4} sales1-sales4;
   array Goal{4} (9000 9300 9600 9900);
   array Achieved{4};
   do i=1 to 4;
      achieved{i}=100*sale{i}/goal{i};
   end;
run;
```

This is an example of a simple table-lookup program. The resulting data set contains the variables that were read from Finance.Qsales, plus the eight variables that were created with ARRAY statements.

SAS Data Set Finance.Report

SalesRep	Sales1	Sales2	Sales3	Sales4	Goal1	Goal2
Britt	8400	8800	9300	9800	9000	9300
Fruchten	9500	9300	9800	8900	9000	9300
Goodyear	9150	9200	9650	11000	9000	9300

Goal3	Goal4	Achieved1	Achieved2	Achieved3	Achieved4
9600	9900	93.333	94.624	96.875	98.990
9600	9900	105.556	100.000	102.083	89.899
9600	9900	101.667	98.925	100.521	111.111

Variables to which initial values are assigned in an ARRAY statement are automatically retained.

The variables Goal1 through Goal4 should not be stored in the data set, because they are needed only to calculate the values of Achieved1 through Achieved4. The next example shows you how to create temporary array elements.

Creating Temporary Array Elements

To create temporary array elements for DATA step processing without creating new variables, specify _TEMPORARY_ after the array name and dimension.

```
data finance.report(drop=i);
   set finance.qsales;
   array sale{4} sales1-sales4;
   array goal{4} _temporary_ (9000 9300 9600 9900);
   array Achieved{4};
   do i=1 to 4;
      achieved{i}=100*sale{i}/goal{i};
   end;
run;
```

Temporary array elements do not appear in the resulting data set.

SAS Data Set Finance.Report

SalesRep	Sales1	Sales2	Sales3	Sales4
Britt	8400	8800	9300	9800
Fruchten	9500	9300	9800	8900
Goodyear	9150	9200	9650	11000

Achieved1	Achieved2	Achieved3	Achieved4
93.333	94.624	96.875	98.990
105.556	100.000	102.083	89.899
101.667	98.925	100.521	111.111

Temporary array elements are useful when the array is needed only to perform a calculation. You can improve performance time by using temporary array elements.

Understanding Multidimensional Arrays

So far, you have learned how to group variables into one-dimensional arrays. You can also group variables into table-like structures called multidimensional arrays. This section teaches you how to define and use two-dimensional arrays, which are a common type of multidimensional array.

Suppose you want to write a DATA step to compare responses on a quiz to the correct answers. As long as there is only one correct answer per question, this is a simple one-to-one comparison.

```
Resp1   →   Answer1
Resp2   →   Answer2
Resp3   →   Answer3
Resp4   →   Answer4
```

However, if there is more than one correct answer per question, you must compare each response to each possible correct answer in order to determine whether there is a match.

```
Resp1   →   Answer1   Answer2   Answer3
Resp2   →   Answer4   Answer5   Answer6
Resp3   →   Answer7   Answer8   Answer9
Resp4   →   Answer10  Answer11  Answer12
```

You can process the above data more easily by grouping the Answer variables into a two-dimensional array. Just as you can think of a one-dimensional array as a single row of variables, as in this example ...

```
Answer1 Answer2 Answer3 Answer4 ... Answer9 Answer10 Answer11 Answer12
```

... you can think of a two-dimensional array as multiple rows of variables.

```
Answer1    Answer2    Answer3
Answer4    Answer5    Answer6
Answer7    Answer8    Answer9
Answer10   Answer11   Answer12
```

Defining a Multidimensional Array

To define a multidimensional array, you specify the number of elements in each dimension, separated by a comma. This ARRAY statement defines a two-dimensional array:

```
array new{3,4} x1-x12;
```

In a two-dimensional array, the two dimensions can be thought of as a table of rows and columns.

```
array new{r,c} x1-x12;
```

The first dimension in the ARRAY statement specifies the number of rows.

```
array new{3,4} x1-x12;
```

The second dimension specifies the number of columns.

```
array new{3,4} x1-x12;
```

columns

x1	x2	x3	x4
x5	x6	x7	x8
x9	x10	x11	x12

You can reference any element of the array by specifying the two dimensions. In the example below, you can perform an action on the variable x7 by specifying the array reference new(2,3). You can easily locate the array element in the table by finding the row (2), then the column (3).

```
array new{3,4} x1-x12;
new(2,3)=0;
```

x1	x2	x3	x4
x5	x6	x7	x8
x9	x10	x11	x12

When you define a two-dimensional array, the array elements are grouped in the order in which they are listed in the ARRAY statement. For example, the array elements x1 through x4 can be thought of as the first row of the table.

```
         ↓  ↓  ↓  ↓
array new{3,4} x1 x2 x3 x4 x5 x6 x7 x8 x9 x10 x11 x12;
```

x1	x2	x3	x4
x5	x6	x7	x8
x9	x10	x11	x12

The elements x5 through x8 become the second row of the table, and so on.

```
                  ↓  ↓  ↓  ↓
array new{3,4} x1 x2 x3 x4 x5 x6 x7 x8 x9 x10 x11 x12;
```

x1	x2	x3	x4
x5	x6	x7	x8
x9	x10	x11	x12

Referencing Elements of a Two-Dimensional Array

Multidimensional arrays are typically used with nested DO loops. The next example uses a one-dimensional array, a two-dimensional array, and a nested DO loop to re-structure a set of variables.

Your company's sales figures are stored by month in the SAS data set Finance.Monthly. Your task is to generate a new data set of quarterly sales rather than monthly sales.

Description of Finance.Monthly

Variable	Type	Length
Year	num	8
Month1	num	8
Month2	num	8
Month3	num	8
Month4	num	8
Month5	num	8
Month6	num	8
Month7	num	8
Month8	num	8
Month9	num	8
Month10	num	8
Month11	num	8
Month12	num	8

Defining the array m{4,3} puts the variables Month1 through Month12 into four groups of three months (calendar quarters).

Table Representation of m Array

Month1	Month2	Month3
Month4	Month5	Month6
Month7	Month8	Month9
Month10	Month11	Month12

```
data finance.quarters;
   set finance.monthly;
   array m{4,3} month1-month12;
```

Defining the array Qtr{4} creates the numeric variables Qtr1, Qtr2, Qtr3, Qtr4, which will be used to sum the sales for each quarter.

```
data finance.quarters;
   set finance.monthly;
   array m{4,3} month1-month12;
   array Qtr{4};
```

A nested DO loop is used to reference the values of the variables Month1 through Month12 and to calculate the values of Qtr1 through Qtr4. Because the variables i and j are used only for loop processing, the DROP= option is used to exclude them from the Finance.Quarters data set.

```
data finance.quarters (drop=i j);
   set finance.monthly;
   array m{4,3} month1-month12;
   array Qtr{4};
```

```
  do i=1 to 4;
    qtr{i}=0;
      do j=1 to 3;
        qtr{i}+m{i,j};
      end;
  end;
run;
```

Each element in the Qtr array represents the sum of one row in the m array. The number of elements in the Qtr array should match the first dimension of the m array (that is, the number of rows in the m array). The first DO loop executes once for each of the four elements of the Qtr array.

The assignment statement, qtr{i}=0, sets the value of qtr{i} to zero after each iteration of the first DO loop. Without the assignment statement, the values of Qtr1, Qtr2, Qtr3, and Qtr4 would accumulate across iterations of the DATA step due to the qtr{i}+m{i,j} sum statement within the DO loop.

```
data finance.quarters(drop=i j);
   set finance.monthly;
   array m{4,3} month1-month12;
   array Qtr{4};
   do i=1 to 4;
     qtr{i}=0;
       do j=1 to 3;
         qtr{i}+m{i,j};
       end;
   end;
run;
```

The second DO loop executes the same number of times as the second dimension of the m array (that is, the number of columns in each row of the m array).

```
data finance.quarters(drop=i j);
   set finance.monthly;
   array m{4,3} month1-month12;
   array Qtr{4};
     do i=1 to 4;
       qtr{i}=0;
         do j=1 to 3;
           qtr{i}+m{i,j};
         end;
     end;
run;
```

To see how the nested DO loop processes these arrays, let's examine the execution of this DATA step.

When this DATA step is compiled, the program data vector is created. The PDV contains the variables Year, Month1 through Month12, and the new variables Qtr1 through Qtr4. (Only the beginning and ending portions of the program data vector are represented here.)

```
data finance.quarters(drop=i j);
   set finance.monthly;
   array m{4,3} month1-month12;
   array Qtr{4};
   do i=1 to 4;
     qtr{i}=0;
       do j=1 to 3;
```

```
            qtr{i}+m{i,j};
        end;
    end;
run;
```

Program Data Vector

N	Year	Month1	Month2	Month3		Qtr1	Qtr2	Qtr3	Qtr4	i	j
•	•	•	•	•		•	•	•	•	•	•

During the first execution of the DATA step, the values of the first observation of Finance.Monthly are read into the program data vector. When the first DO loop executes the first time, the index variable i is set to 1.

```
data finance.quarters(drop=i j);
    set finance.monthly;
    array m{4,3} month1-month12;
    array Qtr{4};
    do i=1 to 4;              i=1
        qtr{i}=0;
        do j=1 to 3;
            qtr{i}+m{i,j};
        end;
    end;
run;
```

Program Data Vector

N	Year	Month1	Month2	Month3		Qtr1	Qtr2	Qtr3	Qtr4	i	j
1	1989	23000	21500	24600		•	•	•	•	1	•

During the first iteration of the nested DO loop, the value of Month1, which is referenced by m(i,j), is added to Qtr1.

```
data finance.quarters(drop=i j);
    set finance.monthly;
    array m{4,3} month1-month12;
    array Qtr{4};
    do i=1 to 4;            i=1
        qtr{i}=0;
      > do j=1 to 3;          j=1
            qtr(1)+m(1,1);
        end;
    end;
run;
```

Program Data Vector

N	Year	Month1	Month2	Month3		Qtr1	Qtr2	Qtr3	Qtr4	i	j
1	1989	23000	21500	24600		23000	•	•	•	1	1

During the second iteration of the nested DO loop, the value of Month2, which is referenced by m (i,j), is added to Qtr1.

```
data finance.quarters(drop=i j);
    set finance.monthly;
```

```
    array m{4,3} month1-month12;
    array Qtr{4};
    do i=1 to 4;          i=1
      qtr{i}=0;
      > do j=1 to 3;       j=2
          qtr{1}+m{1,2};
        end;
    end;
run;
```

Program Data Vector

N	Year	Month1	Month2	Month3		Qtr1	Qtr2	Qtr3	Qtr4	i	j
1	1989	23000	21500	24600		44500	•	•	•	1	2

The nested DO loop continues to execute until the index variable j exceeds the stop value, **3**. When the nested DO loop completes execution, the total sales for the first quarter, Qtr1, have been computed.

```
data finance.quarters(drop=i j);
  set finance.monthly;
  array m{4,3} month1-month12;
  array Qtr{4};
  do i=1 to 4;          i=1
    qtr{i}=0;
    > do j=1 to 3;       j=3
        qtr{1}+m{1,3};
      end;
  end;
run;
```

Program Data Vector

N	Year	Month1	Month2	Month3		Qtr1	Qtr2	Qtr3	Qtr4	i	j
1	1989	23000	21500	24600		69100	•	•	•	1	4

The outer DO loop increments i to 2, and the process continues for the array element Qtr2 and the m array elements Month4 through Month6.

```
data finance.quarters(drop=i j);
  set finance.monthly;
  array m{4,3} month1-month12;
  array Qtr{4};
 > do i=1 to 4;          i=2
    qtr{i}=0;
      do j=1 to 3;       j=1
          qtr{i}+m{i,j};
      end;
  end;
run;
```

Program Data Vector

N		Month2	Month3	Month4		Qtr1	Qtr2	Qtr3	Qtr4	i	j
1		21500	24600	23300		69100	23300	•	•	2	1

After the outer DO loop completes execution, the end of the DATA step is reached, and the first observation is written to the data set Finance.Quarters.

```
data finance.quarters(drop=i j);
   set finance.monthly;
   array m{4,3} month1-month12;
   array Qtr{4};
 > do i=1 to 4;            i=5 (loop ends)
     qtr{i}=0;
      do j=1 to 3;
         qtr{i}+m{i,j};
      end;
   end;
run;
```

Program Data Vector

N		Month2	Month3	Month4			Qtr1	Qtr2	Qtr3	Qtr4	i	j
1		21500	24600	23300			69100	64400	69200	71800	5	4

What you have seen so far represents the first iteration of the DATA step. All observations in the data set Finance.Monthly are processed in the same manner. Below is a portion of the resulting data set, which contains the sales figures grouped by quarters.

SAS Data Set Finance.Quarters (Partial Listing)

Year	Qtr1	Qtr2	Qtr3	Qtr4
1989	69100	64400	69200	71800
1990	73100	72000	83200	82800
1991	73400	81800	85200	87800

Additional Features

You've seen a number of uses for arrays, including creating variables, performing repetitive calculations, and performing table lookups. You can also use arrays for rotating (transposing) a SAS data set.

When you rotate a SAS data set, you change variables to observations or observations to variables. For example, suppose you want to rotate the Finance.Funddrive data set to create four output observations from each input observation.

SAS Data Set Finance.Funddrive

LastName	Qtr1	Qtr2	Qtr3	Qtr4
ADAMS	18	18	20	20
ALEXANDE	15	18	15	10
APPLE	25	25	25	25
ARTHUR	10	25	20	30
AVERY	15	15	15	15
BAREFOOT	20	20	20	20
BAUCOM	25	20	20	30
BLAIR	10	10	5	10
BLALOCK	5	10	10	15
BOSTIC	20	25	30	25
BRADLEY	12	16	14	18
BRADY	20	20	20	20
BROWN	18	18	18	18
BRYANT	16	18	20	18
BURNETTE	10	10	10	10
CHEUNG	30	30	30	30
LEHMAN	20	20	20	20
VALADEZ	14	18	40	25

The following program rotates the data set and lists the first 16 observations in the new data set.

```
data work.rotate(drop=qtr1-qtr4);
   set finance.funddrive;
   array contrib{4} qtr1-qtr4;
   do Qtr=1 to 4;
      Amount=contrib{qtr};
      output;
   end;
run;
proc print data=rotate(obs=16) noobs;
run;
```

Figure 15.1 *The First 16 Observations in the New, Rotated Data Set*

LastName	Qtr	Amount
ADAMS	1	18
ADAMS	2	18
ADAMS	3	20
ADAMS	4	20
ALEXANDER	1	15
ALEXANDER	2	18
ALEXANDER	3	15
ALEXANDER	4	10
APPLE	1	25
APPLE	2	25
APPLE	3	25
APPLE	4	25
ARTHUR	1	10
ARTHUR	2	25
ARTHUR	3	20
ARTHUR	4	30

Chapter Summary

Text Summary

Purpose of SAS Arrays

An array is a temporary grouping of variables under a single name. This can reduce the number of statements needed to process variables and can simplify the maintenance of DATA step programs.

Defining an Array

To group variables into an array, use an ARRAY statement that specifies the array's name; its dimension enclosed in braces, brackets, or parentheses; and the elements to include. For example: array sales{4} qtr1 qtr2 qtr3 qtr4;

Variable Lists as Array Elements

You can use a variable list to specify array elements. Depending on the form of the variable list, it can specify all numeric or all character variables, or a numbered range of variables.

Referencing Elements of an Array

When you define an array in a DATA step, a subscript is assigned to each element. During execution, you can use an array reference to perform actions on specific array elements. When used in a DO loop, for example, the index variable of the iterative DO statement can reference each element of the array.

The DIM Function

When using DO loops to process arrays, you can also use the DIM function to specify the TO clause of the iterative DO statement. When you use the DIM function, you do not have to re-specify the stop value of a DO statement if you change the dimension of the array.

Creating Variables with the ARRAY Statement

If you don't specify array elements in an ARRAY statement, SAS automatically creates variable names for you by concatenating the array name and the numbers 1, 2, 3 ... up to the array dimension. To create an array of character variables, add a dollar sign ($) after the array dimension. By default, all character variables that are created with an ARRAY statement are assigned a length of 8; however, you can specify a different length after the dollar sign.

Assigning Initial Values to Arrays

To assign initial values in an ARRAY statement, place the values in parentheses after the array elements, specifying one initial value for each array element and separating each value with a comma or blank. To assign initial values to character variables, enclose each value in quotation marks.

Creating Temporary Array Elements

You can create temporary array elements for DATA step processing without creating additional variables. Just specify _TEMPORARY_ after the array name and dimension. This is useful when the array is needed only to perform a calculation.

Multidimensional Arrays

To define a multidimensional array, specify the number of elements in each dimension, separated by a comma. For example, `array new{3,4} x1-x12;` defines a two-dimensional array, with the first dimension specifying the number of rows (3) and the second dimension specifying the number of columns (4).

Referencing Elements of a Two-Dimensional Array

Multidimensional arrays are typically used with nested DO loops. If a DO loop processes a two-dimensional array, you can reference any element within the array by specifying the two dimensions.

Rotating Data Sets

You can use arrays to rotate a data set. Rotating a data set changes variables to observations or observations to variables.

Syntax

ARRAY *array-name{dimension} <elements>*;

array-name(subscript)

DIM(*array-name*)

Sample Programs

```
data work.report(drop=i);
   set master.temps;
   array wkday{7} mon tue wed thr fri sat sun;
```

```
      do i=1 to 7;
         wkday{i}=5*(wkday{i}-32)/9;
      end;
run;

data hrd.convert(drop=i);
   set hrd.fitclass;
   array wt{6} weight1-weight6;
   do i=1 to dim(wt);
      wt{i}=wt{i}*2.2046;
   end;
run;

data hrd.diff(drop=i);
   set hrd.convert;
   array wt{6} weight1-weight6;
   array WgtDiff{5};
   do i=1 to 5;
      wgtdiff{i}=wt{i+1}-wt{i};
   end;
run;

data finance.report(drop=i);
   set finance.qsales;
   array sale{4} sales1-sales4;
   array goal{4} _temporary_ (9000 9300 9600 9900);
   array Achieved{4};
   do i=1 to 4;
      achieved{i}=100*sale{i}/goal{i};
   end;
run;

data finance.quarters(drop=i j);
   set finance.monthly;
   array m{4,3} month1-month12;
   array Qtr{4};
   do i=1 to 4;
     qtr{i}=0;
       do j=1 to 3;
          qtr{i}+m{i,j};
       end;
   end;
run;
```

Points to Remember

- A SAS array exists only for the duration of the DATA step.

- Do not give an array the same name as a variable in the same DATA step. Also, avoid using the name of a SAS function as an array name; the array will be correct, but you won't be able to use the function in the same DATA step, and a warning will be written to the SAS log.

- You can indicate the dimension of a one-dimensional array with an asterisk (*) as long as you specify the elements of the array.

- When referencing array elements, be careful not to confuse variable names with the array references.

Chapter Quiz

Select the best answer for each question. After completing the quiz, check your answers using the answer key in the appendix.

1. Which statement is *false* regarding an ARRAY statement?

    a. It is an executable statement.

    b. It can be used to create variables.

    c. It must contain either all numeric or all character elements.

    d. It must be used to define an array before the array name can be referenced.

2. What belongs within the braces of this ARRAY statement?

    ```
    array contrib{?} qtr1-qtr4;
    ```

    a. quarter

    b. quarter*

    c. 1-4

    d. 4

3. For the program below, select an iterative DO statement to process all elements in the contrib array.

    ```
    data work.contrib;
       array contrib{4} qtr1-qtr4;
          ...
             contrib{i}=contrib{i}*1.25;
       end;
    run;
    ```

    a. do i=4;

    b. do i=1 to 4;

    c. do until i=4;

    d. do while i le 4;

4. What is the value of the index variable that references Jul in the statements below?

    ```
    array quarter{4} Jan Apr Jul Oct;
    do i=1 to 4;
       yeargoal=quarter{i}*1.2;
    end;
    ```

    a. 1

    b. 2

    c. 3

    d. 4

5. Which DO statement would *not* process all the elements in the factors array shown below?

    ```
    array factors{*} age height weight bloodpr;
    ```

a. `do i=1 to dim(factors);`

b. `do i=1 to dim(*);`

c. `do i=1,2,3,4;`

d. `do i=1 to 4;`

6. Which statement below is *false* regarding the use of arrays to create variables?

    a. The variables are added to the program data vector during the compilation of the DATA step.

    b. You do not need to specify the array elements in the ARRAY statement.

    c. By default, all character variables are assigned a length of eight.

    d. Only character variables can be created.

7. For the first observation, what is the value of diff{i} at the end of the second iteration of the DO loop?

Weight1	Weight2	Weight3
192	200	215
137	130	125
220	210	213

```
array wt{*} weight1-weight10;
array diff{9};
do i=1 to 9;
   diff{i}=wt{i+1}-wt{i};
end;
```

a. `15`

b. `10`

c. `8`

d. `-7`

8. Finish the ARRAY statement below to create temporary array elements that have initial values of **9000**, **9300**, **9600**, and **9900**.

    `array goal{4} ... ;`

    a. `_temporary_ (9000 9300 9600 9900)`

    b. `temporary (9000 9300 9600 9900)`

    c. `_temporary_ 9000 9300 9600 9900`

    d. `(temporary) 9000 9300 9600 9900`

9. Based on the ARRAY statement below, select the array reference for the array element q50.

    `array ques{3,25} q1-q75;`

    a. `ques{q50}`

    b. `ques{1,50}`

    c. `ques{2,25}`

    d. `ques{3,0}`

10. Select the ARRAY statement that defines the array in the following program.

```
    data coat;
        input category high1-high3 / low1-low3;
        array compare{2,3} high1-high3 low1-low3;
        do i=1 to 2;
            do j=1 to 3;
                compare{i,j}=round(compare{i,j}*1.12);
            end;
        end;
datalines;
5555 9 8 7 6
4  3  2  1
8888 21 12 34 64
13 14 15 16
;
run;
```

a. `array compare{1,6} high1-high3 low1-low3;`

b. `array compare{2,3} high1-high3 low1-low3;`

c. `array compare{3,2} high1-high3 low1-low3;`

d. `array compare{3,3} high1-high3 low1-low3;`

Chapter 16

Reading Raw Data in Fixed Fields

Overview

Introduction

Raw data can be organized in several ways.

```
1---+----10---+----20---+----
BIRD FEEDER    LG088  3 $29.95
GLASS MUGS     SB082  6 $25.00
GLASS TRAY     BQ049 12 $39.95
PADDED HANGRS  MN256 15 $9.95
JEWELRY BOX    AJ498 23 $45.00
RED APRON      AQ072  9 $6.50
CRYSTAL VASE   AQ672 27 $29.95
PICNIC BASKET  LS930 21 $15.00
```

This external file contains data that is arranged in columns or fixed fields. You can specify a beginning and ending column for each field. However, this file contains nonstandard data, because one of the variable's values includes a special character, the dollar sign ($).

This external file contains no special characters, but its data is free-format, meaning that it is not arranged in columns. Notice that the values for a particular field do not begin and end in the same columns.

```
1---+----10---+----20---+--
BARNES NORTHWEST 36098.45
FARLSON SOUTHWEST 24394.09
LAWRENCE NORTHEAST 19504.26
NELSON SOUTHEAST 16930.84
STEWART MIDWEST 23845.13
TAYLOR MIDWEST 12354.42
TREADWAY SOUTHWEST 41092.84
WALSTON SOUTHEAST 28938.71
```

How your data is organized and what type of data you have determine which input style you should use to read the data. SAS provides three primary input styles: column input, formatted input, and list input. This chapter teaches you how to use column input and formatted input to read standard and nonstandard data that is arranged in fixed fields.

Objectives

In this chapter, you learn to

- distinguish between standard and nonstandard numeric data

- read standard fixed-field data

- read nonstandard fixed-field data.

Review of Column Input

Overview

In "Creating SAS Data Sets from External Files" on page 152, you learned how to use column input to read raw data that is stored in an external file.

You can use column input to read the values for Item, IDnum, InStock, and BackOrd from the raw data file that is referenced by the fileref Invent.

```
input Item $ 1-13 IDnum $ 15-19 InStock 21-22
      BackOrd 24-25;
```

```
            Raw Data File Invent
1---+----10---+----20---+--
BIRD FEEDER   LG088  3 20
GLASS MUGS    SB082  6 12
GLASS TRAY    BQ049 12  6
PADDED HANGRS MN256 15 20
JEWELRY BOX   AJ498 23  0
RED APRON     AQ072  9 12
CRYSTAL VASE  AQ672 27  0
PICNIC BASKET LS930 21  0
```

Notice that the INPUT statement lists the variables with their corresponding column locations in order from left to right. However, one of the features of column input is the ability to read fields in any order.

For example, you could have read the values for InStock and BackOrd before the values for Item and IDnum.

```
input InStock 21-22 BackOrd 24-25 Item $ 1-13
      IDnum $ 15-19;
```

When you print a report that is based on this data set, the variables will be listed in the order in which they were created by default.

InStock	BackOrd	Item	IDnum
3	20	BIRD FEEDER	LG088
6	12	GLASS MUGS	SB082
12	6	GLASS TRAY	BQ049
15	20	PADDED HANGRS	MN256
23	0	JEWELRY BOX	AJ498
9	12	RED APRON	AQ072
27	0	CRYSTAL VASE	AQ672
21	0	PICNIC BASKET	LS930

Column Input Features

Column input has several features that make it useful for reading raw data.

- It can be used to read character variable values that contain embedded blanks.

```
input Name $ 1-25;
```

```
1---+----10---+----20---+-
JOSEPH PAUL THACKERY JR.
```

- No placeholder is required for missing data. A blank field is read as missing and does not cause other fields to be read incorrectly.

```
input Item $ 1-13 IDnum $ 15-19
      InStock 21-22 BackOrd 24-25;
```

```
1---+----10---+----20---+--
PADDED HANGRS MN256 15 20
JEWELRY BOX    AJ498     0
RED APRON      AQ072  9 12
```

- Fields or parts of fields can be re-read.

```
input Item $ 1-13 IDnum $ 15-19 Supplier $ 15-16
      InStock 21-22 BackOrd 24-25;
```

```
1---+----10---+----20---+--
PADDED HANGRS MN256 15 20
JEWELRY BOX    AJ498 23  0
RED APRON      AQ072  9 12
```

- Fields do not have to be separated by blanks or other delimiters.

```
input Item $ 1-13 IDnum $ 14-18 InStock 19-20
      BackOrd 21-22;
```

```
1---+----10---+----20---+--
PADDED HANGRSMN2561520
JEWELRY BOX    AJ49823 0
RED APRON      AQ072 912
```

Identifying Nonstandard Numeric Data

Standard Numeric Data

Standard numeric data values can contain only

- numbers

- decimal points

- numbers in scientific, or E, notation (23E4)

- minus signs and plus signs.

Some examples of standard numeric data are 15, -15, 15.4, +.05, 1.54E3, and -1.54E-3.

Nonstandard Numeric Data

Nonstandard numeric data includes

- values that contain special characters, such as percent signs (%), dollar signs ($), and commas (,)

- date and time values

- data in fraction, integer binary, real binary, and hexadecimal forms.

The external file referenced by the fileref Empdata contains the personnel information for the technical writing department of a small computer manufacturer. The fields contain values for each employee's last name, first name, job title, and annual salary.

```
        Raw Data File Empdata
1---+----10---+----20---+---
EVANS    DONNY 112 29,996.63
HELMS    LISA  105 18,567.23
HIGGINS  JOHN  111 25,309.00
LARSON   AMY   113 32,696.78
MOORE    MARY  112 28,945.89
POWELL   JASON 103 35,099.50
RILEY    JUDY  111 25,309.00
RYAN     NEAL  112 28,180.00
WILSON   HENRY 113 31,875.46
WOODS    CHIP  105 17,098.71
```

Notice that the values for Salary contain commas. So, the values for Salary are considered to be nonstandard numeric values.

Choosing an Input Style

Nonstandard data values require an input style that has more flexibility than column input.

You can use formatted input, which combines the features of column input with the ability to read both standard and nonstandard data.

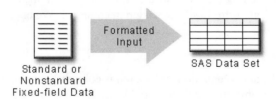

Whenever you encounter raw data that is organized into fixed fields, you can use

- column input to read standard data only

- formatted input to read both standard and nonstandard data.

Using Formatted Input

The General Form of Formatted Input

Formatted input is a very powerful method for reading both standard and nonstandard data in fixed fields.

General form, INPUT statement using formatted input:

INPUT *<pointer-control> variable informat.*;

where

- *pointer-control* positions the input pointer on a specified column
- *variable* is the name of the variable
- *informat* is the special instruction that specifies how SAS reads the raw data.

In this chapter, you'll work with two column pointer controls.

- The *@n* moves the input pointer to a specific column number.
- The *+n* moves the input pointer forward to a column number that is relative to the current position.

Let's first look at the *@n* column pointer control.

Using the @n Column Pointer Control

The *@n* is an absolute pointer control that moves the input pointer to a specific column number. The *@* moves the pointer to column *n*, which is the first column of the field that is being read.

General form, INPUT statement using formatted input and the *@n* pointer control:

INPUT *@n variable informat.*;

- *variable* is the name of the variable
- *informat* is a special instruction that specifies how SAS reads the raw data.

Let's use the *@n* pointer control to locate variable values in the external file Empdata. As you can see, the values for LastName begin in column 1. We could start with the *@1* pointer control.

```
input @1 LastName $7.
```

However, the default column pointer location is column 1, so you do not need to use a column pointer control to read the first field.

```
input LastName $7.
```

```
v---+----10---+----20---+---
EVANS    DONNY 112 29,996.63
HELMS    LISA  105 18,567.23
HIGGINS  JOHN  111 25,309.00
LARSON   AMY   113 32,696.78
MOORE    MARY  112 28,945.89
POWELL   JASON 103 35,099.50
RILEY    JUDY  111 25,309.00
```

Next, the values for FirstName begin in column 9. To point to column 9, use an @ sign and the column number in the INPUT statement:

```
input LastName $7. @9 FirstName $5.
```

```
1---+---v10---+----20---+---
EVANS    DONNY 112 29,996.63
HELMS    LISA  105 18,567.23
HIGGINS  JOHN  111 25,309.00
LARSON   AMY   113 32,696.78
MOORE    MARY  112 28,945.89
POWELL   JASON 103 35,099.50
RILEY    JUDY  111 25,309.00
```

The $7. and $5. informats are explained later in this chapter.

Reading Columns in Any Order

Column pointer controls are very useful. For example, you can use the @*n* to move a pointer forward or backward when reading a record.

In this INPUT statement, the value for FirstName is read first, starting in column 9.

```
input @9 FirstName $5.
```

```
1---+---v10---+----20---+---
EVANS    DONNY 112 29,996.63
HELMS    LISA  105 18,567.23
HIGGINS  JOHN  111 25,309.00
```

Now let's read the values for LastName, which begin in the first column. Here you must use the @*n* pointer control to move the pointer back to column 1.

```
input @9 FirstName $5. @1 LastName $7.
```

```
v---+----10---+----20---+---
EVANS    DONNY 112 29,996.63
HELMS    LISA  105 18,567.23
HIGGINS  JOHN  111 25,309.00
```

The rest of the INPUT statement specifies the column locations of the raw data value for JobTitle and Salary.

```
input @9 FirstName $5. @1 LastName $7. @15 JobTitle 3.
     @19 Salary comma9.;
```

```
1---+----10---v---v20---+---
EVANS     DONNY 112 29,996.63
HELMS     LISA  105 18,567.23
HIGGINS JOHN    111 25,309.00
```

The $5., $7., 3., and comma9. informats are explained later in this chapter.

The +n Pointer Control

The +*n* pointer control moves the input pointer forward to a column number that is relative to the current position. The + moves the pointer forward *n* columns.

General form, INPUT statement using formatted input and the +*n* pointer control:

INPUT +*n variable informat.*;

* *variable* is the name of the variable

* *informat* is a special instruction that specifies how SAS reads the raw data.

In order to count correctly, it is important to understand where the column pointer is located after each data value is read. Let's look at an example.

Suppose you want to read the data from Empdata in the following order: LastName, FirstName, Salary, JobTitle. Because the values for LastName begin in column 1, a column pointer control is not needed.

```
input LastName $7.
```

```
v---+----10---+----20---+---
EVANS     DONNY 112 29,996.63
HELMS     LISA  105 18,567.23
HIGGINS JOHN    111 25,309.00
```

With formatted input, the column pointer moves to the first column following the field that was just read. In this example, after LastName is read, the pointer moves to column 8.

```
1---+--v-10---+----20---+---
EVANS     DONNY 112 29,996.63
HELMS     LISA  105 18,567.23
HIGGINS JOHN    111 25,309.00
```

To start reading FirstName, which begins in column 9, you move the column pointer ahead 1 column with +1.

```
input LastName $7. +1 FirstName $5.
```

```
1---+---v10---+----20---+---
EVANS     DONNY 112 29,996.63
HELMS     LISA  105 18,567.23
HIGGINS JOHN    111 25,309.00
```

After reading FirstName, the column pointer moves to column 14. Now you want to skip over the values for JobTitle and read the values for Salary, which begin in column 19. So you move the column pointer ahead 5 columns from column 14.

```
input LastName $7. +1 FirstName $5. +5 Salary comma9.
```

```
                    12345
1---+----10---v---v20---+---
EVANS    DONNY 112 29,996.63
HELMS    LISA  105 18,567.23
HIGGINS  JOHN  111 25,309.00
```

The last field to be read contains the values for JobTitle. You can use the @n column pointer control to return to column 15.

```
input LastName $7. +1 FirstName $5. +5 Salary comma9.
      @15 JobTitle 3.;
```

```
1---+----10---v----20---+---
EVANS    DONNY 112 29,996.63
HELMS    LISA  105 18,567.23
HIGGINS  JOHN  111 25,309.00
```

The $7., $5., comma9., and 3. informats are explained later in this chapter.

You can use the notation +(-*n*) to move the +*n* pointer backwards. For more information about the +(-*n*) notation, see the SAS documentation.

Using Informats

Overview

Remember that the general form of the INPUT statement for formatted input is:

INPUT *<pointer-control> variable informat.*;

An informat is an instruction that tells SAS how to read raw data. SAS provides many informats for reading standard and nonstandard data values. Here is a small sample.

Table 16.1   *Selected Informats for Reading Data*

PERCENT*w.d*	DATE*w.*	NENGO*w.*
$BINARY*w.*	DATETIME*w.*	PD*w.d*
$VARYING*w.*	HEX*w.*	PERCENT*w.*
$*w.*	JULIAN*w.*	TIME*w.*
COMMA*w.d*	MMDDYY*w.*	*w.d*

Note that

- each informat contains a w value to indicate the width of the raw data field

- each informat also contains a period, which is a required delimiter

- for some informats, the optional d value specifies the number of implied decimal places

- informats for reading character data always begin with a dollar sign ($).

For a complete list of informats, see the SAS documentation.

Reading Character Values

The $w. informat enables you to read character data. The w represents the field width of the data value (the total number of columns that contain the raw data field).

In the example below, the $ indicates that FirstName is a character variable, the 5 indicates a field width of five columns, and a period ends the informat.

```
input @9 FirstName $5.;
```

```
              12345
1---+---v10---+----20---+---
EVANS    DONNY 112 29,996.63
HELMS    LISA  105 18,567.23
HIGGINS  JOHN  111 25,309.00
LARSON   AMY   113 32,696.78
MOORE    MARY  112 28,945.89
POWELL   JASON 103 35,099.50
RILEY    JUDY  111 25,309.00
RYAN     NEAL  112 28,180.00
WILSON   HENRY 113 31,875.46
WOODS    CHIP  105 17,098.71
```

Reading Standard Numeric Data

The informat for reading standard numeric data is the w.d informat.

The w specifies the field width of the raw data value, the period serves as a delimiter, and the d optionally specifies the number of implied decimal places for the value. The w.d informat ignores any specified d value if the data already contains a decimal point.

For example, the raw data value shown below contains 6 digits (4 are decimals) and 1 decimal point. Therefore, the w. informat requires a field width of only 7 to correctly read the raw data value.

Raw Data Value		w. Informat		Variable Value
34.0008	▶	7.	▶	34.0008

In the example shown below, the values for JobTitle in columns 15-17 contain only numbers. Remember that standard numeric data values can contain only numbers, decimal points, scientific notation, and plus and minus signs.

A *d* value is not necessary to read the values for JobTitle. Simply move the column pointer forward 7 spaces to column 15, name the variable, and specify a field width of 3.

```
input @9 FirstName $5. @1 LastName $7. +7 JobTitle 3.;
```

```
                    123
1---+----10---v----20---+---
EVANS    DONNY  112 29,996.63
HELMS    LISA   105 18,567.23
HIGGINS  JOHN   111 25,309.00
LARSON   AMY    113 32,696.78
MOORE    MARY   112 28,945.89
POWELL   JASON  103 35,099.50
RILEY    JUDY   111 25,309.00
```

 Be certain to specify the period in the informat name.

Reading Nonstandard Numeric Data

The COMMA*w.d* informat is used to read numeric values and to remove embedded

- blanks

- commas

- dashes

- dollar signs

- percent signs

- right parentheses

- left parentheses, which are interpreted as minus signs.

The COMMA*w.d* informat has three parts:

1.	the informat name	COMMA
2.	a value that specifies the width of the field to be read (including dollar signs, decimal places, or other special characters), followed by a period	*w.*
3.	an optional value that specifies the number of implied decimal places for a value (not necessary if the value already contains decimal places).	*d*

In the example below, the values for Salary contain commas, which means that they are nonstandard numeric values.

The values for Salary begin in column 19, so use the *@n* or *+n* pointer control to point to column 19, and then name the variable.

```
data sasuser.empinfo;
   infile empdata;
   input @9 FirstName $5. @1 LastName $7. +7 JobTitle 3.
      @19 Salary;
```

```
1---+----10---+---v20---+---
EVANS    DONNY 112 29,996.63
HELMS    LISA  105 18,567.23
HIGGINS  JOHN  111 25,309.00
LARSON   AMY   113 32,696.78
MOORE    MARY  112 28,945.89
POWELL   JASON 103 35,099.50
RILEY    JUDY  111 25,309.00
```

Now add the COMMA*w.d* informat and specify the field width. The values end in column 27, so the field width is 9 columns. Add a RUN statement to complete the DATA step.

```
data sasuser.empinfo;
   infile empdata;
   input @9 FirstName $5. @1 LastName $7. +7 JobTitle 3.
        @19 Salary comma9.;
 run;
```

```
                     123456789
1---+----10---+---v20---+-v-
EVANS    DONNY 112 29,996.63
HELMS    LISA  105 18,567.23
HIGGINS  JOHN  111 25,309.00
LARSON   AMY   113 32,696.78
MOORE    MARY  112 28,945.89
POWELL   JASON 103 35,099.50
RILEY    JUDY  111 25,309.00
```

If you use PROC PRINT to display the data set, the commas are removed from the values for Salary in the resulting output.

```
data sasuser.empinfo;
  infile empdata;
  input @9 FirstName $5. @1 LastName $7. +7 JobTitle 3.
        @19 Salary comma9.;
run;
proc print data=sasuser.empinfo;
run;
```

Figure 16.1 *Output from the PRINT Procedure*

Obs	FirstName	LastName	JobTitle	Salary
1	DONNY	EVANS	112	29996.63
2	ALISA	HELMS	105	18567.23
3	JOHN	HIGGINS	111	25309.00
4	AMY	LARSON	113	32696.78
5	MARY	MOORE	112	28945.89
6	JASON	POWELL	103	35099.50
7	JUDY	RILEY	111	25309.00

Thus, the COMMA*w.d* informat does more than simply read the raw data values. It removes special characters such as commas from numeric data and stores only numeric values in a SAS data set.

DATA Step Processing of Informats

Let's place our INPUT statement in a DATA step and submit it for processing. Remember that after the DATA step is submitted, it is compiled and then executed.

```
data sasuser.empinfo;
   infile empdata;
   input @9 FirstName $5. @1 LastName $7. +7 JobTitle 3.
         @19 Salary comma9.;
run;
```

During the compile phase, the character variables in the program data vector are defined with the exact length specified by the informat. But notice that the lengths that are defined for JobTitle and Salary in the program data vector are different from the lengths that are specified by their informats.

```
                            123 123456789
1---+----10---V-V-V20---+-V-
EVANS    DONNY 112 29,996.63
HELMS    LISA  105 18,567.23
HIGGINS  JOHN  111 25,309.00
```

Program Data Vector

N_	ERROR_	FirstName	LastName	JobTitle	Salary
		$5.	$7.	8.	8.
		DONNY	EVANS	112	29996.6

Remember, by default, SAS stores numeric values (no matter how many digits the value contains) as floating-point numbers in 8 bytes of storage. The length of a stored numeric variable is not affected by an informat's width nor by other column specifications in an INPUT statement.

However, it is still necessary to specify the actual width of a raw data field in an INPUT statement. Otherwise, if you specify a default field width of 8 for all numeric values, SAS will read inappropriate variable values when the program executes.

In the following example, the values for JobTitle would contain embedded blanks, which represent invalid numeric values.

```
data sasuser.empinfo;
   infile empdata;
   input @9 FirstName $5. @1 LastName $7.
         +7 JobTitle 8. @19 Salary comma8.;
run;
```

```
            12345678
              12345678
1---+----10---V---V20V--+V--
EVANS   DONNY 112 29,996.63
HELMS   LISA  105 18,567.23
HIGGINS JOHN  111 25,309.00
```

Program Data Vector

N	ERROR	FirstName	LastName	JobTitle	Salary
		$5.	$7.	8.	8.
		DONNY	EVANS	•	29996.6

Remember that the w value of the informat represents the width of the field in the raw data file. The values for JobTitle have a width of only 3 in the raw data file. However, because they are numeric values, SAS stores them with a default length of 8.

Record Formats

The record format of an external file might affect how data is read with column input and formatted input. A record format specifies how records are organized in a file. Two common record formats are fixed-length records and variable-length records.

Fixed-Length Records

External files that have a fixed-length record format have an end-of-record marker after a predetermined number of columns. A typical record length is 80 columns.

```
123456789...............................80
< field1 >< field2 >< field3 >..unused space    record 1
123456789...............................80
< field1 >< field2 >< field3 >< field4 >....    record 2
123456789...............................80
< field1 >< field2 >....  unused space ......    record 3
```

Variable-Length Records

Files that have a variable-length record format have an end-of-record marker after the last field in each record.

As you can see, the length of each record varies.

```
123456789...................*
< field1 >< field2 >< field3 >    record 1
123456789.........................*
< field1 >< field2 >< field3 >< field4 >    record 2
123456789..........*
< field1 >< field2 >    record 3
```

Reading Variable-Length Records

When you work with variable-length records that contain fixed-field data, you might have values that are shorter than others or that are missing. This can cause problems when you try to read the raw data into your SAS data set.

For example, notice that the following INPUT statement specifies a field width of 8 columns for Receipts. In the third record, the input pointer encounters an end-of-record marker before the 8th column.

```
input Dept $ 1-11 @13 Receipts comma8.;
```

The asterisk symbolizes the end-of-record marker and is not part of the data.

```
                   12345678
1---+----10---+---v20--
BED/BATH    1,354.93*
HOUSEWARES  2,464.05*
GARDEN       923.34*
GRILL        598.34*
SHOES       1,345.82*
SPORTS*
TOYS        6,536.53*
```

The input pointer moves down to the next record in an attempt to complete the value for Receipts. However, **GRILL** is a character value, and Receipts is a numeric variable. Thus, an invalid data error occurs, and Receipts is set to missing.

```
v---+----10---+----20--
BED/BATH     1,354.93*
HOUSEWARES   2,464.05*
GARDEN        923.34*
GRILL         598.34*
SHOES        1,345.82*
SPORTS*
TOYS         6,536.53*
```

The PAD Option

When reading variable-length records that contain fixed-field data, you can avoid problems by using the PAD option in the INFILE statement. The PAD option pads each record with blanks so that all data lines have the same length.

```
infile receipts pad;
```

```
1---+----10---+----20--
BED/BATH    1,354.93
HOUSEWARES  2,464.05
GARDEN      923.34
GRILL       598.34
SHOES       1,345.82
SPORTS
TOYS        6,536.53
```

When you use column input or formatted input to read fixed-field data in variable-length records, remember to determine whether or not you need to use the PAD option. For more information about the PAD option, see the SAS documentation.

⚠ The PAD option is useful only when missing data occurs at the end of a record or when SAS encounters an end-of-record marker before all fields are completely read.

The default value of the maximum record length is determined by your operating system. If you encounter unexpected results when reading many variables, you might need to change the maximum record length by specifying the LRECL=option in the INFILE statement. For more information about the LRECL= option, see the SAS documentation for your operating environment.

Chapter Summary

Text Summary

Review of Column Input

When data is arranged in columns or fixed fields, you can use column input to read them. With column input, the beginning and ending column are specified for each field. Character variables are identified by a dollar ($) sign.

Column input has several features.

- Fields can be read in any order.

- It can be used to read character variables that contain embedded blanks.

- No placeholder is required for missing data. A blank field is read as missing and does not cause other fields to be read incorrectly.

- Fields or parts of fields can be re-read.

- Fields do not have to be separated by blanks or other delimiters

- It can be used to read standard character and numeric data.

Identifying Nonstandard Numeric Data

Standard numeric data values are values that contain only numbers, scientific notation, decimal points, and plus and minus signs. When numeric data contains characters such as commas or dollar signs, the data is considered to be nonstandard.

Nonstandard numeric data includes

- values that contain special characters, such as percent signs, dollar signs, and commas

- date and time values

- data in fraction, integer binary, real binary, and hexadecimal forms.

Choosing an Input Style

SAS provides two input styles for reading data in fixed fields: column input and formatted input. You can use

- column input to read standard data only

- formatted input to read both standard and nonstandard data.

Using Formatted Input

Formatted input uses column pointer controls to position the input pointer on a specified column. A column pointer control is optional when the first variable begins in the first column.

The @*n* is an absolute pointer control that moves the input pointer to a specific column number. You can read columns in any order with the @*n* column pointer control.

The +*n* is a relative pointer control that moves the input pointer forward to a column number that is relative to the current position. The +*n* pointer control cannot move backwards. However, you can use the notation +(-*n*) to move the pointer control backwards.

Using Informats

An informat tells SAS how to read raw data. There are informats for reading standard and nonstandard character and numeric values.

Informats always contain a *w* value to indicate the width of the raw data field. A period (.) ends the informat or separates the *w* value from the optional *d* value, which specifies the number of implied decimal places.

Record Formats

A record format specifies how records are organized in a file. The two most common are fixed-length records and variable-length records.

When you read variable-length records that contain fixed-field data into a SAS data set, there might be values that are shorter than others or that are missing. The PAD option pads each record with blanks so that all data lines have the same length.

Syntax

LIBNAME *libref 'SAS-data-library'*;

FILENAME *fileref 'filename'*;

DATA *SAS-data-set*;

    INFILE *file-specification*;

    INPUT *<pointer-control> variable informat.*;

RUN;

PROC PRINT DATA=*SAS-data-set*;

RUN;

Sample Program

```
libname sasuser 'c:\data\sales';
filename vandata 'c:\records\vans.dat';
data sasuser.vansales;
   infile vandata;
   input +12 Quarter 1. @1 Region $9.
         +6 TotalSales comma11.;
run;
proc print data=sasuser.vansales;
run;
```

Points to Remember

- When you use column input or formatted input, the input pointer stops on the column following the last column that was read.

- When you use informats, you do not need to specify a *d* value if the data values already contain decimal places.

- Column input can be used to read standard character or numeric data only.

- Formatted input can be used to read both standard and nonstandard data.

- When reading variable-length records that contain fixed-field data, you can avoid problems by using the PAD option in the INFILE statement.

Chapter Quiz

Select the best answer for each question. After completing the quiz, check your answers using the answer key in the appendix.

1. Which SAS statement correctly uses column input to read the values in the raw data file below in this order: Address (4th field), SquareFeet (second field), Style (first field), Bedrooms (third field)?

```
1---+----10---+----20---+----30
2STORY 1810 4 SHEPPARD AVENUE
CONDO  1200 2 RAND STREET
RANCH  1550 3 MARKET STREET
```

a. ```
   input Address 15-29 SquareFeet 8-11 Style 1-6
         Bedrooms 13;
   ```

b. ```
   input $ 15-29 Address 8-11 SquareFeet $ 1-6 Style
         13 Bedrooms;
   ```

c. ```
   input Address $ 15-29 SquareFeet 8-11 Style $ 1-6
         Bedrooms 13;
   ```

d. ```
   input Address 15-29 $ SquareFeet 8-11 Style 1-6
         $ Bedrooms 13;
   ```

2. Which is *not* an advantage of column input?

a. It can be used to read character variables that contain embedded blanks.

b.   No placeholder is required for missing data.

c.   Standard as well as nonstandard data values can be read.

d.   Fields do not have to be separated by blanks or other delimiters.

3.   Which is an example of standard numeric data?

a.   -34.245

b. $24,234.25

c.   1/2

d.   50%

4.   Formatted input can be used to read

a.   standard free-format data

b.   standard data in fixed fields

c.   nonstandard data in fixed fields

d.   both standard and nonstandard data in fixed fields

5.   Which informat should you use to read the values in column 1-5?

```
1---+----10---+----20---+----30
2STORY 1810 4 SHEPPARD AVENUE
CONDO  1200 2 RAND STREET
RANCH  1550 3 MARKET STREET
```

a.   *w.*

b.   *$w.*

c.   *w.d*

d.   COMMA*w.d*

6.   The COMMA*w.d* informat can be used to read which of the following values?

a.   **12,805**

b.   **$177.95**

c.   **18%**

d.   all of the above

7.   Which INPUT statement correctly reads the values for ModelNumber (first field) after the values for Item (second field)? Both Item and ModelNumber are character variables.

```
1---+----10---+----20---+----30
DG345 CD PLAYER   $174.99
HJ756 VCR         $298.99
AS658 CAMCORDER $1,195.99
```

a.   input +7 Item $9. @1 ModelNumber $5.;

b.   input +6 Item $9. @1 ModelNumber $5.;

c.   input @7 Item $9. +1 ModelNumber $5.;

d. `input @7 Item $9 @1 ModelNumber 5.;`

8. Which INPUT statement correctly reads the numeric values for Cost (third field)?

```
1---+----10---+----20---+----30
DG345 CD PLAYER    $174.99
HJ756 VCR          $298.99
AS658 CAMCORDER $1,195.99
```

a. `input @17 Cost 7.2;`

b. `input @17 Cost 9.2.;`

c. `input @17 Cost comma7.;`

d. `input @17 Cost comma9.;`

9. Which SAS statement correctly uses formatted input to read the values in this order: Item (first field), UnitCost (second field), Quantity (third field)?

```
1---+----10---+----20---+
ENVELOPE  $13.25 500
DISKETTES $29.50  10
BANDS      $2.50 600
RIBBON    $94.20  12
```

a. `input @1 Item $9. +1 UnitCost comma6.`
   `     @18 Quantity 3.;`

b. `input Item $9. @11 UnitCost comma6.`
   `     @18 Quantity 3.;`

c. `input Item $9. +1 UnitCost comma6.`
   `     @18 Quantity 3.;`

d. all of the above

10. Which raw data file requires the PAD option in the INFILE statement in order to correctly read the data using either column input or formatted input?

a.

```
1---+----10---+----20---+
JONES    M 48 128.6
LAVERNE  M 58 158
JAFFE    F 33 115.5
WILSON   M 28 130
```

b.

```
1---+----10---+----20---+
JONES    M 48 128.6
LAVERNE  M 58 158.0
JAFFE    F 33 115.5
WILSON   M 28 130.0
```

c.

```
1---+----10---+----20---+
JONES     M 48 128.6
LAVERNE   M 58    158
JAFFE     F 33 115.5
WILSON    M 28    130
```

d.

```
1---+----10---+----20---+
  JONES     M 48 128.6
LAVERNE   M 58 158.0
  JAFFE     F 33 115.5
  WILSON    M 28 130.0
```

Chapter 17
Reading Free-Format Data

Overview

Introduction

Raw data can be organized in several ways.

This external file contains data that is arranged in columns, or fixed fields. You can specify a beginning and ending column for each field.

Figure 17.1 *Raw Data in Columns*

```
1---+----10---+----20---+----30--
BIRD FEEDER    LG088    3   20
GLASS MUGS     SB082    6   12
GLASS TRAY     BQ049   12    6
PADDED HANGRS  MN256   15   20
JEWELRY BOX    AJ498   23    0
RED APRON      AQ072    9   12
CRYSTAL VASE   AQ672   27    0
PICNIC BASKET  LS930   21    0
```

By contrast, the following external file contains data that is free-format, meaning data that is not arranged in columns. Notice that the values for a particular field do not begin and end in the same columns.

Figure 17.2 *Raw Data in Free-Format*

```
1---+----10---+----20---+----30--
ABRAMS L.MARKETING $18,209.03
BARCLAY M.MARKETING $18,435.71
COURTNEY W.MARKETING $20,006.16
FARLEY J.PUBLICATIONS $21,305.89
HEINS W.PUBLICATIONS $20,539.23
```

How your data is organized determines which input style you should use to read the data. SAS provides three primary input styles: column, formatted, and list input. This chapter teaches you how to use list input to read free-format data that is not arranged in fixed fields.

Objectives

In this chapter, you learn to use the INPUT statement with list input to read

- free-format data (data that is not organized in fixed fields)

- free-format data that is separated by nonblank delimiters, such as commas

- free-format data that contains missing values

- character values that exceed eight characters

- nonstandard free-format data

- character values that contain embedded blanks.

  In addition, you learn how to mix column, formatted, and list input styles in a single INPUT statement.

Free-Format Data

So far, we have worked with raw data that is in fixed fields. In doing so, we used column input to read standard data values in fixed fields. We have also used formatted input to read both standard and nonstandard data in fixed fields.

Suppose you have raw data that is free-format. That is, it is not arranged in fixed fields. The fields are often separated by blanks or by some other delimiter, as shown below. In this case, column input and formatted input that you might have used before to read standard and nonstandard data in fixed fields will not enable you to read all of the values in the raw data file.

Figure 17.3   *Raw Data in Free-Format That Is Separated by Blanks*

```
1---+----10---+----20---+----30
ABRAMS#L.#MARKETING#$8,209
BARCLAY#M.#MARKETING#$8,435
COURTNEY#W.#MARKETING#$9,006
FARLEY#J.#PUBLICATIONS#$8,305
HEINS#W.#PUBLICATIONS#$9,539
```

Using List Input

Overview

List input is a powerful tool for reading both standard and nonstandard free-format data.

General form, INPUT statement using list input:

INPUT *variable* <$>;

where

- *variable* specifies the variable whose value the INPUT statement is to read

- $ specifies that the variable is a character variable.

Suppose you have an external data file like the one shown below. The file, which is referenced by the fileref Credit, contains the results of a survey on the use of credit cards by males and females in the 18-39 age range.

Figure 17.4 *Raw Data File Creditcard*

```
        Raw Data File Credit
1---+----10---+----20
MALE 27 1 8 0 0
FEMALE 29 3 14 5 10
FEMALE 34 2 10 3 3
MALE 35 2 12 4 8
FEMALE 36 4 16 3 7
MALE 21 1 5 0 0
MALE 25 2 9 2 1
FEMALE 21 1 4 2 6
MALE 38 3 11 4 3
FEMALE 30 3 5 1 0
```

You need to read the data values for

- gender

- age

- number of bank credit cards

- bank card use per month

- number of department store credit cards

- department store card use per month.

List input might be the easiest input style to use because, as shown in the INPUT statement below, you simply list the variable names in the same order as the corresponding raw data fields. Remember to distinguish character variables from numeric variables with dollar signs.

```
input Gender $ Age Bankcard FreqBank Deptcard FreqDept;
```

Because list input, by default, does not specify column locations,

- all fields must be separated by at least one blank or other delimiter

- fields must be read in order from left to right

- you cannot skip or re-read fields.

Processing List Input

It's important to remember that list input causes SAS to scan the input lines for values rather than reading from specific columns. When the INPUT statement is submitted for processing, the input pointer is positioned at column 1 of the raw data file, as shown below.

```
data sasuser.creditsurvey;
   infile creditcard.dat;
   input Gender $ Age Bankcard FreqBank Deptcard
       FreqDept;
run;
```

Figure 17.5 *Raw Data File with the Cursor Positioned at Column 1*

```
V---+----10---+----20
MALE 27 1 8 0 0
FEMALE 29 3 14 5 10
FEMALE 34 2 10 3 3
```

SAS reads the first field until it encounters a blank space. The blank space indicates the end of the field, and the data value is assigned to the program data vector for the first variable in the INPUT statement.

Figure 17.6 *Raw Data File with the Cursor Positioned at the First Blank Space*

```
1---V----10---+----20
MALE 27 1 8 0 0
FEMALE 29 3 14 5 10
FEMALE 34 2 10 3 3
```

Next, SAS scans the record until the next nonblank space is found, and the second value is read until another blank is encountered. Then the value is assigned to its corresponding variable in the program data vector.

Figure 17.7 *Raw Data File with the Cursor Positioned at the Second Blank Space*

```
1---+--V-10---+----20
MALE 27 1 8 0 0
FEMALE 29 3 14 5 10
FEMALE 34 2 10 3 3
```

This process of scanning ahead to the next nonblank column, reading the data value until a blank is encountered, and assigning the value to a variable in the program data vector continues until all of the fields have been read and values have been assigned to variables in the program data vector.

Figure 17.8 *Program Data Vector*

Program Data Vector

N	Gender	Age	Bankcard	FreqBank	Deptcard	FreqDept
1	MALE	27	1	8	0	0

When the DATA step has finished executing, you can display the data set with the PRINT procedure. The following code produces the output below.

```
proc print data=sasuser.creditsurvey;
run;
```

Figure 17.9 Data Set Displayed by PROC PRINT

Obs	Gender	Age	Bankcard	FreqBank	Deptcard	FreqDept
1	MALE	27	1	8	0	0
2	FEMALE	29	3	14	5	10
3	FEMALE	34	2	10	3	3
4	MALE	35	2	12	4	8
5	FEMALE	36	4	16	3	7
6	MALE	21	1	5	0	0
7	MALE	25	2	9	2	1

Working with Delimiters

Most free-format data fields are clearly separated by blanks and are easy to imagine as variables and observations. But fields can also be separated by other delimiters, such as commas, as shown below.

Figure 17.10 Raw Data File with Comma Delimiters

```
      Raw Data File Credit
1---+----10---+----20
MALE,27,1,8,0,0
FEMALE,29,3,14,5,10
FEMALE,34,2,10,3,3
MALE,35,2,12,4,8
FEMALE,36,4,16,3,7
MALE,21,1,5,0,0
MALE,25,2,9,2,1
FEMALE,21,1,4,2,6
MALE,38,3,11,4,3
FEMALE,30,3,5,1,0
```

When characters other than blanks are used to separate the data values, you can tell SAS which field delimiter to use. Use the DLM= option in the INFILE statement to specify a delimiter other than a blank (the default).

General form, DLM= option:

DLM=_delimiter(s)_

where _delimiter(s)_ specifies a delimiter for list input in either of the following forms:

- _'list-of-delimiting-characters'_ specifies one or more characters (up to 200) to read as delimiters. The list of characters must be enclosed in quotation marks.

- _character-variable_ specifies a character variable whose value becomes the delimiter.

- DELIMITER is an alias for the DLM option.

Example

The following program creates the output shown below.

```
data sasuser.creditsurvey;
   infile creditcardcomma.dat dlm=',';
   input Gender $ Age Bankcard FreqBank
         Deptcard FreqDept;
run;
proc print data=sasuser.creditsurvey;
run;
```

Figure 17.11 *Output from the DLM= Option*

Obs	Gender	Age	Bankcard	FreqBank	Deptcard	FreqDept
1	MALE	27	1	8	0	0
2	FEMALE	29	3	14	5	10
3	FEMALE	34	2	10	3	3
4	MALE	35	2	12	4	8
5	FEMALE	36	4	16	3	7
6	MALE	21	1	5	0	0
7	MALE	25	2	9	2	1
8	FEMALE	21	1	4	2	6
9	MALE	38	3	11	4	3
10	FEMALE	30	3	5	1	0

⚠ The field delimiter must *not* be a character that occurs in a data value. For example, the following raw data file contains values for LastName and Salary. Notice that the values for Salary contain commas.

Figure 17.12 *Raw Data File with Commas in Values*

```
1---+----10---+----20
BROWN 24,456.09
JOHNSON 25,467.17
McABE 21,766.36
```

If the field delimiter is also a comma, the fields are identified incorrectly, as shown below.

Figure 17.13 *Raw Data File with Incorrect Use of Commas*

```
1---+----10---+----20
BROWN,24,456.09
JOHNSON,25,467.17
McABE,21,766.36
```

Figure 17.14 *Output When Commas Are Used Incorrectly*

SAS Data Set

Obs	LastName	Salary
1	BROWN	24
2	JOHNSON	25
3	McABE	21

Later in this chapter you'll learn how to work with data values that contain delimiters.

Reading a Range of Variables

When the variable values in the raw data file are sequential and are separated by a blank (or by another delimiter), you can specify a range of variables in the INPUT statement. This is especially useful if your data contains similar variables, such as the answers to a questionnaire.

For example, the following INPUT statement creates five new numeric variables and assigns them the names Ques1, Ques2, Ques3, and so on. You can also specify a range in the VAR statement with the PROC PRINT step to list a range of specific variables.

```
data sasuser.phonesurvey;
   infile phonesurvey;
   input IDnum $ Ques1-Ques5;
run;
proc print data=sasuser.phonesurvey;
   var ques1-ques3;
run;
```

Figure 17.15 *Raw Data File with Sequential Variables*

Raw Data File Survey

```
1---+----10---+----20
1000 23 94 56 85 99
1001 26 55 49 87 85
1002 33 99 54 82 94
1003 71 33 22 44 92
1004 88 49 29 57 83
```

Figure 17.16 *Output Using Sequential Variable Names*

Obs	Ques1	Ques2	Ques3
1	23	94	56
2	26	55	49
3	33	99	54
4	71	33	22
5	88	49	29

If you specify a range of character variables, both the variable list and the $ sign must be enclosed in parentheses.

```
data sasuser.stores;
   infile stordata;
   input Age (Store1-Store3) ($);
run;
proc print data=sasuser.stores;
run;
```

You can also specify a range of variables using formatted input. If you specify a range of variables using formatted input, both the variable list and the informat must be enclosed in parentheses, regardless of the variable's type.

```
data sasuser.scores;
   infile group3;
   input Age (Score1-Score4) (6.);
run;
```

Limitations of List Input

In its default form, list input places several restrictions on the types of data that can be read:

- Although the width of a field can be greater than eight columns, both character and numeric variables have a default length of 8. Character values that are longer than eight characters will be truncated.

- Data must be in standard numeric or character format.

- Character values cannot contain embedded delimiters.

- Missing numeric and character values must be represented by a period or some other character

There are ways to work around these limitations using modified list input, which will be discussed later in this chapter.

Reading Missing Values

Reading Missing Values at the End of a Record

Suppose that the third person represented in the raw data file below did not answer the questions about how many department store credit cards she has and about how often she uses them.

Figure 17.17 *Raw Data File with Missing Values at the End of a Record*

```
1---+----10---+----20
MALE 27 1 8 0 0
FEMALE  3 14 5 10
FEMALE 34 2 10
MALE 35 2 12 4 8
FEMALE 36 4 16 3 7
MALE 21 1 5 0 0
MALE 25 2 9 2 1
FEMALE 21 1 4 2 6
MALE 38 3 11 4 3
FEMALE 30 3 5 1 0
```

Because the missing values occur at the *end* of the record, you can use the MISSOVER option in the INFILE statement to assign the missing values to variables with missing data at the end of a record. The MISSOVER option prevents SAS from reading the next record if, when using list input, it does not find values in the current line for all the INPUT statement variables. At the *end* of the current record, values that are expected but not found are set to missing.

For the raw data file shown above, the MISSOVER option prevents the fields in the fourth record from being read as values for Deptcard and FreqDept in the third observation. Note that Deptcard and FreqDept are set to missing.

```
data sasuser.creditsurvey;
   infile creditcard missover;
   input Gender $ Age Bankcard FreqBank
         Deptcard FreqDept;
run;
proc print data=sasuser.creditsurvey;
run;
```

Figure 17.18 *Output Showing Missing Values*

Obs	Gender	Age	Bankcard	FreqBank	Deptcard	FreqDept
1	MALE	27	1	8	0	0
2	FEMALE	29	3	14	5	10
3	FEMALE	34	2	10	.	.
4	MALE	35	2	12	4	8
5	FEMALE	36	4	16	3	7
6	MALE	21	1	5	0	0

⚠ The MISSOVER option works only for missing values that occur at the *end* of the record.

Reading Missing Values at the Beginning or Middle of a Record

Remember that the MISSOVER option works only for missing values that occur at the end of the record. A different method is required when you use list input to read raw data that contains missing values at the beginning or middle of a record. Let's see what happens when a missing value occurs at the beginning or middle of a record.

Suppose the value for Age is missing in the first record.

Figure 17.19   *Raw Data File with Missing Values at the Beginning or Middle of a Record*

```
  Raw Data File Credit2
1---+----10---+----20
MALE,,1,8,0,0
FEMALE,29,3,14,5,10
FEMALE,34,2,10,3,3
MALE,35,2,12,4,8
FEMALE,36,4,16,3,7
```

When the program below executes, each field in the raw data file is read one by one. The INPUT statement tells SAS to read six data values from each record. However, the first record contains only five values.

```
data sasuser.creditsurvey;
   infile credit2 dlm=',';
   input Gender $ Age Bankcard FreqBank
         Deptcard FreqDept;
run;
proc print data=sasuser.creditsurvey;
run;
```

The two commas in the first record are interpreted as one delimiter. The incorrect value (**1**) is read for Age. The program continues to read subsequent incorrect values for Bankcard (**8**), FreqBank (**0**), and Deptcard (**0**). The program then attempts to read the character field FEMALE, at the beginning of the second record, as the value for the numeric variable FreqDept. This causes the value of FreqDept in the first observation to be interpreted as missing. The input pointer then moves down to the third record to begin reading values for the second observation. Therefore, the first observation in the data set contains incorrect values and values from the second record in the raw data file are not included.

Figure 17.20   *Output with Missing Data Records*

Obs	Gender	Age	Bankcard	FreqBank	Deptcard	FreqDept
1	MALE	1	8	0	0	.
2	FEMALE	34	2	10	3	3
3	MALE	35	2	12	4	8
4	FEMALE	36	4	16	3	7

You can use the Delimiter Sensitive Data (DSD) option in the INFILE statement to correctly read the raw data. The DSD option changes how SAS treats delimiters when list input is used. Specifically, the DSD option

- sets the default delimiter to a comma

- treats two consecutive delimiters as a missing value

- removes quotation marks from values.

When the following program reads the raw data file, the DSD option sets the default delimiter to a comma and treats the two consecutive delimiters as a missing value. Therefore, the data is read correctly.

```
data sasuser.creditsurvey;
   infile creditcardcomma.dat dsd;
```

```
      input Gender $ Age Bankcard FreqBank
            Deptcard FreqDept;
run;
proc print data=sasuser.creditsurvey;
run;
```

Figure 17.21 *DSD Raw Data and Output*

Raw Data File Credit2

```
1---+----10---+----20
MALE,,1,8,0,0
FEMALE,29,3,14,5,10
FEMALE,34,2,10,3,3
MALE,35,2,12,4,8
FEMALE,36,4,16,3,7
```

Obs	Gender	Age	Bankcard	FreqBank	Deptcard	FreqDept
1	MALE	.	1	8	0	0
2	FEMALE	29	3	14	5	10
3	FEMALE	34	2	10	3	3
4	MALE	35	2	12	4	8
5	FEMALE	36	4	16	3	7

If the data uses multiple delimiters or a single delimiter other than a comma, simply specify the delimiter value(s) with the DLM= option. In the following example, an asterisk (*) is used as a delimiter. However, the data is still read correctly because of the DSD option.

```
data sasuser.creditsurvey;
   infile credit3.dat dsd dlm='*';
   input Gender $ Age Bankcard FreqBank
         Deptcard FreqDept;
run;
proc print data=sasuser.creditsurvey;
run;
```

Figure 17.22 *Raw Data with Multiple Delimiters and Output*

Raw Data File Credit3

```
1---+----10---+----20
MALE**1*8*0*0
FEMALE*29*3*14*5*10
FEMALE*34*2*10*3*3
MALE*35*2*12*4*8
FEMALE*36*4*16*3*7
```

Obs	Gender	Age	Bankcard	FreqBank	Deptcard	FreqDept
1	MALE	.	1	8	0	0
2	FEMALE	29	3	14	5	10
3	FEMALE	34	2	10	3	3
4	MALE	35	2	12	4	8
5	FEMALE	36	4	16	3	7

The DSD option can also be used to read raw data when there is a missing value at the beginning of a record, as long as a delimiter precedes the first value in the record.

```
data sasuser.creditsurvey;
   infile credit4.dat dsd;
```

```
        input Gender $ Age Bankcard FreqBank
              Deptcard FreqDept;
run;
proc print data=sasuser.creditsurvey;
run;
```

Figure 17.23   *Raw Data with Missing Data and Output*

```
          Raw Data File Credit4
1---+----10---+----20
,27,1,8,0,0
FEMALE,29,3,14,5,10
FEMALE,34,2,10,3,3
MALE,35,2,12,4,8
FEMALE,36,4,16,3,7
```

Obs	Gender	Age	Bankcard	FreqBank	Deptcard	FreqDept
1		27	1	8	0	0
2	FEMALE	29	3	14	5	10
3	FEMALE	34	2	10	3	3
4	MALE	35	2	12	4	8
5	FEMALE	36	4	16	3	7

You can also use the DSD and DLM= options to read fields that are delimited by blanks.

```
data sasuser.creditsurvey;
   infile credit5.dat dsd dlm=' ';
   input Gender $ Age Bankcard FreqBank
         Deptcard FreqDept;
run;
```

Later in this chapter, you'll learn how to use the DSD option to remove quotation marks from values in raw data.

Specifying the Length of Character Variables

Overview

Remember that when you use list input to read raw data, character variables are assigned a default length of 8. Let's see what happens when list input is used to read character variables whose values are longer than 8.

The raw data file referenced by the fileref Citydata contains 1970 and 1980 population figures for several large U.S. cities. Notice that some city names are rather long.

Figure 17.24 Raw Data File with Character Values That Are Longer than 8

```
1---+----10---+----20---+----
ANCHORAGE 48081 174431
ATLANTA 495039 425022
BOSTON 641071 562994
CHARLOTTE 241420 314447
CHICAGO 3369357 3005072
DALLAS 844401 904078
DENVER 514678 492365
DETROIT 1514063 1203339
MIAMI 334859 346865
PHILADELPHIA 1949996 1688210
SACRAMENTO 257105 275741
```

The longer character values are truncated when they are read into the program data vector.

Figure 17.25 Program Data Vector

Program Data Vector

N	Rank	City	Pop70	Pop80
1	1	ANCHORAG	48081	174431

PROC PRINT output shows the truncated values for City.

```
data sasuser.growth;
   infile citydata.dat;
   input City $ Pop70 Pop80;
run;
proc print data=sasuser.growth;
run;
```

Figure 17.26 Output with Truncated Values

Obs	City	Pop70	Pop80
1	ANCHORAG	48081	174431
2	ATLANTA	495039	425022
3	BOSTON	641071	562994
4	CHARLOTT	241420	314447
5	CHICAGO	3369357	3005072
6	DALLAS	844401	904078
7	DENVER	514678	492365
8	DETROIT	1514063	1203339
9	MIAMI	334859	346865
10	PHILADEL	1949996	1688210
11	SACRAMEN	257105	275741

The LENGTH Statement

Remember, variable attributes are defined when the variable is first encountered in the DATA step. In the program below, the LENGTH statement precedes the INPUT statement and defines both the length and type of the variable City. A length of 12 has been assigned to accommodate **PHILADELPHIA**, which is the longest value for City.

```
data sasuser.growth;
   infile citydata.dat;
   length City $ 12;
   input city $ Pop70 Pop80;
run;
proc print data=sasuser.growth;
run;
```

Figure 17.27   Raw Data File with Character Values That Are Longer than 8

```
1---+----10---+----20---+----
ANCHORAGE 48081 174431
ATLANTA 495039 425022
BOSTON 641071 562994
CHARLOTTE 241420 314447
CHICAGO 3369357 3005072
DALLAS 844401 904078
DENVER 514678 492365
DETROIT 1514063 1203339
MIAMI 334859 346865
PHILADELPHIA 1949996 1688210
SACRAMENTO 257105 275741
```

Using this method, you do not need to specify City's type in the INPUT statement. However, leaving the $ in the INPUT statement will not produce an error. Your output should now display the complete values for City.

Figure 17.28   Output Using Length Statement

Obs	City	Pop70	Pop80
1	ANCHORAGE	48081	174431
2	ATLANTA	495039	425022
3	BOSTON	641071	562994
4	CHARLOTTE	241420	314447
5	CHICAGO	3369357	3005072
6	DALLAS	844401	904078
7	DENVER	514678	492365
8	DETROIT	1514063	1203339
9	MIAMI	334859	346865
10	PHILADELPHIA	1949996	1688210
11	SACRAMENTO	257105	275741

Because variable attributes are defined when the variable is first encountered in the DATA step, a variable that is defined in a LENGTH statement (if it precedes an INPUT statement) will appear first in the data set, regardless of the order of the variables in the INPUT statement.

Modifying List Input

Overview

You can make list input more versatile by using modified list input. There are two modifiers that can be used with list input.

- The ampersand (&) modifier is used to read character values that contain embedded blanks.

- The colon (:) modifier is used to read nonstandard data values and character values that are longer than eight characters, but which contain no embedded blanks.

You can use modified list input to read the file shown below. This file contains the names of the ten largest U.S. cities ranked in order based on their 1986 estimated population figures.

Notice that some of the values for city names contain embedded blanks and are followed by two blanks. Also, note that the values representing the population of each city are nonstandard numeric values (they contain commas).

Figure 17.29 *Raw Data File Topten*

```
              Raw Data File Topten
1---+----10---+----20---+--
 1 NEW YORK  7,262,700
 2 LOS ANGELES  3,259,340
 3 CHICAGO  3,009,530
 4 HOUSTON  1,728,910
 5 PHILADELPHIA  1,642,900
 6 DETROIT  1,086,220
 7 SAN DIEGO  1,015,190
 8 DALLAS  1,003,520
 9 SAN ANTONIO  914,350
10 PHOENIX  894,070
```

In the following sections you will learn how to use the ampersand (&) modifier to read the values for city (`City`). Then you will learn how the colon (:) modifier can be used to read the nonstandard numeric values that represent population (`Pop86`).

Reading Values That Contain Embedded Blanks

The ampersand (&) modifier enables you to read character values that contain single embedded blanks. The & indicates that a character value that is read with list input might contain one or more single embedded blanks. The value is read until two or more consecutive blanks are encountered. The & modifier precedes a specified informat if one is used.

```
input Rank City &;
```

In the raw data file shown below, each value of City is followed by two consecutive blanks. There are two ways that you can use list input to read the values of City.

Using the & Modifier with a LENGTH Statement

As shown below, you can use a LENGTH statement to define the length of `City`, and then add an & modifier to the INPUT statement to indicate that the values contain embedded blanks.

```
data sasuser.cityrank;
   infile topten.dat;
   length City $ 12;
   input Rank city &;
```

Figure 17.30 *Raw Data File Topten*

```
            Raw Data File Topten
1---+----10---+----20---+--
 1 NEW YORK   7,262,700
 2 LOS ANGELES   3,259,340
 3 CHICAGO   3,009,530
 4 HOUSTON   1,728,910
 5 PHILADELPHIA   1,642,900
 6 DETROIT   1,086,220
 7 SAN DIEGO   1,015,190
 8 DALLAS   1,003,520
 9 SAN ANTONIO   914,350
10 PHOENIX   894,070
```

Using the & Modifier with an Informat

You can also read the values for City with the & modifier followed by the $w. informat, which reads standard character values, as shown below. When you do this, the w value in the informat determines the variable's length and should be large enough to accommodate the longest value.

Figure 17.31 *Raw Data File Topten*

```
            Raw Data File Topten
1---+----10---+----20---+--
 1 NEW YORK   7,262,700
 2 LOS ANGELES   3,259,340
 3 CHICAGO   3,009,530
 4 HOUSTON   1,728,910
 5 PHILADELPHIA   1,642,900
 6 DETROIT   1,086,220
 7 SAN DIEGO   1,015,190
 8 DALLAS   1,003,520
 9 SAN ANTONIO   914,350
10 PHOENIX   894,070
```

Remember that you must use two consecutive blanks as delimiters when you use the & modifier. You cannot use any other delimiter to indicate the end of each field.

Reading Nonstandard Values

The colon (:) modifier enables you to read nonstandard data values and character values that are longer than eight characters, but which contain no embedded blanks. The : indicates that values are read until a blank (or other delimiter) is encountered, and then an informat is applied. If an informat for reading character values is specified, the *w* value specifies the variable's length, overriding the default length.

Notice the values representing the 1986 population of each city in the raw data file below. Because they contain commas, these values are nonstandard numeric values.

Figure 17.32   *Raw Data File Topten*

```
           Raw Data File Topten
1---+----10---+----20---+--
 1 NEW YORK  7,262,700
 2 LOS ANGELES  3,259,340
 3 CHICAGO  3,009,530
 4 HOUSTON  1,728,910
 5 PHILADELPHIA  1,642,900
 6 DETROIT  1,086,220
 7 SAN DIEGO  1,015,190
 8 DALLAS  1,003,520
 9 SAN ANTONIO  914,350
10 PHOENIX  894,070
```

In order to read these values, you can modify list input with the colon (:) modifier, followed by the COMMA*w.d* informat, as shown in the program below. Notice that the COMMA*w.d* informat does *not* specify a *w* value.

```
data sasuser.cityrank;
   infile topten.dat;
   input Rank City & $12.
        Pop86 : comma.;
```

Remember that list input reads each value until the next blank is detected. The default length of numeric variables is 8, so you don't need to specify a w value to indicate the length of a numeric variable.

This is different from using a numeric informat with formatted input. In that case, you must specify a *w* value in order to indicate the number of columns to be read.

Processing the DATA Step

At compile time, the informat $12. in the example below sets the length of City to 12 and stores this information in the descriptor portion of the data set. During the execution phase, however, the *w* value of 12 does not determine the number of columns that are read. This is different from the function of informats in the formatted input style.

```
data sasuser.cityrank;
   infile topten.dat;
   input Rank City & $12.
```

```
        Pop86 : comma.;
run;
```

Figure 17.33 *Reading Raw Data File with Character Values That Are Longer than 8*

```
1---+----10---+----20---+--
1 NEW YORK  7,262,700
2 LOS ANGELES  3,259,340
3 CHICAGO  3,009,530
4 HOUSTON  1,728,910
5 PHILADELPHIA  1,642,900
```

The & modifier indicates that the values for City should be read until two consecutive blanks are encountered. Therefore, the value **NEW YORK** is read from column 4 to column 11, a total of only 8 columns. When blanks are encountered in both columns 12 and 13, the value **NEW YORK** is written to the program data vector.

```
data sasuser.cityrank;
   infile topten.dat;
   input Rank City & $12.
        Pop86 : comma.;
run;
```

Figure 17.34 *Reading City Value from Raw Data File*

```
    12345678
1---+----10v--+----20---+--
1 NEW YORK  7,262,700
2 LOS ANGELES  3,259,340
3 CHICAGO  3,009,530
4 HOUSTON  1,728,910
5 PHILADELPHIA  1,642,900
```

Program Data Vector

N	Rank	City	Pop86
1	1	New York	.

The input pointer moves forward to the next nonblank column, which is column 14 in the first record. Now the values for Pop86 are read from column 14 until the next blank is encountered. The COMMA$w.d$ informat removes the commas, and the value is written to the program data vector.

```
data sasuser.cityrank;
   infile topten.dat  input Rank City & $12.
        Pop86 : comma.;
run;
```

Figure 17.35   Reading Raw Data File POP86 Value

```
1---+----10---+----20-v-+--
1 NEW YORK  7,262,700
2 LOS ANGELES  3,259,340
3 CHICAGO  3,009,530
4 HOUSTON  1,728,910
5 PHILADELPHIA  1,642,900
```

Program Data Vector

N	Rank	City	Pop86
1	1	New York	7262700

Notice that the *character* values for City and Pop86 are stored correctly in the data set.

Figure 17.36   SAS Data Set Cityrank

SAS Data Set Perm.Cityrank

Rank	City	Pop86
1	NEW YORK	7262700
2	LOS ANGELES	3259340
3	CHICAGO	3009530
4	HOUSTON	1728910
5	PHILADELPHIA	1642900
6	DETROIT	1086220
7	SAN DIEGO	1015190
8	DALLAS	1003520
9	SAN ANTONIO	914350
10	PHOENIX	894070

Comparing Formatted Input and Modified List Input

As you have seen, informats work differently in modified list input than they do in formatted input. With formatted input, the informat determines both the length of character variables and the number of columns that are read. The same number of columns are read from each record.

```
input @3 City $12.;
```

Figure 17.37   *Raw Data Showing That the Same Number of Columns Are Read from Each Record*

```
   123456789--12              123456789--12
1---+----10---+----20---+--  1---+----10---+----20---+--
1 NEW YORK       7,262,700   1 NEW YORK       7,262,700
2 LOS ANGELES   3,259,340    2 LOS ANGELES   3,259,340
3 CHICAGO       3,009,530    3 CHICAGO       3,009,530
4 HOUSTON       1,728,910    4 HOUSTON       1,728,910
5 PHILADELPHIA 1,642,900     5 PHILADELPHIA 1,642,900
```

The informat in modified list input determines only the length of the variable, *not* the number of columns that are read. Here, the raw data values are read until two consecutive blanks are encountered.

```
input City & $12.;
```

Figure 17.38   *Raw Data Showing That Values Are Read until Two Consecutive Blanks Are Encountered*

```
   12345678                   123456789--12
1---+----10---+----20---+--  1---+----10---+----20---+--
1 NEW YORK   7,262,700       1 NEW YORK   7,262,700
2 LOS ANGELES  3,259,340     2 LOS ANGELES  3,259,340
3 CHICAGO  3,009,530         3 CHICAGO  3,009,530
4 HOUSTON  1,728,910         4 HOUSTON  1,728,910
5 PHILADELPHIA  1,642,900    5 PHILADELPHIA  1,642,900
```

Creating Free-Format Data

Overview

"Creating SAS Data Sets from External Files" on page 152 explained how the PUT statement can be used with column output to write observations from a SAS data set to a raw data file. The PUT statement can also be used with list output to create free-format raw data files.

List output is similar to list input. With list output, you simply list the names of the variables whose values you want to write. The PUT statement writes a value, leaves a blank, and then writes the next value.

General form, PUT statement using list output:

PUT *variable* <: *format*>;

where

- *variable* specifies the variable whose value you want to write

- : precedes a format

- *format* specifies a format to use for writing the data values.

The following program creates the raw data file Findat, using the SAS data set sasuser.Finance. The DATE*w.* format is used to write the value of Date in the form *DDMMYYYY*.

```
data _null_;
   set sasuser.finance;
   file 'c:\data\findat.txt';
   put ssn name salary date : date9.;
run;
```

Figure 17.39 *SAS Data Set Finance*

SAS Data Set Finance

SSN	Name	Salary	Date
074-53-9892	Vincent	35000	05/22/97
776-84-5391	Phillipon	29750	12/15/96
929-75-0218	Gunter	27500	04/30/97
446-93-2122	Harbinger	33900	07/08/96
228-88-9649	Benito	28000	03/04/96
029-46-9261	Rudelich	35000	02/15/95
442-21-8075	Sirignano	5000	11/22/95

Figure 17.40 *Raw Data File Findat*

Raw Data File Findat

```
1---+----10---+----20---+----30---+----40
074-53-9892 Vincent 35000 22MAY1997
776-84-5391 Phillipon 29750 15DEC1996
929-75-0218 Gunter 27500 30APR1997
446-93-2122 Harbinger 33900 08JUL1996
228-88-9649 Benito 28000 04MAR1996
029-46-9261 Rudelich 35000 15FEB1995
442-21-8075 Sirignano 5000 22NOV1995
```

Specifying a Delimiter

You can use the DLM= option with a FILE statement to create a character-delimited raw data file.

```
data _null_;
   set sasuser.finance;
   file 'c:\data\findat2' dlm=',';
   put ssn name salary date : date9.;
run;
```

Figure 17.41 *SAS Data Set Finance*

SAS Data Set Finance

SSN	Name	Salary	Date
074-53-9892	Vincent	35000	05/22/97
776-84-5391	Phillipon	29750	12/15/96
929-75-0218	Gunter	27500	04/30/97
446-93-2122	Harbinger	33900	07/08/96
228-88-9649	Benito	28000	03/04/96
029-46-9261	Rudelich	35000	02/15/95
442-21-8075	Sirignano	5000	11/22/95

Figure 17.42 *Raw Data File Created by the DLM= Option*

Raw Data File Findat2

```
1---+----10---+----20---+----30---+----40
074-53-9892,Vincent,35000,22MAY1997
776-84-5391,Phillipon,29750,15DEC1996
929-75-0218,Gunter,27500,30APR1997
446-93-2122,Harbinger,33900,08JUL1996
228-88-9649,Benito,28000,04MAR1996
029-46-9261,Rudelich,35000,15FEB1995
442-21-8075,Sirignano,5000,22NOV1995
```

For creating a simple raw data file, an alternative to the DATA step is the EXPORT procedure.

General form, PROC EXPORT:

PROC EXPORT DATA=*SAS-data-set*;

OUTFILE=*filename* <DELIMITER=*'delimiter'*>;

RUN;

where

- *SAS-data-set* names the input SAS data set
- *filename* specifies the complete path and filename of the output
- *delimiter* specifies the delimiter to separate columns of data in the output file.

For more information about the EXPORT procedure, see the SAS documentation.

Using the DSD Option

What happens if you need to create a comma-delimited file that requires the use of a format that writes out values using commas?

If you used the following program, the resulting raw data file would contain five fields rather than four.

```
data _null_;
   set sasuser.finance;
   file 'c:\data\findat2' dlm=',';
```

```
   put ssn name salary : comma6. date : date9.;
run;
```

Figure 17.43   *SAS Data Set Finance*

SAS Data Set Finance

SSN	Name	Salary	Date
074-53-9892	Vincent	35000	05/22/97
776-84-5391	Phillipon	29750	12/15/96
929-75-0218	Gunter	27500	04/30/97
446-93-2122	Harbinger	33900	07/08/96
228-88-9649	Benito	28000	03/04/96
029-46-9261	Rudelich	35000	02/15/95
442-21-8075	Sirignano	5000	11/22/95

Figure 17.44   *Raw Data Created with the DLM Option*

```
Raw Data File Findat2
1---+----10---+----20---+----30---+----40
074-53-9892,Vincent,35,000,22MAY1997
776-84-5391,Phillipon,29,750,15DEC1996
929-75-0218,Gunter,27,500,30APR1997
446-93-2122,Harbinger,33,900,08JUL1996
228-88-9649,Benito,28,000,04MAR1996
029-46-9261,Rudelich,35,000,15FEB1995
442-21-8075,Sirignano,5,000,22NOV1995
```

You can use the DSD option in the FILE statement to specify that data values containing commas should be enclosed in quotation marks. Remember that the DSD option uses a comma as a delimiter, so a DLM= option isn't necessary here.

```
data _null_;
   set sasuser.finance;
   file 'c:\data\findat2' dsd;
   put ssn name salary : comma. date : date9.;
run;
```

Figure 17.45   *Raw Data Created with the DSD Option*

```
Raw Data File Findat2
1---+----10---+----20---+----30---+----40
074-53-9892,Vincent,"35,000",22MAY1997
776-84-5391,Phillipon,"29,750",15DEC1996
929-75-0218,Gunter,"27,500",30APR1997
446-93-2122,Harbinger,"33,900",08JUL1996
228-88-9649,Benito,"28,000",04MAR1996
029-46-9261,Rudelich,"35,000",15FEB1995
442-21-8075,Sirignano,"5,000",22NOV1995
```

Reading Values That Contain Delimiters within a Quoted String

You can also use the DSD option in an INFILE statement to read values that contain
delimiters within a quoted string. As shown in the following PROC PRINT output, the
INPUT statement correctly interprets the commas within the values for Salary, does not
interpret them as delimiters, and removes the quotation marks from the character strings
before the value is stored.

```
data work.finance2;
   infile findat2 dsd;
   length SSN $ 11 Name $ 9;
   input ssn name Salary : comma. Date : date9.;
run;
proc print data=work.finance2;
   format date date9.;
run;
```

Figure 17.46   *Raw Data File Findat2*

```
                      Raw Data File Findat2
1---+----10---+----20---+----30---+----40
074-53-9892,Vincent,"35,000",22MAY1997
776-84-5391,Phillipon,"29,750",15DEC1996
929-75-0218,Gunter,"27,500",30APR1997
446-93-2122,Harbinger,"33,900",08JUL1996
228-88-9649,Benito,"28,000",04MAR1996
029-46-9261,Rudelich,"35,000",15FEB1995
442-21-8075,Sirignano,"5,000",22NOV1995
```

Figure 17.47   *Output Created with PROC PRINT*

Obs	SSN	Name	Salary	Date
1	074-53-9892	Vincent	35000	22MAY1997
2	776-84-5391	Phillipon	29750	15DEC1996
3	929-75-0218	Gunter	27500	30APR1997
4	446-93-2122	Harbinger	33900	08JUL1996
5	228-88-9649	Benito	28000	04MAR1996
6	029-46-9261	Rudelich	35000	15FEB1995
7	442-21-8075	Sirignano	5000	22NOV1995

Mixing Input Styles

Evaluating your raw data and choosing the most appropriate input style is a very
important task. You have already worked with three input styles for reading raw data.

Table 17.1 *Input Styles and the Types of Information They Read*

Input Style	Reads
Column	standard data values in fixed fields
Formatted	standard and nonstandard data values in fixed fields
List	data values that are not arranged in fixed fields, but are separated by blanks or other delimiters

With some file layouts, you might need to mix input styles in the same INPUT statement in order to read the data correctly.

Look at the raw data file below and think about how to combine input styles to read these values.

Figure 17.48 *Raw Data Showing Mixed Input Styles*

```
1---+----10---+----20---+----30---+----40-
209-20-3721 07JAN78 41,983 SALES 2896
223-96-8933 03MAY86 27,356 EDUCATION 2344
232-18-3485 17AUG81 33,167 MARKETING 2674
251-25-9392 08SEP84 34,033 RESEARCH 2956
```

- Column input is an appropriate choice for the first field because the values can be read as standard character values and are located in fixed columns.

- The next two fields are also located in fixed columns, but the values require an informat. So, formatted input is a good choice here.

- Values in the fourth field begin in column 28 but do not end in the same column. List input is appropriate here, but notice that some values are longer than eight characters. You need to use the : format modifier with an informat to read these values.

- The last field does not always begin or end in the same column, so list input is the best input style for those values.

Table 17.2 *Input Styles to Read the Raw Data*

Field Description	Starting Column	Field Width	Data Type	Input Style
Social Security #	1	11	character	column
Date of Hire	13	7	date	formatted
Annual Salary	21	6	numeric	formatted
Department	28	5 to 9	character	list
Phone Extension	??	4	character	list

The INPUT statement to read the data should look like this:

```
data sasuser.mixedstyles;
   infile rawdata.dat;
   input SSN $ 1-11 @13 HireDate date7.
         @21 Salary comma6. Department : $9. Phone $;
run;
proc print data=sasuser.mixedstyles;
run;
```

When you submit the PRINT procedure, the output displays values for each variable.

Figure 17.49 *Output Created by PROC PRINT*

Obs	SSN	HireDate	Salary	Department	Phone
1	209-20-3721	6581	41983	SALES	2896
2	223-96-8933	9619	27356	EDUCATION	2344
3	232-18-3485	7899	33167	MARKETING	2674
4	251-25-9392	9017	34033	RESEARCH	2956

Additional Features

Writing Character Strings and Variable Values

You can use a PUT statement to write both character strings and variable values to a raw data file. To write out a character string, simply add a character string, enclosed in quotation marks, to the PUT statement. It's a good idea to include a blank space as the last character in the string to avoid spacing problems.

```
filename totaldat 'c:\records\junsales.txt';
data _null_;
   set work.totals;
   file totaldat;
   put 'Sales for salesrep ' salesrep
       'totaled ' sales : dollar9.;
run;
```

Figure 17.50 *SAS Data Set Work.Totals*

SAS Data Set Work.Totals

Obs	SalesRep	Sales
1	Friedman	$14,893
2	Keane	$14,324
3	Schuster	$13,914
4	Davidson	$13,674

Figure 17.51 *Raw Data File Created Using the PUT Statement*

Raw Data File Totaldat

```
1---+----10---+----20---+----30---+----40---+
Sales for salesrep Friedman totaled $14,893
Sales for salesrep Keane totaled $14,324
Sales for salesrep Schuster totaled $13,914
Sales for salesrep Davidson totaled $13,674
```

🖱️ For more information about using the PUT statement to write character strings, see the SAS documentation.

Chapter Summary

Text Summary

Free-Format Data

External files can contain raw data that is free-format. That is, the data is not arranged in fixed fields. The fields can be separated by blanks, or by some other delimiter, such as commas.

Using List Input

Free-format data can easily be read with list input because you do not need to specify column locations of the data. You simply list the variable names in the same order as the corresponding raw data fields. You must distinguish character variables from numeric variables by using the dollar ($) sign.

When characters other than blanks are used to separate the data values, you can specify the field delimiter by using the DLM= option in the INFILE statement.

You can also specify a range of variables in the INPUT statement when the variable values in the raw data file are sequential and are separated by blanks (or by some other delimiter). This is especially useful if your data contains similar variables, such as the answers to a questionnaire.

In its simplest form, list input places several limitations on the types of data that can be read.

Reading Missing Values

If your data contains missing values at the end of a record, you can use the INFILE statement with the MISSOVER option to prevent SAS from reading the next record to find the missing values.

If your data contains missing values at the beginning or in the middle of a record, you might be able to use the DSD option in the INFILE statement to correctly read the raw data. The DSD option sets the default delimiter to a comma and treats two consecutive delimiters as a missing value.

If the data uses multiple delimiters or a single delimiter other than a comma, you can use both the DSD option and the DLM= option in the INFILE statement.

The DSD option can also be used to read raw data when there is a missing value at the beginning of a record, as long as a delimiter precedes the first value in the record.

Specifying the Length of Character Values

You can specify the length of character variables by using the LENGTH statement. The LENGTH statement enables you to use list input to read character values that are longer than eight characters without truncating them.

Because variable attributes are defined when the variable is first encountered in the DATA step, the LENGTH statement precedes the INPUT statement and defines both the length and the type of the variable.

When you use the LENGTH statement, you do not need to specify the variable type again in the INPUT statement.

Modifying List Input

Modified list input can be used to read values that contain embedded blanks and nonstandard values. Modified list input uses two format modifiers:

- the ampersand (&) modifier enables you to read character values that contain single embedded blanks

- the colon (:) modifier enables you to read nonstandard data values and character values that are longer than eight characters, but which contain no embedded blanks.

Remember that informats work differently in modified list input than they do in formatted input.

Creating Free-Format Data

You can create a raw data file using list output. With list output, you simply list the names of the variables whose values you want to write. The PUT statement writes a value, leaves a blank, and then writes the next value.

You can use the DLM= option with a FILE statement to create a delimited raw data file. You can use the DSD option in a FILE statement to specify that data values containing commas should be enclosed in quotation marks. You can also use the DSD option to read values that contain delimiters within a quoted string.

Mixing Input Styles

With some file layouts, you might need to mix input styles in the same INPUT statement in order to read the data correctly.

Syntax

Reading Free-Format Data

LIBNAME *libref 'SAS-data-library'*;

FILENAME *fileref 'filename'*;

DATA *SAS-data-set*;

    INFILE *file-specification* <DLM *'delimiter'*> <MISSOVER><DSD>;

    LENGTH *variable* **$** *length*;

    INPUT *variable*<$> <&|:><*informat*>;

RUN;

PROC PRINT DATA==*SAS-data-set*;

RUN;

Creating Free-Format Data

LIBNAME *libref* '*SAS-data-library*';

DATA _NULL_;

    SET *SAS-data-set*;

    FILE *file-specification* <DLM '*delimiter*'> <DSD>;

    PUT *variable*<: *format*>;

RUN;

Sample Programs

Reading Free-Format Data

```
libname sasuser 'c:\records\data';
filename credit 'c:\records\credit.dat';
data sasuser.carduse;
   infile creditcard.dat dlm='#' missover;
   length LastName $ 14;
   input lastname $ Gender $ Age CardType $
       Total : comma.;
run;
proc print data=sasuser.carduse;
run;
```

Creating Raw Data Using List Output

```
libname sasuser 'c:\records\data';
data _null_;
   set sasuser.finance;
   file 'c:\accounts\newdata.txt' dsd;
   put ssn name salary : comma. date : date9.;
run;
```

Points to Remember

- When you use list input,

  - fields must be separated by at least one blank or other delimiter.

  - fields must be read in order, left to right. You cannot skip or re-read fields.

  - use a LENGTH statement to avoid truncating character values that are longer than eight characters.

- In formatted input, the informat determines both the length of character variables and the number of columns that are read. The same number of columns are read from each record.

- The informat in modified list input determines only the length of the variable value, not the number of columns that are read.

Chapter Quiz

Select the best answer for each question. After completing the quiz, check your answers using the answer key in the appendix.

1. The raw data file referenced by the fileref Students contains data that is

   Figure 17.52 *Raw Data File Students*

   ```
   Raw Data File Students
   1---+----10---+----20---+
   FRED JOHNSON 18 USC 1
   ASHLEY FERRIS 20 NCSU 3
   BETH ROSEMONT 21 UNC 4
   ```

   a. arranged in fixed fields

   b. free-format

   c. mixed-format

   d. arranged in columns

2. Which input style should be used to read the values in the raw data file that is referenced by the fileref Students?

   Figure 17.53 *Raw Data File Students*

   ```
   Raw Data File Students
   1---+----10---+----20---+
   FRED JOHNSON 18 USC 1
   ASHLEY FERRIS 20 NCSU 3
   BETH ROSEMONT 21 UNC 4
   ```

   a. column

   b. formatted

   c. list

   d. mixed

3. Which SAS program was used to create the raw data file Teamdat from the SAS data set Work.Scores?

Figure 17.54 *SAS Data Set Students and Raw Data File Teamdat*

SAS Data Set Work.Scores

Obs	Name	HighScore	Team
1	Joe	87	Blue Beetles, Durham
2	Dani	79	Raleigh Racers, Raleigh
3	Lisa	85	Sand Sharks, Cary
4	Matthew	76	Blue Beetles, Durham

Raw Data File Teamdat
```
1---+----10---+----20---+----30---+
Joe,87,"Blue Beetles, Durham"
Dani,79,"Raleigh Racers, Raleigh"
Lisa,85,"Sand Sharks, Cary"
Matthew,76,"Blue Beetles, Durham"
```

a.
```
data _null_;
    set work.scores;
    file 'c:\data\teamdat' dlm=',';
    put name highscore team;
run;
```

b.
```
data _null_;
    set work.scores;
    file 'c:\data\teamdat' dlm=' ';
    put name highscore team;
run;
```

c.
```
data _null_;
    set work.scores;
    file 'c:\data\teamdat' dsd;
    put name highscore team;
run;
```

d.
```
data _null_;
    set work.scores;
    file 'c:\data\teamdat';
    put name highscore team;
run;
```

4. Which SAS statement reads the raw data values in order and assigns them to the variables shown below?

Variables: FirstName (character), LastName (character), Age (numeric), School (character), Class (numeric)

Figure 17.55 *Raw Data File Students*

Raw Data File Students
```
1---+----10---+----20---+
FRED JOHNSON 18 USC 1
ASHLEY FERRIS 20 NCSU 3
BETH ROSEMONT 21 UNC 4
```

a. `input FirstName $ LastName $ Age School $ Class;`

b. `input FirstName LastName Age School Class;`

c. ```
input FirstName $ 1-4 LastName $ 6-12 Age 14-15
      School $ 17-19 Class 21;
```

d. ```
input FirstName 1-4 LastName 6-12 Age 14-15
      School 17-19 Class 21;
```

5. Which SAS statement should be used to read the raw data file that is referenced by the fileref Salesrep?

Figure 17.56 *Raw Data File Salesrep*

```
Raw Data File Salesrep
1---+----10---+----20---+----30
ELAINE:FRIEDMAN:WILMINGTON:2102
JIM:LLOYD:20:RALEIGH:38392
JENNIFER:WU:21:GREENSBORO:1436
```

a. ```
infile salesrep;
```

b. ```
infile salesrep ':';
```

c. ```
infile salesrep dlm;
```

d. ```
infile salesrep dlm=':';
```

6. Which of the following raw data files can be read by using the MISSOVER option in the INFILE statement? Spaces for missing values are highlighted with colored blocks.

a.

Figure 17.57 *Raw Data File with Blank Spaces at the End of Line Three*

```
1---+----10---+----20---+----
ORANGE SUNNYDALE 20 10
PINEAPPLE ALOHA 7 10
GRAPE FARMFRESH 3 ▓
APPLE FARMFRESH 16 5
GRAPEFRUIT SUNNYDALE 12 8
```

b.

Figure 17.58 *Raw Data File with Blank Spaces in the Middle of Line Three*

```
1---+----10---+----20---+----
ORANGE SUNNYDALE 20 10
PINEAPPLE ALOHA 7 10
GRAPE FARMFRESH ▓ 17
APPLE FARMFRESH 16 5
GRAPEFRUIT SUNNYDALE 12 8
```

c.

Figure 17.59 *Raw Data File with Blank Spaces in the Middle of Line Three*

```
1---+----10---+----20---+----
ORANGE SUNNYDALE 20 10
PINEAPPLE ALOHA 7 10
GRAPE ▓▓▓▓▓ 3 17
APPLE FARMFRESH 16 5
GRAPEFRUIT SUNNYDALE 12 8
```

d.

Figure 17.60 *Raw Data File with Blank Spaces at the Beginning of Line Three*

```
1---+----10---+----20---+----
ORANGE SUNNYDALE 20 10
PINEAPPLE ALOHA 7 10
        FARMFRESH 3
APPLE FARMFRESH 16 5
GRAPEFRUIT SUNNYDALE 12 8
```

7. Which SAS program correctly reads the data in the raw data file that is referenced by the fileref Volunteer?

Figure 17.61 *Raw Data File Volunteer*

Raw Data File Volunteer

```
1---+----10---+----20---+----30
ARLENE BIGGERSTAFF 19 UNC 2
JOSEPH CONSTANTINO 21 CLEM 2
MARTIN FIELDS 18 UNCG 1
```

a.
```
data sasuser.contest;
    infile volunteer;
    input FirstName $ LastName $ Age
        School $ Class;
run;
```

b.
```
data sasuser.contest;
    infile volunteer;
    length LastName $ 11;
    input FirstName $ lastname $ Age
        School $ Class;
run;
```

c.
```
data sasuser.contest;
    infile volunteer;
    input FirstName $ lastname $ Age
        School $ Class;  length LastName $ 11;
run;
```

d.
```
data sasuser.contest;
    infile volunteer;
    input FirstName $ LastName $ 11. Age
        School $ Class;
run;
```

8. Which type of input should be used to read the values in the raw data file that is referenced by the fileref University?

Figure 17.62 *Raw Data File University*

```
Raw Data File University
1---+----10---+----20---+----30
UNC ASHEVILLE  2,712
UNC CHAPEL HILL  24,189
UNC CHARLOTTE  15,031
UNC GREENSBORO  12,323
```

    a. column

    b. formatted

    c. list

    d. modified list

9. Which SAS statement correctly reads the values for Flavor and Quantity? Make sure the length of each variable can accommodate the values shown.

Figure 17.63 *Raw Data File Cookies*

```
Raw Data File Cookies
1---+----10---+----20---+----30
CHOCOLATE CHIP  10,453
OATMEAL  12,187
PEANUT BUTTER  11,546
SUGAR  12,331
```

    a. `input Flavor & $9. Quantity : comma.;`

    b. `input Flavor & $14. Quantity : comma.;`

    c. `input Flavor : $14. Quantity & comma.;`

    d. `input Flavor $14. Quantity : comma.;`

10. Which SAS statement correctly reads the raw data values in order and assigns them to these corresponding variables: Year (numeric), School (character), Enrolled (numeric)?

Figure 17.64 *Raw Data File Founding*

```
Raw Data File Founding
1---+----10---+----20---+----30---+----40
1868 U OF CALIFORNIA BERKELEY  31,612
1906 U OF CALIFORNIA DAVIS  21,838
1965 U OF CALIFORNIA IRVINE  15,874
1919 U OF CALIFORNIA LOS ANGELES  35,730
```

    a. `input Year School & $27.`
          `Enrolled : comma.;`

    b. `input Year 1-4 School & $27.`
          `Enrolled : comma.;`

    c. `input @1 Year 4. +1 School & $27.`
          `Enrolled : comma.;`

    d. all of the above

Chapter 18
Reading Date and Time Values

Overview

Introduction

SAS provides many informats for reading raw data values in various forms. "Reading Raw Data in Fixed Fields" on page 514 explained how informats can be used to read standard and nonstandard data. In this chapter, you learn how to use a special category of SAS informats called date and time informats. These informats enable you to read a variety of common date and time expressions. After you read date and time values, you can also perform calculations with them.

```
options yearcutoff=1920;
data perm.aprbills;
   infile aprdata;
   input LastName $8. @10 DateIn mmddyy8. +1 DateOut
      mmddyy8. +1 RoomRate 6. @35 EquipCost 6.;
   Days=dateout-datein+1;
   RoomCharge=days*roomrate;
   Total=roomcharge+equipcost;
run;
```

Figure 18.1   *Program Data Vector*

Program Data Vector

LastName	DateIn	DateOut	RoomRate	EquipCost	Days	RoomCharge	Total
Akron	14339	14343	175.00	298.45	5	875.00	1173.45

+1

Objectives

In this chapter, you learn how

- SAS stores date and time values

- to use SAS informats to read common date and time expressions

- to handle two-digit year values

- to calculate time intervals by subtracting two dates

- to multiply a time interval by a rate

- to display various date and time values.

How SAS Stores Date Values

Before you read date or time values into a SAS data set or use those values in calculations, you should understand how SAS stores date and time values.

When you use a SAS informat to read a date, SAS converts it to a numeric date value. A SAS date value is the number of days from January 1, 1960, to the given date.

Figure 18.2   *SAS Calculation of Date Values*

Jan. 1, 1959	Jan. 1, 1960	Jan. 1, 1961
← ⁻365	0	366 →

Here are some examples of how the appropriate SAS informat can convert different expressions for the date January 2, 2000, to a single SAS date value:

Date Expression	SAS Date Informat	SAS Date Value
02Jan00	DATE*w*.	14611
01-02-2000	MMDDYY*w*.	14611
02/01/00	DDMMYY*w*.	14611
2000/01/02	YYMMDD*w*.	14611

Storing dates and times as numeric values enables you to use dates and times in calculations much as you would use any other number.

How SAS Stores Time Values

SAS stores time values similar to the way it stores date values. A SAS time value is stored as the number of seconds since midnight.

Figure 18.3 *SAS Calculation of Time Values*

A SAS datetime is a special value that combines both date and time information. A SAS datetime value is stored as the number of seconds between midnight on January 1, 1960, and a given date and time.

Figure 18.4 *SAS Calculation of Date and Time Values*

More about SAS Date and Time Values

As you use SAS date and time values, remember that

- SAS date values are based on the Gregorian calendar, and they are valid for dates from A.D. 1582 through A.D. 20,000. Use caution when working with historical dates. Most of Europe started to use the Gregorian calendar in 1582. Great Britain and the American colonies adopted it in 1752. Check the adoption date for other parts of the world before making important calculations.

- SAS makes adjustments for leap years but ignores leap seconds.

- SAS does not make adjustments for daylight saving time.

Reading Dates and Times with Informats

Overview

You use SAS date and time informats to read date and time expressions and convert them to SAS date and time values. Like other SAS informats, date and time informats are composed of

- an informat name
- a field width
- a period delimiter.

SAS informat names indicate the form of date expression that can be read using that particular informat. Here are some examples of common date and time informats:

- DATE*w.*
- DATETIME*w.*
- MMDDYY*w.*
- TIME*w.*

As you know, there are several ways to represent a date. For example, all the following expressions represent the date October 15, 1999. Each of these common date expressions can be read using the appropriate SAS date informat.

Table 18.1 *Date Expressions and Corresponding SAS Date Informats*

Date Expression	SAS Date Informat
10/15/99	MMDDYY*w.*
15Oct99	DATE*w.*
10-15-99	MMDDYY*w.*
99/10/15	YYMMDD*w.*

Specifying Informats

Using the INPUT statement with an informat after a variable name is the simplest way to read date and time values into a variable.

General form, INPUT statement with an informat:

INPUT *<pointer-control> variable informat.*;

where

- *pointer-control* specifies the absolute or relative position to move the pointer.
- *variable* is the name of the variable being read.
- *informat.* is any valid SAS informat. Note that the informat includes a final period.

For example, the following INPUT statement uses two informats:

```
input @15 Style $3. @21 Price 5;
```

The *$w.* character informat ($3.) reads values, starting at column 15 of the raw data, into the variable Style. The *w.d* numeric informat (5) reads values, starting at column 21, into the variable Price.

Now let's look at some specific date and time informats you can use.

MMDDYYw. Informat

You can tell by its name that the informat MMDDYY*w*. reads date values in the form 10/15/99.

General form, values read with MMDDYY*w*. informat:

mmddyy or *mmddyyyy*

where

- *mm* is an integer between 01 and 12, representing the month
- *dd* is an integer between 01 and 31, representing the day
- *yy* or *yyyy* is an integer that represents the year.

In the MMDDYY*w*. informat, the month, day, and year fields can be separated by blanks or delimiters such as - or /. If delimiters are present, they must occur between all fields in the values. Remember to specify a field width that includes not only the month, day, and year values, but any delimiters as well. Here are some date expressions that you can read using the MMDDYY*w*. informat:

Table 18.2 *Date Expressions and Corresponding SAS Date Informats*

Date Expression	SAS Date Informat
101599	MMDDYY6.
10/15/99	MMDDYY8.
10 15 99	MMDDYY8.
10-15-1999	MMDDYY10.

DATEw. Informat

The DATE*w.* informat reads date values in the form 30May2000.

General form, values read with DATE*w.* informat:

ddmmmyy or *ddmmmyyyy*

where

- *dd* is an integer from 01 to 31, representing the day
- *mmm* is the first three letters of the month's name
- *yy* or *yyyy* is an integer that represents the year.

Blanks or other special characters can appear between the day, month, and year, as long as you increase the width of the informat to include these delimiters. Here are some date expressions that you can read using the DATEw. informat:

Table 18.3 *Date Expressions and Corresponding SAS Date Informats*

Date Expression	SAS Date Informat
30May00	`DATE7.`
30May2000	`DATE9.`
30-May-2000	`DATE11.`

TIMEw. Informat

The TIME*w.* informat reads values in the form *hh:mm:ss.ss*.

General form, values read with TIME*w.* informat:

hh:mm:ss.ss

where

- *hh* is an integer from 00 to 23, representing the hour
- *mm* is an integer from 00 to 59, representing the minute
- *ss.ss* is an optional field that represents seconds and hundredths of seconds.

If you do not enter a value for *ss.ss*, a value of zero is assumed. Here are some examples of time expressions that you can read using the TIME*w.* informat:

Table 18.4 *Time Expressions and Corresponding SAS Time Informats*

TimeExpression	SAS Time Informat
17:00:01.34	`TIME11.`

TimeExpression	SAS Time Informat
17:00	**TIME5.**
2:34	**TIME5.**

⚠️ Notice the last example. The field is only 4 columns wide, but a *w* value of **5** is specified. Five is the minimum acceptable field width for the TIME*w*. informat. If you specify a *w* value less than **5**, you'll receive the following error message in the SAS log:

Figure 18.5 *SAS Log*

```
                              SAS Log
 ┌──────────────────────────────────────────────────────────┐
 │ ERROR 29 - 85: Width specified for informat               │
 │                TIME is invalid.                           │
 └──────────────────────────────────────────────────────────┘
```

DATETIMEw. Informat

The DATETIME*w*. informat reads expressions that are composed of two parts, a date value and a time value, in the form: *ddmmmyy hh:mm:ss.ss*.

General form, values read with DATETIME*w*. informat:

ddmmmyy hh:mm:ss.ss

where

- *ddmmmyy* is the date value, the same form as for the DATE*w*. informat
- the time value must be in the form *hh:mm:ss.ss*
- *hh* is an integer from 00 to 23, representing the hour
- *mm* is an integer from 00 to 59, representing the minute
- *ss.ss* is an optional field that represents seconds and hundredths of seconds
- the date value and time value are separated by a blank or other delimiter.

If you do not enter a value for *ss.ss*, a value of **zero** is assumed.

Here are some examples of the DATETIME*w*. informat. Note that in the time value, you must use delimiters to separate the values for hour, minutes, and seconds.

Table 18.5 *Date and Time Expressions and Corresponding SAS Datetime Informats*

Date and Time Expression	SAS Datetime Informat
30May2000:10:03:17.2	**DATETIME20.**
30May00 10:03:17.2	**DATETIME18.**
30May2000/10:03	**DATETIME15.**

YEARCUTOFF= SAS System Option

Recall from "Referencing Files and Setting Options" on page 42 that the value of the YEARCUTOFF= system option affects only two-digit year values. A date value that contains a four-digit year value will be interpreted correctly even if it does not fall within the 100-year span set by the YEARCUTOFF= system option.

Table 18.6 Date Expressions with Corresponding Date Informats and Interpretations

Date Expression	SAS Date Informat	Interpreted As
06Oct59	**date7.**	06Oct1959
17Mar1783	**date9.**	17Mar1783

However, if you specify an inappropriate field width, you will receive incorrect results. Notice that the date expression in the table below contains a four-digit year value. The informat specifies a *w* value that is too small to read the entire value, so the last two digits of the year are truncated.

Table 18.7 Date Expressions with Corresponding Date Informats and Interpretations

Date Expression	SAS Date Informat	Interpreted As
17Mar1783	**date7.**	17Mar2017

Another problem arises if you use the wrong informat to read a date or time expression. The SAS log displays an invalid data message, and the variable's values are set to missing.

Figure 18.6 SAS Log

```
1---+----10
03/23/98
```

Figure 18.7 SAS Log Showing Invalid Data Message

```
                        SAS Log

3   input birthday date8.;
4   run;
NOTE: Invalid data for BIRTHDAY in line 3 1-8.
RULE: ----+----1----+----3----+----4----+----5
3      03/23/98
BIRTHDAY=. _ERROR_=1 _N_=1
```

When you work with date and time values,

- check the default value of the YEARCUTOFF= system option, and change it if necessary. The default YEARCUTOFF= value is **1920**.

- specify the proper informat for reading a date value.

- specify the correct field width so that the entire date value is read.

Using Dates and Times in Calculations

In this chapter so far, you've learned how date and time informats read common date and time expressions in specific forms. Now you will see how converting date and time expressions to numeric SAS date values can be useful, particularly for determining time intervals or performing calculations.

Suppose you work in the billing department of a small community hospital. It's your job to create a SAS data set from the raw data file that is referenced by the fileref Aprdata. A portion of the raw data file below shows data values that represent each patient's

- last name

- date checked in

- date checked out

- daily room rate

- equipment cost.

Figure 18.8 *Raw Data File Aprdata*

```
                  Raw Data File Aprdata
1---+----10---+----20---+----30---+----40
Akron      04/05/99 04/09/99 175.00 298.45
Brown      04/12/99 05/01/99 125.00 326.78
Carnes     04/27/99 04/29/99 125.00 174.24
Denison    04/11/99 04/12/99 175.00  87.41
Fields     04/15/99 04/22/99 175.00 378.96
Jamison    04/16/99 04/23/99 125.00 346.28
```

The data set that you create must also include variable values that represent how many days each person stayed in the hospital, the total room charges, and the total of all expenses that each patient incurred. When building the SAS program, you must first name the data set, identify the raw data file Aprdata, and use formatted input to read the data.

The following example is shown with the YEARCUTOFF= system option. When you work with two-digit year data, remember to check the default value of the YEARCUTOFF= option, and change it if necessary.

```
options yearcutoff=1920;
data perm.aprbills;
   infile aprdata;
   input LastName $8.
```

Notice that the values in the second and third fields are in the form *mmddyy*. To complete the INPUT statement, add instructions to read the values for RoomRate (fourth field) and EquipCost (fifth field), and add a semicolon.

```
options yearcutoff=1920;
data perm.aprbills;
   infile aprdata;
   input LastName $8. @10 DateIn mmddyy8. +1 DateOut
        mmddyy8. +1 RoomRate 6. @35 EquipCost 6.;
```

Now that the INPUT statement is complete, calculate how many days each patient was hospitalized. Because DateIn and DateOut are numeric variables, you can simply subtract to find the difference. But because the dates should be inclusive (patients are charged for both the first and last days), you must add 1 to the difference. Call this new variable Days.

```
options yearcutoff=1920;
data perm.aprbills;
   infile aprdata;
   input LastName $8. @10 DateIn mmddyy8. +1 DateOut
        mmddyy8. +1 RoomRate 6. @35 EquipCost 6.;
   Days=dateout-datein+1;
```

You can calculate a total room charge by multiplying Days times RoomRate.

```
options yearcutoff=1920;
data perm.aprbills;
   infile aprdata;
   input LastName $8. @10 DateIn mmddyy8. +1 DateOut
        mmddyy8. +1 RoomRate 6. @35 EquipCost 6.;
   Days=dateout-datein+1;
   RoomCharge=days*roomrate;
```

Calculating the total cost for each patient is easy. Create a variable named Total whose value is the sum of RoomCharge and EquipCost. Then add a PROC PRINT step and a RUN statement to view the new data.

```
options yearcutoff=1920;
data perm.aprbills;
   infile aprdata;
   input LastName $8. @10 DateIn mmddyy8. +1 DateOut
   mmddyy8. +1 RoomRate 6. @35 EquipCost 6.;
   Days=dateout-datein+1;
   RoomCharge=days*roomrate;
   Total=roomcharge+equipcost;
run;
proc print data=perm.aprbills;
run;
```

Figure 18.9 *Table Created with PROC PRINT*

Obs	LastName	DateIn	DateOut	RoomRate	EquipCost	Days	RoomCharge	Total
1	Akron	14339	14343	175	298.45	5	875	1173.45
2	Brown	14346	14365	125	326.78	20	2500	2826.78
3	Carnes	14361	14363	125	174.24	3	375	549.24
4	Denison	14345	14346	175	87.41	2	350	437.41
5	Fields	14349	14356	175	378.96	8	1400	1778.96
6	Jamison	14350	14357	125	346.28	8	1000	1346.28

If the values for DateIn and DateOut look odd to you, remember that these are SAS date values. Applying a format such as MMDDYY displays them as they appeared in Aprdata. You'll work with some other date and time formats later in this chapter.

Follow the execution of the program below. When the DATA step executes, the values for DateIn and DateOut are converted to SAS date values.

```
options yearcutoff=1920;
data perm.aprbills;
   infile aprdata;
   input LastName $8. @10 DateIn mmddyy8. +1 DateOut
         mmddyy8. +1 RoomRate 6. @35 EquipCost 6.;
   Days=dateout-datein+1;
   RoomCharge=days*roomrate;
   Total=roomcharge+equipcost;
run;
```

Figure 18.10 *Raw Data File Aprdata and Program Data Vector*

```
                 Raw Data File Aprdata
1---+----10---+----20---+----30---+----40
Akron    04/05/99 04/09/99 175.00 298.45
Brown    04/12/99 05/01/99 125.00 326.78
Carnes   04/27/99 04/29/99 125.00 174.24
Denison  04/11/99 04/12/99 175.00  87.41
Fields   04/15/99 04/22/99 175.00 378.96
Jamison  04/16/99 04/23/99 125.00 346.28
```

Program Data Vector

LastName	DateIn	DateOut	RoomRate	EquipCost	Days	RoomCharge	Total
Akron	14339	14343	•	•	•	•	•

After the INPUT statement, Days is created by subtracting DateIn from DateOut and adding 1.

```
options yearcutoff=1920;
data perm.aprbills;
   infile aprdata;
   input LastName $8. @10 DateIn mmddyy8. +1 DateOut
      mmddyy8. +1 RoomRate 6. @35 EquipCost 6.;
   Days=dateout-datein+1;
   RoomCharge=days*roomrate;
   Total=roomcharge+equipcost;
run;
```

Figure 18.11 *Program Data Vector*

Program Data Vector

LastName	DateIn	DateOut	RoomRate	EquipCost	Days	RoomCharge	Total
Akron	14339	14343	175.00	298.45	5	•	•

The value for RoomCharge is calculated next. RoomCharge is the product of Days and RoomRate.

```
options yearcutoff=1920;
data perm.aprbills;
   infile aprdata;
   input LastName $8. @10 DateIn mmddyy8. +1 DateOut
      mmddyy8. +1 RoomRate 6. @35 EquipCost 6.;
   Days=dateout-datein+1;
   RoomCharge=days*roomrate;
   Total=roomcharge+equipcost;
run;
```

Figure 18.12 *Raw Data File Aprdata and Program Data Vector*

Raw Data File Aprdata

```
1---+----10---+----20---+----30---+----40
Akron     04/05/99 04/09/99 175.00 298.45
Brown     04/12/99 05/01/99 125.00 326.78
Carnes    04/27/99 04/29/99 125.00 174.24
Denison   04/11/99 04/12/99 175.00  87.41
Fields    04/15/99 04/22/99 175.00 378.96
Jamison   04/16/99 04/23/99 125.00 346.28
```

Program Data Vector

LastName	DateIn	DateOut	RoomRate	EquipCost	Days	RoomCharge	Total
Akron	14339	14343	175.00	298.45	5	875.00	•

The value for Total is the final calculation. Total is the sum of EquipCost and RoomCharge.

```
options yearcutoff=1920;
data perm.aprbills;
   infile aprdata;
   input LastName $8. @10 DateIn mmddyy8. +1 DateOut
      mmddyy8. +1 RoomRate 6. @35 EquipCost 6.;
   Days=dateout-datein+1;
   RoomCharge=days*roomrate;
   Total=roomcharge+equipcost;
run;
```

Figure 18.13   *Raw Data File Aprdata and Program Data Vector*

Raw Data File Aprdata

```
1---+----10---+----20---+----30---+----40
Akron      04/05/99 04/09/99 175.00 298.45
Brown      04/12/99 05/01/99 125.00 326.78
Carnes     04/27/99 04/29/99 125.00 174.24
Denison    04/11/99 04/12/99 175.00  87.41
Fields     04/15/99 04/22/99 175.00 378.96
Jamison    04/16/99 04/23/99 125.00 346.28
```

Program Data Vector

LastName	DateIn	DateOut	RoomRate	EquipCost	Days	RoomCharge	Total
Akron	14339	14343	175.00	298.45	5	875.00	1173.45

Using Date and Time Formats

Overview

Remember that when Perm.Aprbills is printed, the values for DateIn and DateOut appear as SAS date values.

Raw Data File Aprdata

```
1---+----10---+----20---+----30---+----40
Akron      04/05/99 04/09/99 175.00 298.45
Brown      04/12/99 05/01/99 125.00 326.78
Carnes     04/27/99 04/29/99 125.00 174.24
Denison    04/11/99 04/12/99 175.00  87.41
Fields     04/15/99 04/22/99 175.00 378.96
Jamison    04/16/99 04/23/99 125.00 346.28
```

```
options yearcutoff=1920;
data perm.aprbills;
   infile aprdata;
   input LastName $8. @10 DateIn mmddyy8. +1 DateOut
      mmddyy8. +1 RoomRate 6. @35 EquipCost 6.;
   Days=dateout-datein+1;
   RoomCharge=days*roomrate;
   Total=roomcharge+equipcost;
run;
proc print data=perm.aprbills;
run;
```

Figure 18.14 PROC PRINT Output for the above Example

Obs	LastName	DateIn	DateOut	RoomRate	EquipCost	Days	RoomCharge	Total
1	Akron	14339	14343	175	298.45	5	875	1173.45
2	Brown	14346	14365	125	326.78	20	2500	2826.78
3	Carnes	14361	14363	125	174.24	3	375	549.24
4	Denison	14345	14346	175	87.41	2	350	437.41
5	Fields	14349	14356	175	378.96	8	1400	1778.96
6	Jamison	14350	14357	125	346.28	8	1000	1346.28

SAS provides many specialized date and time formats that enable you to specify how date and time values are displayed. Let's look at two date formats: WEEKDATE*w.* and WORDDATE*w.*

The WEEKDATE*w.* Format

You can use the WEEKDATE*w.* format to write date values in a format that displays the day of the week, month, day, and year.

General form, WEEKDATE*w.* format:

WEEKDATE*w.*

The WEEKDATE*w.* format writes date values in the form *day-of-week*, *month-name dd*, *yy* (or *yyyy*).

where

- *dd* is an integer between 01 and 31, representing the day

- *yy* or *yyyy* is an integer that represents the year.

Tip If the *w* value is too small to write the complete day of the week and month, SAS abbreviates as needed.

```
proc print data=perm.aprbills;
   format datein dateout weekdate17.;
run;
```

Figure 18.15 PROC PRINT Output for the above Example

Obs	LastName	DateIn	DateOut	RoomRate	EquipCost	Days	RoomCharge	To
1	Akron	Mon, Apr 5, 1999	Fri, Apr 9, 1999	175	298.45	5	875	1173
2	Brown	Mon, Apr 12, 1999	Sat, May 1, 1999	125	326.78	20	2500	2826
3	Carnes	Tue, Apr 27,1999	Thu, Apr 29, 1999	125	174.24	3	375	549
4	Denison	Sun, Apr 11, 1999	Mon, Apr 12, 1999	175	87.41	2	350	437
5	Fields	Thu, Apr 15, 1999	Thu, Apr 22, 1999	175	378.96	8	1400	1778
6	Jamison	Fri, Apr 16, 1999	Fri, Apr 23, 1999	125	346.28	8	1000	1346

You can vary the results by changing the w value in the format.

FORMAT Statement	Result
format datein weekdate3.;	Mon
format datein weekdate6.;	Monday
format datein weekdate17.;	Mon, Apr 5, 1999
format datein weekdate21.;	Monday, April 5, 1999

The WORDDATEw. Format

The WORDDATE*w*. format is similar to the WEEKDATE*w*. format, but it does not display the day of the week or the two-digit year values.

General form, WORDDATE*w*. format:

WORDDATE*w*.

The WORDDATE*w*. format writes date values in the form *month-name dd, yyyy*.

where

- *dd* is an integer between 01 and 31, representing the day

- *yyyy* is an integer that represents the year.

Tip 🗒 If the *w* value is too small to write the complete month, SAS abbreviates as needed.

```
proc print data=perm.aprbills;
   format datein dateout worddate12.;
run;
```

Figure 18.16 *PROC PRINT Output for the above Example*

Obs	LastName	DateIn	DateOut	RoomRate	EquipCost	Days	RoomCharge	Total
1	Akron	Apr 5, 1999	Apr 9, 1999	175	298.45	5	875	1173.45
2	Brown	Apr 12, 1999	May 1, 1999	125	326.78	20	2500	2826.78
3	Carnes	Apr 27,1999	Apr 29, 1999	125	174.24	3	375	549.24
4	Denison	Apr 11, 1999	Apr 12, 1999	175	87.41	2	350	437.41
5	Fields	Apr 15, 1999	Apr 22, 1999	175	378.96	8	1400	1778.96
6	Jamison	Apr 16, 1999	Apr 23, 1999	125	346.28	8	1000	1346.28

You can vary the results by changing the *w* value in the format.

Table 18.8 *FORMAT Statements and Corresponding Results*

FORMAT Statement	Result
format datein worddate3.;	Apr
format datein worddate5.;	April

FORMAT Statement	Result
`format datein worddate14.;`	April 15, 1999

Remember that you can permanently assign a format to variable values by including a FORMAT statement in the DATA step.

```
options yearcutoff=1920;
data work.aprbills;
   infile aprdata;
   input LastName $8. @10 DateIn mmddyy8. +1 DateOut
      mmddyy8. +1 RoomRate 6. @35 EquipCost 6.;
   Days=dateout-datein+1;
   RoomCharge=days*roomrate;
   Total=roomcharge+equipcost;
   format datein dateout worddate12.;
run;
proc print data=work.aprbills;
run;
```

Chapter Summary

Text Summary

How SAS Stores Date and Time Values

SAS stores dates as numeric SAS date values, which represent the number of days from January 1, 1960. SAS time values are the number of seconds since midnight.

Reading Dates and Times with Informats

Use SAS informats to read date and time expressions and convert them to SAS date and time values.

- MMDDYY*w.* reads dates such as 053090, 05/30/90, or 05 30 1990.

- DATE*w.* reads dates such as 30May1990, 30May90, or 30-May-1990.

- TIME*w.* reads times such as 17:00, 17:00:01.34, or 2:34.

- DATETIME*w.* reads dates and times such as 30May1990:10:03:17.2, 30May90 10:03:17.2, or 30May1990/10:03.

Two-digit year values require special consideration. When a two-digit year value is read, SAS defaults to a year within a 100-year span that is determined by the YEARCUTOFF= system option. The default value of YEARCUTOFF= is **1920**. You can check or reset the value of this option in your SAS session to use a different 100-year span for date informats.

Using Dates and Times in Calculations

Date and time values can be used in calculations like other numeric values. In addition to tracking time intervals, SAS date and time values can be used with SAS functions and with complex calculations.

Using Date and Time Formats

SAS provides many specialized date and time formats that enable you to specify how date and time values are displayed. You can use the WEEKDATE*w.* format to write date values in the form *day-of-week, month-name dd, yy* (or *yyyy*). You can use the WORDDATE*w.* format to write date values in the form *month-name dd, yyyy*.

Syntax

OPTIONS YEARCUTOFF=*yyyy*;
DATA *SAS-data-set*;
    INFILE *file-specification*;
    INPUT *<pointer-control> variable informat.*;
RUN;
PROC PRINT DATA=*SAS-data-set*;
    FORMAT *variable format.*;
RUN;

Sample Program

```
options yearcutoff=1920;
data perm.aprbills;
   infile aprdata;
   input LastName $8. @10 DateIn mmddyy8.
       +1 DateOut mmddyy8. +1 RoomRate 6.
       @35 EquipCost 6.;
   Days=dateout-datein+1;
   RoomCharge=days*roomrate;
   Total=roomcharge+equipcost;
run;
proc print data=perm.aprbills;
   format datein dateout worddate12.;
run;
```

Points to Remember

- SAS makes adjustments for leap years, but not for leap seconds or daylight saving time.

- The minimum acceptable field width for the TIME*w.* informat is 5. If you specify a *w* value less than **5**, you'll receive an error message in the SAS log.

- The default value of the YEARCUTOFF= option is **1920**. When you work with two-digit year data, remember to check the default value of the YEARCUTOFF= option, and change it if necessary.

- The value of the YEARCUTOFF= system option does not affect four-digit year values. Four-digit values are always read correctly.

- Be sure to specify the proper informat for reading a date value, and specify the correct field width so that the entire value is read.

- If SAS date values appear in your program output, use a date format to display them in legible form.

Chapter Quiz

Select the best answer for each question. After completing the quiz, you can check your answers using the answer key in the appendix.

1. SAS date values are the number of days since which date?

   a. January 1, 1900

   b. January 1, 1950

   c. January 1, 1960

   d. January 1, 1970

2. A great advantage of storing dates and times as SAS numeric date and time values is that

   a. they can easily be edited.

   b. they can easily be read and understood.

   c. they can be used in text strings like other character values.

   d. they can be used in calculations like other numeric values.

3. SAS does not automatically make adjustments for daylight saving time, but it *does* make adjustments for:

   a. leap seconds

   b. leap years

   c. Julian dates

   d. time zones

4. An input data file has date expressions in the form 10222001. Which SAS informat should you use to read these dates?

   a. DATE6.

   b. DATE8.

   c. MMDDYY6.

   d. MMDDYY8.

5. The minimum width of the TIME*w*. informat is:

   a. 4

   b. 5

   c. 6

   d. 7

6. Shown below are date and time expressions and corresponding SAS datetime informats. Which date and time expression *cannot* be read by the informat that is shown beside it?

   a. 30May2000:10:03:17.2 `DATETIME20.`

   b. 30May00 10:03:17.2 `DATETIME18.`

   c. 30May2000/10:03 `DATETIME15.`

    d. 30May2000/1003 **DATETIME14.**

7. What is the default value of the YEARCUTOFF= system option?

    a. **1920**

    b. **1910**

    c. **1900**

    d. **1930**

8. Suppose your input data file contains the date expression 13APR2009. The YEARCUTOFF= system option is set to 1910. SAS will read the date as:

    a. 13APR1909

    b. 13APR1920

    c. 13APR2009

    d. 13APR2020

9. Suppose the YEARCUTOFF= system option is set to 1920. An input file contains the date expression 12/08/1925, which is being read with the MMDDYY8. informat. Which date will appear in your data?

    a. 08DEC1920

    b. 08DEC1925

    c. 08DEC2019

    d. 08DEC2025

10. Suppose your program creates two variables from an input file. Both variables are stored as SAS date values: FirstDay records the start of a billing cycle, and LastDay records the end of that cycle. The code for calculating the total number of days in the cycle would be:

    a. `TotDays=lastday-firstday;`

    b. `TotDays=lastday-firstday+1;`

    c. `TotDays=lastday/firstday;`

    d. You cannot use date values in calculations.

Chapter 19

Creating a Single Observation from Multiple Records

Overview

Introduction

Information for one observation can be spread out over several raw data file records.
You can write multiple INPUT statements when multiple input records comprise a single
observation, as shown below.

```
input Lname $ 1-8 Fname $ 10-15;
input Department $ 1-12 JobCode $ 15-19;
input Salary comma10.;
```

Figure 19.1 *Multiple Records Comprising Each Observation*

```
1---+----10---+----
ABRAMS    THOMAS
MARKETING      SR01
$25,209.03
BARCLAY   ROBERT
EDUCATION      IN01
$24,435.71
COURTNEY MARK
PUBLICATIONS   TW01
$24,006.16
```

Alternatively, you can write one INPUT statement that contains a line pointer control to specify the record(s) from which values are to be read.

```
input #1 Lname $ 1-8 Fname $ 10-15
      #2 Department $ 1-12 JobCode $ 15-19
      #3 Salary comma10.;
```

Figure 19.2 *Multiple Records Comprising Each Observation*

```
1---+----10---+----
ABRAMS    THOMAS
MARKETING      SR01
$25,209.03
BARCLAY   ROBERT
EDUCATION      IN01
$24,435.71
COURTNEY MARK
PUBLICATIONS   TW01
$24,006.16
```

Objectives

In this chapter, you learn to

- read multiple records sequentially and create a single observation

- read multiple records non-sequentially and create a single observation.

Using Line Pointer Controls

You know that as SAS reads raw data values, it keeps track of its position with an input pointer. You have used column pointer controls and column specifications to determine the column placement of the input pointer.

Table 19.1 *Input Statements for Column Specifications and Column Pointer Controls*

Column Specifications	`input Name $ 1-12 Age 15-16 Gender $ 18;`
Column Pointer Controls	`input Name $12. @15 Age 2. @18 Gender $1.;`

But you can also position the input pointer on a specific record by using a line pointer control in the INPUT statement.

Table 19.2 *INPUT Statement and Raw Data File Admit*

```
input #2 Name $ 1-12 Age 15-16
Gender $ 18;
```

```
        Raw Data File Admit
1---+----10---+----
S. Thompson    37 M
L. Rochester   31 F
M. Sabatello   43 M
```

There are two types of line pointer controls.

- The forward slash (/) specifies a line location that is relative to the current one.

- The #*n* specifies the absolute number of the line to which you want to move the pointer.

First we'll look at the forward slash (/). Later in the chapter, you'll learn how to use the #*n*, and you will see how these two controls can be combined.

Reading Multiple Records Sequentially

The Forward Slash (/) Line Pointer Control

Use the forward slash (/) line pointer control to read multiple records sequentially. The / advances the input pointer to the next record. The / line pointer control moves the input pointer forward only and must be specified *after* the instructions for reading the values in the current record.

The single INPUT statement below reads the values for Lname and Fname in the first record, followed by the values for Department and JobCode in the second record. Then the value for Salary is read in the third record.

```
input Lname $ 1-8 Fname $ 10-15 /
      Department $ 1-12 JobCode $ 15-19 /
      Salary comma10.;
```

Figure 19.3   Multiple Records Comprising Each Observation

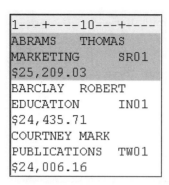

```
1---+----10---+----
ABRAMS   THOMAS
MARKETING      SR01
$25,209.03
BARCLAY  ROBERT
EDUCATION      IN01
$24,435.71
COURTNEY MARK
PUBLICATIONS   TW01
$24,006.16
```

Using the / Line Pointer Control

Look at the forward slash (/) line pointer control in the following example.

The raw data file Memdata contains the mailing list of a professional organization. Your task is to combine the information for each member into a single observation. We'll begin by reading each member's name, followed by the street address, and finally the city, state, and zip code.

- As you write the instructions to read the values for Fname and Lname, notice that not all of the values for Lname begin in the same column. So, you should use standard list input to read these values.

```
data perm.members;
   infile memdata;
   input Fname $ Lname $
```

Figure 19.4   Raw Data File Memdata

```
1---+----10---+----20
LEE ATHNOS
1215 RAINTREE CIRCLE
PHOENIX   AZ 85044
HEIDIE BAKER
1751 DIEHL ROAD
VIENNA   VA 22124
MYRON BARKER
131 DONERAIL DRIVE
ATLANTA  GA 30363
JOYCE BENEFIT
85 MAPLE AVENUE
MENLO PARK  CA 94025
```

- Now you want to read the values for Address from the second record. The / line pointer control advances the input pointer to the next record. At this point the INPUT statement is incomplete, so you should not place a semicolon after the line pointer control.

```
data perm.members;
   infile memdata;
   input Fname $ Lname $ /
```

Figure 19.5 *Raw Data File Memdata*

```
1---+----10---+----20
LEE ATHNOS
1215 RAINTREE CIRCLE
PHOENIX   AZ 85044
HEIDIE BAKER
1751 DIEHL ROAD
VIENNA   VA 22124
MYRON BARKER
131 DONERAIL DRIVE
ATLANTA   GA 30363
JOYCE BENEFIT
85 MAPLE AVENUE
MENLO PARK   CA 94025
```

- You can use column input to read the values in the next record as one variable named Address. Then add a line pointer control to move the input pointer to the next record.

```
data perm.members;
   infile memdata;
   input Fname $ Lname $ /
        Address $ 1-20 /
```

Figure 19.6 *Line Pointer Control Advanced One Record*

```
1---+----10---+----20
LEE ATHNOS
1215 RAINTREE CIRCLE
PHOENIX   AZ 85044
HEIDIE BAKER
1751 DIEHL ROAD
VIENNA   VA 22124
MYRON BARKER
131 DONERAIL DRIVE
ATLANTA   GA 30363
JOYCE BENEFIT
85 MAPLE AVENUE
MENLO PARK   CA 94025
```

- As you write the statements to read the values for City, notice that some of the values are longer than eight characters and contain embedded blanks. Also note that each value is followed by two consecutive blanks. To read these values, you should use modified list input with the ampersand (&) modifier.

The values for State and the values for Zip do not begin in the same column. Therefore, you should use list input to read these values.

```
data perm.members;
   infile memdata;
   input Fname $ Lname $ /
        Address $ 1-20 /
        City & $10. State $ Zip $;
run;
```

Figure 19.7 *Line Pointer Control Advanced Another Record*

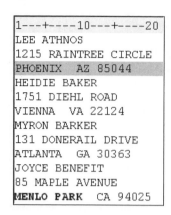

Sequential Processing of Multiple Records in the DATA Step

Now that you've learned the basics of using the / line pointer control, let's look at the sequential processing of multiple records in the DATA step.

The values in the first record are read, and the / line pointer control moves the input pointer to the second record.

Figure 19.8 *Reading the First Record of the First Observation*

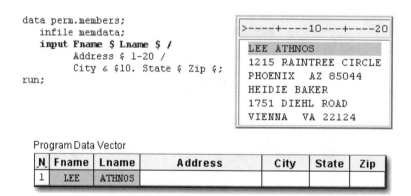

The values for Address are read, and the second / line pointer control advances the input pointer to the third record.

Figure 19.9 *Reading the Second Record of the First Observation*

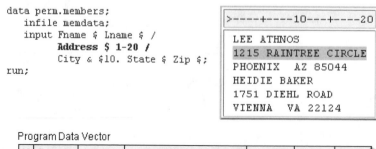

The values for City, State, and Zip are read. The INPUT statement is complete.

Figure 19.10   *Reading the Third Record of the First Observation*

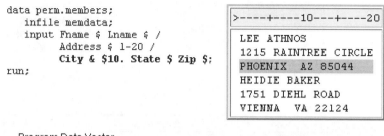

The values in the program data vector are written to the data set as the first observation.

Figure 19.11   *The Program Data Vector and the SAS Data Set Perm.Members*

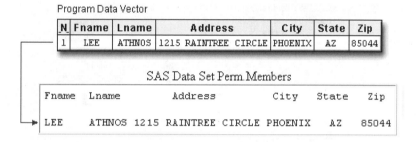

Control returns to the top of the DATA step. The variable values are reinitialized to missing.

Figure 19.12   *Reinitializing the Variable Values to Missing*

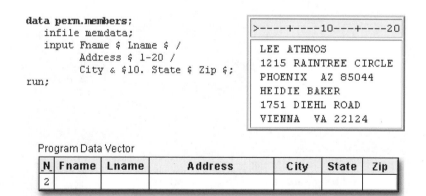

During the second iteration, values for Fname and Lname are read beginning in column 1 of the fourth record.

Figure 19.13   Reading the First Record of the Second Observation

The values for Address are read. The / line pointer control advances the input pointer to the fifth record.

Figure 19.14   Reading the Second Record of the Second Observation

The values for City, State, and Zip are read. The INPUT statement is complete.

Figure 19.15   Reading the Third Record of the First Observation

The values in the program data vector are written to the data set as the second observation.

Figure 19.16 *The Program Data Vector and the SAS Data Set Perm.Members*

Program Data Vector

N.	Fname	Lname	Address	City	State	Zip
2	HEIDIE	BAKER	1751 DIEHL ROAD	VIENNA	VA	22124

SAS Data Set Perm.Members

Fname	Lname	Address	City	State	Zip
LEE	ATHNOS	1215 RAINTREE CIRCLE	PHOENIX	AZ	85044
HEIDIE	BAKER	1751 DIEHL ROAD	VIENNA	VA	22124

After the data set is complete, PROC PRINT output for Perm.Members shows that each observation contains the complete information for one member.

```
proc print data=perm.members;
run;
```

Figure 19.17 *PROC PRINT Output of the Complete Perm.Members Data Set*

Obs	Fname	Lname	Address	City	State	Zip
1	LEE	ATHNOS	1215 RAINTREE CIRCLE	PHOENIX	AZ	85044
2	HEIDIE	BAKER	1751 DIEHL ROAD	VIENNA	VA	22124
3	MYRON	BARKER	131 DONERAIL DRIVE	ATLANTA	GA	30363
4	JOYCE	BENEFIT	85 MAPLE AVENUE	MENLO PARK	CA	94025

Number of Records per Observation

Note that the raw data file must contain the same number of records for each observation.

Suppose there are only two records for the second member. Remember, the INPUT statement is set up to read three records.

```
data perm.members;
   infile memdata;
   input Fname $ Lname $ /
Address $ 1-20 /
City & $10. State $ Zip $;
```

Figure 19.18 *Raw Data File Memdata*

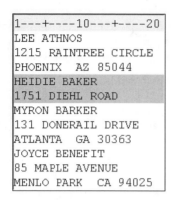

```
1---+----10---+----20
LEE ATHNOS
1215 RAINTREE CIRCLE
PHOENIX   AZ 85044
HEIDIE BAKER
1751 DIEHL ROAD
MYRON BARKER
131 DONERAIL DRIVE
ATLANTA   GA 30363
JOYCE BENEFIT
85 MAPLE AVENUE
MENLO PARK   CA 94025
```

The second member's name and address are read and assigned to corresponding variables. Then the input pointer advances to the next record, as directed by the INPUT statement, and the third member's name is read as a value for City.

The INPUT statement looks for a value for State and Zip, so the input pointer advances to the next record and reads the member's address.

The PROC PRINT output for this data set illustrates the problem.

Obs	Fname	Lname	Address	City	State	Zip
1	LEE	ATHNOS	1215 RAINTREE CIRCLE	PHOENIX	AZ	85044
2	HEIDIE	BAKER	1751 DIEHL ROAD	MYRON BARK	131	DONERAIL
3	ATLANTA GA		JOYCE BENEFIT	85 MAPLE A	MENLO	PARK

Before you write the INPUT statement, check whether the raw data file contains the same number of records for each observation. In this raw data file there are now three records for each observation.

Figure 19.19 *Verifying the Number of Records for Each Observation*

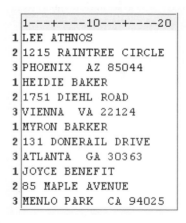

```
 1---+----10---+----20
1 LEE ATHNOS
2 1215 RAINTREE CIRCLE
3 PHOENIX   AZ 85044
1 HEIDIE BAKER
2 1751 DIEHL ROAD
3 VIENNA   VA 22124
1 MYRON BARKER
2 131 DONERAIL DRIVE
3 ATLANTA   GA 30363
1 JOYCE BENEFIT
2 85 MAPLE AVENUE
3 MENLO PARK   CA 94025
```

For more information about working with raw data files that contain missing records, see the SAS documentation.

Reading Multiple Records Non-Sequentially

The #n Line Pointer Control

You already know how to read multiple records sequentially by using the / line pointer control. Now let's look at reading multiple records non-sequentially by using the #*n* line pointer control.

The #*n* specifies the absolute number of the line to which you want to move the input pointer. The #*n* pointer control can read records in any order; therefore, it must be specified *before* the instructions for reading values in a specific record.

The INPUT statement below first reads the values for Department and JobCode in the second record, then the values for Lname and Fname in the first record. Finally, it reads the value for Salary in the third record.

```
input #2 Department $ 1-12 JobCode $ 15-19
      #1 Lname $ Fname $
      #3 Salary comma10.;
```

Figure 19.20 *The Records of the First Observation*

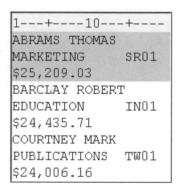

```
1---+----10---+----
ABRAMS THOMAS
MARKETING      SR01
$25,209.03
BARCLAY ROBERT
EDUCATION      IN01
$24,435.71
COURTNEY MARK
PUBLICATIONS   TW01
$24,006.16
```

Using the #n Line Pointer Control

Look at the #*n* line pointer control in the following example.

The raw data file Patdata contains information about the patients of a small group of general surgeons.

The first three records contain a patient's name, address, city, state, and zip code. The fourth record contains the patient's ID number followed by the name of the primary physician.

Figure 19.21 Observation Records in a Raw Data File

```
1---+----10---+----20---
1 ALEX BEDWAN
2 609 WILTON MEADOW DRIVE
3 GARNER NC 27529
4 XM034 FLOYD
  ALISON BEYER
  8521 HOLLY SPRINGS ROAD
  APEX NC 27502
  XF124 LAWSON
```

Suppose you want to read each patient's information in the following order:

1. ID number (ID)

2. first name (Fname)

3. last name (Lname)

4. address (Address)

5. city (City)

6. state (State)

7. zip (Zip)

8. doctor (Doctor)

- To read the values for ID in the fourth record, specify #4 before naming the variable and defining its attributes.

```
data perm.patients;
   infile patdata;
   input #4 ID $5.
```

Figure 19.22 Specifying the ID Value in the Fourth Record of an Observation

```
1---+----10---+----20---
ALEX BEDWAN
609 WILTON MEADOW DRIVE
GARNER NC 27529
XM034 FLOYD
```

- To read the values for Fname and Lname in the first record, specify #1 before naming the variables and defining their attributes.

```
data perm.patients;
   infile patdata;
   input #4 ID $5.
         #1 Fname $ Lname $
```

Figure 19.23 Specifying the First Record of an Observation

```
1---+----10---+----20---
ALEX BEDWAN
609 WILTON MEADOW DRIVE
GARNER NC 27529
XM034 FLOYD
```

- Use the *#n* line pointer control to move the input pointer to the second record and read the value for Address.

```
data perm.patients;
   infile patdata;
   input #4 ID $5.
        #1 Fname $ Lname $
        #2 Address $23.
```

Figure 19.24 *Specifying the Second Record of an Observation*

```
1---+----10---+----20---
ALEX BEDWAN
609 WILTON MEADOW DRIVE
GARNER NC 27529
XM034 FLOYD
```

- Now move the input pointer to the third record and read the values for City, State, and Zip, in that order.

In this raw data file, the values for City contain eight characters or less and do not contain embedded blanks, so you can use standard list input to read these values.

```
data perm.patients;
   infile patdata;
   input #4 ID $5.
        #1 Fname $ Lname $
        #2 Address $23.
        #3 City $ State $ Zip $
```

Figure 19.25 *Specifying the Third Record of an Observation*

```
1---+----10---+----20---
ALEX BEDWAN
609 WILTON MEADOW DRIVE
GARNER NC 27529
XM034 FLOYD
```

- Now you need to move the input pointer down to the fourth record to read the values for Doctor, which begin in column 7. Don't forget to add a semicolon at the end of the INPUT statement. A RUN statement completes the program.

```
data perm.patients;
   infile patdata;
   input #4 ID $5.
        #1 Fname $ Lname $
        #2 Address $23.
        #3 City $ State $ Zip $
        #4 @7 Doctor $6.;
run;
```

Figure 19.26 *Specifying the Doctor Value in the Fourth Record of an Observation*

```
1---+----10---+----20---
ALEX BEDWAN
609 WILTON MEADOW DRIVE
GARNER NC 27529
XM034 FLOYD
```

Execution of the DATA Step

The *#n* pointer controls in the program below cause four records to be read for each execution of the DATA step.

```
data perm.patients;
   infile patdata;
   input #4 ID $5.
         #1 Fname $ Lname $
         #2 Address $23.
         #3 City $ State $ Zip $
         #4 @7 Doctor $6.;
run;
```

The first time the DATA step executes, the first four records are read, and an observation is written to the data set.

Figure 19.27 *Raw Data File with the First Four Records Highlighted*

```
1---+----10---+----20---
ALEX BEDWAN
609 WILTON MEADOW DRIVE
GARNER NC 27529
XM034 FLOYD
ALISON BEYER
8521 HOLLY SPRINGS ROAD
APEX NC 27502
XF124 LAWSON
```

During the second iteration, the next four records are read, and the second observation is written to the data set, and so on.

Figure 19.28 *Raw Data File with the Next Four Records Highlighted*

```
1---+----10---+----20---
ALEX BEDWAN
609 WILTON MEADOW DRIVE
GARNER NC 27529
XM034 FLOYD
ALISON BEYER
8521 HOLLY SPRINGS ROAD
APEX NC 27502
XF124 LAWSON
```

The PROC PRINT output of the data set shows how information that was spread over several records has been condensed into one observation.

```
proc print data=perm.patients noobs;
run;
```

Figure 19.29 PROC PRINT Output of Data Set Perm.Patients

ID	Fname	Lname	Address	City	State	Zip	Doctor
XM034	ALEX	BEDWAN	609 WILTON MEADOW DRIVE	GARNER	NC	27529	FLOYD
XF124	ALISON	BEYER	8521 HOLLY SPRINGS ROAD	APEX	NC	27502	LAWSON
XF232	LISA	BONNER	109 BRAMPTON AVENUE	CARY	NC	27511	LAWSON
XM065	GEORGE	CHESSON	3801 WOODSIDE COURT	GARNER	NC	27529	FLOYD

Combining Line Pointer Controls

The forward slash (/) line pointer control and the #*n* line pointer control can be used together in a SAS program to read multiple records both sequentially and non-sequentially.

For example, you could use both the / line pointer control and the #*n* line pointer control to read the variables in the raw data file Patdata in the following order:

1. ID
2. Fname
3. Lname
4. Address
5. City
6. State
7. Zip
8. Doctor

```
data perm.patients;
   infile patdata;
   input #4 ID $5.
         #1 Fname $ Lname $ /
            Address $23. /
            City $ State $ Zip $ /
            @7 Doctor $6.;
run;
```

Figure 19.30 Raw Data File with the First Four Records Highlighted

```
1---+----10---+----20---
ALEX BEDWAN
609 WILTON MEADOW DRIVE
GARNER NC 27529
XM034 FLOYD
ALISON BEYER
8521 HOLLY SPRINGS ROAD
APEX NC 27502
XF124 LAWSON
```

- To read the values for ID in the fourth record, specify #4 before naming the variable and defining its attributes.
- Specify #1 to move the input pointer back to the first record, where the values for Fname and Lname are read.

- Because the next record to be read is sequential, you can use the / line pointer control after the variable Lname to move the input pointer to the second record, where the value for Address is read.

- The / line pointer control in the next line directs the input pointer to the third record, where the values for City, State, and Zip are read.

- The final / line pointer control moves the input pointer back to the fourth record, where the value for Doctor is read.

Alternatively, you can use just the #*n* line pointer control (as shown earlier in this chapter and below) to read the variables in the order shown above.

```
data perm.patients;
   infile patdata;
   input #4 ID $5.
         #1 Fname $ Lname $
         #2 Address $23.
         #3 City $ State $ Zip $
         #4 @7 Doctor $6.;
run;
```

Figure 19.31   Raw Data File with the First Four Records Highlighted

```
1---+----10---+----20---
ALEX BEDWAN
609 WILTON MEADOW DRIVE
GARNER NC 27529
XM034 FLOYD
ALISON BEYER
8521 HOLLY SPRINGS ROAD
APEX NC 27502
XF124 LAWSON
```

Chapter Summary

Text Summary

Multiple Records per Observation

Information for one observation can be spread out over several records. You can write one INPUT statement that contains line pointer controls to specify the record(s) from which values are read.

Reading Multiple Records Sequentially

The forward slash (/) line pointer control is used to read multiple records sequentially. Each time a / pointer is encountered, the input pointer advances to the next line.

Reading Multiple Records Non-Sequentially

The #*n* line pointer control is used to read multiple records non-sequentially. The #*n* specifies the absolute number of the line to which you want to move the pointer.

Combining Line Pointer Controls

The / line pointer control and the *#n* line pointer control can be combined within a SAS program to read multiple records both sequentially and non-sequentially.

Syntax

> **LIBNAME** *libref 'SAS-data-library'*;
>
> **FILENAME** *fileref 'filename'*;
>
> **DATA** *SAS-data-set*;
>
> > **INFILE** *file-specification*;
> >
> > **INPUT** *#n variable ...*
> >
> > > *#n @n variable ... / variable ... / variable ...* ;
>
> **RUN;**
>
> **PROC PRINT DATA**=*SAS-data-set*;
>
> **RUN;**

Sample Program

```
libname perm 'c:\records\empdata';
filename personel 'c:\records\empdata\new.dat';
data perm.emplist3;
   infile personel;
   input #2 Department $ 5-16
         #1 @16 ID $4. @1 Name $14. /
             JobCode 3. /
             Salary comma9.;
run;
proc print data=perm.emplist3;
run;
```

Points to Remember

- When a file contains multiple records per observation, depending on the program, the file might need to contain the same number of records for each observation.

- Because the / pointer control can move forward only, the pointer control is specified *after* the values in the current record are read.

- The *#n* pointer control can read records in any order and must be specified *before* the variable names are defined.

- A semicolon should be placed at the end of the *complete* INPUT statement.

Chapter Quiz

Select the best answer for each question. After completing the quiz, check your answers using the answer key in the appendix.

1. You can position the input pointer on a specific record by using

   a. column pointer controls.

    b. column specifications.

    c. line pointer controls.

    d. line hold specifiers.

2. Which pointer control is used to read multiple records sequentially?

    a. *@n*

    b. *+n*

    c. /

    d. all of the above

3. Which pointer control can be used to read records non-sequentially?

    a. *@n*

    b. *#n*

    c. *+n*

    d. /

4. Which SAS statement correctly reads the values for Fname, Lname, Address, City, State and Zip in order?

    Figure 19.32   Raw Data File

```
1---+----10---+----20---
LAWRENCE CALDWELL
1010 LAKE STREET
ANAHEIM CA 94122
RACHEL CHEVONT
3719 OLIVE VIEW ROAD
HARTFORD CT 06183
```

    a.
```
input Fname $ Lname $ /
      Address $20. /
      City $ State $ Zip $;
```

    b.
```
input Fname $ Lname $ /;
      Address $20. /;
      City $ State $ Zip $;
```

    c.
```
input / Fname $ Lname $
      / Address $20.
      City $ State $ Zip $;
```

    d.
```
input / Fname $ Lname $;
      / Address $20.;
      City $ State $ Zip $;
```

5. Which INPUT statement correctly reads the values for ID in the fourth record, and then returns to the first record to read the values for Fname and Lname?

Figure 19.33 *Raw Data File*

```
1---+----10---+----20---
GEORGE CHESSON
3801 WOODSIDE COURT
GARNER NC 27529
XM065 FLOYD
JAMES COLDWELL
123-A TARBERT
APEX NC 27529
XM065 LAWSON
```

a. input #4 ID $5.
          #1 Fname $ Lname $;

b. input #4 ID $ 1-5
          #1 Fname $ Lname $;

c. input #4 ID $
          #1 Fname $ Lname $;

d. all of the above

6. How many records will be read for each execution of the DATA step?

Figure 19.34 *Raw Data File*

```
1---+----10---+----20---
SKIRT BLACK
COTTON
036499 $44.98
SKIRT NAVY
LINEN
036899 $51.50
DRESS RED
SILK
037299 $76.98
```

```
data spring.sportswr;
   infile newitems;
   input #1 Item $ Color $
         #3 @8 Price comma6.
         #2 Fabric $
         #3 SKU $ 1-6;
run;
```

a. one

b. two

c. three

d. four

7. Which INPUT statement correctly reads the values for City, State, and Zip?

Figure 19.35 Raw Data File

```
1---+----10---+----20---
DINA   FIELDS
904   MAPLE   CIRCLE
DURHAM   NC   27713
ELIZABETH   GARRISON
1293   OAK   AVENUE
CHAPEL HILL   NC   27614
DAVID   HARRINGTON
2426   ELMWOOD   LANE
RALEIGH   NC   27803
```

a. input #3 City $ State $ Zip $;

b. input #3 City & $11. State $ Zip $;

c. input #3 City $11. +2 State $2. + 2 Zip $5.;

d. all of the above

8. Which program does *not* read the values in the first record as a variable named Item and the values in the second record as two variables named Inventory and Type?

Figure 19.36 Raw Data File

```
1---+----10---+----20---
COLORED PENCILS
12 BOXES
WATERCOLOR PAINT
8 PALETTES
DRAWING PAPER
15 PADS
```

a.
```
data perm.supplies;
     infile instock pad;
     input Item & $16. /
          Inventory 2. Type $8.;
run;
```

b.
```
data perm.supplies;
     infile instock pad;
     input Item & $16.
          / Inventory 2. Type $8.;
 run;
```

c.
```
data perm.supplies;
     infile instock pad;
     input #1 Item & $16.
          Inventory 2. Type $8.;
run;
```

d.
```
data perm.supplies;
     infile instock pad;
     input Item & $16.
          #2 Inventory 2. Type $8.;
run;
```

9. Which INPUT statement reads the values for Lname, Fname, Department, and Salary (in that order)?

Figure 19.37 *Raw Data File*

```
1---+----10---+----20---
ABRAMS THOMAS
SALES $25,209.03
BARCLAY ROBERT
MARKETING $29,180.36
COURTNEY MARK
PUBLICATIONS $24,006.16
```

a. `input #1 Lname $ Fname $ /`
      `Department $12. Salary comma10.;`

b. `input #1 Lname $ Fname $ /`
      `Department : $12. Salary : comma.;`

c. `input #1 Lname $ Fname $`
      `#2 Department : $12. Salary : comma.;`

d. both b and c

10. Which raw data file poses potential problems when you are reading multiple records for each observation?

   a.

   Figure 19.38 *Raw Data File*

   ```
   1---+----10---+----20---
   LAWRENCE CALDWELL
   1010 LAKE STREET
   ANAHEIM CA 94122
   RACHEL CHEVONT
   3719 OLIVE VIEW ROAD
   HARTFORD CT 06183
   ```

   b.

   Figure 19.39 *Raw Data File*

   ```
   1---+----10---+----20---
   SHIRT LT BLUE SOLID
   SKU 128699
   $38.99
   SHIRT DK BLUE STRIPE
   SKU 128799
   $41.99
   ```

   c.

   Figure 19.40 *Raw Data File*

   ```
   1---+----10---+----20---
   MARCUS JONES
   SR01 $26,134.00
   MARY ROBERTSON
   COURTNEY NEILS
   TWO1 $28,342.00
   ```

   d.

Figure 19.41 Raw Data File

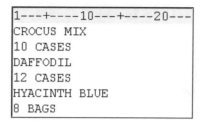

```
1---+----10---+----20---
CROCUS MIX
10 CASES
DAFFODIL
12 CASES
HYACINTH BLUE
8 BAGS
```

Chapter 20

Creating Multiple Observations from a Single Record

Overview

Introduction

Sometimes raw data files contain data for several observations in one record. Data can be stored in this manner to reduce the size of the entire data file.

Each record can contain

- repeating blocks of data that represent separate observations

Figure 20.1 *Repeating Blocks of Data*

```
1---+----10---+----20---+----30--
01APR90 68 02APR90 67 03APR90 70
04APR90 74 05APR90 72 06APR90 73
07APR90 71 08APR90 75 09APR90 76
```

- an ID field followed by an equal number of repeating fields that represent separate observations

Figure 20.2 *ID Field with Repeating Fields*

```
1---+----10---+----20---+----30--
001 WALKING AEROBICS CYCLING
002 SWIMMING CYCLING SKIING
003 TENNIS SWIMMING AEROBICS
```

- an ID field followed by a varying number of repeating fields that represent separate observations.

Figure 20.3 *ID Field with Varying Number of Repeating Fields*

```
1---+----10---+----20---+----30--
001 WALKING
002 SWIMMING CYCLING SKIING
003 TENNIS SWIMMING
```

This chapter shows you several ways of creating multiple observations from a single record.

Objectives

In this chapter, you learn to

- create multiple observations from a single record that contains repeating blocks of data

- create multiple observations from a single record that contains one ID field followed by the *same* number of repeating fields

- create multiple observations from a single record that contains one ID field followed by a *varying* number of repeating fields.

Additionally, you learn to

- hold the current record across iterations of the DATA step

- hold the current record for the next INPUT statement

- execute SAS statements based on a variable's value

- explicitly write an observation to a data set

- execute SAS statements while a condition is true.

Reading Repeating Blocks of Data

Overview

Each record in the file Tempdata contains three blocks of data. Each block contains a date followed by the day's high temperature in a small city located in the southern United States.

Figure 20.4   *Raw Data File Tempdata*

```
            Raw Data File Tempdata
1---+----10---+----20---+----30--
01APR90 68 02APR90 67 03APR90 70
04APR90 74 05APR90 72 06APR90 73
07APR90 71 08APR90 75 09APR90 76
10APR90 78 11APR90 70 12APR90 69
13APR90 71 14APR90 72 15APR90 74
16APR90 73 17APR90 71 18APR90 75
19APR90 75 20APR90 73 21APR90 75
22APR90 77 23APR90 78 24APR90 80
25APR90 78 26APR90 77 27APR90 79
28APR90 81 29APR90 81 30APR90 84
```

You could write a DATA step that reads each record and creates three different Date and Temp variables.

Figure 20.5   *Three Date and Temp Variables*

SAS Data Set

Date1	Temp1	Date2	Temp2	Date3	Temp3
11048	68	11049	67	11050	70

Alternatively, you could create a separate observation for each block of data in a record. This data set is better structured for analysis and reporting with SAS procedures.

Figure 20.6   *Separate Observations for Each Block of Data in a Record*

SAS Data Set

Date	HighTemp
11048	68
11049	67
11050	70

Holding the Current Record with a Line-Hold Specifier

You need to hold the current record so the input statement can read, and SAS can output, repeated blocks of data on the same record. This is easily accomplished by using a line-hold specifier in the INPUT statement.

SAS provides two line-hold specifiers.

- The trailing at sign (@) holds the input record for the execution of the next INPUT statement.

- The double trailing at sign (@@) holds the input record for the execution of the next INPUT statement, even across iterations of the DATA step.

The term *trailing* indicates that the @ or @@ must be the *last* item specified in the INPUT statement. For example,

```
input Name $20. @; or input Name $20. @@;
```

This chapter teaches you how the trailing @@ can be used to hold a record across multiple iterations of the DATA step.

Using the Double Trailing At Sign (@@) to Hold the Current Record

Normally, each time a DATA step executes, the INPUT statement reads the next record. But when you use the trailing @@, the INPUT statement continues reading from the same record.

Table 20.1 *Example with the Double Trailing At Sign*

`Input Score;`
`Input Score @@;`

Figure 20.7 *Raw Data File and Program Data Vector*

Figure 20.8 *Raw Data File and Program Data Vector*

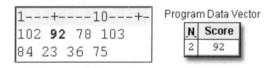

The double trailing at sign (@@)

- works like the trailing @ except it holds the data line in the input buffer across multiple executions of the DATA step

- typically is used to read multiple SAS observations from a single data line

- should not be used with the @ pointer control, with column input, nor with the MISSOVER option.

A record that is held by the double trailing at sign (@@) is not released until either of the following events occurs:

- the input pointer moves past the end of the record. Then the input pointer moves down to the next record.

Figure 20.9 *Raw Data File*

```
1---+----10--▼+-|
102 92 78 103
84 23 36 75
```

- an INPUT statement that has no trailing at sign executes.

```
input ID $ @@;
.

.

.

input Department 5.;
```

The following example requires only one INPUT statement to read the values for Date and HighTemp, but the INPUT statement must execute three times for each record.

The INPUT statement reads a block of values for Date and HighTemp, and holds the current record by using the trailing @@. The values in the program data vector are written to the data set as an observation, and control returns to the top of the DATA step.

```
data perm.april90;
   infile tempdata;
   input Date : date. HighTemp @@;
```

Figure 20.10 *Control Returned to the Top of the DATA Step*

```
        Raw Data File Tempdata
1---+----1▼---+----20---+----30--
01APR90 68 02APR90  67 03APR90  70
04APR90 74 05APR90  72 06APR90  73
07APR90 71 08APR90  75 09APR90  76
```

In the next iteration, the INPUT statement reads the next block of values for Date and HighTemp from the same record.

Figure 20.11 *Date and High Temp*

```
        Raw Data File Tempdata
1---+----10---+----2▼---+----30--
01APR90 68 02APR90 67 03APR90  70
04APR90 74 05APR90 72 06APR90  73
07APR90 71 08APR90 75 09APR90  76
```

Completing the DATA Step

You can add a FORMAT statement to the DATA step to display the date or time values with a specified format. The FORMAT statement below uses the DATE*w.* format to display the values for Date in the form *ddmmmyyyy*.

```
data perm.april90;
   infile tempdata;
   input Date : date. HighTemp @@;
   format date date9.;
run;
```

Figure 20.12 *Displaying Dates in a Specified Format*

Raw Data File Tempdata

```
1---+----10---+----20---+----30--
01APR90 68 02APR90 67 03APR90 70
04APR90 74 05APR90 72 06APR90 73
07APR90 71 08APR90 75 09APR90 76
```

DATA Step Processing of Repeating Blocks of Data

Complete DATA Step

```
data perm.april90;
   infile tempdata;
   input Date : date. HighTemp @@;
   format date date9.;
run;
```

Example

As the execution phase begins, the input pointer rests on column 1 of record 1.

Figure 20.13 *Input Pointer on Column 1 of Record 1*

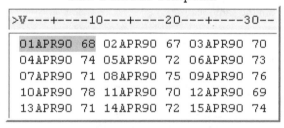

During the first iteration of the DATA step, the first block of values for Date and High Temp are read into the program data vector.

Figure 20.14 *Reading the First Block of Values*

Raw Data File Tempdata

```
>----+----1V---+----20---+----30--
01APR90 68  02APR90 67  03APR90 70
04APR90 74  05APR90 72  06APR90 73
07APR90 71  08APR90 75  09APR90 76
10APR90 78  11APR90 70  12APR90 69
13APR90 71  14APR90 72  15APR90 74
```

Program Data Vector

N	Date	HighTemp
1	11048	68

The first observation is written to the data set.

Figure 20.15 *Control Returns to the Top of the DATA Step*

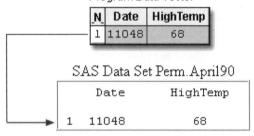

Raw Data File Tempdata

```
>----+----1V---+----20---+----30--
01APR90 68  02APR90 67  03APR90 70
04APR90 74  05APR90 72  06APR90 73
07APR90 71  08APR90 75  09APR90 76
10APR90 78  11APR90 70  12APR90 69
13APR90 71  14APR90 72  15APR90 74
```

Program Data Vector

N	Date	HighTemp
1	11048	68

SAS Data Set Perm.April90

	Date	HighTemp
1	11048	68

Control returns to the top of the DATA step, and the values are reset to missing.

Figure 20.16 *Reset Values*

Program Data Vector

N	Date	HighTemp
2	•	•

During the second iteration, the @@ prevents the input pointer from moving down to the next record. Instead, the INPUT statement reads the second block of values for Date and HighTemp from the first record.

Figure 20.17 *Reading the Second Block of Values from the First Record*

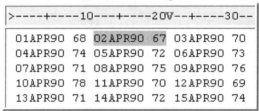

The second observation is written to the data set, and control returns to the top of the DATA step.

Figure 20.18 *Writing the Second Observation to the Data Set*

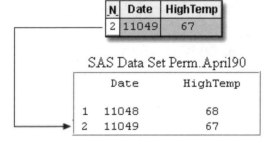

During the third iteration, the last block of values is read and written to the data set as the third observation.

Figure 20.19 *Writing the Third Observation to the Data Set*

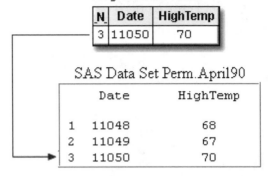

Raw Data File Tempdata

```
>----+----10---+----20---+----30-V
01APR90  68  02APR90  67  03APR90  70
04APR90  74  05APR90  72  06APR90  73
07APR90  71  08APR90  75  09APR90  76
10APR90  78  11APR90  70  12APR90  69
13APR90  71  14APR90  72  15APR90  74
```

Program Data Vector

N	Date	HighTemp
3	11050	70

SAS Data Set Perm.April90

	Date	HighTemp
1	11048	68
2	11049	67
3	11050	70

During the fourth iteration, the first block of values in the second record is read and written as the fourth observation.

Figure 20.20 *Writing the Fourth Observation to the Data Set*

Raw Data File Tempdata

```
>----+----1V---+----20---+----30--
01APR90  68  02APR90  67  03APR90  70
04APR90  74  05APR90  72  06APR90  73
07APR90  71  08APR90  75  09APR90  76
10APR90  78  11APR90  70  12APR90  69
13APR90  71  14APR90  72  15APR90  74
```

Program Data Vector

N	Date	HighTemp
4	11051	74

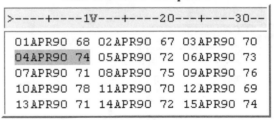

SAS Data Set Perm.April90

	Date	HighTemp
1	11048	68
2	11049	67
3	11050	70
4	11051	74

The execution phase continues until the last block of data is read.

Figure 20.21 *Writing the Last Observation to the Data Set*

Raw Data File Tempdata

```
>----+----10---+----20---+----30-V

01APR90 68 02APR90 67 03APR90 70
04APR90 74 05APR90 72 06APR90 73
07APR90 71 08APR90 75 09APR90 76
10APR90 78 11APR90 70 12APR90 69
13APR90 71 14APR90 72 15APR90 74
```

SAS Data Set Perm.April90

```
        Date      HighTemp

   1    11048        68
          .
          .
          .
  13    11060        71
  14    11061        72
  15    11062        74
```

You can display the data set with the PRINT procedure.

```
proc print data=perm.april90;
run;
```

Figure 20.22 *PROC PRINT Output of the Data Set.*

Obs	Date	HighTemp
1	01APR1990	68
2	02APR1990	67
3	03APR1990	70
4	04APR1990	74
5	05APR1990	72
6	06APR1990	73
7	07APR1990	71
8	08APR1990	75
9	09APR1990	76
10	10APR1990	78
11	11APR1990	70
12	12APR1990	69
13	13APR1990	71
14	14APR1990	71
15	15APR1990	74

Reading the Same Number of Repeating Fields

Overview

So far, you have created multiple observations from a single record by executing the DATA step once for each block of data in a record.

Now, look at another file that is organized differently.

Each record in the file Data97 contains a sales representative's ID number, followed by four repeating fields that represent his or her quarterly sales totals for 1997.

You want to pair each employee ID number with one quarterly sales total to produce a single observation. Four observations are generated from each record.

Figure 20.23   *Multiple Fields for the Same ID*

```
                  Raw Data File Data97
1---+----10---+----20---+----30---+----40
0734 1,323.34 2,472.85 3,276.65 5,345.52
0943 1,908.34 2,560.38 3,472.09 5,290.86
1009 2,934.12 3,308.41 4,176.18 7,581.81
1043 1,295.38 5,980.28 8,876.84 6,345.94
1190 2,189.84 5,023.57 2,794.67 4,243.35
1382 3,456.34 2,065.83 3,139.08 6,503.49
1734 2,345.83 3,423.32 1,034.43 1,942.28
```

ID	Quarter	Sales
0734	1	1323.34
0734	2	2472.85
0734	3	3276.65
0734	4	5345.52
0943	1	1908.34
0943	2	2560.38
0943	3	3472.09
0943	4	5290.86

To accomplish this, you must execute the DATA step once for each record, repetitively reading and writing values in one iteration.

This means that a DATA step must

- read the value for ID and hold the current record

- create a new variable named Quarter to identify the fiscal quarter for each sales figure

- read a new value for Sales and write the values to the data set as an observation

- continue reading a new value for Sales and writing values to the data set three more times.

Using the Single Trailing At Sign (@) to Hold the Current Record

First, you need to read the value for ID and hold the record so that subsequent values for Sales can be read.

```
data perm.sales97;
   infile data97;
   input ID $
```

Figure 20.24   Holding a Record

```
                    Raw Data File Data97
1---V----10---+----20---+----30---+----40
0734 1,323.34 2,472.85 3,276.65 5,345.52
0943 1,908.34 2,560.38 3,472.09 5,290.86
1009 2,934.12 3,308.41 4,176.18 7,581.81
```

You are already familiar with the trailing @@, which holds the current record across multiple iterations of the DATA step.

However, in this case, you want to hold the record with the trailing @, so that a second INPUT statement can read the multiple sales values from a single record within the same iteration of the DATA step. Like the trailing @@, the single trailing @

- enables the next INPUT statement to continue reading from the same record

- releases the current record when a subsequent INPUT statement executes without a line-hold specifier.

It's easy to distinguish between the trailing @@ and the trailing @ by remembering that

- the double trailing at sign (@@) holds a record across multiple iterations of the DATA step until the end of the record is reached.

- the single trailing at sign (@) releases a record when control returns to the top of the DATA step.

In this example, the first INPUT statement reads the value for ID and uses the trailing @ to hold the current record for the next INPUT statement in the DATA step.

```
data perm.sales97;
   infile data97;
   input ID $ @;
   input Sales : comma. @;
output;
```

Figure 20.25   Reading the Value for ID

```
                    Raw Data File Data97
1---V----10---+----20---+----30---+----40
0734 1,323.34 2,472.85 3,276.65 5,345.52
0943 1,908.34 2,560.38 3,472.09 5,290.86
1009 2,934.12 3,308.41 4,176.18 7,581.81
```

The second INPUT statement reads a value for Sales and holds the record. The COMMA*w.d* informat in the INPUT statement reads the numeric value for Sales and removes the embedded commas. An OUTPUT statement writes the observation to the SAS data set, and the DATA step continues processing.

Notice that the COMMA*w.d* informat does *not* specify a *w* value. Remember that list input reads values until the next blank is detected. The default length of numeric variables is 8 bytes, so you don't need to specify a *w* value to determine the length of a numeric variable.

When all of the repeating fields have been read and output, control returns to the top of the DATA step, and the record is released.

```
data perm.sales97;
   infile data97;
   input ID $ @;
   input Sales : comma. @;
   output;
   input Sales : comma. @;
   output;
   input Sales : comma. @;
   output;
   input Sales : comma. @;
   output;
run;
```

Figure 20.26   *Reading the Value for Sales*

```
                   Raw Data File Data97
1---+----10--V+----20---+----30---+----40
0734 1,323.34 2,472.85 3,276.65 5,345.52
0943 1,908.34 2,560.38 3,472.09 5,290.86
1009 2,934.12 3,308.41 4,176.18 7,581.81
```

More Efficient Programming

Each record contains four different values for the variable Sales, so the INPUT statement must execute four times. Rather than writing four INPUT statements, you can execute one INPUT statement repeatedly in an iterative DO loop.

Each time the loop executes, you need to write the values for ID, Quarter, and Sales as an observation to the data set. This is easily accomplished by using the OUTPUT statement.

```
data perm.sales97;
   infile data97;
   input ID $ @;
   do Quarter=1 to 4;
      input Sales : comma. @;
      output;
   end;
run;
```

By default, every DATA step contains an implicit OUTPUT statement at the end of the step. Placing an explicit OUTPUT statement in a DATA step overrides the automatic output, and SAS adds an observation to a data set only when the explicit OUTPUT statement is executed.

Processing a DATA Step That Contains an Iterative DO Loop

Now that the program is complete, let's see how SAS processes a DATA step that
contains an iterative DO loop.

```
data perm.sales97;
   infile data97;
   input ID $ @;
   do Quarter=1 to 4;
      input Sales : comma. @;
      output;
   end;
run;
```

During the first iteration, the value for ID is read and Quarter is initialized to **1** as the
loop begins to execute.

Figure 20.27 *Reading the Value for ID and Initializing Quarter*

Raw Data File Data97

```
>----V----10---+----20---+----30---+----40-
 0734 1,323.34 2,472.85 3,276.65 5,345.52
 0943 1,908.34 2,560.38 3,472.09 5,290.86
 1009 2,934.12 3,308.41 4,176.18 7,581.81
```

Program Data Vector

N	ID	Quarter	Sales
1	0734	1	.

The INPUT statement reads the first repeating field and assigns the value to Sales in the
program data vector. The @ holds the current record.

Figure 20.28 *Results of the INPUT Statement*

Raw Data File Data97

```
>----+----10--V+----20---+----30---+----40-
 0734 1,323.34 2,472.85 3,276.65 5,345.52
 0943 1,908.34 2,560.38 3,472.09 5,290.86
 1009 2,934.12 3,308.41 4,176.18 7,581.81
```

Program Data Vector

N	ID	Quarter	Sales
1	0734	1	1323.34

The OUTPUT statement writes the values in the program data vector to the data set as
the first observation.

Figure 20.29   *Results of the OUTPUT Statement*

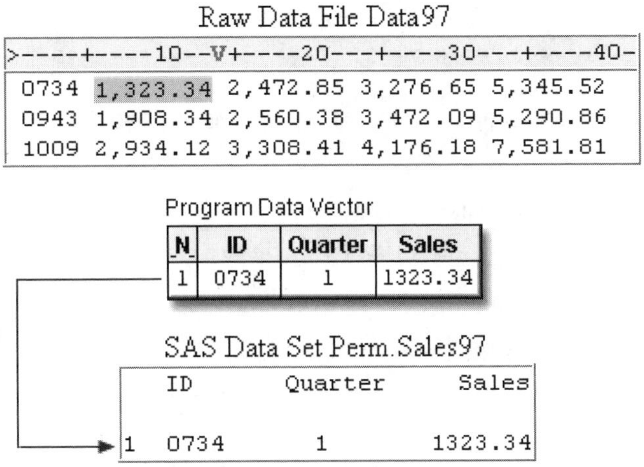

Raw Data File Data97

```
>----+----10--V+----20---+----30---+----40-
 0734 1,323.34 2,472.85 3,276.65 5,345.52
 0943 1,908.34 2,560.38 3,472.09 5,290.86
 1009 2,934.12 3,308.41 4,176.18 7,581.81
```

Program Data Vector

N	ID	Quarter	Sales
1	0734	1	1323.34

SAS Data Set Perm.Sales97

	ID	Quarter	Sales
1	0734	1	1323.34

The END statement indicates the bottom of the loop, but control returns to the DO statement, not to the top of the DATA step. Now the value of Quarter is incremented to 2.

Display 20.1   *Results of the END Statement*

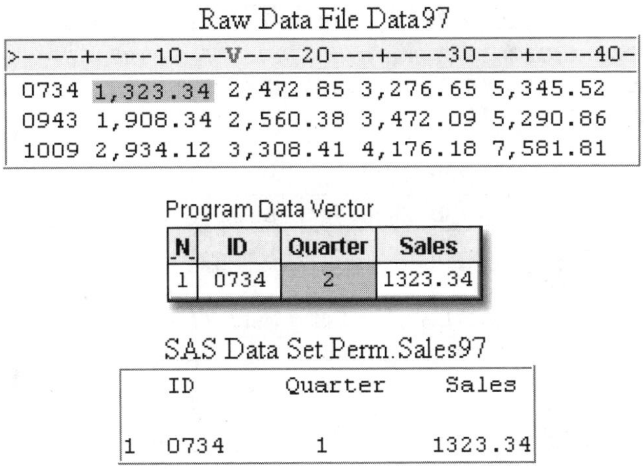

Raw Data File Data97

```
>----+----10---V----20---+----30---+----40-
 0734 1,323.34 2,472.85 3,276.65 5,345.52
 0943 1,908.34 2,560.38 3,472.09 5,290.86
 1009 2,934.12 3,308.41 4,176.18 7,581.81
```

Program Data Vector

N	ID	Quarter	Sales
1	0734	2	1323.34

SAS Data Set Perm.Sales97

	ID	Quarter	Sales
1	0734	1	1323.34

The INPUT statement executes again, reading the second repeating field and storing the value for Sales in the program data vector.

Figure 20.30 *Results of the Second Reading of the INPUT Statement*

Raw Data File Data97

```
>----+----10---+----20-V-+----30---+----40-
 0734 1,323.34 2,472.85 3,276.65 5,345.52
 0943 1,908.34 2,560.38 3,472.09 5,290.86
 1009 2,934.12 3,308.41 4,176.18 7,581.81
```

Program Data Vector

N	ID	Quarter	Sales
1	0734	2	2472.85

SAS Data Set Perm.Sales97

	ID	Quarter	Sales
1	0734	1	1323.34

The OUTPUT statement writes the values in the program data vector as the second observation.

Figure 20.31 *Results of the Second Reading of the OUTPUT Statement*

Raw Data File Data97

```
>----+----10---+----20-V-+----30---+----40-
 0734 1,323.34 2,472.85 3,276.65 5,345.52
 0943 1,908.34 2,560.38 3,472.09 5,290.86
 1009 2,934.12 3,308.41 4,176.18 7,581.81
```

Program Data Vector

N	ID	Quarter	Sales
1	0734	2	2472.85

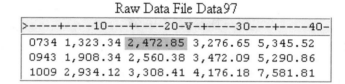

SAS Data Set Perm.Sales97

	ID	Quarter	Sales
1	0734	1	1323.34
2	0734	2	2472.85

The loop continues executing while the value for Quarter is **3**, and then **4**. In the process, the third and fourth observations are written.

Figure 20.32 *Writing the Third and Fourth Observations*

Raw Data File Data97

```
>----+----10---+----20---+----30---+----4V-
 0734 1,323.34 2,472.85 3,276.65 5,345.52
 0943 1,908.34 2,560.38 3,472.09 5,290.86
 1009 2,934.12 3,308.41 4,176.18 7,581.81
```

Program Data Vector

N	ID	Quarter	Sales
1	0734	4	5345.52

SAS Data Set Perm.Sales97

	ID	Quarter	Sales
1	0734	1	1323.34
2	0734	2	2472.85
3	0734	3	3276.65
4	0734	4	5345.52

After the fourth observation is written, Quarter is incremented to **5** at the bottom of the DO loop and control returns to the top of the loop. The loop does not execute again because the value of Quarter is now greater than **4**.

Raw Data File Data97

```
>----+----10---+----20---+----30---+----4V-
 0734 1,323.34 2,472.85 3,276.65 5,345.52
 0943 1,908.34 2,560.38 3,472.09 5,290.86
 1009 2,934.12 3,308.41 4,176.18 7,581.81
```

Program Data Vector

N	ID	Quarter	Sales
1	0734	5	5345.52

SAS Data Set Perm.Sales97

	ID	Quarter	Sales
1	0734	1	1323.34
2	0734	2	2472.85
3	0734	3	3276.65
4	0734	4	5345.52

Control returns to the top of the DATA step, and the input pointer moves to column 1 of the next record. The variable values in the program data vector are reset to missing.

Figure 20.33 *Returning Control to the Top of the DATA Step*

Raw Data File Data97

```
>V---+----10---+----20---+----30---+----40-
 0734 1,323.34 2,472.85 3,276.65 5,345.52
 0943 1,908.34 2,560.38 3,472.09 5,290.86
 1009 2,934.12 3,308.41 4,176.18 7,581.81
```

Program Data Vector

N	ID	Quarter	Sales
2	•	•	•

SAS Data Set Perm.Sales97

```
     ID       Quarter      Sales

1   0734        1         1323.34
2   0734        2         2472.85
3   0734        3         3276.65
4   0734        4         5345.52
```

When the execution phase is complete, you can display the data set with the PRINT procedure.

```
proc print data=perm.sales97;
run;
```

Figure 20.34 *Output of PROC PRINT*

Obs	ID	Quarter	Sales
1	0734	1	1323.34
2	0734	2	2472.85
3	0734	3	3276.65
4	0734	4	5345.52
5	0943	1	1908.34
6	0943	2	2560.38
7	0943	3	3472.09
8	0943	4	5290.86
9	1009	1	2934.12
10	1009	2	3308.41
11	1009	3	4176.18
12	1009	4	7581.81

Reading a Varying Number of Repeating Fields

Overview

So far, each record in the file Data97 has contained the same number of repeating fields.

Figure 20.35 *Raw Data File Data97*

```
                Raw Data File Data97
1---+----10---+----20---+----30---+----40
0734 1,323.34 2,472.85 3,276.65 5,345.52
0943 1,908.34 2,560.38 3,472.09 5,290.86
1009 2,934.12 3,308.41 4,176.18 7,581.81
1043 1,295.38 5,980.28 8,876.84 6,345.94
1190 2,189.84 5,023.57 2,794.67 4,243.35
1382 3,456.34 2,065.83 3,139.08 6,503.49
1734 2,345.83 3,423.32 1,034.43 1,942.28
```

But suppose that some of the employees quit at various times. Their records might not contain sales totals for the second, third, or fourth quarter. These records contain a variable number of repeating fields.

Figure 20.36 *Raw Data File Data97 Showing Empty Records*

```
                Raw Data File Data97
1---+----10---+----20---+----30---+----40
1824 1,323.34 2,472.85
1943 1,908.34
2046 1,423.52 1,673.46 3,276.65
2063 2,345.34 2,452.45 3,523.52 2,983.01
```

The DATA step that you just wrote won't work with a variable number of repeating fields because now the value of Quarter is not constant for every record.

```
data perm.sales97;
   infile data97;
   input ID $ @;
   do Quarter=1 to 4;
      input Sales : comma. @;
      output;
   end;
run;
```

Using the MISSOVER Option

You can adapt the DATA step to accommodate a varying number of Sales values.

Like the previous example with the same number of repeating fields, your DATA step must read the same record repeatedly. However, you need to prevent the input pointer from moving to the next record when there are missing Sales values.

You can use the MISSOVER option in an INFILE statement to prevent SAS from reading the next record when missing values are encountered at the end of a record. Essentially, records that have a varying number of repeating fields are records that contain missing values, so you need to specify the MISSOVER option here as well.

Because there is at least one value for the repeating field, Sales, in each record, the first INPUT statement reads *both* the value for ID and the first Sales value for each record. The trailing @ holds the record so that any subsequent repeating fields can be read.

```
data perm.sales97;
   infile data97 missover;
   input ID $ Sales : comma. @;
```

Figure 20.37   Holding a Record

```
                     Raw Data File Data97
1---+----10--V+----20---+----30---+----40
1824 1,323.34 2,472.85
1943 1,908.34
2046 1,423.52 1,673.46 3,276.65
2063 2,345.34 2,452.45 3,523.52 2,983.01
```

SAS provides several options to control reading past the end of a line. You've seen the MISSOVER option for setting remaining INPUT statement variables to missing values if the pointer reaches the end of a record. You can also use other options such as the TRUNCOVER option, which reads column or formatted input when the last variable that is read by the INPUT statement contains varying-length data. The TRUNCOVER option assigns the contents of the input buffer to a variable when the field is shorter than expected.

Other related options include FLOWOVER (the default), STOPOVER, and SCANOVER. For more information about TRUNCOVER and related options, see the SAS documentation.

Executing SAS Statements While a Condition Is True

Now consider how many times to read each record. Earlier, you created an index variable named Quarter whose value ranged from **1** to **4** because there were four repeating fields.

Now you want to read the record only while a value for Sales exists. Use a DO WHILE statement instead of the iterative DO statement, enclosing the expression in parentheses. In the example below, the DO WHILE statement executes while the value of Sales is not equal to a missing value (which is represented by a period).

```
data perm.sales97;
   infile data97 missover;
   input ID $ Sales : comma. @;
   do while (sales ne .);
```

Creating a Counter Variable

Because the DO WHILE statement does not create an index variable, you can create your own "counter" variable. You can use a sum statement to increment the value of the counter variable each time the DO WHILE loop executes.

In the example below, the assignment statement that precedes the loop creates the counter variable Quarter and assigns it an initial value of zero. Each time the DO WHILE loop executes, the sum statement increments the value of Quarter by one.

```
data perm.sales97;
   infile data97 missover;
   input ID $ Sales : comma. @;
   Quarter=0;
   do while (sales ne .);
      quarter+1;
```

Completing the DO WHILE Loop

Now look at the other statements that should be executed in the DO WHILE loop. First, you need an OUTPUT statement to write the current observation to the data set. Then, another INPUT statement reads the next value for Sales and holds the record. You complete the DO WHILE loop with an END statement.

```
data perm.sales97;
   infile data97 missover;
   input ID $ Sales : comma. @;
   Quarter=0;
   do while (sales ne .);
      quarter+1;
      output;
      input sales : comma. @;
   end;
run;
```

Processing a DATA Step That Has a Varying Number of Repeating Fields

This example uses the following DATA step:

```
data perm.sales97;
   infile data97 missover;
   input ID $ Sales : comma. @;
   Quarter=0;
   do while (sales ne .);
      quarter+1;
      output;
      input sales : comma. @;
   end;
run;
```

During the first iteration of the DATA step, values for ID and Sales are read. Quarter is initialized to *0*.

Figure 20.38   *Initializing the Value of Quarter to 0*

Raw Data File Data97

```
>----+----10--V+----20---+----30---+----40-
1824 1,323.34 2,472.85
1943 1,908.34
2046 1,423.52 1,673.46 3,276.65
2063 2,345.34 2,452.45 3,523.52 2,983.01
```

Program Data Vector

N	ID	Sales	Quarter
1	1824	1323.34	0

The DO WHILE statement checks to see if Sales has a value, which it does, so the other statements in the DO loop execute. The Value of Quarter is incremented by 1 and the current observation is written to the data set.

Figure 20.39   *Executing the DO Loop*

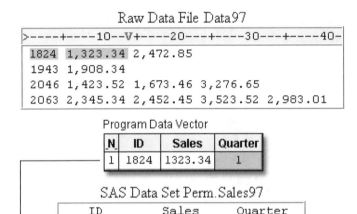

Raw Data File Data97

```
>----+----10--V+----20---+----30---+----40-
1824 1,323.34 2,472.85
1943 1,908.34
2046 1,423.52 1,673.46 3,276.65
2063 2,345.34 2,452.45 3,523.52 2,983.01
```

Program Data Vector

N	ID	Sales	Quarter
1	1824	1323.34	1

SAS Data Set Perm.Sales97

	ID	Sales	Quarter
1	1824	1323.34	1

The INPUT statement reads the next value for Sales, the end of the loop is reached, and control returns to the DO WHILE statement.

Figure 20.40   *Returning Control to the DO WHILE Statement*

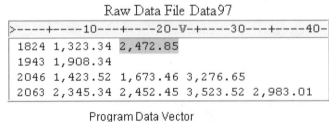

Raw Data File Data97

```
>----+----10---+----20-V-+----30---+----40-
1824 1,323.34 2,472.85
1943 1,908.34
2046 1,423.52 1,673.46 3,276.65
2063 2,345.34 2,452.45 3,523.52 2,983.01
```

Program Data Vector

N	ID	Sales	Quarter
1	1824	2472.85	1

SAS Data Set Perm.Sales97

	ID	Sales	Quarter
1	1824	1323.34	1

The condition is checked and Sales still has a value, so the loop executes again.

Figure 20.41 *Executing the Loop Again*

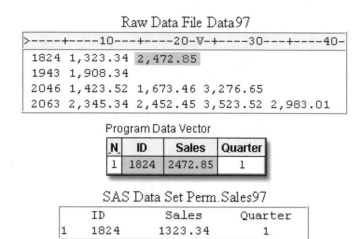

Quarter is incremented to 2, and the values in the program data vector are written as the second observation.

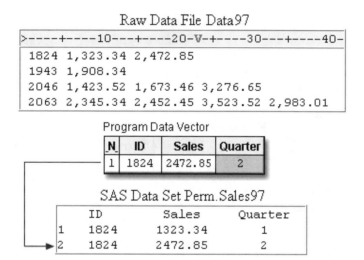

The INPUT statement executes again. The MISSOVER option prevents the input pointer from moving to the next record in search of another value for Sales. Therefore, SALEs receives a missing value.

Raw Data File Data97

```
>----+----10---+----20-V-+----30---+----40-
1824 1,323.34 2,472.85
1943 1,908.34
2046 1,423.52 1,673.46 3,276.65
2063 2,345.34 2,452.45 3,523.52 2,983.01
```

Program Data Vector

N	ID	Sales	Quarter
1	1824	•	2

SAS Data Set Perm.Sales97

	ID	Sales	Quarter
1	1824	1323.34	1
2	1824	2472.85	2

The end of the loop is reached, and control returns to the DO WHILE statement. Because the condition is now false, the statements in the loop are not executed and the values in the PDV are not output.

Figure 20.42 *Returning Control the DO WHILE Statement*

Raw Data File Data97

```
>----+----10---+----20-V-+----30---+----40-
1824 1,323.34 2,472.85
1943 1,908.34
2046 1,423.52 1,673.46 3,276.65
2063 2,345.34 2,452.45 3,523.52 2,983.01
```

Program Data Vector

N	ID	Sales	Quarter
1	1824	•	2

SAS Data Set Perm.Sales97

	ID	Sales	Quarter
1	1824	1323.34	1
2	1824	2472.85	2

Instead, control returns to the top of the DATA step, the values in the program data vector are reset to missing, and the input statement reads the next record. The DATA step continues executing until the end of the file.

Figure 20.43 *Returning Control to the Top of the DATA Step*

Raw Data File Data97

```
>V---+----10---+----20---+----30---+----40-
 1824 1,323.34 2,472.85
 1943 1,908.34
 2046 1,423.52 1,673.46 3,276.65
 2063 2,345.34 2,452.45 3,523.52 2,983.01
```

Program Data Vector

N	ID	Sales	Quarter
2	•	•	•

SAS Data Set Perm.Sales97

	ID	Sales	Quarter
1	1824	1323.34	1
2	1824	2472.85	2

PROC PRINT output for the data set shows a varying number of observations for each employee.

```
proc print data=perm.sales97;
run;
```

Figure 20.44 *Output of PROC PRINT*

Obs	ID	Sales	Quarter
1	1824	1323.34	1
2	1824	2472.85	2
3	1943	2199.23	1
4	2046	3598.48	1
5	2046	4697.98	2
6	2046	4598.45	3
7	2063	4963.87	1
8	2063	3434.42	2
9	2063	2241.64	3
10	2063	2759.11	4

Chapter Summary

Text Summary

File Formats

One raw data record might contain data for several observations. Data might be stored in this manner in order to reduce the size of the entire file. The data can be organized into

- repeating blocks of data

- an ID field followed by the same number of repeating fields

- an ID field followed by a varying number of repeating fields.

Reading Repeating Blocks of Data

To create multiple observations from a record that contains repeating blocks of data, the INPUT statement needs to hold the current record until each block of data has been read and written to the data set as an observation. The DATA step should include statements that

1. read the first block of values and hold the current record with the double trailing at sign (@@) line-hold specifier

2. optionally add a FORMAT statement to display date or time values with a specified format

3. write the first block of values as an observation

4. execute the DATA step until all repeating blocks have been read.

Reading the Same Number of Repeating Fields

To create multiple observations from a record that contains an ID field and the same number of repeating fields, you must execute the DATA step once for each record, repetitively reading and writing values in one iteration. The DATA step should include statements that

1. read the ID field and hold the current record with the single trailing at sign (@) line-hold specifier

2. execute SAS statements using an iterative DO loop. The iterative DO loop repetitively processes statements that

   - read the next value of the repeating field and hold the record with the @ line-hold specifier

   - explicitly write an observation to the data set by using an OUTPUT statement.

3. complete the iterative DO loop with an END statement.

Reading a Varying Number of Repeating Fields

To create multiple observations from a record that contains an ID field and a varying number of repeating fields, you must execute the DATA step once for each record, repetitively reading and writing values in one iteration while the value of the repeating field is nonmissing. The DATA step should include statements that

1. prevent SAS from reading the next record if missing values were encountered in the current record. The MISSOVER option on the INFILE statement prevents reading the next record.

2. read the ID field and the first repeating field, and then hold the record with the single trailing at sign (@) line-hold specifier

3. optionally create a counter variable

4. execute SAS statements while a condition is true, using a DO WHILE loop. A DO WHILE loop repetitively processes statements that

   - optionally increment the value of the counter variable by using a sum statement

   - explicitly add an observation to the data set by using an OUTPUT statement

- read the next value of the repeating field and hold the record with the single trailing at sign (@) line-hold specifier.

5. complete the DO WHILE loop with an END statement.

Syntax

Repeating Blocks of Data

LIBNAME *libref 'SAS-data-library'*;

FILENAME *fileref 'filename'*;

DATA SAS-data-set;

    INFILE *file-specification*;

    INPUT *variables* **@@**;

    FORMAT *date/time-variable format*;

RUN;

An ID Field Followed by the Same Number of Repeating Fields

LIBNAME *libref 'SAS-data-library'*;

FILENAME *fileref 'filename'*;

DATA *SAS-data-set*;

    INFILE *file-specification*;

    INPUT *id-variable* **@**;

    DO *index-variable specification*;

        INPUT *repeating-variable* **@**;

        OUTPUT;

    END;

RUN;

An ID Field Followed by a Varying Number of Repeating Fields

LIBNAME *libref 'SAS-data-library'*;

FILENAME *fileref 'filename'*;

DATA *SAS-data-set*;

    INFILE *file-specification* **MISSOVER**;

    INPUT *id-variable repeating-variable* **@**;

    counter-variable=0;

    DO WHILE *(expression)*;

        counter-variable+1;

        OUTPUT;

        INPUT *repeating-variable* **@**;

    END;

RUN;

Sample Programs

Repeating Blocks of Data

```
libname perm 'c:\records\weather';
filename tempdata 'c:\records\weather\tempdata';
data perm.april90;
   infile tempdata;
   input Date : date. HighTemp @@;
   format date date9.;
run;
```

An ID Field Followed by the Same Number of Repeating Fields

```
libname perm 'c:\records\sales';
filename data97 'c:\records\sales\1997.dat';
data perm.sales97;
   infile data97;
   input ID $ @;
   do Quarter=1 to 4;
      input Sales : comma. @;
      output;
   end;
run;
```

An ID Field Followed by a Varying Number of Repeating Fields

```
libname perm 'c:\records\sales';
filename data97 'c:\records\sales\1997.dat';
data perm.sales97;
   infile data97 missover;
   input ID $ Sales : comma. @;
   Quarter=0;
   do while (sales ne .);
      quarter+1;
      output;
      input sales : comma. @;
   end;
run;
```

Points to Remember

- The double trailing at sign (@@) holds a record across multiple iterations of the DATA step until the end of the record is reached.

- The single trailing at sign (@) releases a record when control returns to the top of the DATA step.

- Use an END statement to complete DO loops and DO WHILE loops.

Chapter Quiz

Select the best answer for each question. After completing the quiz, check your answers using the answer key in the appendix.

1. Which is true for the double trailing at sign (@@)?

   a. It enables the next INPUT statement to read from the current record across multiple iterations of the DATA step.

   b. It must be the last item specified in the INPUT statement.

   c. It is released when the input pointer moves past the end of the record.

   d. all of the above

2. A record that is being held by a single trailing at sign (@) is automatically released when

   a. the input pointer moves past the end of the record.

   b. the next iteration of the DATA step begins.

   c. another INPUT statement that has an @ executes.

   d. another value is read from the observation.

3. Which SAS program correctly creates a separate observation for each block of data?

   Figure 20.45 *Raw Data File*

   ```
   1---+----10---+----20---+----30---+----40---+
   1001 apple 1002 banana 1003 cherry
   1004 guava 1005 kiwi 1006 papaya
   1007 pineapple 1008 raspberry 1009 strawberry
   ```

   a.
   ```
   data perm.produce;
       infile fruit;
       input Item $ Variety : $10.;
   run;
   ```

   b.
   ```
   data perm.produce;
       infile fruit;
       input Item $ Variety : $10. @;
   run;
   ```

   c.
   ```
   data perm.produce;
       infile fruit;
       input Item $ Variety : $10. @@;
   run;
   ```

   d.
   ```
   data perm.produce;
       infile fruit @@;
       input Item $ Variety : $10.;
   run;
   ```

4. Which SAS program reads the values for ID and holds the record for each value of Quantity, so that three observations are created for each record?

Figure 20.46 *Raw Data File*

```
1---+----10---+----20---+----30
2101 21,208 19,047 22,890
2102 18,775 20,214 22,654
2103 19,763 22,927 21,862
```

a.
```
data work.sales;
    infile unitsold;
    input ID $;
    do week=1 to 3;
        input Quantity : comma.;
        output;
    end;
run;
```

b.
```
data work.sales;
    infile unitsold;
    input ID $ @@;
    do week=1 to 3;
        input Quantity : comma.;
        output;
    end;
run;
```

c.
```
data work.sales;
    infile unitsold;
    input ID $ @;
    do week=1 to 3;
        input Quantity : comma.;
        output;
    end;
run;
```

d.
```
data work.sales;
    infile unitsold;
    input ID $ @;
    do week=1 to 3;
        input Quantity : comma. @;
        output;
    end;
run;
```

5. Which SAS statement repetitively executes several statements when the value of an index variable named count ranges from **1** to **50**, incremented by 5?

   a. `do count=1 to 50 by 5;`

   b. `do while count=1 to 50 by 5;`

   c. `do count=1 to 50 + 5;`

   d. `do while (count=1 to 50 + 5);`

6. Which option below, when used in a DATA step, writes an observation to the data set after each value for Activity has been read?

   a.
   ```
   do choice=1 to 3;
       input Activity : $10. @;
       output;
   ```

```
      end;
      run;

b.  do choice=1 to 3;
        input Activity : $10. @;
    end;
    output;
    run;

c.  do choice=1 to 3;
        input Activity : $10. @;
    end;
    run;
```

    d. both a and b

7. Which SAS statement repetitively executes several statements while the value of Cholesterol is greater than **200**?

    a. `do cholesterol > 200;`

    b. `do cholesterol gt 200;`

    c. `do while (cholesterol > 200);`

    d. `do while cholesterol > 200;`

8. Which choice below is an example of a sum statement?

    a. `totalpay=1;`

    b. `totalpay+1;`

    c. `totalpay*1;`

    d. `totalpay by 1;`

9. Which program creates the SAS data set Perm.Topstore from the raw data file shown below?

Figure 20.47 *Raw Data File*

```
1---+----10---+----20---+----30---+
1001 77,163.19 76,804.75 74,384.27
1002 76,612.93 81,456.34 82,063.97
1003 82,185.16 79,742.33
```

Figure 20.48 *Output from PROC PRINT*

SAS Data Set Perm.Topstore

Store	Sales	Month
1001	77163.19	1
1001	76804.75	2
1001	74384.27	3
1002	76612.93	1
1002	81456.34	2
1002	82063.97	3
1003	82185.16	1
1003	79742.33	2

a.
```
data perm.topstores;
    infile sales98 missover;
    input Store Sales : comma. @;
    do while (sales ne .);
        month + 1;
        output;
        input sales : comma. @;
    end;
run;
```

b.
```
data perm.topstores;
    infile sales98 missover;
    input Store Sales : comma. @;
    do while (sales ne .);
        Month=0;
        month + 1;
        output;
        input sales : comma. @;
    end;
run;
```

c.
```
data perm.topstores;
    infile sales98 missover;
    input Store Sales : comma.
    Month @;
    do while (sales ne .);
        month + 1;
        input sales : comma. @;
    end;
    output;
run;
```

d.
```
data perm.topstores;
    infile sales98 missover;
    input Store Sales : comma. @;
    Month=0;
    do while (sales ne .);
        month + 1;
        output;
        input sales : comma. @;
```

```
        end;
    run;
```

10. How many observations are produced by the DATA step that reads this external file?

Figure 20.49 *Raw Data File*

```
1---+----10---+----20---+----30---+----40
01 CHOCOLATE VANILLA RASPBERRY
02 VANILLA PEACH
03 CHOCOLATE
04 RASPBERRY PEACH CHOCOLATE
05 STRAWBERRY VANILLA CHOCOLATE
```

```
data perm.choices;
    infile icecream missover;
    input Day $ Flavor : $10. @;
    do while (flavor ne ' ');
        output; input flavor : $10. @;
    end;
run;
```

a. 3

b. 5

c. 12

d. 15

Chapter 21
Reading Hierarchical Files

Overview

Introduction

Raw data files can be hierarchical in structure, consisting of a header record and one or more detail records. Typically, each record contains a field that identifies the record type.

Here, the P indicates a header record that contains a patient's ID number. The C indicates a detail record that contains the date of the patient's appointment and the charges that the patient has incurred.

Raw Data File

```
1---+----10---+----20
P   1095
C   01-08-89   $45.0
C   01-17-89   $37.5
P   1096
C   01-09-89   $156.5
P   1097
C   01-02-89   $109.0
P   1099
C   01-03-89   $45.0
C   01-05-89   $45.0
P   1201
C   01-05-89   $37.0
C   01-10-89   $45.0
```

You can build a SAS data set from a hierarchical file by creating one observation per detail record, retaining the patient ID from the header record.

SAS Data Set

Obs	ID	Date	Amount
1	1095	01/08/89	$45.00
2	1095	01/17/89	$37.50
3	1096	01/09/89	$156.50
4	1097	01/02/89	$109.00
5	1099	01/03/89	$45.00
6	1099	01/05/89	$45.00
7	1201	01/05/89	$37.00
8	1201	01/10/89	$45.00

You can also build a SAS data set from a hierarchical file by creating one observation per header record and combining the information from detail records into summary variables.

SAS Data Set

Obs	ID	Total
1	1095	$82.50
2	1096	$156.50
3	1097	$109.00
4	1099	$90.00
5	1201	$82.00

In this chapter, you learn how to read from a hierarchical file and create a SAS data set that contains either one observation per detail record or one observation per header record.

Objectives

In this chapter, you learn to

- retain the value of a variable

- conditionally execute a SAS statement

- determine when the last observation is being processed

- conditionally execute multiple SAS statements.

You can also review how to

- use a line-hold specifier to hold the current record

- explicitly write an observation to a data set.

Creating One Observation per Detail Record

Overview

In order to create one observation per detail record, it is necessary to distinguish between header and detail records. Having a field that identifies the type of the record makes this task easier.

In the partial raw data file Census, shown below, H indicates a header record that contains a street address, and P indicates a detail record that contains information about a person who lives at that address.

```
          Raw Data File
1---+----10---+----
H 321 S. MAIN ST
P MARY E      21 F
P WILLIAM M 23 M
P SUSAN K     3 F
H 324 S. MAIN ST
P THOMAS H   79 M
P WALTER S   46 M
P ALICE A    42 F
P MARYANN A 20 F
P JOHN S     16 M
H 325A S. MAIN ST
```

Let's see how you can create a data set that contains one observation for each person who lives at a specific address.

SAS Data Set

Obs	Address	Name	Age	Gender
1	321 S. MAIN ST	MARY E	21	F
2	321 S. MAIN ST	WILLIAM M	23	M
3	321 S. MAIN ST	SUSAN K	3	F
4	324 S. MAIN ST	THOMAS H	79	M
5	324 S. MAIN ST	WALTER S	46	M
6	324 S. MAIN ST	ALICE A	42	F
7	324 S. MAIN ST	MARYANN A	20	F
8	324 S. MAIN ST	JOHN S	16	M
9	325A S. MAIN ST	JAMES L	34	M
10	325A S. MAIN ST	LIZA A	31	F
11	325B S. MAIN ST	MARGO K	27	F

Retaining the Values of Variables

As you write the DATA step to read this file, remember that you want to keep the header record as a part of each observation until the next header record is encountered. To do this, you need to use a RETAIN statement to retain the values for Address across iterations of the DATA step.

```
data perm.people;
   infile census;
   retain Address;
```

```
1---+----10---+----
H 321 S. MAIN ST
P MARY E     21 F
P WILLIAM M 23 M
P SUSAN K    3 F
```

Retain the variable Address so that its value will be available to subsequent iterations of the DATA step.

Next, read the first field in each record, which identifies the record's type. You also need to use the @ line-hold specifier to hold the current record so that the other values in the record can be read later.

```
data perm.people;
   infile census;
   retain Address;
   input type $1. @;
```

```
1---+----10---+----
H 321 S. MAIN ST
P MARY E     21 F
P WILLIAM M 23 M
P SUSAN K    3 F
```

Conditionally Executing SAS Statements

You can use the value of type to identify each record. If type is **H**, you need to execute an INPUT statement to read the values for Address. However, if type is **P**, then execute an INPUT statement to read the values for Name, Age, and Gender.

You can tell SAS to perform a given task based on a specific condition by using an IF-THEN statement.

```
data perm.people;
   infile census;
   retain Address;
   input type $1. @;
   if type='H' then
      input @3 address $15.;
```

```
v---+----10---+----
H 321 S. MAIN ST
P MARY E      21 F
P WILLIAM M 23 M
P SUSAN K     3 F
```

Expressions in conditional statements usually involve some kind of comparison. In the example shown above, a variable is compared to a constant. When the condition is met, the expression is evaluated as true, and the statement that follows the keyword THEN is executed.

The expression defines a condition so that when the value of type is **H**, the INPUT statement reads the values for Address. However, when the value of type is not **H**, the expression is evaluated as false, and the INPUT statement is not executed. Notice that the value is enclosed in quotation marks because it is a character value.

⚠ When you compare values, be sure to express the values exactly as they are coded in the data. For example, the expression below would evaluate to false because the values in the data are stored in uppercase letters.

```
if type='h' then ... ;
```

Reading a Detail Record

Now think about what needs to happen when a detail record is read. Remember, you want to write an observation to the data set only when the value of type is **P**.

```
v---+----10---+----
H 321 S. MAIN ST
P MARY E      21 F
P WILLIAM M 23 M
P SUSAN K     3 F
```

You can use a subsetting IF statement to check for the condition that type is **P**. The remaining DATA step statements execute only when the condition is true. If type is not **P**, then the values for Name, Age, and Gender are not read, the values in the program data vector are not written to the data set as an observation, and control returns to the top of the DATA step. However, Address is retained.

If type is **P**, Name, Age, and Gender are read, and an observation is written to the data set. Remember that you want to create an observation for detail records only.

```
data perm.people;
   infile census;
   retain Address;
   input type $1. @;
   if type='H' then input @3 address $15.;
   if type='P';
   input @3 Name $10. @13 Age 3. @16 Gender $1.;
run;
```

Dropping Variables

Because type is useful only for identifying a record's type, drop the variable from the data set. The DROP= option in the DATA statement shown here prevents the variable type from being written to the data set.

```
data perm.people (drop=type);
   infile census;
   retain Address;
   input type $1. @;
   if type='H' then input @3 address $15.;
   if type='P';
   input @3 Name $10. @13 Age 3. @16 Gender $1.;
run;
```

Processing a DATA Step That Creates One Observation per Detail Record

Let's see how this DATA step is processed.

```
data perm.people (drop=type);
   infile census;
   retain Address;
   input type $1. @;
   if type='H' then input @3 address $15.;
   if type='P';
   input @3 Name $10. @13 Age 3. @16 Gender $1.;
run;
```

At compile time, the variable type is flagged so that its values are not written to the data set. Address is flagged so that its value is retained across iterations of the DATA step.

```
>----+----10---+----
H 321 S. MAIN ST
P MARY E      21 F
P WILLIAM M 23 M
P SUSAN K      3 F
H 324 S. MAIN ST
```

		Retain	Drop			
N	Address		type	Name	Age	Gender
•					•	

As the DATA step begins to execute, the INPUT statement reads the value for type and holds the first record.

```
>----+----10---+----
H 321 S. MAIN ST
P MARY E      21 F
P WILLIAM M 23 M
P SUSAN K      3 F
H 324 S. MAIN ST
```

		Retain	Drop			
N	Address		type	Name	Age	Gender
1			H		•	

The condition type='H' is checked and found to be true, so the INPUT statement reads the value for Address in the first record.

```
>----+----10---+----
H 321 S. MAIN ST
P MARY E      21 F
P WILLIAM M 23 M
P SUSAN K      3 F
H 324 S. MAIN ST
```

		Retain	Drop			
N	Address		type	Name	Age	Gender
1	321 S. MAIN ST		H		•	

Next, the subsetting IF statement checks for the condition type='P'. Because the condition is not true, the remaining statements are not executed and control returns to the top of the DATA step. The PDV is initialized but Address is retained.

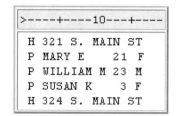

As the second iteration begins, the input pointer moves to the next record and a new value for type is read. The condition expressed in the IF-THEN statement is not true, so the statement following the THEN keyword is not executed.

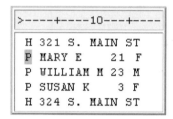

Now the subsetting IF statement checks for the condition type='P'. In this iteration, the condition is true, so the final INPUT statement reads the values for Name, Age, and Gender.

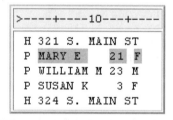

Then the values in the program data vector are written as the first observation, and control returns to the top of the DATA step. Notice that the values for type are not included.

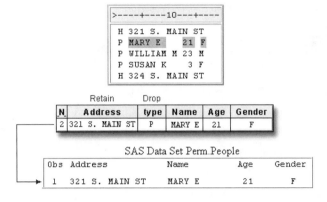

As execution continues, observations are produced from the third and fourth records. However, notice that the fifth record is a header record. During the fifth iteration, the condition type='H' is true, so a new Address is read into the program data vector, overlaying the previous value.

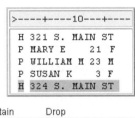

Displaying Your Results

When the execution phase is complete, you can display the data set by using the PRINT procedure. The first 10 observations are displayed.

Figure 21.1 *Output from the PRINT Procedure*

Obs	Address	Name	Age	Gender
1	321 S. MAIN ST	MARY E	21	F
2	321 S. MAIN ST	WILLIAM M	23	M
3	321 S. MAIN ST	SUSAN K	3	F
4	324 S. MAIN ST	THOMAS H	79	M
5	324 S. MAIN ST	WALTER S	46	M
6	324 S. MAIN ST	ALICE A	42	F
7	324 S. MAIN ST	MARYANN A	20	F
8	324 S. MAIN ST	JOHN S	16	M
9	325A S. MAIN ST	JAMES L	34	M
10	325A S. MAIN ST	LIZA A	31	F

Creating One Observation per Header Record

Overview

In the previous example, you learned how to create one observation per detail record. But suppose you only want to know how many people reside at each address. You can create a data set that reads each detail record, counts the number of people, and stores this number in a summary variable.

In the example below, this summary variable, Total, is computed for each address. As you can see, creating one observation per header record condenses a large amount of information into a concise data set.

```
          Raw Data File
1---+----10---+----20
H 321 S. MAIN ST
P MARY E     21 F
P WILLIAM M 23 M
P SUSAN K     3 F
H 324 S. MAIN ST
P THOMAS H   79 M
P WALTER S   46 M
P ALICE A    42 F
P MARYANN A 20 F
P JOHN S     16 M
H 325A S. MAIN ST
P JAMES L 34 M
P LIZA A 31 F
H 325B S. MAIN ST
P MARGO K 27 F
P WILLIAM R 27 M
P ROBERT W 1 M
```

SAS Data Set

Address	Total
321 S. MAIN ST	3
324 S. MAIN ST	5
325A S. MAIN ST	2
325B S. MAIN ST	3

As you write the DATA step to read this file, you need to think about performing several tasks. First, the value of Address must be retained as detail records are read and summarized.

```
data perm.people;
   infile census;
   retain Address;
```

```
1---+----10---+----
H 321 S. MAIN ST
P MARY E     21 F
P WILLIAM M 23 M
P SUSAN K    3 F
```

Next, the value of type must be read in order to determine whether the current record is a header record or a detail record. Add an @ to hold the record so that another INPUT statement can read the remaining values.

```
data perm.people;
   infile census;
   retain Address;
   input type $1. @;
```

```
V---+----10---+----
H 321 S. MAIN ST
P MARY E      21 F
P WILLIAM M 23 M
P SUSAN K     3 F
```

When the value of type indicates a header record, several statements need to be executed. When the value of type indicates a detail record, you need to define an alternative set of actions. Let's look at executing different sets of statements for each value of type.

DO Group Actions for Header Records

To execute multiple SAS statements based on the value of a variable, you can use a simple DO group with an IF-THEN statement. When the condition type='H' is true, several statements need to be executed.

```
data perm.residnts;
   infile census;
   retain Address;
   input type $1. @;
   if type='H' then do;
```

- First, you need to determine whether this is the first header record in the external file. You do not want the first header record to be written as an observation until the related detail records have been read and summarized.

  _N_ is an automatic variable whose value is the number of times the DATA step has begun to execute. The expression _n_ > 1 defines a condition where the DATA step has executed more than once. Use this expression in conjunction with the previous IF-THEN statement to check for these two conditions:

  1. The current record is a header record.

  2. The DATA step has executed more than once.

  ```
  data perm.residnts;
     infile census;
     retain Address;
     input type $1. @;
     if type='H' then do;
        if _n_ > 1
  ```

- When the conditions type='H' and _n_ > 1 are true, an OUTPUT statement is executed. Thus, each header record except for the first one causes an observation to be written to the data set.

  ```
  data perm.residnts;
     infile census;
     retain Address;
     input type $1. @;
     if type='H' then do;
        if _n_ > 1 then output;
  ```

- An assignment statement creates the summary variable Total and sets its value to **0**.

  ```
  data perm.residnts;
     infile census;
     retain Address;
     input type $1. @;
     if type='H' then do;
        if _n_ > 1 then output;
        Total=0;
  ```

- An INPUT statement reads the values for Address.

  ```
  data perm.residnts;
     infile census;
     retain Address;
     input type $1. @;
     if type='H' then do;
        if _n_ > 1 then output;
        Total=0;
        input address $ 3-17;
  ```

- An END statement closes the DO group.

  ```
  data perm.residnts;
     infile census;
     retain Address;
     input type $1. @;
     if type='H' then do;
        if _n_ > 1 then output;
        Total=0;
  ```

```
      input address $ 3-17;
   end;
```

Reading Detail Records

When the value of type is not **H**, you need to define an alternative action. You can do this by adding an ELSE statement after the DO group.

Remember that the IF-THEN statement executes a SAS statement when the condition specified in the IF clause is true. By adding an ELSE statement after the IF-THEN statement, you define an alternative action to be performed when the IF condition is false.

```
data perm.residnts;
   infile census;
   retain Address;
   input type $1. @;
   if type='H' then do;
      if _n_ > 1 then output;
      Total=0;
      input address $ 3-17;
   end;
   else
```

The only other type of record is a detail record, represented by a P. You want to count each person who is represented by a detail record and store the accumulated value in the summary variable Total. You do not need to read the values for Name, Age, and Gender.

```
data perm.residnts;
   infile census;
   retain Address;
   input type $1. @;
   if type='H' then do;
      if _n_ > 1 then output;
      Total=0;
      input address $ 3-17;
   end;
   else if type='P' then
```

```
1---+----10---+----20
H 321 S. MAIN ST
P MARY E     21 F
P WILLIAM M 23 M
P SUSAN K    3 F
H 324 S. MAIN ST
P THOMAS H  79 M
P WALTER S  46 M
P ALICE A   42 F
P MARYANN A 20 F
P JOHN S    16 M
```

At this point, the value of Total to **0** has already been initialized each time a header record is read. Now, as each detail record is read, you can increment the value of Total by using a sum statement. In this example you're counting the number of detail records for each header record, so you increment the value of Total by 1 when the value of type is **P**.

```
data perm.residnts;
   infile census;
   retain Address;
   input type $1. @;
   if type='H' then do;
      if _n_ > 1 then output;
      Total=0;
      input address $ 3-17;
   end;
   else if type='P' then total+1;
```

⚠ Remember that a sum statement enables you to add any valid SAS expression to an accumulator variable.

```
else if type='B' then total+cost;
```

⚠ Remember also that the value generated by a sum statement is automatically retained throughout the DATA step. That is why it is important to set the value of Total to **0** each time a header record is read.

Determining the End of the External File

Your program writes an observation to the data set only when another header record is read and the DATA step has executed more than once. But after the last detail record is read, there are no more header records to cause the last observation to be written to the data set.

```
data perm.residnts;
   infile census;
   retain Address;
   input type $1. @;
   if type='H' then do;
      if _n_ > 1 then
         output;
      Total=0;
      input address $ 3-17;
   end;
   else if type='P'
         then total+1;
```

```
1---+----10---+----20
H 321 S. MAIN ST
P MARY E     21 F
P WILLIAM M 23 M
P SUSAN K    3 F
H 324 S. MAIN ST
P THOMAS H  79 M
P WALTER S  46 M
P ALICE A   42 F
P MARYANN A 20 F
P JOHN S    16 M
H 325A S. MAIN ST
P JAMES L 34 M
P LIZA A 31 F
H 325B S. MAIN ST
P MARGO K 27 F
P WILLIAM R 27 M
P ROBERT W 1 M
```

You need to determine when the last record in the file is read so that you can then execute another explicit OUTPUT statement. You can determine when the current record is the last record in an external file by specifying the END= option in the INFILE statement. (You learned how to use the END= option with a SET statement in "Reading SAS Data Sets" on page 334.)

General form, INFILE statement with the END= option:

INFILE *file-specification* **END=**variable**;**

where *variable* is a temporary numeric variable whose value is **0** until the last line is read and **1** after the last line is read.

Like automatic variables, the END= variable is not written to the data set.

In the following example, the END= variable is defined in the INFILE statement as last. When last has a value other than **0**, the OUTPUT statement writes the final observation to the data set.

```
data perm.residnts;
   infile census end=last;
   retain Address;
   input type $1. @;
   if type='H' then do;
    if _n_ > 1 then output;
       Total=0;
      input address $ 3-17;
   end;
   else if type='P' then total+1;
   if last then output;
```

A DROP= option in the DATA statement drops the variable type from the data set, and a RUN statement completes the DATA step.

```
data perm.residnts (drop=type);
   infile census end=last;
   retain Address;
   input type $1. @;
```

```
      if type='H' then do;
         if _n_ > 1 then output;
            Total=0;
         input address $ 3-17;
      end;
      else if type='P' then total+1;
      if last then output;
run;
```

Processing a DATA Step That Creates One Observation per Header Record

During the compile phase, the variable type is flagged so that later it can be dropped. The value for Address and Total (SUM statement) are retained.

```
data perm.people (drop=type);
   infile census end=last;
   retain Address;
   input type $1. @;
   if type='H'then do;
      if _n_ > 1 then output;
      Total=0;
      input Address $3-17;
   end;
   else if type='P' then total+1;
   if last then output;
run;
```

```
>----+----10---+----

H 321 S. MAIN ST
P MARY E      21 F
P WILLIAM M 23 M
P SUSAN K     3 F
H 324 S. MAIN ST
```

		Retain	Drop	Retain
		Retain		Retain
N	**last**	**Address**	**type**	**Total**
•	•			•

As the execution begins, _N_ is 1 and last is 0. Total is 0 because of the sum statement.

```
data perm.people (drop=type);
   infile census end=last;
   retain Address;
   input type $1. @;
   if type='H'then do;
      if _n_ > 1 then output;
      Total=0;
      input Address $3-17;
   end;
   else if type='P' then total+1;
   if last then output;
run;
```

```
>V---+----10---+----

H 321 S. MAIN ST
P MARY E      21 F
P WILLIAM M 23 M
P SUSAN K     3 F
H 324 S. MAIN ST
```

		Retain	Drop	Retain
N	**last**	**Address**	**type**	**Total**
1	0			0

Now the value for type is read, the condition type='H' is true, and therefore the statements in the DO group execute.

```
data perm.people (drop=type);
   infile census end=last;
   retain Address;
   input type $1. @;
   if type='H'then do;
      if _n_ > 1 then output;
      Total=0;
      input Address $3-17;
   end;
   else if type='P' then total+1;
   if last then output;
run;
```

```
>-V--+----10---+----

H 321 S. MAIN ST
P MARY E      21 F
P WILLIAM M 23 M
P SUSAN K     3 F
H 324 S. MAIN ST
```

		Retain	Drop	Retain
N	last	Address	type	Total
1	0		H	0

The condition N>1 is not true, so the OUTPUT statement is not executed. However, Total is assigned the value of 0 and the value for Address is read.

```
data perm.people (drop=type);
   infile census end=last;
   retain Address;
   input type $1. @;
   if type='H'then do;
      if _n_ > 1 then output;
      Total=0;
      input Address $3-17;
   end;
   else if type='P' then total+1;
   if last then output;
run;
```

```
>----+----10---+--V-

H 321 S. MAIN ST
P MARY E      21 F
P WILLIAM M 23 M
P SUSAN K     3 F
H 324 S. MAIN ST
```

		Retain	Drop	Retain
N	last	Address	type	Total
1	0	321 S. MAIN ST	H	0

The END statement closes the DO group. The alternative condition expressed in the ELSE statement is not checked because the first condition, type='H', was true.

```
data perm.people (drop=type);
   infile census end=last;
   retain Address;
   input type $1. @;
   if type='H'then do;
      if _n_ > 1 then output;
      Total=0;
      input Address $3-17;
   end;
   else if type='P' then total+1;
   if last then output;
run;
```

```
>----+----10---+----

H 321 S. MAIN ST
P MARY E      21 F
P WILLIAM M 23 M
P SUSAN K     3 F
H 324 S. MAIN ST
```

		Retain	Drop	Retain
N	last	Address	type	Total
1	0	321 S. MAIN ST	H	0

The value of last is still **0**, so the OUTPUT statement is not executed. Control returns to the top of the DATA step.

```
data perm.people (drop=type);
   infile census end=last;
   retain Address;
   input type $1. @;
   if type='H'then do;
      if _n_ > 1 then output;
      Total=0;
      input Address $3-17;
   end;
   else if type='P' then total+1;
   if last then output;
run;
```

```
> ---+----10---+-V--

H 321 S. MAIN ST
P MARY E    21 F
P WILLIAM M 23 M
P SUSAN K    3 F
H 324 S. MAIN ST
```

		Retain	Drop	Retain
N	last	Address	type	Total
2	0	321 S. MAIN ST		0

During the second iteration, the value of type is `'P'` and Total is incremented by 1. Again, the value of last is 0, so control returns to the top of the DATA step.

```
data perm.people (drop=type);
   infile census end=last;
   retain Address;
   input type $1. @;
   if type='H'then do;
      if _n_ > 1 then output;
      Total=0;
      input Address $3-17;
   end;
   else if type='P' then total+1;
   if last then output;
run;
```

```
>-V--+----10---+----

H 321 S. MAIN ST
P MARY E    21 F
P WILLIAM M 23 M
P SUSAN K    3 F
H 324 S. MAIN ST
```

		Retain	Drop	Retain
N	last	Address	type	Total
2	0	321 S. MAIN ST	P	1

During the fifth iteration, the value of type is `'H'` and _N_ is greater than 1, so the values for Address and Total are written to the data set as the first observation.

```
data perm.people (drop=type);
   infile census end=last;
   retain Address;
   input type $1. @;
   if type='H'then do;
      if _n_ > 1 then output;
      Total=0;
      input Address $3-17;
   end;
   else if type='P' then total+1;
   if last then output;
run;
```

```
>-V--+----10---+----

H 321 S. MAIN ST
P MARY E    21 F
P WILLIAM M 23 M
P SUSAN K    3 F
H 324 S. MAIN ST
```

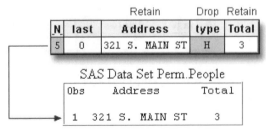

		Retain	Drop	Retain
N	last	Address	type	Total
5	0	321 S. MAIN ST	H	3

SAS Data Set Perm.People

Obs	Address	Total
1	321 S. MAIN ST	3

As the last record in the file is read, the variable last is set to **1**. Now that the condition for last is true, the values in the program data vector are written to the data set as the final observation.

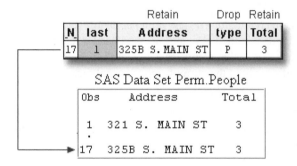

```
data perm.people (drop=type);
   infile census end=last;
   retain Address;
   input type $1. @;
   if type='H'then do;
      if _n_ > 1 then output;
      Total=0;
      input Address $3-17;
   end;
   else if type='P' then total+1;
   if last then output;
run;
```

```
                          >-V--+----10---+----

                          P  LIZA A     31 F
                          H  325B S.  MAIN ST
                          P  MARGO K    27 F
                          P  WILLIAM R 27 F
                          P  ROBERT W   1 M
```

	Retain		Drop	Retain
N	**last**	**Address**	**type**	**Total**
17	1	325B S.MAIN ST	P	3

SAS Data Set Perm.People

```
Obs      Address        Total

1    321 S. MAIN ST      3
.
17   325B S. MAIN ST     3
```

Chapter Summary

Text Summary

Hierarchical Raw Data Files

Raw data files can be hierarchical in structure, consisting of a header record and one or more detail records. You can build a SAS data set from a hierarchical file by creating one observation

- per detail record, storing header data with each observation

- per header record, combining the information from detail records into summary variables.

Creating One Observation per Detail Record

To create one observation per detail record, it is necessary to distinguish between header and detail records. Having a field that identifies the type of the record makes this task easier.

As you write the DATA step, use a RETAIN statement to retain header data in the PDV until the next header record is encountered.

Next, you need to read the field in each record that identifies the record's type. Remember to use the @ line-hold specifier to hold the current value of each record type so that the other values in the record can be read by subsequent INPUT statements.

Use an IF-THEN statement to check for the condition that the record is a header record. If the record is a header record, you need to execute an INPUT statement to read values for that record.

You can use a subsetting IF statement to check for the condition that the record is a detail record. If the record is a detail record, use another INPUT statement to read values in that record.

Use the DROP= option to exclude the variable that identifies each record's type from the output data set.

Creating One Observation per Header Record

Creating one observation per header record condenses a large amount of information into a concise data set. As you write the DATA step, you need to think about performing several tasks.

As with creating one observation per detail record, use a RETAIN statement to retain header data in the PDV until the next header record is encountered. Then read the field in each record that identifies the record's type. Remember to use the @ line-hold specifier to hold the current record so that the other values in the record can be read by subsequent INPUT statements.

When the record is a header record, multiple statements need to be executed. You can do this by adding a simple DO group to an IF-THEN statement. Within the DO group, you need to

1. determine whether this is the first header record in the external file by using the automatic variable _N_

2. use an OUTPUT statement to write each header record except for the first one to the data set

3. use an assignment statement to create a summary variable, and set its value to *0*

4. add an INPUT statement to read values in the header record

5. close the DO group with an END statement.

When the record is a detail record, you need to define an alternative set of actions. You can do this by adding an ELSE statement after the DO group. As each detail record is read, increment the value of the summary variable by using a sum statement.

After the last detail record is read, there are no more header records to cause the last observation to be written to the data set. Determine when the current record is the last record in an external file by specifying the END= option in the INFILE statement. Again, you can use the DROP= option to exclude the variable that identifies each record's type from the output data set.

Syntax

Syntax to Create One Observation for Each Detail Record
LIBNAME libref *'SAS-data-library'*;

FILENAME *fileref 'filename'*;

DATA *SAS-data-set* (**DROP=**variable);

> **INFILE** *file-specification*;
> **RETAIN** *variable*;
> **INPUT** *variable*;
> **IF** *variable='condition'* **THEN** *SAS statement*;
> **IF** *variable='condition'*;
>   *SAS-statement*;
> **RUN**;

Syntax to Create One Observation for Each Header Record

LIBNAME *libref 'SAS-data-library'*;
FILENAME fileref *'filename'*;
DATA *SAS-data-set* **(DROP=** *variable***)**;
> **INFILE** *file-specification* **END=***variable*;
> **RETAIN** *variable*;
> **INPUT** *variable*;
> **IF** *variable='condition'* **THEN DO**;
>   **IF** _N_ > 1 **THEN OUTPUT**;
>   *summary-variable*=0;
>   **INPUT** *variable*;
> **END**;
> ELSE **IF** *variable='condition'* **THEN**
>   *summary-variable+expression*;
> **IF** *variable* **THEN OUTPUT**;
RUN;

Sample Programs

Program to Create One Observation for Each Detail Record

```
libname perm 'c:\records\census2k';
filename census 'c:\records\census2k\survey.dat';
data perm.people(drop=type);
   infile census;
   retain Address;
   input type $1. @;
   if type='H' then input @3 address $15.;
   if type='P';
   input @3 Name $10. @13 Age 3. @16 Gender $1.;
run;
```

Program to Create One Observation for Each Header Record

```
libname perm 'c:\records\census2k';
filename census 'c:\records\census2k\survey.dat';
data perm.residnts (drop=type);
   infile census end=last;
   retain Address;
   input type $1. @;
   if type='H' then do;
      if _n_ > 1 then output;
```

```
          Total=0;
          input address $ 3-17;
       end;
       else if type='P' then total+1;
       if last then output;
   run;
```

Points to Remember

- As with automatic variables, the END= variable is not written to the data set.

- Values are automatically retained when using a sum statement. Therefore, it may be necessary to set the value of the counter variable back to 0 when a new header is encountered.

Chapter Quiz

Select the best answer for each question. After completing the quiz, check your answers using the answer key in the appendix.

1. When you write a DATA step to create one observation per detail record you need to

   a. distinguish between header and detail records.

   b. keep the header record as a part of each observation until the next header record is encountered.

   c. hold the current value of each record type so that the other values in the record can be read.

   d. all of the above

2. Which SAS statement reads the value for code (in the first field), and then holds the value until an INPUT statement reads the remaining value in each observation in the same iteration of the DATA step?

   ```
   1---+----10---+----20---+----30
   01 Office Supplies
   02 Paper Clips
   02 Tape
   01 Art Supplies
   02 Crayons
   02 Finger Paint
   02 Construction Paper
   ```

   a. `input code $2. @;`

   b. `input code $2. @@;`

   c. `retain code;`

   d. none of the above

3. Which SAS statement checks for the condition that Record equals C and executes a single statement to read the values for Amount?

   a. `if record=c then input @3 Amount comma7.;`

b. `if record='C' then input @3 Amount comma7.;`

c. `if record='C' then do input @3 Amount comma7.;`

d. `if record=C then do input @3 Amount comma7.;`

4. After the value for code is read in the sixth iteration, which illustration of the program data vector is correct?

```
1---+----10---+----20---+----30
H Lettuce
P Green Leaf    Quality Growers
P Iceberg       Pleasant Farm
P Romaine       Quality Growers
H Squash
P Yellow        Tasty Acres
P Zucchini      Pleasant Farm
```

```
data perm.produce (drop=code);
   infile orders;
   retain Vegetable;
   input code $1. @;
   if code='H' then input @3 vegetable $6.;
   if code='P';
   input @3 Variety : $10. @15 Supplier : $15.;
run;
proc print data=perm.produce;
run;
```

a.

_N_	Vegetable	code	Variety	Supplier
6		P		

b.

_N_	Vegetable	code	Variety	Supplier
6	Squash	P		

c.

_N_	Vegetable	code	Variety	Supplier
6	Squash	H		

d.

_N_	Vegetable	code	Variety	Supplier
6	Squash	H	Yellow	

5. What happens when the fourth iteration of the DATA step is complete?

```
1---+----10---+----20---+----30
F Apples
V Gala                $2.50
V Golden Delicious $1.99
V Rome                $2.35
F Oranges
V Navel               $2.79
V Temple              $2.99
```

```
data perm.orders (drop=type);
   infile produce;
   retain Fruit;
   input type $1. @;
   if type='F' then input @3 fruit $7.;
   if type='V';
   input @3 Variety : $16. @20 Price comma5.;
run;
```

a. All of the values in the program data vector are written to the data set as the third observation.

b. All of the values in the program data vector are written to the data set as the fourth observation.

c. The values for Fruit, Variety, and Price are written to the data set as the third observation.

d. The values for Fruit, Variety, and Price are written to the data set as the fourth observation.

6. Which SAS statement indicates that several other statements should be executed when Record has a value of **A**?

```
1---+----10---+----20---+----30
A 124153-01
C $153.02
D $20.00
D $20.00
```

a. `if record='A' then do;`

b. `if record=A then do;`

c. `if record='A' then;`

d. `if record=A then;`

7. Which is true for the following statements (X indicates a header record)?

```
if code='X' then do;
   if _n_ > 1 then output;
   Total=0;
   input Name $ 3-20;
end;
```

a. _N_ equals the number of times the DATA step has begun to execute.

b. When code='X' and _n_ > 1 are true, an OUTPUT statement is executed.

c. Each header record causes an observation to be written to the data set.

d. a and b

8. What happens when the condition type='P' is false?

```
if type='P' then input @3 ID $5. @9 Address $20.;
else if type='V' then input @3 Charge 6.;
```

   a. The values for ID and Address are read.

   b. The values for Charge are read.

   c. type is assigned the value of **v**.

   d. The ELSE statement is executed.

9. What happens when last has a value other than zero?

```
data perm.househld (drop=code);
   infile citydata end=last;
   retain Address;
   input type $1. @;
   if code='A' then do;
      if _n_ > 1 then output;
      Total=0;
      input address $ 3-17;
   end;
   else if code='N' then total+1;
   if last then output;
run;
```

   a. last has a value of **1**.

   b. The OUTPUT statement writes the last observation to the data set.

   c. The current value of last is written to the DATA set.

   d. a and b

10. Based on the values in the program data vector, what happens next?

_N_	last	Department	Extension	type	Total	Amount
5	0	Accounting	x3808	D	16.50	.

```
1---+----10---+----20---+----30
D Accounting   x3808
S Paper Clips  $2.50
S Paper        $4.95
S Binders      $9.05
D Personnel    x3810
S Markers      $8.98
S Pencils      $3.35
```

```
data work.supplies (drop=type amount);
   infile orders end=last;
   retain Department Extension;
   input type $1. @;
   if type='D' then do;
      if _n_ > 1 then output;
      Total=0;
      input @3 department $10. @16 extension $5.;
   end;
   else if type='S' then do;
      input @16 Amount comma5.;
```

```
            total+amount;
            if last then output; end;
run;
```

a. All the values in the program data vector are written to the data set as the first observation.

b. The values for Department, Total, and Extension are written to the data set as the first observation.

c. The values for Department, Total, and Extension are written to the data set as the fourth observation.

d. The value of last changes to `1`.

Appendix 1
Quiz Answer Keys

Chapter 1: Base Programming

1. How many observations and variables does the data set below contain?

Name	Sex	Age
Picker	M	32
Fletcher		28
Romano	F	.
Choi	M	42

   a. 3 observations, 4 variables

   b. 3 observations, 3 variables

   c. 4 observations, 3 variables

   d. can't tell because some values are missing

Correct answer: c

Rows in the data set are called observations, and columns are called variables. Missing values don't affect the structure of the data set.

2. How many program steps are executed when the program below is processed?

```
data user.tables;
    infile jobs;
    input date name $ job $;
run;
proc sort data=user.tables;
    by name;
run;
proc print data=user.tables;
run;
```

   a. three

   b. four

   c. five

   d. six

Correct answer: a

When it encounters a DATA, PROC, or RUN statement, SAS stops reading statements and executes the previous step in the program. The program above contains one DATA step and two PROC steps, for a total of three program steps.

3. What type of variable is the variable AcctNum in the data set below?

AcctNum	Gender
3456_1	M
2451_2	
Romano	F
Choi	M

   a. numeric

   b. character

   c. can be either character or numeric

d. can't tell from the data shown

Correct answer: b

It must be a character variable, because the values contain letters and underscores, which are not valid characters for numeric values.

4. What type of variable is the variable Wear in the data set below, assuming that there is a missing value in the data set?

Brand	Wear
Acme	43
Ajax	34
Atlas	.

a. numeric

b. character

c. can be either character or numeric

d. can't tell from the data shown

Correct answer: a

It must be a numeric variable, because the missing value is indicated by a period rather than by a blank.

5. Which of the following variable names is valid?

a. 4BirthDate

b. $Cost

c. _Items_

d. Tax-Rate

Correct answer: c

Variable names follow the same rules as SAS data set names. They can be 1 to 32 characters long, must begin with a letter (A-Z, either uppercase or lowercase) or an underscore, and can continue with any combination of numerals, letters, or underscores.

6. Which of the following files is a permanent SAS file?

a. Sashelp.PrdSale

b. Sasuser.MySales

c. Profits.Quarter1

d. all of the above

Correct answer: d

To store a file permanently in a SAS data library, you assign it a libref other than the default Work. For example, by assigning the libref Profits to a SAS data library, you specify that files within the library are to be stored until you delete them. Therefore, SAS files in the Sashelp and Sasuser and Profits libraries are permanent files.

7.   In a DATA step, how can you reference a temporary SAS data set named Forecast?

   a.   Forecast

   b.   Work.Forecast

   c.   Sales.Forecast (after assigning the libref Sales)

   d.   only a and b above

Correct answer: d

To reference a temporary SAS file in a DATA step or PROC step, you can specify the one-level name of the file (for example, Forecast) or the two-level name using the libref Work (for example, Work.Forecast).

8.   What is the default length for the numeric variable Balance?

Name	Balance
Adams	105.73
Geller	107.89
Martinez	97.45
Noble	182.50

   a.   5

   b.   6

   c.   7

   d.   8

Correct answer: d

The numeric variable Balance has a default length of 8. Numeric values (no matter how many digits they contain) are stored in 8 bytes of storage unless you specify a different length.

9.   How many statements does the following SAS program contain?

```
proc print data=new.prodsale
              label double;
   var state day price1 price2; where state='NC';
   label state='Name of State'; run;
```

   a.   three

   b.   four

   c.   five

   d.   six

Correct answer: c

The five statements are: 1) the PROC PRINT statement (two lines long); 2) the VAR statement; 3) the WHERE statement (on the same line as the VAR statement); 4) the LABEL statement; and 5) the RUN statement (on the same line as the LABEL statement).

10.   What is a SAS library?

a. collection of SAS files, such as SAS data sets and catalogs

b. in some operating environments, a physical collection of SAS files

c. a group of SAS files in the same folder or directory.

d. all of the above

Correct answer: d

Every SAS file is stored in a SAS library, which is a collection of SAS files, such as SAS data sets and catalogs. In some operating environments, a SAS library is a physical collection of files. In others, the files are only logically related. In the Windows and UNIX environments, a SAS library is typically a group of SAS files in the same folder or directory.

Chapter 2: Referencing Files and Setting Options

1. If you submit the following program, how does the output look?

```
options pagesize=55 nonumber;
proc tabulate data=clinic.admit;
   class actlevel;
   var age height weight;
   table actlevel,(age height weight)*mean;
run;
options linesize=80;
proc means data=clinic.heart min max maxdec=1;
   var arterial heart cardiac urinary;
   class survive sex;
run;
```

a. The PROC MEANS output has a print line width of 80 characters, but the PROC TABULATE output has no print line width.

b. The PROC TABULATE output has no page numbers, but the PROC MEANS output has page numbers.

c. Each page of output from both PROC steps is 55 lines long and has no page numbers, and the PROC MEANS output has a print line width of 80 characters.

d. The date does not appear on output from either PROC step.

Correct answer: c

When you specify a system option, it remains in effect until you change the option or end your SAS session, so both PROC steps generate output that is printed 55 lines per page with no page numbers. If you don't specify a system option, SAS uses the default value for that system option.

2. How can you create SAS output in HTML format on any SAS platform?

a. by specifying system options

b. by using programming statements

c. by using SAS windows to specify the result format

d. you can't create HTML output on all SAS platforms

Correct answer: b

You can create HTML output using programming statements on any SAS platform. In addition, on all except mainframe platforms, you can use SAS windows to specify HTML as a result format.

3. In order for the date values `05May1955` and `04Mar2046` to be read correctly, what value must the YEARCUTOFF= option have?

   a. a value between 1947 and 1954, inclusive

   b. 1955 or higher

   c. 1946 or higher

   d. any value

Correct answer: d

As long as you specify an informat with the correct field width for reading the entire date value, the YEARCUTOFF= option doesn't affect date values that have four-digit years.

4. When you specify an engine for a library, you always specify

   a. the file format for files that are stored in the library.

   b. the version of SAS that you are using.

   c. access to other software vendors' files.

   d. instructions for creating temporary SAS files.

Correct answer: a

A SAS engine is a set of internal instructions that SAS uses for writing to and reading from files in a SAS library. Each engine specifies the file format for files that are stored in the library, which in turn enables SAS to access files with a particular format. Some engines access SAS files, and other engines support access to other vendors' files.

5. Which statement prints a summary of all the files stored in the library named Area51?

   a. `proc contents data=area51._all_ nods;`

   b. `proc contents data=area51 _all_ nods;`

   c. `proc contents data=area51 _all_ noobs`

   d. `proc contents data=area51 _all_.nods;`

Correct answer: a

To print a summary of library contents with the CONTENTS procedure, use a period to append the _ALL_ option to the libref. Adding the NODS option suppresses detailed information about the files.

6. The following PROC PRINT output was created immediately after PROC TABULATE output. Which system options were specified when the report was created?

```
                                                        1
                        10:03 Friday, March 17, 2000

                                        Act
          Obs   ID  Height Weight  Level     Fee

            1   2458   72      168   HIGH    85.20
            2   2462   66      152   HIGH   124.80
            3   2501   61      123   LOW    149.75
            4   2523   63      137   MOD    149.75
            5   2539   71      158   LOW    124.80
            6   2544   76      193   HIGH   124.80
            7   2552   67      151   MOD    149.75
            8   2555   70      173   MOD    149.75
            9   2563   73      154   LOW    124.80
```

    a. OBS=, DATE, and NONUMBER

    b. NUMBER, PAGENO=1, and DATE

    c. NUMBER and DATE only

    d. none of the above

Correct answer: b

Clearly, the DATE and NUMBER (page number) options are specified. Because the page number on the output is 1, even though PROC TABULATE output was just produced, PAGENO=1 must also have been specified. If you don't specify PAGENO=, all output in the Output window is numbered sequentially throughout your SAS session.

7. Which of the following programs correctly references a SAS data set named SalesAnalysis that is stored in a permanent SAS library?

    a.
```
data saleslibrary.salesanalysis;
    set mydata.quarter1sales;
    if sales>100000;
run;
```

    b.
```
data mysales.totals;
    set sales_99.salesanalysis;
    if totalsales>50000;
run;
```

    c.
```
aproc print data=salesanalysis.quarter1;
    var sales salesrep month;
run;
```

    d.
```
proc freq data=1999data.salesanalysis;
    tables quarter*sales;
run;
```

Correct answer: b

Librefs must be 1 to 8 characters long, must begin with a letter or underscore, and can contain only letters, numerals, or underscores. After you assign a libref, you specify it as the first level in the two-level name for a SAS file.

8. Which time span is used to interpret two-digit year values if the YEARCUTOFF= option is set to 1950?

    a. 1950-2049

    b. 1950-2050

c. 1949-2050

d. 1950-2000

Correct answer: a

The YEARCUTOFF= option specifies which 100-year span is used to interpret two-digit year values. The default value of YEARCUTOFF= is **1920**. However, you can override the default and change the value of YEARCUTOFF= to the first year of another 100-year span. If you specify YEARCUTOFF=**1950**, then the 100-year span will be from **1950** to **2049**.

9. Assuming you are using SAS code and not special SAS windows, which one of the following statements is *false*?

   a. LIBNAME statements can be stored with a SAS program to reference the SAS library automatically when you submit the program.

   b. When you delete a libref, SAS no longer has access to the files in the library. However, the contents of the library still exist on your operating system.

   c. Librefs can last from one SAS session to another.

   d. You can access files that were created with other vendors' software by submitting a LIBNAME statement.

Correct answer: c

The LIBNAME statement is global, which means that librefs remain in effect until you modify them, cancel them, or end your SAS session. Therefore, the LIBNAME statement assigns the libref for the current SAS session only. You must assign a libref before accessing SAS files that are stored in a permanent SAS data library.

10. What does the following statement do?

```
libname osiris spss 'c:\myfiles\sasdata\data.spss';
```

   a. defines a library called Spss using the OSIRIS engine

   b. defines a library called Osiris using the SPSS engine

   c. defines two libraries called Osiris and Spss using the default engine

   d. defines the default library using the OSIRIS and SPSS engines

Correct answer: b

In the LIBNAME statement, you specify the library name before the engine name. Both are followed by the path.

Chapter 3: Editing and Debugging SAS Programs

1. As you write and edit SAS programs it's a good idea to

   a. begin DATA and PROC steps in column one.

   b. indent statements within a step.

   c. begin RUN statements in column one.

   d. do all of the above.

Correct answer: d

Although you can write SAS statements in almost any format, a consistent layout enhances readability and enables you to understand the program's purpose. It's a good idea to begin DATA and PROC steps in column one, to indent statements within a step, to begin RUN statements in column one, and to include a RUN statement after every DATA step or PROC step.

2. Suppose you have submitted a SAS program that contains spelling errors. Which set of steps should you perform, in the order shown, to revise and resubmit the program?

   a. • Correct the errors.

      • Clear the Log window.

      • Resubmit the program.

      • Check the Log window.

   b. • Correct the errors.

      • Resubmit the program.

      • Check the Output window.

      • Check the Log window.

   c. • Correct the errors.

      • Clear the Log window.

      • Resubmit the program.

      • Check the Output window.

   d. • Correct the errors.

      • Clear the Output window.

      • Resubmit the program.

      • Check the Output window.

   Correct answer: a

   To modify programs that contain errors, if you use the Program Editor window, you usually need to recall the submitted statements from the recall buffer to the Program Editor window, where you can correct the problems. After correcting the errors, you can resubmit the revised program. However, before doing so, it's a good idea to clear the messages from the Log window so that you don't confuse the old error messages with the new messages. Remember to check the Log window again to verify that your program ran correctly.

3. What happens if you submit the following program?

```
proc sort data=clinic.stress out=maxrates;
   by maxhr;
run;
proc print data=maxrates label double noobs;
   label rechr='Recovery Heart Rate;
   var resthr maxhr rechr date;
   where toler='I' and resthr>90;
   sum fee;
run;
```

   a. Log messages indicate that the program ran successfully.

b. A "PROC SORT running" message appears at the top of the active window, and a log message may indicate an error in a statement that seems to be valid.

c. A log message indicates that an option is not valid or not recognized.

d. A "PROC PRINT running" message appears at the top of the active window, and a log message may indicate that a quoted string has become too long or that the statement is ambiguous.

Correct answer: d

The missing quotation mark in the LABEL statement causes SAS to misinterpret the statements in the program. When you submit the program, SAS is unable to resolve the PROC step, and a "PROC PRINT running" message appears at the top of the active window.

4. What generally happens when a syntax error is detected?

a. SAS continues processing the step.

b. SAS continues to process the step, and the Log window displays messages about the error.

c. SAS stops processing the step in which the error occurred, and the Log window displays messages about the error.

d. SAS stops processing the step in which the error occurred, and the Output window displays messages about the error.

Correct answer: c

Syntax errors generally cause SAS to stop processing the step in which the error occurred. When a program that contains an error is submitted, messages regarding the problem also appear in the Log window. When a syntax error is detected, the Log window displays the word ERROR, identifies the possible location of the error, and gives an explanation of the error.

5. A syntax error occurs when

a. Some data values are not appropriate for the SAS statements that are specified in a program.

b. the form of the elements in a SAS statement is correct, but the elements are not valid for that usage.

c. program statements do not conform to the rules of the SAS language.

d. none of the above

Correct answer: c

Syntax errors are common types of errors. Some SAS system options and features of the code editing window can help you identify syntax errors. Other types of errors include data errors and logic errors.

6. How can you tell whether you have specified an invalid option in a SAS program?

a. A log message indicates an error in a statement that seems to be valid.

b. A log message indicates that an option is not valid or not recognized.

c. The message "PROC running" or "DATA step running" appears at the top of the active window.

d. You can't tell until you view the output from the program.

Correct answer: b

When you submit a SAS statement that contains an invalid option, a log message notifies you that the option is not valid or not recognized. You should recall the program, remove or replace the invalid option, check your statement syntax as needed, and resubmit the corrected program.

7. Which of the following programs contains a syntax error?

   a.
   ```
   proc sort data=sasuser.mysales;
       by region;
   run;
   ```

   b.
   ```
   dat sasuser.mysales;
       set mydata.sales99;
   run;
   ```

   c.
   ```
   proc print data=sasuser.mysales label;
       label region='Sales Region';
   run;
   ```

   d. none of the above

Correct answer: b

The DATA step contains a misspelled keyword (dat instead of data). However, this is such a common (and easily interpretable) error that SAS produces only a warning message, not an error.

8. What should you do after submitting the following program in the Windows or UNIX operating environment?

   ```
   proc print data=mysales;
       where state='NC;
   run;
   ```

   a. Submit a RUN statement to complete the PROC step.

   b. Recall the program. Then add a quotation mark and resubmit the corrected program.

   c. Cancel the submitted statements. Then recall the program, add a quotation mark, and resubmit the corrected program.

   d. Recall the program. Then replace the invalid option and resubmit the corrected program.

Correct answer: c

This program contains an unbalanced quotation mark. When you have an unbalanced quotation mark, SAS is often unable to detect the end of the statement in which it occurs. Simply adding a quotation mark and resubmitting your program usually does not solve the problem. SAS still considers the quotation marks to be unbalanced. To correct the error, you need to resolve the unbalanced quotation mark before you recall, correct, and resubmit the program.

9. Which of the following commands opens a file in the code editing window?

   a. `file 'd:\programs\sas\newprog.sas'`

   b. `include 'd:\programs\sas\newprog.sas'`

   c. `open 'd:\programs\sas\newprog.sas'`

   d. all of the above

Correct answer: b

One way of opening a file in the code editing window is by using the INCLUDE command. Using the INCLUDE command enables you to open a single program or combine stored programs in a single window. To save a SAS program, you can use the FILE command.

10. Suppose you submit a short, simple DATA step. If the active window displays the message "DATA step running" for a long time, what probably happened?

    a. You misspelled a keyword.

    b. You forgot to end the DATA step with a RUN statement.

    c. You specified an invalid data set option.

    d. Some data values weren't appropriate for the SAS statements that you specified.

Correct answer: b

Without a RUN statement (or a following DATA or PROC step), the DATA step doesn't execute, so it continues to run. Unbalanced quotation marks can also cause the "DATA step running" message if relatively little code follows the unbalanced quotation mark. The other three problems above generate errors in the Log window.

Chapter 4: Creating List Reports

1. Which PROC PRINT step below creates the following output

Date	On	Changed	Flight
04MAR99	232	18	219
05MAR99	160	4	219
06MAR99	163	14	219
07MAR99	241	9	219
08MAR99	183	11	219
09MAR99	211	18	219
10MAR99	167	7	219

    a.
```
proc print data=flights.laguardia noobs;
    var on changed flight;
    where on>=160;
run;
```

    b.
```
proc print data=flights.laguardia;
    var date on changed flight;
    where changed>3;
run;
```

    c.
```
proc print data=flights.laguardia label;
    id date;
    var boarded transferred flight;
    label boarded='On' transferred='Changed';
    where flight='219';
run;
```

d. ```
   proc print flights.laguardia noobs;
      id date;
      var date on changed flight;
      where flight='219';
   run;
   ```

Correct answer: c

The DATA= option specifies the data set that you are listing, and the ID statement replaces the Obs column with the specified variable. The VAR statement specifies variables and controls the order in which they appear, and the WHERE statement selects rows based on a condition. The LABEL option in the PROC PRINT statement causes the labels specified in the LABEL statement to be displayed.

2. Which of the following PROC PRINT steps is correct if labels are not stored with the data set?

   a. ```
      proc print data=allsales.totals label;
         label region8='Region 8 Yearly Totals';
      run;
      ```

   b. ```
      proc print data=allsales.totals;
         label region8='Region 8 Yearly Totals';
      run;
      ```

   c. ```
      proc print data allsales.totals label noobs;
      run;
      ```

   d. ```
      proc print allsales.totals label;
      run;
      ```

Correct answer: a

You use the DATA= option to specify the data set to be printed. The LABEL option specifies that variable labels appear in output instead of variable names.

3. Which of the following statements selects from a data set only those observations for which the value of the variable Style is **RANCH**, **SPLIT**, or **TWOSTORY**?

   a. `where style='RANCH' or 'SPLIT' or 'TWOSTORY';`

   b. `where style in 'RANCH' or 'SPLIT' or 'TWOSTORY';`

   c. `where style in (RANCH, SPLIT, TWOSTORY);`

   d. `where style in ('RANCH','SPLIT','TWOSTORY');`

Correct answer: d

In the WHERE statement, the IN operator enables you to select observations based on several values. You specify values in parentheses and separated by spaces or commas. Character values must be enclosed in quotation marks and must be in the same case as in the data set.

4. If you want to sort your data and create a temporary data set named Calc to store the sorted data, which of the following steps should you submit?

   a. ```
      proc sort data=work.calc out=finance.dividend;
      run;
      ```

   b. ```
      proc sort dividend out=calc;
         by account;
      run;
      ```
      ```

c. ```
proc sort data=finance.dividend out=work.calc;
   by account;
run;
```

d. ```
proc sort from finance.dividend to calc;
   by account;
run;
```

Correct answer: c

In a PROC SORT step, you specify the DATA= option to specify the data set to sort. The OUT= option specifies an output data set. The required BY statement specifies the variable(s) to use in sorting the data.

5. Which options are used to create the following PROC PRINT output?

```
                              13:27 Monday, March 22, 1999

   Patient     Arterial     Heart      Cardiac     Urinary

     203          88          95          66          110

      54          83         183          95            0

     664          72         111         332           12

     210          74          97         369            0

     101          80         130         291            0
```

a. the DATE system option and the LABEL option in PROC PRINT

b. the DATE and NONUMBER system options and the DOUBLE and NOOBS options in PROC PRINT

c. the DATE and NONUMBER system options and the DOUBLE option in PROC PRINT

d. the DATE and NONUMBER system options and the NOOBS option in PROC PRINT

Correct answer: b

The DATE and NONUMBER system options cause the output to appear with the date but without page numbers. In the PROC PRINT step, the DOUBLE option specifies double spacing, and the NOOBS option removes the default Obs column.

6. Which of the following statements can you use in a PROC PRINT step to create this output?

Month	Instructors	AerClass	WalkJogRun	Swim
01	1	37	91	83
02	2	41	102	27
03	1	52	98	19
04	1	61	118	22
05	3	49	88	29
	8	240	497	180

a. ```
var month instructors;
sum instructors aerclass walkjogrun swim;
```
```

b.
```
var month;
   sum instructors aerclass walkjogrun swim;
```

c.
```
var month instructors aerclass;
   sum instructors aerclass walkjogrun swim;
```

d. all of the above

Correct answer: d

You do not need to name the variables in a VAR statement if you specify them in the SUM statement, but you can. If you choose not to name the variables in the VAR statement as well, then the SUM statement determines their order in the output.

7. What happens if you submit the following program?

```
proc sort data=clinic.diabetes;
run;
proc print data=clinic.diabetes;
   var age height weight pulse;
   where sex='F';
run
```

a. The PROC PRINT step runs successfully, printing observations in their sorted order.

b. The PROC SORT step permanently sorts the input data set.

c. The PROC SORT step generates errors and stops processing, but the PROC PRINT step runs successfully, printing observations in their original (unsorted) order.

d. The PROC SORT step runs successfully, but the PROC PRINT step generates errors and stops processing.

Correct answer: c

The BY statement is required in PROC SORT. Without it, the PROC SORT step fails. However, the PROC PRINT step prints the original data set as requested.

8. If you submit the following program, which output does it create?

```
proc sort data=finance.loans out=work.loans;
   by months amount;
run;
proc print data=work.loans noobs;
   var months;
   sum amount payment;
   where months<360;
run;
```

a.

Months	Amount	Payment
12	$3,500	$308.52
24	$8,700	$403.47
36	$10,000	$325.02
48	$5,000	$128.02
	$27,200	$1,165.03

b.

Months	Amount	Payment
12	$3,500	$308.52
24	$8,700	$403.47
36	$10,000	$325.02
48	$5,000	$128.02
	27,200	1,165.03

c.

Months	Amount	Payment
12	$3,500	$308.52
48	$5,000	$128.02
24	$8,700	$403.47
36	$10,000	$325.02
	$27,200	$1,165.03

d.

Months	Amount	Payment
12	$3,500	$308.52
24	$8,700	$403.47
36	$10,000	$325.02
48	$5,000	$128.02
		$1,165.03

Correct answer: a

Column totals appear at the end of the report in the same format as the values of the variables, so b is incorrect. Work.Loans is sorted by Month and Amount, so c is incorrect. The program sums both Amount and Payment, so d is incorrect.

9. Choose the statement below that selects rows in which

   - the amount is less than or equal to $5000

   - the account is 101-1092 or the rate equals 0.095.

   a. ```
   where amount <= 5000 and
         account='101-1092' or rate = 0.095;
   ```

   b. ```
   where (amount le 5000 and account='101-1092')
         or rate = 0.095;
   ```

   c. ```
   where amount <= 5000 and
         (account='101-1092' or rate eq 0.095);
   ```

   d. ```
   where amount <= 5000 or account='101-1092'
         and rate = 0.095;
   ```

Correct answer: c

To ensure that the compound expression is evaluated correctly, you can use parentheses to group

```
account='101-1092' or rate eq 0.095
```

```
OBS  Account    Amount     Rate   Months  Payment

 1   101-1092   $22,000   10.00%      60   $467.43
 2   101-1731  $114,000    9.50%     360   $958.57
 3   101-1289   $10,000   10.50%      36   $325.02
 4   101-3144    $3,500   10.50%      12   $308.52
 5   103-1135    $8,700   10.50%      24   $403.47
 6   103-1994   $18,500   10.00%      60   $393.07
 7   103-2335    $5,000   10.50%      48   $128.02
 8   103-3864   $87,500    9.50%     360   $735.75
 9   103-3891   $30,000    9.75%     360   $257.75
```

For example, from the data set above, a and b above select observations 2 and 8 (those that have a rate of 0.095); c selects no observations; and d selects observations 4 and 7 (those that have an amount less than or equal to 5000).

10. What does PROC PRINT display by default?

   a. PROC PRINT does not create a default report; you must specify the rows and columns to be displayed.

   b. PROC PRINT displays all observations and variables in the data set. If you want an additional column for observation numbers, you can request it.

   c. PROC PRINT displays columns in the following order: a column for observation numbers, all character variables, and all numeric variables.

   d. PROC PRINT displays all observations and variables in the data set, a column for observation numbers on the far left, and variables in the order in which they occur in the data set.

Correct answer: d

You can remove the column for observation numbers. You can also specify the variables you want, and you can select observations according to conditions.

Chapter 5: Creating SAS Data Sets from External Files

1. Which SAS statement associates the fileref Crime with the raw data file `C:\States\Data\Crime.dat`?

   a. `filename crime 'c:\states\data\crime.dat';`

   b. `filename crime c:\states\data\crime.dat;`

   c. `fileref crime 'c:\states\data\crime.dat';`

   d. `filename 'c:\states\data\crime' crime.dat;`

Correct answer: a

You assign a fileref by using a FILENAME statement in the same way that you assign a libref by using a LIBNAME statement.

2. Filerefs remain in effect until . . .

a. you change them.

b. you cancel them.

c. you end your SAS session.

d. all of the above

Correct answer: d

Like LIBNAME statements, FILENAME statements are global; they remain in effect until you change them, cancel them, or end your SAS session.

3. Which statement identifies a raw data file to be read with the fileref Products and specifies that the DATA step read only records 1-15?

a. `infile products obs 15;`

b. `infile products obs=15;`

c. `input products obs=15;`

d. `input products 1-15;`

Correct answer: b

You use an INFILE statement to specify the raw data file to be read. You can specify a fileref or an actual filename (in quotation marks). The OBS= option in the INFILE statement enables you to process only records 1 through *n*.

4. Which of the following programs correctly writes the observations from the data set below to a raw data file?

SAS data set Work.Patients

ID	Sex	Age	Height	Weight	Pulse
2304	F	16	61	102	100
1128	M	43	71	218	76
4425	F	48	66	162	80
1387	F	57	64	142	70
9012	F	39	63	157	68
6312	M	52	72	240	77
5438	F	42	62	168	83
3788	M	38	73	234	71
9125	F	56	64	159	70
3438	M	15	66	140	67

a.
```
data _null_;
    set work.patients;
    infile 'c:\clinic\patients\referals.dat';
    input id 1-4 sex 6 age 8-9 height 11-12
        weight 14-16 pulse 18-20;
run;
```

b.
```
data referals.dat;
    set work.patients;
    input id 1-4 sex 6 age 8-9 height 11-12
        weight 14-16 pulse 18-20;
run;
```

c. ```
data _null_;
    set work.patients;
    file c:\clinic\patients\referals.dat;
    put id 1-4 sex 6 age 8-9 height 11-12
        weight 14-16 pulse 18-20;
run;
```

d. ```
data _null_;
    set work.patients;
    file 'c:\clinic\patients\referals.dat';
    put id 1-4 sex 6 age 8-9 height 11-12
        weight 14-16 pulse 18-20;
run;
```

Correct answer: d

The keyword _NULL_ in the DATA statement enables you to use the power of the DATA step without actually creating a SAS data set. You use FILE and PUT statements to write observations from a SAS data set to a raw data file. The FILE statement specifies the raw data file and the PUT statement describes the lines to write to the raw data file. The filename and location specified in the FILE statement must be enclosed in quotation marks.

5. Which raw data file can be read using column input?

   a.

   ```
   1---+----10---+----20---+
   Henderson CA 26 ADM
   Josephs SC 33 SALES
   Williams MN 40 HRD
   Rogan NY RECRTN
   ```

   b.

   ```
   1---+----10---+----20---+----30
   2803 Deborah Campos  173.97
   2912 Bill    Marin   205.14
   3015 Helen   Stinson 194.08
   3122 Nicole  Terry   187.65
   ```

   c.

   ```
   1---+----10---+----20---+
   Avery John       $601.23
   Davison Sherrill $723.15
   Holbrook Grace   $489.76
   Jansen Mike      $638.42
   ```

   d. all of the above.

Correct answer: b

Column input is appropriate only in some situations. When you use column input, your data must be standard character or numeric values, and they must be in fixed fields. That is, values for a particular variable must be in the same location in all records.

6. Which program creates the PROC PRINT output shown below?

```
1---+----10---+----20---+----30
3427 Chen    Steve Raleigh
1436 Davis   Lee   Atlanta
2812 King    Vicky Memphis
1653 Sanchez Jack  Atlanta
```

Obs	ID	LastName	FirstName	City
1	3427	Chen	Steve	Raleigh
2	1436	Davis	Lee	Atlanta
3	2812	King	Vicky	Memphis
4	1653	Sanchez	Jack	Atlanta

a.
```
data work.salesrep;
    infile empdata;
    input ID $ 1-4 LastName $ 6-12
        FirstName $ 14-18 City $ 20-29;
run;
proc print data=work.salesrep;
run;
```

b.
```
data work.salesrep;
    infile empdata;
    input ID $ 1-4 Name $ 6-12
        FirstName $ 14-18 City $ 20-29;
run;
proc print data=work.salesrep;
run;
```

c.
```
data work.salesrep;
    infile empdata;
    input ID $ 1-4 name1 $ 6-12
        name2 $ 14-18 City $ 20-29;
run;
proc print data=work.salesrep;
run;
```

d. all of the above.

Correct answer: a

The INPUT statement creates a variable using the name that you assign to each field. Therefore, when you write an INPUT statement, you need to specify the variable names exactly as you want them to appear in the SAS data set.

7. Which statement correctly reads the fields in the following order: StockNumber, Price, Item, Finish, Style?

Field Name	Start Column	End Column	Data Type
StockNumber	1	3	character
Finish	5	9	character
Style	11	18	character
Item	20	24	character

Field Name	Start Column	End Column	Data Type
Price	27	32	numeric

```
1---+----10---+----20---+----30---+
310 oak    pedestal table  329.99
311 maple pedestal table  369.99
312 brass floor    lamp    79.99
313 glass table    lamp    59.99
313 oak    rocking  chair  153.99
```

a. input StockNumber $ 1-3 Finish $ 5-9 Style $ 11-18
          Item $ 20-24 Price 27-32;

b. input StockNumber $ 1-3 Price 27-32
          Item $ 20-24 Finish $ 5-9 Style $ 11-18;

c. input $ StockNumber 1-3 Price 27-32 $
          Item   20-24 $ Finish 5-9 $ Style 11-18;

d. input StockNumber $ 1-3 Price $ 27-32
          Item $ 20-24 Finish $ 5-9 Style $ 11-18;

Correct answer: b

You can use column input to read fields in any order. You must specify the variable name, identify character variables with a $, and specify the correct starting and ending column for each field.

8. Which statement correctly re-defines the values of the variable Income as 100 percent higher?

a. income=income*1.00;

b. income=income+(income*2.00);

c. income=income*2;

d. income=*2;

Correct answer: c

To re-define the values of the variable Income in an assignment statement, you specify the variable name on the left side of the equal sign and an appropriate expression including the variable name on the right side of the equal sign.

9. Which program correctly reads instream data?

a. data finance.newloan;
        input datalines;
        if country='JAPAN';
        MonthAvg=amount/12;
     1998 US   CARS   194324.12
     1998 US   TRUCKS 142290.30
     1998 CANADA CARS   10483.44
     1998 CANADA TRUCKS 93543.64
     1998 MEXICO CARS   22500.57
     1998 MEXICO TRUCKS 10098.88
     1998 JAPAN CARS   15066.43
     1998 JAPAN TRUCKS 40700.34

     ;

b.
```
data finance.newloan;
    input Year 1-4 Country $ 6-11
        Vehicle $ 13-18 Amount 20-28;
    if country='JAPAN';
    MonthAvg=amount/12;
    datalines;
run;
```

c.
```
data finance.newloan;
    input Year 1-4 Country 6-11
        Vehicle 13-18 Amount 20-28;
    if country='JAPAN';
    MonthAvg=amount/12;
    datalines;
1998 US      CARS   194324.12
1998 US      TRUCKS 142290.30
1998 CANADA CARS     10483.44
1998 CANADA TRUCKS  93543.64
1998 MEXICO CARS     22500.57
1998 MEXICO TRUCKS  10098.88
1998 JAPAN  CARS     15066.43
1998 JAPAN  TRUCKS  40700.34
;
```

d.
```
data finance.newloan;
    input Year 1-4 Country $ 6-11
        Vehicle $ 13-18 Amount 20-28;
    if country='JAPAN';
    MonthAvg=amount/12;
    datalines;
1998 US      CARS   194324.12
1998 US      TRUCKS 142290.30
1998 CANADA CARS     10483.44
1998 CANADA TRUCKS  93543.64
1998 MEXICO CARS     22500.57
1998 MEXICO TRUCKS  10098.88
1998 JAPAN  CARS     15066.43
1998 JAPAN  TRUCKS  40700.34
;
```

Correct answer: d

To read instream data, you specify a DATALINES statement and data lines, followed by a null statement (single semicolon) to indicate the end of the input data. Program a contains no DATALINES statement, and the INPUT statement doesn't specify the fields to read. Program b contains no data lines, and the INPUT statement in program c doesn't specify the necessary dollar signs for the character variables Country and Vehicle.

10. Which SAS statement subsets the raw data shown below so that only the observations in which Sex (in the second field) has a value of **F** are processed?

```
----+----10---+----20---+--
Alfred   M 14 69.0 112.5
Becka    F 13 65.3 98.0
Gail     F 14 64.3 90.0
Jeffrey  M 13 62.5 84.0
John     M 12 59.0 99.5
Karen    F 12 56.3 77.0
Mary     F 15 66.5 112.0
Philip   M 16 72.0 150.0
Sandy    F 11 51.3 50.5
Tammy    F 14 62.8 102.5
William  M 15 66.5 112.0
```

a. if sex=f;

b. if sex=F;

c. if sex='F';

d. a or b

Correct answer: c

To subset data, you can use a subsetting IF statement in any DATA step to process only those observations that meet a specified condition. Because Sex is a character variable, the value **F** must be enclosed in quotation marks and must be in the same case as in the data set.

Chapter 6: Understanding DATA Step Processing

1. Which of the following is *not* created during the compilation phase?

   a. the data set descriptor

   b. the first observation

   c. the program data vector

   d. the _N_ and _ERROR_ automatic variables

   Correct answer: b

   During the compilation phase, the program data vector is created. The program data vector includes the two automatic variables _N_ and _ERROR_. The descriptor portion of the new SAS data set is created at the end of the compilation phase. The descriptor portion includes the name of the data set, the number of observations and variables, and the names and attributes of the variables. Observations are not written until the execution phase.

2. During the compilation phase, SAS scans each statement in the DATA step, looking for syntax errors. Which of the following is *not* considered a syntax error?

   a. incorrect values and formats

   b. invalid options or variable names

   c. missing or invalid punctuation

   d. missing or misspelled keywords

Correct answer: a

Syntax checking can detect many common errors, but it cannot verify the values of variables or the correctness of formats.

3. Unless otherwise directed, the DATA step executes...

    a. once for each compilation phase.

    b. once for each DATA step statement.

    c. once for each record in the input file.

    d. once for each variable in the input file.

Correct answer: c

The DATA step executes once for each record in the input file, unless otherwise directed.

4. At the beginning of the execution phase, the value of _N_ is 1, the value of _ERROR_ is 0, and the values of the remaining variables are set to:

    a. 0

    b. 1

    c. undefined

    d. missing

Correct answer: d

The remaining variables are initialized to missing. Missing numeric values are represented by periods, and missing character values are represented by blanks.

5. Suppose you run a program that causes three DATA errors. What is the value of the automatic variable _ERROR_ when the observation that contains the third error is processed?

    a. 0

    b. 1

    c. 2

    d. 3

Correct answer: b

The default value of _ERROR_ is 0, which means there is no data error. When an error occurs, whether one error or multiple errors, the value is set to 1.

6. Which of the following actions occurs at the beginning of an iteration of the DATA step?

    a. The automatic variables _N_ and _ERROR_ are incremented by one.

    b. The DATA step stops execution.

    c. The descriptor portion of the data set is written.

    d. The values of variables created in programming statements are re-set to missing in the program data vector.

Correct answer: d

By default, at the end of the DATA step, the values in the program data vector are written to the data set as an observation, control returns to the top of the DATA step, the value of the automatic variable _N_ is incremented by one, and the values of variables created in programming statements are reset to missing. The automatic variable _ERROR_ is reset to 0 if necessary.

7. Based on the DATA step shown below, in what order will the variables be stored in the new data set?

```
data perm.update;
   infile invent;
   input IDnum $ Item $ 1-13 Instock 21-22
        BackOrd 24-25;
   Total=instock+backord;
run;
```

   a. IDnum Item InStock BackOrd Total

   b. Item IDnum InStock BackOrd Total

   c. Total IDnum Item InStock BackOrd

   d. Total Item IDnum InStock BackOrd

   Correct answer: a

   The order in which variables are defined in the DATA step determines the order in which the variables are stored in the data set.

8. If SAS detects syntax errors, then...

   a. data set variables will contain missing values.

   b. the DATA step does not compile.

   c. the DATA step still compiles, but it does not execute.

   d. the DATA step still compiles and executes.

   Correct answer: c

   When SAS cannot detect syntax errors, the DATA step compiles, but it does not execute.

9. What is wrong with this program?

```
data perm.update;
   infile invent
   input Item $ 1-13 IDnum $ 15-19 Instock 21-22
        BackOrd 24-25;
   total=instock+backord;
run;
```

   a. missing semicolon on second line

   b. missing semicolon on third line

   c. incorrect order of variables

   d. incorrect variable type

   Correct answer: a

   A semicolon is missing from the second line. It will cause an error because the INPUT statement will be interpreted as invalid INFILE statement options.

10. Look carefully at this section of a SAS session log. Based on the note, what was the most likely problem with the DATA step?

```
NOTE: Invalid data for IDnum in line 7 15-19.
RULE: ----+----1----+----2----+----3----+----4
7          Bird Feeder LG088 3 20
Item=Bird Feeder IDnum=. InStock=3 BackOrd=20
Total=23 _ERROR_=1 _N_=1
```

a. A keyword was misspelled in the DATA step.

b. A semicolon was missing from the INFILE statement.

c. A variable was misspelled in the INPUT statement.

d. A dollar sign was missing in the INPUT statement.

Correct answer: d

The third line of the log displays the value for IDnum, which is clearly a character value. The fourth line displays the values in the program data vector and shows a period, the symbol for a missing numeric value, for IDnum, the other values are correctly assigned. Thus, it appears that numeric values were expected for IDnum. A dollar sign, to indicate character values, is missing from the INPUT statement.

Chapter 7: Creating and Applying User-Defined Formats

1. If you don't specify the LIBRARY= option on the PROC FORMAT statement, your formats are stored in Work.Formats, and they exist ...

a. only for the current procedure.

b. only for the current DATA step.

c. only for the current SAS session.

d. permanently.

Correct answer: c

If you do not specify the LIBRARY= option, formats are stored in a default format catalog named Work.Formats. As the libref Work implies, any format that is stored in Work.Formats is a temporary format that exists only for the current SAS session.

2. Which of the following statements will store your formats in a permanent catalog?

a. ```
libname library 'c:\sas\formats\lib';
proc format lib=library
   ...;
```

b. ```
libname library 'c:\sas\formats\lib';
format lib=library
   ...;
```

c.
```
library='c:\sas\formats\lib';
proc format library
    ...;
```

d.
```
library='c:\sas\formats\lib';
proc library
    ...;
```

Correct answer: a

To store formats in a permanent catalog, you first write a LIBNAME statement to associate the libref with the SAS data library in which the catalog will be stored. Then add the LIB= (or LIBRARY=) option to the PROC FORMAT statement, specifying the name of the catalog.

3. When creating a format with the VALUE statement, the new format's name

   - cannot end with a number

   - cannot end with a period

   - cannot be the name of a SAS format, and...

   a. cannot be the name of a data set variable.

   b. must be at least two characters long.

   c. must be at least eight characters long.

   d. must begin with a dollar sign ($) if used with a character variable.

Correct answer: d

The name of a format that is created with a VALUE statement must begin with a dollar sign ($) if it applies to a character variable.

4. Which of the following FORMAT procedures is written correctly?

   a.
   ```
   proc format lib=library
        value colorfmt;
              1='Red'
              2='Green'
              3='Blue'
      run;
   ```

   b.
   ```
   proc format lib=library;
        value colorfmt
              1='Red'
              2='Green'
              3='Blue';
      run;
   ```

   c.
   ```
   proc format lib=library;
        value colorfmt;
              1='Red'
              2='Green'
              3='Blue'
      run;
   ```

   d.
   ```
   proc format lib=library;
        value colorfmt
              1='Red';
              2='Green';
   ```

```
                    3='Blue';
        run;
```

Correct answer: b

A semicolon is needed after the PROC FORMAT statement. The VALUE statement begins with the keyword VALUE and ends with a semicolon after all the labels have been defined.

5. Which of these is *false*? Ranges in the VALUE statement can specify...

    a. a single value, such as **24** or **'S'**.

    b. a range of numeric values, such as **0-1500**.

    c. a range of character values, such as **'A'-'M'**.

    d. a list of numeric and character values separated by commas, such as **90,'B', 180,'D',270**.

Correct answer: d

You can list values separated by commas, but the list must contain either all numeric values or all character values. Data set variables are either numeric or character.

6. How many characters can be used in a label?

    a. 40

    b. 96

    c. 200

    d. 256

Correct answer: d

When specifying a label, enclose it in quotation marks and limit the label to 256 characters.

7. Which keyword can be used to label missing numeric values as well as any values that are not specified in a range?

    a. LOW

    b. MISS

    c. MISSING

    d. OTHER

Correct answer: d

MISS and MISSING are invalid keywords, and LOW does not include missing numeric values. The keyword OTHER can be used in the VALUE statement to label missing values as well as any values that are not specifically included in a range.

8. You can place the FORMAT statement in either a DATA step or a PROC step. What happens when you place it in a DATA step?

    a. You temporarily associate the formats with variables.

    b. You permanently associate the formats with variables.

    c. You replace the original data with the format labels.

    d. You make the formats available to other data sets.

Correct answer: b

By placing the FORMAT statement in a DATA step, you permanently associate the defined format with variables.

9. The format JOBFMT was created in a FORMAT procedure. Which FORMAT statement will apply it to the variable JobTitle in the program output?

    a. `format jobtitle jobfmt;`

    b. `format jobtitle jobfmt.;`

    c. `format jobtitle=jobfmt;`

    d. `format jobtitle='jobfmt';`

    Correct answer: b

    To associate a user-defined format with a variable, place a period at the end of the format name when it is used in the FORMAT statement.

10. Which keyword, when added to the PROC FORMAT statement, will display all the formats in your catalog?

    a. CATALOG

    b. LISTFMT

    c. FMTCAT

    d. FMTLIB

    Correct answer: d

    Adding the keyword FMTLIB to the PROC FORMAT statement displays a list of all the formats in your catalog, along with descriptions of their values.

Chapter 8: Producing Descriptive Statistics

1. The default statistics produced by the MEANS procedure are *n*, mean, minimum, maximum, and...

    a. median

    b. range

    c. standard deviation

    d. standard error of the mean.

    Correct answer: c

    By default, the MEANS procedure produces the n, mean, minimum, maximum, and standard deviation.

2. Which statement will limit a PROC MEANS analysis to the variables Boarded, Transfer, and Deplane?

    a. `by boarded transfer deplane;`

    b. `class boarded transfer deplane;`

    c. `output boarded transfer deplane;`

d. `var boarded transfer deplane;`

Correct answer: d

To specify the variables that PROC MEANS analyzes, add a VAR statement and list the variable names.

3. The data set Survey.Health includes the following variables. Which is a poor candidate for PROC MEANS analysis?

   a. `IDnum`

   b. `Age`

   c. `Height`

   d. `Weight`

Correct answer: a

Unlike Age, Height, or Weight, the values of IDnum are unlikely to yield any useful statistics.

4. Which of the following statements is true regarding BY group processing?

   a. BY variables must be either indexed or sorted.

   b. Summary statistics are computed for BY variables.

   c. BY group processing is preferred when you are categorizing data that contains few variables.

   d. BY group processing overwrites your data set with the newly grouped observations.

Correct answer: a

Unlike CLASS processing, BY group processing requires that your data already be indexed or sorted in the order of the BY variables. You might need to run the SORT procedure before using PROC MEANS with a BY group.

5. Which group processing statement produced the PROC MEANS output shown below?

The MEANS Procedure

Survive	Sex	N Obs	Variable	N	Mean	Std Dev	Minimum	Maximum
DIED	1	4	Arterial	4	92.5000000	10.4721854	83.0000000	103.0000000
			Heart	4	111.0000000	53.4103610	54.0000000	183.0000000
			Cardiac	4	176.7500000	75.2257713	95.0000000	260.0000000
			Urinary	4	98.0000000	186.1343601	0	377.0000000
	2	6	Arterial	6	94.1666667	27.3160514	72.0000000	145.0000000
			Heart	6	103.6666667	16.6573307	81.0000000	130.0000000
			Cardiac	6	318.3333333	102.6034437	156.0000000	424.0000000
			Urinary	6	100.3333333	155.7134120	0	405.0000000
SURV	1	5	Arterial	5	77.2000000	12.1942609	61.0000000	88.0000000
			Heart	5	109.0000000	31.9687347	77.0000000	149.0000000
			Cardiac	5	298.0000000	139.8499196	66.0000000	410.0000000
			Urinary	5	100.8000000	60.1722527	44.0000000	200.0000000
	2	5	Arterial	5	78.8000000	6.8337398	72.0000000	87.0000000
			Heart	5	100.0000000	13.3790882	84.0000000	111.0000000
			Cardiac	5	330.2000000	86.9839066	256.0000000	471.0000000
			Urinary	5	111.2000000	152.4096454	12.0000000	377.0000000

a. `class sex survive;`

b. `class survive sex;`

c. `by sex survive;`

d. `by survive sex;`

Correct answer: b

A CLASS statement produces a single large table, whereas BY group processing creates a series of small tables. The order of the variables in the CLASS statement determines their order in the output table.

6. Which program can be used to create the following output?

Sex	N Obs	Variable	N	Mean	Std Dev	Minimum	Maximum
F	11	Age	11	48.9090909	13.3075508	16.0000000	63.0000000
		Height	11	63.9090909	2.1191765	61.0000000	68.0000000
		Weight	11	150.4545455	18.4464828	102.0000000	168.0000000
M	9	Age	9	44.0000000	12.3895117	15.0000000	54.0000000
		Height	9	70.6666667	2.6457513	66.0000000	75.0000000
		Weight	9	204.2222222	30.2893454	140.0000000	240.0000000

a.
```
proc means data=clinic.diabetes;
    var age height weight;
    class sex;
    output out=work.sum_gender
        mean=AvgAge AvgHeight AvgWeight;
run;
```

b.
```
proc summary data=clinic.diabetes print;
    var age height weight; class sex;
    output out=work.sum_gender
        mean=AvgAge AvgHeight AvgWeight;
run;
```

c.
```
proc means data=clinic.diabetes noprint;
    var age height weight;
    class sex;
    output out=work.sum_gender
        mean=AvgAge AvgHeight AvgWeight;
run;
```

d. Both a and b.

Correct answer: d

You can use either PROC MEANS or PROC SUMMARY to create the table. Adding a PRINT option to the PROC SUMMARY statement produces the same report as if you used PROC MEANS.

7. By default, PROC FREQ creates a table of frequencies and percentages for which data set variables?

a. character variables

b. numeric variables

c. both character and numeric variables

d. none: variables must always be specified

Correct answer: c

By default, PROC FREQ creates a table for all variables in a data set.

8. Frequency distributions work best with variables that contain

    a. continuous values.

    b. numeric values.

    c. categorical values.

    d. unique values.

Correct answer: c

Both continuous values and unique values can result in lengthy, meaningless tables. Frequency distributions work best with categorical values.

9. Which PROC FREQ step produced this two-way table?

The FREQ Procedure

Frequency Percent Row Pct Col Pct	Table of Weight by Height			
		Height		
Weight	< 5'5"	5'5-10"	> 5'10"	Total
< 140	2	0	0	2
	10.00	0.00	0.00	10.00
	100.00	0.00	0.00	
	28.57	0.00	0.00	
140-180	5	5	0	10
	25.00	25.00	0.00	50.00
	50.00	50.00	0.00	
	71.43	62.50	0.00	
> 180	0	3	5	8
	0.00	15.00	25.00	40.00
	0.00	37.50	62.50	
	0.00	37.50	100.00	
Total	7	8	5	20
	35.00	40.00	25.00	100.00

a. ```
    proc freq data=clinic.diabetes;
       tables height weight;
       format height htfmt. weight wtfmt.;
    run;
    ```

b.  ```
    proc freq data=clinic.diabetes;
       tables weight height;
       format weight wtfmt. height htfmt.;
    run;
    ```

c. ```
    proc freq data=clinic.diabetes;
       tables height*weight;
       format height htfmt. weight wtfmt.;
    run;
    ```

d.  ```
    proc freq data=clinic.diabetes;
       tables weight*height;
       format weight wtfmt. height htfmt.;
    run;
    ```

Correct answer: d

An asterisk is used to join the variables in a two-way TABLES statement. The first variable forms the table rows. The second variable forms the table columns.

10. Which PROC FREQ step produced this table?

The FREQ Procedure

Percent	Table of Sex by Weight			
	Weight			
Sex	< 140	140-180	> 180	Total
F	10.00	45.00	0.00	55.00
M	0.00	5.00	40.00	45.00
Total	2	10	8	20
	10.00	50.00	40.00	100.00

a.
```
proc freq data=clinic.diabetes;
   tables sex weight / list;
   format weight wtfmt.;
run;
```

b.
```
proc freq data=clinic.diabetes;
   tables sex*weight / nocol;
   format weight wtfmt.;
run;
```

c.
```
proc freq data=clinic.diabetes;
   tables sex weight / norow nocol;
   format weight wtfmt.;
run;
```

d.
```
proc freq data=clinic.diabetes;
   tables sex*weight / nofreq norow nocol;
   format weight wtfmt.;
run;
```

Correct answer: d

An asterisk is used to join the variables in crosstabulation tables. The only results shown in this table are cell percentages. The NOFREQ option suppresses cell frequencies, the NOROW option suppresses row percentages, and the NOCOL option suppresses column percentages.

Chapter 9: Producing HTML Output

1. Using ODS statements, how many types of output can you generate at once?

   a. 1 (only HTML output)

   b. 2

   c. 3

   d. as many as you want

Correct answer: d

You can generate any number of output types as long as you open the ODS destination for each type of output you want to create.

2. If ODS is set to its default settings in the SAS windowing environment for Microsoft Windows and UNIX, , what types of output are created by the code below?

```
ods html file='c:\myhtml.htm';
ods pdf file='c:\mypdf.pdf';
```

   a. HTML and PDF

   b. PDF only

   c. HTML, PDF, and listing

   d. No output is created because ODS is closed by default.

Correct answer: b

HTML output is created by default in the SAS windowing environment for Microsoft Windows and UNIX, so these statements create HTML and PDF output.

3. In the SAS windowing environment for Microsoft Windows and UNIX, what is the purpose of closing the HTML destination in the code shown below?

```
ods html close;
ods rtf ... ;
```

   a. It conserves system resources.

   b. It simplifies your program.

   c. It makes your program compatible with other hardware platforms.

   d. It makes your program compatible with previous versions of SAS.

Correct answer: a

By default, in the SAS windowing environment for Microsoft Windows and UNIX, SAS programs produce HTML output. If you want only RTF output, it's a good idea to close the HTML destination before creating RTF output, as an open destination uses system resources.

4. When the code shown below is run, what will the file `D:\Output\body.html` contain?

```
ods html body='d:\output\body.html';
proc print data=work.alpha;
run;
proc print data=work.beta;
run;
ods html close;
```

   a. The PROC PRINT output for Work.Alpha.

   b. The PROC PRINT output for Work.Beta.

   c. The PROC PRINT output for both Work.Alpha and Work.Beta.

   d. Nothing. No output will be written to `D:\Output\body.html`.

Correct answer: c

When multiple procedures are run while HTML output is open, procedure output is appended to the same body file.

5. When the code shown below is run, what file will be referenced by the links in `D:\Output\contents.html`?

```
ods html body='d:\output\body.html'
        contents='d:\output\contents.html'
        frame='d:\output\frame.html';
```

a. **D:\Output\body.html**

b. **D:\Output\contents.html**

c. **D:\Output\frame.html**

d. There are no links from the file **D:\Output\contents.html**.

Correct answer: a

The CONTENTS= option creates a table of contents containing links to the body file, **D:\Output\body.html**.

6. The table of contents created by the CONTENTS= option contains a numbered heading for

a. each procedure.

b. each procedure that creates output.

c. each procedure and DATA step.

d. each HTML file created by your program.

Correct answer: b

The table of contents contains a numbered heading for each procedure that creates output.

7. When the code shown below is run, what will the file **D:\Output\frame.html** display?

```
ods html body='d:\output\body.html'
        contents='d:\output\contents.html'
        frame='d:\output\frame.html';
```

a. The file **D:\Output\contents.html**.

b. The file **D:\Output\frame.html**.

c. The files **D:\Output\contents.html** and **D:\Output\body.html**.

d. It displays no other files.

Correct answer: c

The FRAME= option creates an HTML file that integrates the table of contents and the body file.

8. What is the purpose of the URL= suboptions shown below?

```
ods html body='d:\output\body.html' (url='body.html')
        contents='d:\output\contents.html'
        (url='contents.html')
        frame='d:\output\frame.html';
```

a. To create absolute link addresses for loading the files from a server.

b. To create relative link addresses for loading the files from a server.

c. To allow HTML files to be loaded from a local drive.

d. To send HTML output to two locations.

Correct answer: b

Specifying the URL= suboption in the file specification provides a URL that ODS uses in the links it creates. Specifying a simple (one name) URL creates a relative link address to the file.

9. Which ODS HTML option was used in creating the following table?

Obs	Sex	Age	Height	Weight	ActLevel
1	M	27	72	168	HIGH
2	F	34	66	152	HIGH
3	F	31	61	123	LOW

a. `format=brown`

b. `format='brown'`

c. `style=brown`

d. `style='brown'`

Correct answer: c

You can change the appearance of HTML output by using the STYLE= option in the ODS HTML statement. The style name doesn't need quotation marks.

10. What is the purpose of the PATH= option?

```
ods html path='d:\output' (url=none)
         body='body.html'
         contents='contents.html'
         frame='frame.html';
```

a. It creates absolute link addresses for loading HTML files from a server.

b. It creates relative link addresses for loading HTML files from a server.

c. It allows HTML files to be loaded from a local drive.

d. It specifies the location of HTML file output.

Correct answer: d

You use the PATH= option to specify the location for HTML files to be stored. When you use the PATH= option, you don't need to specify the full path name for the body, contents, or frame files.

Chapter 10: Creating and Managing Variables

1. Which program creates the output shown below?

Raw Data File Furnture

```
1---+----10---+----20---+----30---+
310 oak     pedestal table 329.99
311 maple pedestal table 369.99
312 brass floor     lamp   79.99
313 glass table     lamp   59.99
314 oak     rocking   chair 153.99
```

StockNum	Finish	Style	Item	TotalPrice
310	oak	pedestal	table	329.99
311	maple	pedestal	table	699.98
312	brass	floor	lamp	779.97
313	glass	table	lamp	839.96

a.
```
data test2;
    infile furnture;
    input StockNum $ 1-3 Finish $ 5-9 Style $ 11-18
        Item $ 20-24 Price 26-31;
    if finish='oak' then delete;
    retain TotPrice 100;
    totalprice+price;
    drop price;
run;
proc print data=test2 noobs;
run;
```

b.
```
data test2;
    infile furnture;
    input StockNum $ 1-3 Finish $ 5-9 Style $ 11-18
        Item $ 20-24 Price 26-31;
    if finish='oak' and price<200 then delete;
    TotalPrice+price;
run;
proc print data=test2 noobs;
run;
```

c.
```
data test2(drop=price);
    infile furnture;
    input StockNum $ 1-3 Finish $ 5-9 Style $ 11-18
        Item $ 20-24 Price 26-31;
    if finish='oak' and price<200 then delete;
    TotalPrice+price;
run;
proc print data=test2 noobs;
run;
```

d.
```
data test2;
    infile furnture;
    input StockNum $ 1-3 Finish $ 5-9 Style $ 11-18
        Item $ 20-24 Price 26-31;
    if finish=oak and price<200 then delete price;
    TotalPrice+price;
run;
proc print data=test2 noobs;
run;
```

Correct answer: c

Program c correctly deletes the observation in which the value of Finish is **oak** and the value of Price is less than **200**. It also creates TotalPrice by summing the variable Price down observations, and then drops Price by using the DROP= data set option in the DATA statement.

2. How is the variable Amount labeled and formatted in the PROC PRINT output?

```
data credit;
   infile creddata;
   input Account $ 1-5 Name $ 7-25 Type $ 27
            Transact $ 29-35 Amount 37-50;
   label amount='Amount of Loan';
   format amount dollar12.2;
run;
proc print data=credit label;
   label amount='Total Amount Loaned';
   format amount comma10.;
run;
```

   a. label Amount of Loan, format DOLLAR12.2

   b. label Total Amount Loaned, format COMMA10.

   c. label Amount, default format

   d. The PROC PRINT step does not execute because two labels and two formats are assigned to the same variable.

Correct answer: b

The PROC PRINT output displays the label Total Amount Loaned for the variable Amount and formats this variable using the COMMA10. format. Temporary labels or formats that are assigned in a PROC step override permanent labels or formats that are assigned in a DATA step.

3. Consider the IF-THEN statement shown below. When the statement is executed, which expression is evaluated first?

```
if finlexam>=95
   and (research='A' or
       (project='A' and present='A'))
   then Grade='A+';
```

   a. `finlexam>=95`

   b. `research='A'`

   c. `project='A' and present='A'`

   d. `research='A' or
        (project='A' and present='A')`

Correct answer: c

Logical comparisons that are enclosed in parentheses are evaluated as true or false before they are compared to other expressions. In the example above, the AND comparison within the nested parentheses is evaluated before being compared to the OR comparison.

4. Consider the small raw data file and program shown below. What is the value of Count after the fourth record is processed?

```
data work.newnums;
   infile numbers;
   input Tens 2-3;
   Count+tens;
run;
```

```
1---+----10
  10
  20

  40
  50
```

a. missing

b. 0

c. 30

d. 70

Correct answer: d

The sum statement adds the result of the expression that is on the right side of the plus sign to the numeric variable that is on the left side. The new value is then retained for subsequent observations. The sum statement ignores the missing value, so the value of Count in the fourth observation would be **10+20+0+40**, or **70**.

5. Now consider the revised program below. What is the value of Count after the third observation is read?

```
data work.newnums;
   infile numbers;
   input Tens 2-3;
   retain Count 100;
   count+tens;
run;
```

```
1---+----10
  10
  20

  40
  50
```

a. missing

b. 0

c. 100

d. 130

Correct answer: d

The RETAIN statement assigns an initial value of **100** to the variable Count, so the value of Count in the third observation would be **100+10+20+0**, or **130**.

6. For the observation shown below, what is the result of the IF-THEN statements?

Status	Type	Count	Action	Control
Ok	3	12	E	Go

```
if status='OK' and type=3
   then Count+1;
if status='S' or action='E'
   then Control='Stop';
```

a. `Count = 12     Control = Go`

b. `Count = 13     Control = Stop`

c. `Count = 12     Control = Stop`

d. `Count = 13     Control = Go`

Correct answer: c

You must enclose character values in quotation marks, and you must specify them in the same case in which they appear in the data set. The value **Ok** is not identical to OK, so the value of Count is not changed by the IF-THEN statement.

7. Which of the following can determine the length of a new variable?

a. the length of the variable's first reference in the DATA step

b. the assignment statement

c. the LENGTH statement

d. all of the above

Correct answer: d

The length of a variable is determined by its first reference in the DATA step. When creating a new character variable, SAS allocates as many bytes of storage space as there are characters in the reference to that variable. The first reference to a new variable can also be made with a LENGTH statement or an assignment statement.

8. Which set of statements is equivalent to the code shown below?

```
if code='1' then Type='Fixed';
if code='2' then Type='Variable';
if code^='1' and code^='2' then Type='Unknown';
```

a.
```
if code='1' then Type='Fixed';
   else if code='2' then Type='Variable';
   else Type='Unknown';
```

b.
```
if code='1' then Type='Fixed';
if code='2' then Type='Variable';
   else Type='Unknown';
```

c.
```
if code='1' then type='Fixed';
   else code='2' and type='Variable';
   else type='Unknown';
```

d.
```
if code='1' and type='Fixed';
   then code='2' and type='Variable';
   else type='Unknown';
```

Correct answer: a

You can write multiple ELSE statements to specify a series of mutually exclusive conditions. The ELSE statement must immediately follow the IF-THEN statement in your program. An ELSE statement executes only if the previous IF-THEN/ELSE statement is false.

9. What is the length of the variable Type, as created in the DATA step below?

```
data finance.newloan;
   set finance.records;
   TotLoan+payment;
   if code='1' then Type='Fixed';
   else Type='Variable';
   length type $ 10;
run;
```

a. 5

b. 8

c. 10

d. it depends on the first value of Type

Correct answer: a

The length of a new variable is determined by the first reference in the DATA step, not by data values. In this case, the length of Type is determined by the value Fixed. The LENGTH statement is in the wrong place; it must occur before any other reference to the variable in the DATA step.

10. Which program contains an error?

a.
```
data clinic.stress(drop=timemin timesec);
   infile tests;
   input ID $ 1-4 Name $ 6-25 RestHR 27-29 MaxHR 31-33
         RecHR 35-37 TimeMin 39-40 TimeSec 42-43
         Tolerance $ 45;
   TotalTime=(timemin*60)+timesec;
   SumSec+totaltime;
run;
```

b.
```
proc print data=clinic.stress;
   label totaltime='Total Duration of Test';
   format timemin 5.2;
   drop sumsec;
run;
```

c.
```
proc print data=clinic.stress(keep=totaltime timemin);
   label totaltime='Total Duration of Test';
   format timemin 5.2;
run;
```

d.
```
data clinic.stress;
   infile tests;
   input ID $ 1-4 Name $ 6-25 RestHR 27-29 MaxHR 31-33
         RecHR 35-37 TimeMin 39-40 TimeSec 42-43
         Tolerance $ 45;
   TotalTime=(timemin*60)+timesec;
   keep id totaltime tolerance;
run;
```

Correct answer: b

To select variables, you can use a DROP or KEEP statement in any DATA step. You can also use the DROP= or KEEP= data set options following a data set name in any DATA or PROC step. However, you cannot use DROP or KEEP statements in PROC steps.

Chapter 11: Reading SAS Data Sets

1. If you submit the following program, which variables appear in the new data set?

```
data work.cardiac(drop=age group);
   set clinic.fitness(keep=age weight group);
   if group=2 and age>40;
run;
```

   a. none

   b. Weight

   c. Age, Group

   d. Age, Weight, Group

   Correct answer: b

   The variables Age, Weight, and Group are specified using the KEEP= option in the SET statement. After processing, Age and Group are dropped in the DATA statement.

2. Which of the following programs correctly reads the data set Orders and creates the data set FastOrdr?

   a.
```
   data catalog.fastordr(drop=ordrtime);
      set july.orders(keep=product units price);
      if ordrtime<4;
      Total=units*price;
   run;
```

   b.
```
   data catalog.orders(drop=ordrtime);
      set july.fastordr(keep=product units price);
      if ordrtime<4;
      Total=units*price;
   run;
```

   c.
```
   data catalog.fastordr(drop=ordrtime);
      set july.orders(keep=product units price
                         ordrtime);
      if ordrtime<4;
      Total=units*price;
   run;
```

   d. none of the above

   Correct answer: c

   You specify the data set to be created in the DATA statement. The DROP= data set option prevents variables from being written to the data set. Because you use the variable OrdrTime when processing your data, you cannot drop OrdrTime in the SET statement. If you use the KEEP= option in the SET statement, then you must list OrdrTime as one of the variables to be kept.

3. Which of the following statements is *false* about BY-group processing?

   When you use the BY statement with the SET statement:

   a. The data sets listed in the SET statement must be indexed or sorted by the values of the BY variable(s).

   b. The DATA step automatically creates two variables, FIRST. and LAST., for each variable in the BY statement.

   c. FIRST. and LAST. identify the first and last observation in each BY group, respectively.

   d. FIRST. and LAST. are stored in the data set.

   Correct answer: d

   When you use the BY statement with the SET statement, the DATA step creates the temporary variables FIRST. and LAST. They are not stored in the data set.

4. There are 500 observations in the data set Usa. What is the result of submitting the following program?

   ```
   data work.getobs5;
      obsnum=5;
      set company.usa(keep=manager payroll) point=obsnum;
      stop;
   run;
   ```

   a. an error

   b. an empty data set

   c. continuous loop

   d. a data set that contains one observation

   Correct answer: b

   The DATA step outputs observations at the end of the DATA step. However, in this program, the STOP statement stops processing before the end of the DATA step. An explicit OUTPUT statement is needed in order to produce an observation.

5. There is no end-of-file condition when you use direct access to read data, so how can your program prevent a continuous loop?

   a. Do not use a POINT= variable.

   b. Check for an invalid value of the POINT= variable.

   c. Do not use an END= variable.

   d. Include an OUTPUT statement.

   Correct answer: b

   To avoid a continuous loop when using direct access, either include a STOP statement or use programming logic that executes a STOP statement when the data step encounters an invalid value of the POINT= variable. If SAS reads an invalid value of the POINT= variable, it sets the automatic variable _ERROR_ to 1. You can use this information to check for conditions that cause continuous looping.

6. Assuming that the data set Company.USA has five or more observations, what is the result of submitting the following program?

```
data work.getobs5;
   obsnum=5;
   set company.usa(keep=manager payroll) point=obsnum;
   output;
   stop;
run;
```

   a. an error

   b. an empty data set

   c. a continuous loop

   d. a data set that contains one observation

   Correct answer: d

   By combining the POINT= option with the OUTPUT and STOP statements, your program can output a single observation.

7. Which of the following statements is *true* regarding direct access of data sets?

   a. You cannot specify END= with POINT=.

   b. You cannot specify OUTPUT with POINT=.

   c. You cannot specify STOP with END=.

   d. You cannot specify FIRST. with LAST.

   Correct answer: a

   The END= option and POINT= option are incompatible in the same SET statement. Use one or the other in your program.

8. What is the result of submitting the following program?

```
data work.addtoend;
   set clinic.stress2 end=last;
   if last;
run;
```

   a. an error

   b. an empty data set

   c. a continuous loop

   d. a data set that contains one observation

   Correct answer: d

   This program uses the END= option to name a temporary variable that contains an end-of-file marker. That variable — last — is set to 1 when the SET statement reads the last observation of the data set.

9. At the start of DATA step processing, during the compilation phase, variables are created in the program data vector (PDV), and observations are set to:

   a. blank

   b. missing

   c. 0

d. there are no observations.

Correct answer: d

At the bottom of the DATA step, the compilation phase is complete, and the descriptor portion of the new SAS data set is created. There are no observations because the DATA step has not yet executed.

10. The DATA step executes:

    a. continuously if you use the POINT= option and the STOP statement.

    b. once for each variable in the output data set.

    c. once for each observation in the input data set.

    d. until it encounters an OUTPUT statement.

Correct answer: c

The DATA step executes once for each observation in the input data set. You use the POINT= option with the STOP statement to prevent continuous looping.

Chapter 12: Combining SAS Data Sets

1. Which program will combine Brothers.One and Brothers.Two to produce Brothers.Three?

    a.
```
data brothers.three;
     set brothers.one;
     set brothers.two;
run;
```

    b.
```
data brothers.three;
     set brothers.one brothers.two;
run;
```

    c.
```
data brothers.three;
     set brothers.one brothers.two;
     by varx;
run;
```

    d.
```
data brothers.three;
     merge brothers.one brothers.two;
     by varx;
run;
```

Correct answer: a

This example is a case of one-to-one matching, which requires multiple SET statements. Where same-named variables occur, values that are read from the second data set replace those read from the first data set. Also, the number of observations in the new data set is the number of observations in the smallest original data set.

2. Which program will combine Actors.Props1 and Actors.Props2 to produce Actors.Props3?

Actors.Props1	
Actor	**Prop**
Curly	Anvil
Larry	Ladder
Moe	Poker

+

Actors.Props2	
Actor	**Prop**
Curly	Ladder
Moe	Pliers

=

Actors.Props3	
Actor	**Prop**
Curly	Anvil
Curly	Ladder
Larry	Ladder
Moe	Poker
Moe	Pliers

a.
```
data actors.props3;
    set actors.props1;
    set actors.props2;
run;
```

b.
```
data actors.props3;
    set actors.props1 actors.props2;
run;
```

c.
```
data actors.props3;
    set actors.props1 actors.props2;
    by actor;
run;
```

d.
```
data actors.props3;
    merge actors.props1 actors.props2;
    by actor;
run;
```

Correct answer: c

This is a case of interleaving, which requires a list of data set names in the SET statement and one or more BY variables in the BY statement. Notice that observations in each BY group are read sequentially, in the order in which the data sets and BY variables are listed. The new data set contains all the variables from all the input data sets, as well as the total number of records from all input data sets.

3. If you submit the following program, which new data set is created?

Work.Dataone		
Career	**Supervis**	**Finance**
72	26	9
63	76	7
96	31	7
96	98	6
84	94	6

Work.Datatwo		
Variety	**Feedback**	**Autonomy**
10	11	70
85	22	93
83	63	73
82	75	97
36	77	97

```
data work.jobsatis;
    set work.dataone work.datatwo;
run;
```

a.

Career	Supervis	Finance	Variety	Feedback	Autonomy
72	26	9	.	.	.
63	76	7	.	.	.
96	31	7	.	.	.
96	98	6	.	.	.
84	94	6	.	.	.
.	.	.	10	11	70
.	.	.	85	22	93
.	.	.	83	63	73
.	.	.	82	75	97
.	.	.	36	77	97

b.

Career	Supervis	Finance	Variety	Feedback	Autonomy
72	26	9	10	11	70
63	76	7	85	22	93
96	31	7	83	63	73
96	98	6	82	75	97
84	94	6	36	77	97

c.

Career	Supervis	Finance
72	26	9
63	76	7
96	31	7
96	98	6
84	94	6
10	11	70
85	22	93
83	63	73
82	75	97
36	77	97

d. none of the above

Correct answer: a

Concatenating appends the observations from one data set to another data set. The new data set contains the total number of records from all input data sets, so b is incorrect. All the variables from all the input data sets appear in the new data set, so c is incorrect.

4. If you concatenate the data sets below in the order shown, what is the value of Sale in observation 2 of the new data set?

Sales.Reps		Sales.Close		Sales.Bonus	
ID	**Name**	**ID**	**Sale**	**ID**	**Bonus**
1	Nay Rong	1	$28,000	1	$2,000
2	Kelly Windsor	2	$30,000	2	$4,000
3	Julio Meraz	2	$40,000	3	$3,000
4	Richard Krabill	3	$15,000	4	$2,500
		3	$20,000		
		3	$25,000		
		4	$35,000		

a. missing

b. `$30,000`

c. `$40,000`

d. you cannot concatenate these data sets

Correct answer: a

The concatenated data sets are read sequentially, in the order in which they are listed in the SET statement. The second observation in Sales.Reps does not contain a value for Sale, so a missing value appears for this variable. (Note that if you merge the data sets, the value of Sale for the second observation is `$30,000`.)

5. What happens if you merge the following data sets by the variable SSN?

1st		2nd		
SSN	**Age**	**SSN**	**Age**	**Date**
029-46-9261	39	029-46-9261	37	02/15/95
074-53-9892	34	074-53-9892	32	05/22/97
228-88-9649	32	228-88-9649	30	03/04/96
442-21-8075	12	442-21-8075	10	11/22/95
446-93-2122	36	446-93-2122	34	07/08/96
776-84-5391	28	776-84-5391	26	12/15/96
929-75-0218	27	929-75-0218	25	04/30/97

a. The values of Age in the 1st data set overwrite the values of Age in the 2nd data set.

b. The values of Age in the 2nd data set overwrite the values of Age in the 1st data set.

c. The DATA step fails because the two data sets contain same-named variables that have different values.

d. The values of Age in the 2nd data set are set to missing.

Correct answer: b

If you have variables with the same name in more than one input data set, values of the same-named variable in the first data set in which it appears are overwritten by values of the same-named variable in subsequent data sets.

6. Suppose you merge data sets Health.Set1 and Health.Set2 below:

Health.Set1		
ID	**Sex**	**Age**
1129	F	48
1274	F	50
1387	F	57
2304	F	16
2486	F	63
4425	F	48
4759	F	60
5438	F	42
6488	F	59
9012	F	39
9125	F	56

Health.Set2		
ID	**Height**	**Weight**
1129	61	137
1387	64	142
2304	61	102
5438	62	168
6488	64	154
9012	63	157
9125	64	159

Which output does the following program create?

```
data work.merged;
   merge health.set1(in=in1) health.set2(in=in2);
   by id;
   if in1 and in2;
run;
proc print data=work.merged;
run;
```

a.

Obs	ID	Sex	Age	Height	Weight
1	1129	F	48	61	137
2	1274	F	50	.	.
3	1387	F	57	64	142
4	2304	F	16	61	102
5	2486	F	63	.	.
6	4425	F	48	.	.
7	4759	F	60	.	.
8	5438	F	42	62	168
9	6488	F	59	64	154
10	9012	F	39	63	157
11	9125	F	56	64	159

b.

Obs	ID	Sex	Age	Height	Weight
1	1129	F	48	61	137
2	1387	F	50	64	142
3	2304	F	57	61	102
4	5438	F	16	62	168
5	6488	F	63	64	154
6	9012	F	48	63	157
7	9125	F	60	64	159
8	5438	F	42	.	.
9	6488	F	59	.	.
10	9012	F	39	.	.
11	9125	F	56	.	.

c.

Obs	ID	Sex	Age	Height	Weight
1	1129	F	48	61	137
2	1387	F	57	64	142
3	2304	F	16	61	102
4	5438	F	42	62	168
5	6488	F	59	64	154
6	9012	F	39	63	157
7	9125	F	56	64	159

d. none of the above

Correct answer: c

The DATA step uses the IN= data set option and the subsetting IF statement to exclude unmatched observations from the output data set. So a and b, which contain unmatched observations, are incorrect.

7. The data sets Ensemble.Spring and Ensemble.Summer both contain a variable named Blue. How do you prevent the values of the variable Blue from being overwritten when you merge the two data sets?

a.
```
data ensemble.merged;
    merge ensemble.spring(in=blue)
          ensemble.summer;
    by fabric;
run;
```

b.
```
data ensemble.merged;
    merge ensemble.spring(out=blue)
          ensemble.summer;
    by fabric;
run;
```

c.
```
data ensemble.merged;
    merge ensemble.spring(blue=navy)
          ensemble.summer;
    by fabric;
run;
```

d.
```
data ensemble.merged;
    merge ensemble.spring(rename=(blue=navy))
```

```
        ensemble.summer;
    by fabric;
  run;
```

Correct answer: d

Match-merging overwrites same-named variables in the first data set with same-named variables in subsequent data sets. To prevent overwriting, rename variables by using the RENAME= data set option in the MERGE statement.

8. What happens if you submit the following program to merge Blood.Donors1 and Blood.Donors2, shown below?

```
data work.merged;
   merge blood.donors1 blood.donors2;
   by id;
run
```

Blood.Donors1		
ID	Type	Units
2304	O	16
1129	A	48
1129	A	50
1129	A	57
2486	B	63

Blood.Donors2		
ID	Code	Units
6488	65	27
1129	63	32
5438	62	39
2304	61	45
1387	64	67

a. The Merged data set contains some missing values because not all observations have matching observations in the other data set.

b. The Merged data set contains eight observations.

c. The DATA step produces errors.

d. Values for Units in Blood.Donors2 overwrite values of Units in Blood.Donors1.

Correct answer: c

The two input data sets are not sorted by values of the BY variable, so the DATA step produces errors and stops processing.

9. If you merge Company.Staff1 and Company.Staff2 below by ID, how many observations does the new data set contain?

Company.Staff1			
ID	Name	Dept	Project
000	Miguel	A12	Document
111	Fred	B45	Survey
222	Diana	B45	Document
888	Monique	A12	Document
999	Vien	D03	Survey

Company.Staff2		
ID	Name	Hours
111	Fred	35
222	Diana	40
777	Steve	0
888	Monique	37

a. 4

b. 5

c. 6

d. 9

Correct answer: c

In this example, the new data set contains one observation for each unique value of ID. The merged data set is shown below.

ID	Name	Dept	Project	Hours
000	Miguel	A12	Document	.
111	Fred	B45	Survey	35
222	Diana	B45	Document	40
777	Steve			0
888	Monique	A12	Document	37
999	Vien	D03	Survey	.

10. If you merge data sets Sales.Reps, Sales.Close, and Sales.Bonus by ID, what is the value of Bonus in the third observation in the new data set?

Sales.Reps

ID	Name
1	Nay Rong
2	Kelly Windsor
3	Julio Meraz
4	Richard Krabill

Sales.Close

ID	Sale
1	$28,000
2	$30,000
2	$40,000
3	$15,000
3	$20,000
3	$25,000
4	$35,000

Sales.Bonus

ID	Bonus
1	$2,000
2	$4,000
3	$3,000
4	$2,500

a. $4,000

b. $3,000

c. missing

d. can't tell from the information given

Correct answer: a

In the new data set, the third observation is the second observation for ID number 2 (Kelly Windsor). The value for Bonus is retained from the previous observation because the BY variable value didn't change. The new data set is shown below.

ID	Name	Sale	Bonus
1	Nay Rong	$28,000	$2,000
2	Kelly Windsor	$30,000	$4,000
2	Kelly Windsor	$40,000	$4,000
3	Julio Meraz	$15,000	$3,000
3	Julio Meraz	$20,000	$3,000
3	Julio Meraz	$25,000	$3,000
4	Richard Krabill	$35,000	$2,500

Chapter 13: Transforming Data with SAS Functions

1. Which function calculates the average of the variables Var1, Var2, Var3, and Var4?

   a. `mean(var1,var4)`

   b. `mean(var1-var4)`

   c. `mean(of var1,var4)`

   d. `mean(of var1-var4)`

   Correct answer: d

   Use a variable list to specify a range of variables as the function argument. When specifying a variable list, be sure to precede the list with the word OF. If you omit the word OF, the function argument might not be interpreted as expected.

2. Within the data set Hrd.Temp, PayRate is a character variable and Hours is a numeric variable. What happens when the following program is run?

   ```
   data work.temp;
      set hrd.temp;
      Salary=payrate*hours;
   run;
   ```

   a. SAS converts the values of PayRate to numeric values. No message is written to the log.

   b. SAS converts the values of PayRate to numeric values. A message is written to the log.

   c. SAS converts the values of Hours to character values. No message is written to the log.

   d. SAS converts the values of Hours to character values. A message is written to the log.

   Correct answer: b

   When this DATA step is executed, SAS automatically converts the character values of PayRate to numeric values so that the calculation can occur. Whenever data is automatically converted, a message is written to the SAS log stating that the conversion has occurred.

3. A typical value for the character variable Target is `123,456`. Which statement correctly converts the values of Target to numeric values when creating the variable TargetNo?

   a. `TargetNo=input(target,comma6.);`

   b. `TargetNo=input(target,comma7.);`

   c. `TargetNo=put(target,comma6.);`

   d. `TargetNo=put(target,comma7.)`

   Correct answer: b

You explicitly convert character values to numeric values by using the INPUT function. Be sure to select an informat that can read the form of the values.

4. A typical value for the numeric variable SiteNum is `12.3`. Which statement correctly converts the values of SiteNum to character values when creating the variable Location?

   a. `Location=dept||'/'||input(sitenum,3.1);`

   b. `Location=dept||'/'||input(sitenum,4.1);`

   c. `Location=dept||'/'||put(sitenum,3.1);`

   d. `Location=dept||'/'||put(sitenum,4.1);`

   Correct answer: d

   You explicitly convert numeric values to character values by using the PUT function. Be sure to select a format that can read the form of the values.

5. Suppose the YEARCUTOFF= system option is set to 1920. Which MDY function creates the date value for January 3, 2020?

   a. `MDY(1,3,20)`

   b. `MDY(3,1,20)`

   c. `MDY(1,3,2020)`

   d. `MDY(3,1,2020)`

   Correct answer: c

   Because the YEARCUTOFF= system option is set to 1920, SAS sees the two-digit year value `20` as `1920`. Four-digit year values are always read correctly.

6. The variable Address2 contains values such as `Piscataway, NJ`. How do you assign the two-letter state abbreviations to a new variable named State?

   a. `State=scan(address2,2);`

   b. `State=scan(address2,13,2);`

   c. `State=substr(address2,2);`

   d. `State=substr(address2,13,2);`

   Correct answer: a

   The SCAN function is used to extract words from a character value when you know the order of the words, when their position varies, and when the words are marked by some delimiter. In this case, you don't need to specify delimiters, because the blank and the comma are default delimiters.

7. The variable IDCode contains values such as `123FA` and `321MB`. The fourth character identifies sex. How do you assign these character codes to a new variable named Sex?

   a. `Sex=scan(idcode,4);`

   b. `Sex=scan(idcode,4,1);`

   c. `Sex=substr(idcode,4);`

   d. `Sex=substr(idcode,4,1);`

   Correct answer: d

The SUBSTR function is best used when you know the exact position of the substring to extract from the character value. You specify the position to start from and the number of characters to extract.

8. Due to growth within the 919 area code, the telephone exchange 555 is being reassigned to the 920 area code. The data set Clients.Piedmont includes the variable Phone, which contains telephone numbers in the form 919-555-1234. Which of the following programs will correctly change the values of Phone?

   a.
   ```
   data work.piedmont(drop=areacode exchange);
      set clients.piedmont;
      Areacode=substr(phone,1,3);
      Exchange=substr(phone,5,3);
      if areacode='919' and exchange='555'
         then scan(phone,1,3)='920';
   run;
   ```

   b.
   ```
   data work.piedmont(drop=areacode exchange);
      set clients.piedmont;
      Areacode=substr(phone,1,3);
      Exchange=substr(phone,5,3);
      if areacode='919' and exchange='555'
         then phone=scan('920',1,3);
   run;
   ```

   c.
   ```
   data work.piedmont(drop=areacode exchange);
      set clients.piedmont;
      Areacode=substr(phone,1,3);
      Exchange=substr(phone,5,3);
      if areacode='919' and exchange='555'
         then substr(phone,5,3)='920';
   run;
   ```

   d.
   ```
   data work.piedmont(drop=areacode exchange);
      set clients.piedmont;
      Areacode=substr(phone,1,3);
      Exchange=substr(phone,5,3);
      if areacode='919' and exchange='555'
         then phone=substr('920',1,3);
   run;
   ```

   Correct answer: c

   The SUBSTR function replaces variable values if it is placed on the left side of an assignment statement. When placed on the right side (as in Question 7), the function extracts a substring.

9. Suppose you need to create the variable FullName by concatenating the values of FirstName, which contains first names, and LastName, which contains last names. What's the best way to remove extra blanks between first names and last names?

   a.
   ```
   data work.maillist;
      set retail.maillist;
      length FullName $ 40;
      fullname=trim firstname||' '||lastname;
   run;
   ```

   b.
   ```
   data work.maillist;
      set retail.maillist;
      length FullName $ 40;
   ```

```
        fullname=trim(firstname)||' '||lastname;
    run;
```

c.
```
data work.maillist;
    set retail.maillist;
    length FullName $ 40;
    fullname=trim(firstname)||' '||trim(lastname);
run;
```

d.
```
data work.maillist;
    set retail.maillist;
    length FullName $ 40;
    fullname=trim(firstname||' '||lastname);
run;
```

Correct answer: b

The TRIM function removes trailing blanks from character values. In this case, extra blanks must be removed from the values of FirstName. Although answer c also works, the extra TRIM function for the variable LastName is unnecessary. Because of the LENGTH statement, all values of FullName are padded to 40 characters.

10. Within the data set Furnitur.Bookcase, the variable Finish contains values such as **ash/cherry/teak/matte-black**. Which of the following creates a subset of the data in which the values of Finish contain the string **walnut**? Make the search for the string case-insensitive.

a.
```
data work.bookcase;
    set furnitur.bookcase;
    if index(finish,walnut) = 0;
run;
```

b.
```
data work.bookcase;
    set furnitur.bookcase;
    if index(finish,'walnut') > 0;
run;
```

c.
```
data work.bookcase;
    set furnitur.bookcase;
    if index(lowcase(finish),walnut) = 0;
run;
```

d.
```
data work.bookcase;
    set furnitur.bookcase;
    if index(lowcase(finish),'walnut') > 0;
run;
```

Correct answer: d

Use the INDEX function in a subsetting IF statement, enclosing the character string in quotation marks. Only those observations in which the function locates the string and returns a value greater than 0 are written to the data set.

Chapter 14: Generating Data with DO Loops

1. Which statement is *false* regarding the use of DO loops?

a. They can contain conditional clauses.

b. They can generate multiple observations.

c. They can be used to combine DATA and PROC steps.

d. They can be used to read data.

Correct answer: c

DO loops are DATA step statements and cannot be used in conjunction with PROC steps.

2. During each execution of the following DO loop, the value of Earned is calculated and is added to its previous value. How many times does this DO loop execute?

```
data finance.earnings;
   Amount=1000;
   Rate=.075/12;
   do month=1 to 12;
      Earned+(amount+earned)*rate;
   end;
run;
```

a. 0

b. 1

c. 12

d. 13

Correct answer: c

The number of iterations is determined by the DO statement's stop value, which in this case is **12**.

3. On January 1 of each year, $5000 is invested in an account. Complete the DATA step below to determine the value of the account after 15 years if a constant interest rate of ten percent is expected.

```
data work.invest;
   ...
      Capital+5000;
      capital+(capital*.10);
   end;
run;
```

a. do count=1 to 15;

b. do count=1 to 15 by 10%;

c. do count=1 to capital;

d. do count=capital to (capital*.10);

Correct answer: a

Use a DO loop to perform repetitive calculations starting at 1 and looping 15 times.

4. In the data set Work.Invest, what would be the stored value for Year?

```
data work.invest;
   do year=1990 to 2004;
      Capital+5000;
      capital+(capital*.10);
```

```
      end;
run;
```

a. missing

b. `1990`

c. `2004`

d. `2005`

Correct answer: d

At the end of the fifteenth iteration of the DO loop, the value for Year is incremented to `2005`. Because this value exceeds the stop value, the DO loop ends. At the bottom of the DATA step, the current values are written to the data set.

5. Which of the following statements is *false* regarding the program shown below?

```
data work.invest;
   do year=1990 to 2004;
      Capital+5000;
      capital+(capital*.10);
      output;
   end;
run;
```

a. The OUTPUT statement writes current values to the data set immediately.

b. The last value for Year in the new data set is `2005`.

c. The OUTPUT statement overrides the automatic output at the end of the DATA step.

d. The DO loop performs 15 iterations.

Correct answer: b

The OUTPUT statement overrides the automatic output at the end of the DATA step. On the last iteration of the DO loop, the value of Year, `2004`, is written to the data set.

6. How many observations will the data set Work.Earn contain?

```
data work.earn;
   Value=2000;
   do year=1 to 20;
      Interest=value*.075;
      value+interest;
      output;
   end;
run;
```

a. 0

b. 1

c. 19

d. 20

Correct answer: d

The number of observations is based on the number of times the OUTPUT statement executes. The new data set has 20 observations, one for each iteration of the DO loop.

7. Which of the following would you use to compare the result of investing $4,000 a year for five years in three different banks that compound interest monthly? Assume a fixed rate for the five-year period.

    a. DO WHILE statement

    b. nested DO loops

    c. DO UNTIL statement

    d. a DO group

    Correct answer: b

    Place the monthly calculation in a DO loop within a DO loop that iterates once for each year. The DO WHILE and DO UNTIL statements are not used here because the number of required iterations is fixed. A non-iterative DO group would not be useful.

8. Which statement is *false* regarding DO UNTIL statements?

    a. The condition is evaluated at the top of the loop, before the enclosed statements are executed.

    b. The enclosed statements are always executed at least once.

    c. SAS statements in the DO loop are executed until the specified condition is true.

    d. The DO loop must have a closing END statement.

    Correct answer: a

    The DO UNTIL condition is evaluated at the bottom of the loop, so the enclosed statements are always excecuted at least once.

9. Select the DO WHILE statement that would generate the same result as the program below.

    ```
    data work.invest;
    capital=100000;
       do until(Capital gt 500000);
          Year+1;
          capital+(capital*.10);
       end;
    run;
    ```

    a. `do while(Capital ge 500000);`

    b. `do while(Capital=500000);`

    c. `do while(Capital le 500000);`

    d. `do while(Capital>500000);`

    Correct answer: c

    Because the DO WHILE loop is evaluated at the top of the loop, you specify the condition that must exist in order to execute the enclosed statements.

10. In the following program, complete the statement so that the program stops generating observations when Distance reaches 250 miles or when 10 gallons of fuel have been used.

```
data work.go250;
   set perm.cars;
   do gallons=1 to 10 ... ;
      Distance=gallons*mpg;
      output;
   end;
run;
```

a. `while(Distance<250)`

b. `when(Distance>250)`

c. `over(Distance le 250)`

d. `until(Distance=250)`

Correct answer: a

The WHILE expression causes the DO loop to stop executing when the value of Distance becomes equal to or greater than **250**.

Chapter 15: Processing Variables with Arrays

1. Which statement is *false* regarding an ARRAY statement?

   a. It is an executable statement.

   b. It can be used to create variables.

   c. It must contain either all numeric or all character elements.

   d. It must be used to define an array before the array name can be referenced.

   Correct answer: a

   An ARRAY statement is not an executable statement; it merely defines an array.

2. What belongs within the braces of this ARRAY statement?

   `array contrib{?} qtr1-qtr4;`

   a. `quarter`

   b. `quarter*`

   c. `1-4`

   d. `4`

   Correct answer: d

   The value in parentheses indicates the number of elements in the array. In this case, there are four elements.

3. For the program below, select an iterative DO statement to process all elements in the contrib array.

   ```
   data work.contrib;
      array contrib{4} qtr1-qtr4;
         ...
         contrib{i}=contrib{i}*1.25;
   ```

```
   end;
run;
```

a. `do i=4;`

b. `do i=1 to 4;`

c. `do until i=4;`

d. `do while i le 4;`

Correct answer: b

In the DO statement, you specify the index variable that represents the values of the array elements. Then specify the start and stop positions of the array elements.

4. What is the value of the index variable that references Jul in the statements below?

```
array quarter{4} Jan Apr Jul Oct;
do i=1 to 4;
   yeargoal=quarter{i}*1.2;
end;
```

a. **1**

b. **2**

c. **3**

d. **4**

Correct answer: c

The index value represents the position of the array element. In this case, the third element is Jul.

5. Which DO statement would *not* process all the elements in the factors array shown below?

```
 array factors{*} age height weight bloodpr;
```

a. `do i=1 to dim(factors);`

b. `do i=1 to dim(*);`

c. `do i=1,2,3,4;`

d. `do i=1 to 4;`

Correct answer: b

To process all the elements in an array, you can either specify the array dimension or use the DIM function with the array name as the argument.

6. Which statement below is *false* regarding the use of arrays to create variables?

a. The variables are added to the program data vector during the compilation of the DATA step.

b. You do not need to specify the array elements in the ARRAY statement.

c. By default, all character variables are assigned a length of eight.

d. Only character variables can be created.

Correct answer: d

Either numeric or character variables can be created by an ARRAY statement.

7. For the first observation, what is the value of diff{i} at the end of the second iteration of the DO loop?

Weight1	Weight2	Weight3
192	200	215
137	130	125
220	210	213

```
array wt{*} weight1-weight10;
array diff{9};
do i=1 to 9;
   diff{i}=wt{i+1}-wt{i};
end;
```

a. **15**

b. **10**

c. **8**

d. **-7**

Correct answer: a

At the end of the second iteration, **diff{i}** resolves as follows:

```
diff{2}=wt{2+1}-wt{2};
diff{2}=215-200
```

8. Finish the ARRAY statement below to create temporary array elements that have initial values of **9000**, **9300**, **9600**, and **9900**.

```
array goal{4} ... ;
```

a. `_temporary_  (9000 9300 9600 9900)`

b. `temporary (9000 9300 9600 9900)`

c. `_temporary_  9000 9300 9600 9900`

d. `(temporary) 9000 9300 9600 9900`

Correct answer: a

To create temporary array elements, specify _TEMPORARY_ after the array name and dimension. Specify an initial value for each element, separated by either blanks or commas, and enclose the values in parentheses.

9. Based on the ARRAY statement below, select the array reference for the array element q50.

```
array ques{3,25} q1-q75;
```

a. `ques{q50}`

b. `ques{1,50}`

c. `ques{2,25}`

d. `ques{3,0}`

Correct answer: c

This two-dimensional array would consist of three rows of 25 elements. The first row would contain q1 through q25, the second row would start with q26 and end with q50, and the third row would start with q51 and end with q75.

10. Select the ARRAY statement that defines the array in the following program.

```
data coat;
    input category high1-high3 / low1-low3;
    array compare{2,3} high1-high3 low1-low3;
    do i=1 to 2;
        do j=1 to 3;
            compare{i,j}=round(compare{i,j}*1.12);
        end;
    end;
datalines;
5555 9 8 7 6
4 3 2 1
8888 21 12 34 64
13 14 15 16
;
run;
```

a. `array compare{1,6} high1-high3 low1-low3;`

b. `array compare{2,3} high1-high3 low1-low3;`

c. `array compare{3,2} high1-high3 low1-low3;`

d. `array compare{3,3} high1-high3 low1-low3;`

Correct answer: b

The nested DO loops indicate that the array is named compare and is a two-dimensional array that has two rows and three columns.

Chapter 16: Reading Raw Data in Fixed Fields

1. Which SAS statement correctly uses column input to read the values in the raw data file below in this order: Address (4th field), SquareFeet (second field), Style (first field), Bedrooms (third field)?

```
1---+----10---+----20---+----30
2STORY 1810 4 SHEPPARD AVENUE
CONDO  1200 2 RAND STREET
RANCH  1550 3 MARKET STREET
```

a. `input Address 15-29 SquareFeet 8-11 Style 1-6`
   `     Bedrooms 13;`

b. `input $ 15-29 Address 8-11 SquareFeet $ 1-6 Style`
   `     13 Bedrooms;`

c. `input Address $ 15-29 SquareFeet 8-11 Style $ 1-6`
   `     Bedrooms 13;`

d. `input Address 15-29 $ SquareFeet 8-11 Style 1-6`
   `     $ Bedrooms 13;`

Correct answer: c

Column input specifies the variable's name, followed by a dollar ($) sign if the values are character values, and the beginning and ending column locations of the raw data values.

2. Which is *not* an advantage of column input?

    a. It can be used to read character variables that contain embedded blanks.

    b. No placeholder is required for missing data.

    c. Standard as well as nonstandard data values can be read.

    d. Fields do not have to be separated by blanks or other delimiters.

Correct answer: c

Column input is useful for reading standard values only.

3. Which is an example of standard numeric data?

    a. -34.245

    b. $24,234.25

    c. 1/2

    d. 50%

Correct answer: a

A standard numeric value can contain numbers, scientific notation, decimal points, and plus and minus signs. Nonstandard numeric data includes values that contain fractions or special characters such as commas, dollar signs, and percent signs.

4. Formatted input can be used to read

    a. standard free-format data

    b. standard data in fixed fields

    c. nonstandard data in fixed fields

    d. both standard and nonstandard data in fixed fields

Correct answer: d

Formatted input can be used to read both standard and nonstandard data in fixed fields.

5. Which informat should you use to read the values in column 1-5?

```
1---+----10---+----20---+----30
2STORY 1810 4 SHEPPARD AVENUE
CONDO  1200 2 RAND STREET
RANCH  1550 3 MARKET STREET
```

    a. *w.*

    b. *$w.*

    c. *w.d*

    d. COMMA*w.d*

Correct answer: b

The $w. informat enables you to read character data. The *w* represents the field width of the data value or the total number of columns that contain the raw data field.

6. The COMMA*w.d* informat can be used to read which of the following values?

   a. **12,805**

   b. **$177.95**

   c. **18%**

   d. all of the above

   Correct answer: d

   The COMMAw.d informat strips out special characters, such as commas, dollar signs, and percent signs, from numeric data and stores only numeric values in a SAS data set.

7. Which INPUT statement correctly reads the values for ModelNumber (first field) after the values for Item (second field)? Both Item and ModelNumber are character variables.

   ```
   1---+----10---+----20---+----30
   DG345 CD PLAYER    $174.99
   HJ756 VCR          $298.99
   AS658 CAMCORDER $1,195.99
   ```

   a. `input +7 Item $9. @1 ModelNumber $5.;`

   b. `input +6 Item $9. @1 ModelNumber $5.;`

   c. `input @7 Item $9. +1 ModelNumber $5.;`

   d. `input @7 Item $9 @1 ModelNumber 5.;`

   Correct answer: b

   The +6 pointer control moves the input pointer to the beginning column of Item, and the values are read. Then the @1 pointer control returns to column 1, where the values for ModelNumber are located.

8. Which INPUT statement correctly reads the numeric values for Cost (third field)?

   ```
   1---+----10---+----20---+----30
   DG345 CD PLAYER    $174.99
   HJ756 VCR          $298.99
   AS658 CAMCORDER $1,195.99
   ```

   a. `input @17 Cost 7.2;`

   b. `input @17 Cost 9.2.;`

   c. `input @17 Cost comma7.;`

   d. `input @17 Cost comma9.;`

   Correct answer: d

   The values for Cost contain dollar signs and commas, so you must use the COMMA*w.d* informat. Counting the numbers, dollar sign, comma, and decimal point, the field width is 9 columns. Because the data value contains decimal places, a d value is not needed.

9. Which SAS statement correctly uses formatted input to read the values in this order: Item (first field), UnitCost (second field), Quantity (third field)?

```
1---+----10---+----20---+
ENVELOPE  $13.25 500
DISKETTES $29.50  10
BANDS      $2.50 600
RIBBON    $94.20  12
```

a. `input @1 Item $9. +1 UnitCost comma6.`
   `@18 Quantity 3.;`

b. `input Item $9. @11 UnitCost comma6.`
   `@18 Quantity 3.;`

c. `input Item $9. +1 UnitCost comma6.`
   `@18 Quantity 3.;`

d. all of the above

Correct answer: d

The default location of the column pointer control is column 1, so a column pointer control is optional for reading the first field. You can use the $@n$ or $+n$ pointer controls to specify the beginning column of the other fields. You can use the $w.$ informat to read the values for Item, the COMMA$w.d$ informat for UnitCost, and the $w.d$ informat for Quantity.

10. Which raw data file requires the PAD option in the INFILE statement in order to correctly read the data using either column input or formatted input?

   a.

```
1---+----10---+----20---+
JONES     M 48 128.6
LAVERNE   M 58 158
JAFFE     F 33 115.5
WILSON    M 28 130
```

   b.

```
1---+----10---+----20---+
JONES     M 48 128.6
LAVERNE   M 58 158.0
JAFFE     F 33 115.5
WILSON    M 28 130.0
```

   c.

```
1---+----10---+----20---+
JONES     M 48 128.6
LAVERNE   M 58    158
JAFFE     F 33 115.5
WILSON    M 28    130
```

   d.

```
1---+----10---+----20---+
   JONES     M 48 128.6
 LAVERNE     M 58 158.0
   JAFFE     F 33 115.5
  WILSON     M 28 130.0
```

Correct answer: a

Use the PAD option in the INFILE statement to read variable-length records that contain fixed-field data. The PAD option pads each record with blanks so that all data lines have the same length.

Chapter 17: Reading Free-Format Data

1. The raw data file referenced by the fileref Students contains data that is

Raw Data File Students
```
1---+----10---+----20---+
FRED JOHNSON 18 USC 1
ASHLEY FERRIS 20 NCSU 3
BETH ROSEMONT 21 UNC 4
```

   a. arranged in fixed fields

   b. free-format

   c. mixed-format

   d. arranged in columns

   Correct answer: b

   The raw data file contains data that is free-format, meaning that the data is not arranged in columns or fixed fields.

2. Which input style should be used to read the values in the raw data file that is referenced by the fileref Students?

Raw Data File Students
```
1---+----10---+----20---+
FRED JOHNSON 18 USC 1
ASHLEY FERRIS 20 NCSU 3
BETH ROSEMONT 21 UNC 4
```

   a. column

   b. formatted

   c. list

   d. mixed

   Correct answer: c

   List input should be used to read data that is free-format because you do not need to specify the column locations of the data.

3. Which SAS program was used to create the raw data file Teamdat from the SAS data set Work.Scores?

SAS Data Set Work.Scores

Obs	Name	HighScore	Team
1	Joe	87	Blue Beetles, Durham
2	Dani	79	Raleigh Racers, Raleigh
3	Lisa	85	Sand Sharks, Cary
4	Matthew	76	Blue Beetles, Durham

Raw Data File Teamdat

```
1---+----10---+----20---+----30---+
Joe,87,"Blue Beetles, Durham"
Dani,79,"Raleigh Racers, Raleigh"
Lisa,85,"Sand Sharks, Cary"
Matthew,76,"Blue Beetles, Durham"
```

a.
```
data _null_;
    set work.scores;
    file 'c:\data\teamdat' dlm=',';
    put name highscore team;
run;
```

b.
```
data _null_;
    set work.scores;
    file 'c:\data\teamdat' dlm=' ';
    put name highscore team;
run;
```

c.
```
data _null_;
    set work.scores;
    file 'c:\data\teamdat' dsd;
    put name highscore team;
run;
```

d.
```
data _null_;
    set work.scores;
    file 'c:\data\teamdat';
    put name highscore team;
run;
```

Correct answer: c

You can use the DSD option in the FILE statement to specify that data values containing commas should be enclosed in quotation marks. The DSD option uses a comma as the delimiter by default.

4. Which SAS statement reads the raw data values in order and assigns them to the variables shown below?

Variables: FirstName (character), LastName (character), Age (numeric), School (character), Class (numeric)

Raw Data File Students

```
1---+----10---+----20---+
FRED JOHNSON 18 USC 1
ASHLEY FERRIS 20 NCSU 3
BETH ROSEMONT 21 UNC 4
```

a. `input FirstName $ LastName $ Age School $ Class;`

b. `input FirstName LastName Age School Class;`

c. `input FirstName $ 1-4 LastName $ 6-12 Age 14-15`
    `     School $ 17-19 Class 21;`

d. `input FirstName 1-4 LastName 6-12 Age 14-15`
    `     School 17-19 Class 21;`

Correct answer: a

Because the data is free-format, list input is used to read the values. With list input, you simply name each variable and identify its type.

5. Which SAS statement should be used to read the raw data file that is referenced by the fileref Salesrep?

Raw Data File Salesrep

```
1---+----10---+----20---+----30
ELAINE:FRIEDMAN:WILMINGTON:2102
JIM:LLOYD:20:RALEIGH:38392
JENNIFER:WU:21:GREENSBORO:1436
```

a. `infile salesrep;`

b. `infile salesrep ':';`

c. `infile salesrep dlm;`

d. `infile salesrep dlm=':';`

Correct answer: d

The INFILE statement identifies the location of the external data file. The DLM= option specifies the colon (:) as the delimiter that separates each field.

6. Which of the following raw data files can be read by using the MISSOVER option in the INFILE statement? Spaces for missing values are highlighted with colored blocks.

a.

```
1---+----10---+----20---+----
ORANGE SUNNYDALE 20 10
PINEAPPLE ALOHA 7 10
GRAPE FARMFRESH 3 ▓
APPLE FARMFRESH 16 5
GRAPEFRUIT SUNNYDALE 12 8
```

b.

```
1---+----10---+----20---+----
ORANGE SUNNYDALE 20 10
PINEAPPLE ALOHA 7 10
GRAPE FARMFRESH ▓  17
APPLE FARMFRESH 16 5
GRAPEFRUIT SUNNYDALE 12 8
```

c.

```
1---+----10---+----20---+----
ORANGE SUNNYDALE 20 10
PINEAPPLE ALOHA 7 10
GRAPE ▓▓▓▓▓▓ 3 17
APPLE FARMFRESH 16 5
GRAPEFRUIT SUNNYDALE 12 8
```

d.

```
1---+----10---+----20---+----
ORANGE SUNNYDALE 20 10
PINEAPPLE ALOHA 7 10
▓▓▓ FARMFRESH 3
APPLE FARMFRESH 16 5
GRAPEFRUIT SUNNYDALE 12 8
```

Correct answer: a

You can use the MISSOVER option in the INFILE statement to read the missing values at the end of a record. The MISSOVER option prevents SAS from moving to the next record if values are missing in the current record.

7. Which SAS program correctly reads the data in the raw data file that is referenced by the fileref Volunteer?

Raw Data File Volunteer

```
1---+----10---+----20---+----30|
ARLENE BIGGERSTAFF 19 UNC 2
JOSEPH CONSTANTINO 21 CLEM 2
MARTIN FIELDS 18 UNCG 1
```

a.
```
data perm.contest;
    infile volunteer;
    input FirstName $ LastName $ Age
        School $ Class;
run;
```

b.
```
data perm.contest;
    infile volunteer;
    length LastName $ 11;
    input FirstName $ lastname $ Age
        School $ Class;
run;
```

c.
```
data perm.contest;
    infile volunteer;
    input FirstName $ lastname $ Age
```

```
            School $ Class;  length LastName $ 11;
   run;
```

d.
```
   data perm.contest;
       infile volunteer;
       input FirstName $ LastName $ 11. Age
            School $ Class;
   run;
```

Correct answer: b

The LENGTH statement extends the length of the character variable LastName so that it is large enough to accommodate the data. Variable attributes such as length are defined the first time a variable is named in a DATA step. The LENGTH statement should precede the INPUT statement so that the correct length is defined.

8. Which type of input should be used to read the values in the raw data file that is referenced by the fileref University?

Raw Data File University

```
1---+----10---+----20---+----30
UNC ASHEVILLE  2,712
UNC CHAPEL HILL  24,189
UNC CHARLOTTE  15,031
UNC GREENSBORO  12,323
```

a. column

b. formatted

c. list

d. modified list

Correct answer: d

Notice that the values for School contain embedded blanks, and the values for Enrolled are nonstandard numeric values. Modified list input can be used to read the values that contain embedded blanks and nonstandard values.

9. Which SAS statement correctly reads the values for Flavor and Quantity? Make sure the length of each variable can accommodate the values shown.

Raw Data File Cookies

```
1---+----10---+----20---+----30
CHOCOLATE CHIP  10,453
OATMEAL  12,187
PEANUT BUTTER  11,546
SUGAR  12,331
```

a. `input Flavor & $9. Quantity : comma.;`

b. `input Flavor & $14. Quantity : comma.;`

c. `input Flavor : $14. Quantity & comma.;`

d. `input Flavor $14. Quantity : comma.;`

Correct answer: b

The INPUT statement uses list input with format modifiers and informats to read the values for each variable. The ampersand (&) modifier enables you to read character values that contain single embedded blanks. The colon (:) modifier enables you to read nonstandard data values and character values that are longer than eight characters, but which contain no embedded blanks.

10. Which SAS statement correctly reads the raw data values in order and assigns them to these corresponding variables: Year (numeric), School (character), Enrolled (numeric)?

    Raw Data File Founding
    ```
    1---+----10---+----20---+----30---+----40
    1868 U OF CALIFORNIA BERKELEY  31,612
    1906 U OF CALIFORNIA DAVIS  21,838
    1965 U OF CALIFORNIA IRVINE  15,874
    1919 U OF CALIFORNIA LOS ANGELES  35,730
    ```

    a. ```
       input Year School & $27.
             Enrolled : comma.;
       ```

    b. ```
       input Year 1-4 School & $27.
             Enrolled : comma.;
       ```

    c. ```
       input @1 Year 4. +1 School & $27.
             Enrolled : comma.;
       ```

    d. all of the above

    Correct answer: d

    The values for Year can be read with column, formatted, or list input. However, the values for School and Enrolled are free-format data that contain embedded blanks or nonstandard values. Therefore, these last two variables must be read with modified list input.

# Chapter 18: Reading Date and Time Values

1. SAS date values are the number of days since which date?

   a. January 1, 1900

   b. January 1, 1950

   c. January 1, 1960

   d. January 1, 1970

   Correct answer: c

   A SAS date value is the number of days from January 1, 1960, to the given date.

2. A great advantage of storing dates and times as SAS numeric date and time values is that

   a. they can easily be edited.

   b. they can easily be read and understood.

   c. they can be used in text strings like other character values.
       ```

    d. they can be used in calculations like other numeric values.

Correct answer: d

In addition to tracking time intervals, SAS date and time values can be used in calculations like other numeric values. This lets you calculate values that involve dates much more easily than in other programming languages.

3. SAS does not automatically make adjustments for daylight saving time, but it *does* make adjustments for:

    a. leap seconds

    b. leap years

    c. Julian dates

    d. time zones

Correct answer: b

SAS automatically makes adjustments for leap years.

4. An input data file has date expressions in the form 10222001. Which SAS informat should you use to read these dates?

    a. DATE6.

    b. DATE8.

    c. MMDDYY6.

    d. MMDDYY8.

Correct answer: d

The SAS informat MMDDYYw. reads dates such as 10222001, 10/22/01, or 10-22-01. In this case, the field width is eight.

5. The minimum width of the TIME*w*. informat is:

    a. 4

    b. 5

    c. 6

    d. 7

Correct answer: b

The minimum acceptable field width for the TIMEw. informat is five. If you specify a *w* value less than five, you will receive an error message in the SAS log.

6. Shown below are date and time expressions and corresponding SAS datetime informats. Which date and time expression *cannot* be read by the informat that is shown beside it?

    a. 30May2000:10:03:17.2 **DATETIME20.**

    b. 0May00 10:03:17.2 **DATETIME18.**

    c. 30May2000/10:03 **DATETIME15.**

    d. 30May2000/1003 **DATETIME14.**

Correct answer: d

In the time value of a date and time expression, you must use delimiters to separate the values for hour, minutes, and seconds.

7. What is the default value of the YEARCUTOFF= system option?

    a. `1920`

    b. `1910`

    c. `1900`

    d. `1930`

Correct answer: a

The default value of YEARCUTOFF= is `1920`. This enables you to read two-digit years from 00-19 as the years 2000 through 2019.

8. Suppose your input data file contains the date expression 13APR2009. The YEARCUTOFF= system option is set to 1910. SAS will read the date as:

    a. 13APR1909

    b. 13APR1920

    c. 13APR2009

    d. 13APR2020

Correct answer: c

The value of the YEARCUTOFF= system option does not affect four-digit year values. Four-digit values are always read correctly.

9. Suppose the YEARCUTOFF= system option is set to 1920. An input file contains the date expression 12/08/1925, which is being read with the MMDDYY8. informat. Which date will appear in your data?

    a. 08DEC1920

    b. 08DEC1925

    c. 08DEC2019

    d. 08DEC2025

Correct answer: c

The *w* value of the informat MMDDYY8. is too small to read the entire value, so the last two digits of the year are truncated. The last two digits thus become 19 instead of 25. Because the YEARCUTOFF= system option is set to 1920, SAS interprets this year as 2019. To avoid such errors, be sure to specify an informat that is wide enough for your date expressions.

10. Suppose your program creates two variables from an input file. Both variables are stored as SAS date values: FirstDay records the start of a billing cycle, and LastDay records the end of that cycle. The code for calculating the total number of days in the cycle would be:

    a. `TotDays=lastday-firstday;`

    b. `TotDays=lastday-firstday+1;`

    c. `TotDays=lastday/firstday;`

    d. You cannot use date values in calculations.

Correct answer: b

To find the number of days spanned by two dates, subtract the first day from the last day and add one. Because SAS date values are numeric values, they can easily be used in calculations.

Chapter 19: Creating a Single Observation from Multiple Records

1. You can position the input pointer on a specific record by using

   a. column pointer controls.

   b. column specifications.

   c. line pointer controls.

   d. line hold specifiers.

   Correct answer: c

   Information for one observation can be spread out over several records. You can write one INPUT statement that contains line pointer controls to specify the record(s) from which values are read.

2. Which pointer control is used to read multiple records sequentially?

   a. @*n*

   b. +*n*

   c. /

   d. all of the above

   Correct answer: c

   The forward slash (/) line pointer control is used to read multiple records sequentially. Each time a / pointer is encountered, the input pointer advances to the next line. @*n* and +*n* are column pointer controls.

3. Which pointer control can be used to read records non-sequentially?

   a. @*n*

   b. #*n*

   c. +*n*

   d. /

   Correct answer: b

   The #*n* line pointer control is used to read records non-sequentially. The #*n* specifies the absolute number of the line to which you want to move the pointer.

4. Which SAS statement correctly reads the values for Fname, Lname, Address, City, State and Zip in order?

```
1---+----10---+----20---
LAWRENCE CALDWELL
1010 LAKE STREET
ANAHEIM CA 94122
RACHEL CHEVONT
3719 OLIVE VIEW ROAD
HARTFORD CT 06183
```

a. ```
    input Fname $ Lname $ /
          Address $20. /
          City $ State $ Zip $;
    ```

b.  ```
    input Fname $ Lname $ /;
          Address $20. /;
          City $ State $ Zip $;
    ```

c. ```
    input / Fname $ Lname $
          / Address $20.
          City $ State $ Zip $;
    ```

d.  ```
    input / Fname $ Lname $;
          / Address $20.;
          City $ State $ Zip $;
    ```

Correct answer: a

The INPUT statement uses the / line pointer control to move the input pointer forward from the first record to the second record, and from the second record to the third record. The / line pointer control only moves the input pointer forward and must be specified after the instructions for reading the values in the current record. You should place a semicolon only at the end of a complete INPUT statement.

5. Which INPUT statement correctly reads the values for ID in the fourth record, and then returns to the first record to read the values for Fname and Lname?

```
1---+----10---+----20---
GEORGE CHESSON
3801 WOODSIDE COURT
GARNER NC 27529
XM065 FLOYD
JAMES COLDWELL
123-A TARBERT
APEX NC 27529
XM065 LAWSON
```

a. ```
    input #4 ID $5.
          #1 Fname $ Lname $;
    ```

b.  ```
    input #4 ID $ 1-5
          #1 Fname $ Lname $;
    ```

c. ```
    input #4 ID $
          #1 Fname $ Lname $;
    ```

d.  all of the above

Correct answer: d

The first #*n* line pointer control enables you to read the values for ID from the fourth record. The second #*n* line pointer control moves back to the first record and reads
    ```

the values for Fname and Lname. You can use formatted input, column input, or list input to read the values for ID.

6. How many records will be read for each execution of the DATA step?

```
1---+----10---+----20---
SKIRT BLACK
COTTON
036499 $44.98
SKIRT NAVY
LINEN
036899 $51.50
DRESS RED
SILK
037299 $76.98
```

```
data spring.sportswr;
   infile newitems;
   input #1 Item $ Color $
         #3 @8 Price comma6.
         #2 Fabric $
         #3 SKU $ 1-6; run;
```

a. one

b. two

c. three

d. four

Correct answer: c

The first time the DATA step executes, the first three records are read, and an observation is written to the data set. During the second iteration, the next three records are read, and the second observation is written to the data set. During the third iteration, the last three records are read, and the final observation is written to the data set.

7. Which INPUT statement correctly reads the values for City, State, and Zip?

```
1---+----10---+----20---
DINA   FIELDS
904   MAPLE   CIRCLE
DURHAM   NC   27713
ELIZABETH   GARRISON
1293   OAK   AVENUE
CHAPEL HILL   NC   27614
DAVID   HARRINGTON
2426   ELMWOOD   LANE
RALEIGH   NC   27803
```

a. `input #3 City $ State $ Zip $;`

b. `input #3 City & $11. State $ Zip $;`

c. `input #3 City $11. +2 State $2. + 2 Zip $5.;`

d. all of the above

Correct answer: b

A combination of modified and simple list input can be used to read the values for City, State, and Zip. You need to use modified list input to read the values for City, because one of the values is longer than eight characters and contains an embedded blank. You cannot use formatted input, because the values do not begin and end in the same column in each record.

8. Which program does *not* read the values in the first record as a variable named Item and the values in the second record as two variables named Inventory and Type?

```
1---+----10---+----20---
COLORED PENCILS
12 BOXES
WATERCOLOR PAINT
8 PALETTES
DRAWING PAPER
15 PADS
```

a.
```
data perm.supplies;
    infile instock pad;
    input Item & $16. /
        Inventory 2. Type $8.;
run;
```

b.
```
data perm.supplies;
    infile instock pad;
    input Item & $16.
        / Inventory 2. Type $8.;
  run;
```

c.
```
data perm.supplies;
    infile instock pad;
    input #1 Item & $16.
        Inventory 2. Type $8.;
run;
```

d.
```
data perm.supplies;
    infile instock pad;
    input Item & $16.
        #2 Inventory 2. Type $8.;
run;
```

Correct answer: c

The values for Item in the first record are read, then the following / or #*n* line pointer control advances the input pointer to the second record, to read the values for Inventory and Type.

9. Which INPUT statement reads the values for Lname, Fname, Department and Salary (in that order)?

```
1---+----10---+----20---
ABRAMS THOMAS
SALES $25,209.03
BARCLAY ROBERT
MARKETING $29,180.36
COURTNEY MARK
PUBLICATIONS $24,006.16
```

a. ```
input #1 Lname $ Fname $ /
      Department $12. Salary comma10.;
```

b. ```
input #1 Lname $ Fname $ /
      Department : $12. Salary : comma.;
```

c. ```
input #1 Lname $ Fname $
      #2 Department : $12. Salary : comma.;
```

d. both b and c

Correct answer: d

You can use either the / or #*n* line pointer control to advance the input pointer to the second line, in order to read the values for Department and Salary. The colon (:) modifier is used to read the character values that are longer than eight characters (Department) and the nonstandard data values (Salary).

10. Which raw data file poses potential problems when you are reading multiple records for each observation?

a.

```
1---+----10---+----20---
LAWRENCE CALDWELL
1010 LAKE STREET
ANAHEIM CA 94122
RACHEL CHEVONT
3719 OLIVE VIEW ROAD
HARTFORD CT 06183
```

b.

```
1---+----10---+----20---
SHIRT LT BLUE SOLID
SKU 128699
$38.99
SHIRT DK BLUE STRIPE
SKU 128799
$41.99
```

c.

```
1---+----10---+----20---
MARCUS JONES
SR01 $26,134.00
MARY ROBERTSON
COURTNEY NEILS
TWO1 $28,342.00
```

d.

```
1---+----10---+----20---
CROCUS MIX
10 CASES
DAFFODIL
12 CASES
HYACINTH BLUE
8 BAGS
```
```

Correct answer: c

The third raw data file does not contain the same number of records for each observation, so the output from this data set will show invalid data for the ID and salary information in the fourth line.

Chapter 20: Creating Multiple Observations from a Single Record

1. Which is true for the double trailing at sign (@@)?

    a. It enables the next INPUT statement to read from the current record across multiple iterations of the DATA step.

    b. It must be the last item specified in the INPUT statement.

    c. It is released when the input pointer moves past the end of the record.

    d. all of the above

    Correct answer: d

    The double trailing at sign (@@) enables the next INPUT statement to read from the current record across multiple iterations of the DATA step. It must be the last item specified in the INPUT statement. A record that is being held by the double trailing at sign (@@) is not released until the input pointer moves past the end of the record, or until an INPUT statement that has no line-hold specifier executes.

2. A record that is being held by a single trailing at sign (@) is automatically released when

    a. the input pointer moves past the end of the record.

    b. the next iteration of the DATA step begins.

    c. another INPUT statement that has an @ executes.

    d. another value is read from the observation.

    Correct answer: b

    Unlike the double trailing at sign (@@), the single trailing at sign (@) is automatically released when control returns to the top of the DATA step for the next iteration. The trailing @ does not toggle on and off. If another INPUT statement that has a trailing @ executes, the holding effect is still on.

3. Which SAS program correctly creates a separate observation for each block of data?

```
1---+----10---+----20---+----30---+----40---+
1001 apple 1002 banana 1003 cherry
1004 guava 1005 kiwi 1006 papaya
1007 pineapple 1008 raspberry 1009 strawberry
```

    a. ```
data perm.produce;
    infile fruit;
    input Item $ Variety : $10.;
run;
```
```

b. ```
data perm.produce;
    infile fruit;
    input Item $ Variety : $10. @;
run;
```

c. ```
data perm.produce;
    infile fruit;
    input Item $ Variety : $10. @@;
run;
```

d. ```
data perm.produce;
    infile fruit @@;
    input Item $ Variety : $10.;
run;
```

Correct answer: c

Each record in this file contains three repeating blocks of data values for Item and Variety. The INPUT statement reads a block of values for Item and Variety, and then holds the current record by using the double-trailing at sign (@@). The values in the program data vector are written to the data set as the first observation. In the next iteration, the INPUT statement reads the next block of values for Item and Variety from the same record.

4. Which SAS program reads the values for ID and holds the record for each value of Quantity, so that three observations are created for each record?

```
1---+----10---+----20---+----30
2101 21,208 19,047 22,890
2102 18,775 20,214 22,654
2103 19,763 22,927 21,862
```

a. ```
data work.sales;
    infile unitsold;
    input ID $;
    do week=1 to 3;
       input Quantity : comma.;
       output;
    end;
run;
```

b. ```
data work.sales;
    infile unitsold;
    input ID $ @@;
    do week=1 to 3;
       input Quantity : comma.;
       output;
    end;
run;
```

c. ```
data work.sales;
    infile unitsold;
    input ID $ @;
    do week=1 to 3;
       input Quantity : comma.;
       output;
    end;
run;
```

```
d.  data work.sales;
        infile unitsold;
        input ID $ @;
        do week=1 to 3;
            input Quantity : comma. @;
            output;
        end;
    run;
```

Correct answer: d

This raw data file contains an ID field followed by repeating fields. The first INPUT statement reads the values for ID and uses the @ line-hold specifier to hold the current record for the next INPUT statement in the DATA step. The second INPUT statement reads the values for Quantity. When all of the repeating fields have been read, control returns to the top of the DATA step, and the record is released.

5. Which SAS statement repetitively executes several statements when the value of an index variable named count ranges from **1** to **50**, incremented by 5?

   a. `do count=1 to 50 by 5;`

   b. `do while count=1 to 50 by 5;`

   c. `do count=1 to 50 + 5;`

   d. `do while (count=1 to 50 + 5);`

Correct answer: a

The iterative DO statement begins the execution of a loop based on the value of an index variable. Here, the loop executes when the value of count ranges from **1** to **50**, incremented by 5.

6. Which option below, when used in a DATA step, writes an observation to the data set after each value for Activity has been read?

   a.
```
do choice=1 to 3;
    input Activity : $10. @;
    output;
end;
run;
```

   b.
```
do choice=1 to 3;
    input Activity : $10. @;
end;
output;
run;
```

   c.
```
do choice=1 to 3;
    input Activity : $10. @;
end;
run;
```

   d. both a and b

Correct answer: a

The OUTPUT statement must be in the loop so that each time a value for Activity is read, an observation is immediately written to the data set.

7. Which SAS statement repetitively executes while the value of Cholesterol is greater than **200**?

a. ```
do cholesterol > 200;
```

b. ```
do cholesterol gt 200;
```

c. ```
do while (cholesterol > 200);
```

d. ```
do while cholesterol > 200;
```

Correct answer: c

The DO WHILE statement checks for the condition that Cholesterol is greater than 200. The expression must be enclosed in parentheses. The expression is evaluated at the top of the loop before the loop executes. If the condition is true, the DO WHILE loop executes. If the expression is false the first time it is evaluated, the loop does not execute.

8. Which choice below is an example of a sum statement?

a. ```
totalpay=1;
```

b. ```
totalpay+1;
```

c. ```
totalpay*1;
```

d. ```
totalpay by 1;
```

Correct answer: b

The sum statement adds the result of an expression to an accumulator variable. The + sign is an essential part of the sum statement. Here, the value of TotalPay is incremented by 1.

9. Which program creates the SAS data set Perm.Topstore from the raw data file shown below?

```
1---+----10---+----20---+----30---+
1001 77,163.19 76,804.75 74,384.27
1002 76,612.93 81,456.34 82,063.97
1003 82,185.16 79,742.33
```

SAS Data Set Perm.Topstore

Store	Sales	Month
1001	77163.19	1
1001	76804.75	2
1001	74384.27	3
1002	76612.93	1
1002	81456.34	2
1002	82063.97	3
1003	82185.16	1
1003	79742.33	2

a. ```
data perm.topstores;
   infile sales98 missover;
   input Store Sales : comma. @;
   do while (sales ne .);
      month + 1;
      output;
```
```

```
                input sales : comma. @;
            end;
        run;
```

b.
```
data perm.topstores;
    infile sales98 missover;
    input Store Sales : comma. @;
    do while (sales ne .);
        Month=0;
        month + 1;
        output;
        input sales : comma. @;
    end;
run;
```

c.
```
data perm.topstores;
    infile sales98 missover;
    input Store Sales : comma.
    Month @;
    do while (sales ne .);
        month + 1;
        input sales : comma. @;
    end;
    output;
run;
```

d.
```
data perm.topstores;
    infile sales98 missover;
    input Store Sales : comma. @;
    Month=0;
    do while (sales ne .);
        month + 1;
        output;
        input sales : comma. @;
    end;
run;
```

Correct answer: d

The first input statement reads STORE and SALES and holds the record. Month is initialized to 0. The DO loop executes while SALES is not missing. Inside the loop, month increments, an observation is output, the next SALES value is read, and the record is held. The loop stops when a missing value of SALES is read. Control returns to the top of the DATA step, the held record is released, and the input statement reads the next record.

10. How many observations are produced by the DATA step that reads this external file?

```
1---+----10---+----20---+----30---+----40
01 CHOCOLATE VANILLA RASPBERRY
02 VANILLA PEACH
03 CHOCOLATE
04 RASPBERRY PEACH CHOCOLATE
05 STRAWBERRY VANILLA CHOCOLATE
```

```
data perm.choices;
    infile icecream missover;
    input Day $ Flavor : $10. @;
```

```
        do while (flavor ne ' ');
            output;
        input flavor : $10. @;
        end;
   run;
```

a. 3

b. 5

c. 12

d. 15

Correct answer: c

This DATA step produces one observation for each repeating field. The MISSOVER option in the INFILE statement prevents SAS from reading the next record when missing values occur at the end of a record. Every observation contains one value for Flavor, paired with the corresponding value for ID. Because there are 12 values for Flavor, there are 12 observations in the data set.

Chapter 21: Reading Hierarchical Files

1. When you write a DATA step to create one observation per detail record you need to

   a. distinguish between header and detail records.

   b. keep header data as a part of each observation until the next header record is encountered.

   c. hold the current record so other values in the record can be read.

   d. all of the above

   Correct answer: d

   In order to create one observation per detail record, it is necessary to distinguish between header and detail records. Use a RETAIN statement to keep header data as part of each observation until the next header record is encountered. You also need to use the @ line-hold specifier to hold the current record so other values in the record can be read.

2. Which SAS statement reads the value for code (the first field), and holds the record until an INPUT statement reads the remaining value from the same record in the same iteration of the DATA step?

   ```
   1---+----10---+----20---+----30
   01 Office Supplies
   02 Paper Clips
   02 Tape
   01 Art Supplies
   02 Crayons
   02 Finger Paint
   02 Construction Paper
   ```

   a. `input code $2. @;`

b. `input code $2. @@;`

c. `retain code;`

d. none of the above

Correct answer: a

An INPUT statement is used to read the value for code. The single @ sign at the end of the INPUT statement holds the current record for a later INPUT statement in the same iteration of the DATA step.

3. Which SAS statement checks for the condition that Record equals C and executes a single statement to read the values for Amount?

a. `if record=c then input @3 Amount comma7.;`

b. `if record='C' then input @3 Amount comma7.;`

c. `if record='C' then do input @3 Amount comma7.;`

d. `if record=C then do input @3 Amount comma7.;`

Correct answer: b

The IF-THEN statement defines the condition that Record equals C and executes an INPUT statement to read the value for Amount when the condition is true. C must be enclosed in quotation marks and must be specified exactly as shown because it is a character value.

4. After the value for code is read in the sixth iteration, which illustration of the program data vector is correct?

```
1---+----10---+----20---+----30
H Lettuce
P Green Leaf   Quality Growers
P Iceberg      Pleasant Farm
P Romaine      Quality Growers
H Squash
P Yellow       Tasty Acres
P Zucchini     Pleasant Farm
```

```
data perm.produce (drop=code);
   infile orders;
   retain Vegetable;
   input code $1. @;
   if code='H' then input @3 vegetable $6.;
   if code='P';
   input @3 Variety : $10. @15 Supplier : $15.;
run;
proc print data=perm.produce;
run;
```

a.

_N_	Vegetable	code	Variety	Supplier
6		P		

b.

_N_	Vegetable	code	Variety	Supplier
6	Squash	P		

c.

_N_	Vegetable	code	Variety	Supplier
6	Squash	H		

d.

_N_	Vegetable	code	Variety	Supplier
6	Squash	H	Yellow	

Correct answer: b

The value of Vegetable is retained across iterations of the DATA step. As the sixth iteration begins, the INPUT statement reads the value for code and holds the record, so that the values for Variety and Supplier can be read with an additional INPUT statement.

5. What happens when the fourth iteration of the DATA step is complete?

```
1---+----10---+----20---+----30
F Apples
V Gala              $2.50
V Golden Delicious $1.99
V Rome              $2.35
F Oranges
V Navel             $2.79
V Temple            $2.99
```

```
data perm.orders (drop=type);
   infile produce;
   retain Fruit;
   input type $1. @;
   if type='F' then input @3 fruit $7.;
   if type='V';
   input @3 Variety : $16. @20 Price comma5.;
run;
```

a. All of the values in the program data vector are written to the data set as the third observation.

b. All of the values in the program data vector are written to the data set as the fourth observation.

c. The values for Fruit, Variety, and Price are written to the data set as the third observation.

d. The values for Fruit, Variety, and Price are written to the data set as the fourth observation.

Correct answer: c

This program creates one observation for each detail record. The RETAIN statement retains the value for Fruit as part of each observation until the values for Variety and

Price can be read. The DROP= option in the DATA statement prevents the variable for type from being written to the data set.

6. Which SAS statement indicates that several other statements should be executed when Record has a value of **A**?

```
1---+----10---+----20---+----30
A 124153-01
C $153.02
D $20.00
D $20.00
```

a. `if record='A' then do;`

b. `if record=A then do;`

c. `if record='A' then;`

d. `if record=A then;`

Correct answer: a

The IF-THEN statement defines the condition that Record equals A and specifies a simple DO group. The keyword DO indicates that several executable statements follow until the DO group is closed by an END statement. The value A must be enclosed in quotation marks and specified exactly as shown because it is a character value.

7. Which is true for the following statements (X indicates a header record)?

```
if code='X' then do;
   if _n_ > 1 then output;
   Total=0;
   input Name $ 3-20;
end;
```

a. _N_ equals the number of times the DATA step has begun to execute.

b. When code='X' and _n_ > 1 are true, an OUTPUT statement is executed.

c. Each header record causes an observation to be written to the data set.

d. a and b

Correct answer: d

_N_ is an automatic variable whose value is the number of times the DATA step has begun to execute. The expression _n_ > 1 defines a condition where the DATA step has executed more than once. When the conditions code='X' and _n_ > 1 are true, an OUTPUT statement is executed, and Total is initialized to zero. Thus, each header record except the first one causes an observation to be written to the data set.

8. What happens when the condition type='P' is false?

```
if type='P' then input @3 ID $5. @9 Address $20.;
else if type='V' then input @3 Charge 6.;
```

a. The values for ID and Address are read.

b. The values for Charge are read.

c. type is assigned the value of **v**.

d. The ELSE statement is executed.

Correct answer: d

The condition is false, so the values for ID and Address are not read. Instead, the ELSE statement is executed and defines another condition which may or may not be true.

9. What happens when last has a value other than zero?

```
data perm.househld (drop=code);
   infile citydata end=last;
   retain Address;
   input type $1. @;
   if code='A' then do;
      if _n_ > 1 then output;
      Total=0;
      input address $ 3-17;
   end;
   else if code='N' then total+1;
   if last then output;
run;
```

a. last has a value of **1**.

b. The OUTPUT statement writes the last observation to the data set.

c. The current value of last is written to the DATA set.

d. a and b

Correct answer: d

You can determine when the current record is the last record in an external file by specifying the END= option in the INFILE statement. last is a temporary numeric variable whose value is zero until the last line is read. last has a value of **1** after the last line is read. Like automatic variables, the END= variable is not written to the data set.

10. Based on the values in the program data vector, what happens next?

_N_	last	Department	Extension	type	Total	Amount
5	0	Accounting	x3808	D	16.50	.

```
1---+----10---+----20---+----30
D Accounting   x3808
S Paper Clips  $2.50
S Paper        $4.95
S Binders      $9.05
D Personnel    x3810
S Markers      $8.98
S Pencils      $3.35
```

```
data work.supplies (drop=type amount);
   infile orders end=last;
   retain Department Extension;
   input type $1. @;
   if type='D' then do;
      if _n_ > 1 then output;
      Total=0;
      input @3 department $10. @16 extension $5.;
   end;
```

```
        else if type='S' then do;
           input @16 Amount comma5.;
           total+amount;
           if last then output;
            end;
run;
```

a. All the values in the program data vector are written to the data set as the first observation.

b. The values for Department, Total, and Extension are written to the data set as the first observation.

c. The values for Department, Total, and Extension are written to the data set as the fourth observation.

d. The value of last changes to **1**.

Correct answer: b

This program creates one observation for each header record and combines information from each detail record into the summary variable, Total. When the value of type is **D** and the value of _N_ is greater than **1**, the OUTPUT statement executes, and the values for Department, Total and Extension are written to the data set as the first observation. The variables _N_ , last, type and Amount are not written to the data set.

Glossary

aggregate storage location

a location in an operating system that can contain a group of distinct files. The exact name for this location varies by operating system; for example, directory, folder, or partitioned data set.

analysis variable

a numeric variable that is used to calculate statistics or to display values. Usually an analysis variable contains quantitative or continuous values, but this is not required.

arithmetic expression

See SAS expression

arithmetic operator

in SAS, any of the symbols (+, -, /, *, and **) that are used to perform addition, subtraction, division, multiplication, or exponentiation in arithmetic expressions. In SYSTEM 2000 software only, ** is not supported.

array

in the SAS programming language, a temporary grouping of SAS variables that have the same data type and that are arranged in a particular order and identified by an array name. The array exists only for the duration of the current DATA step.

array name

a name that is selected to identify a group of variables or temporary data elements. It must be a valid SAS name that is not the name of a variable in the same DATA step or SCL (SAS Component Language) program.

array reference

a reference to an element to be processed in an array.

attribute

See variable attribute

Base SAS

the core product that is part of SAS Foundation and is installed with every deployment of SAS software. Base SAS provides an information delivery system for accessing, managing, analyzing, and presenting data.

Base SAS software

See Base SAS

BY group

a group of observations or rows that have the same value for a variable that is specified in a BY statement. If more than one variable is specified in a BY statement, then the BY group is a group of observations that have a unique combination of values for those variables.

BY group variable

See BY variable

BY value

the value of a BY variable.

BY variable

a variable that is named in a BY statement and whose values define groups of observations to process.

BY-group processing

the process of using the BY statement to process observations that are ordered, grouped, or indexed according to the values of one or more variables. Many SAS procedures and the DATA step support BY-group processing. For example, you can use BY-group processing with the PRINT procedure to print separate reports for different groups of observations in a single SAS data set.

catalog

See SAS catalog

character format

a set of instructions that tell SAS to use a specific pattern for writing character data values.

character function

a type of function that enables you to manipulate, compare, evaluate, or analyze character strings.

character informat

a set of instructions that tell SAS to use a specific pattern for reading character data values into character variables.

character value

a value that can contain alphabetic characters, the numeric characters 0 through 9, and other special characters.

character variable

a variable whose values can consist of alphabetic characters and special characters as well as numeric characters.

class variable

See classification variable

classification variable

a variable whose values are used to group (or classify) the observations in a data set into different groups that are meaningful for analysis. A classification variable can have either character or numeric values. Classification variables include group, subgroup, category, and BY variables.

code editing window

a generic term for any window into which users can type or paste program code, or in which they can make changes to existing programs.

column input

in the DATA step, a style of input in which column numbers are included in the INPUT statement to tell SAS which columns contain the values for each variable. This style of input is useful when the values for each variable are in the same location in all records.

comment

See comment statement

comment statement

information that is embedded in a SAS program and that serves as explanatory text. SAS ignores comments during processing but writes them to the SAS log. Comment syntax has several forms. For example, a comment can appear as a statement that begins with an asterisk and ends with a semicolon, as in * message ;.

comparison operator

in programming languages, a symbol or mnemonic code that is used in expressions to test for a particular relationship between two values or text strings. For example, the symbol < and its corresponding mnemonic, LT, are used to determine whether one value is less than another.

compilation

See program compilation

compound expression

an expression that contains more than one operator.

concatenate

to join the contents of two or more elements, end to end, forming a separate element. Examples of elements are character values, tables, external files, SAS data sets, and SAS libraries.

condition

in a SAS program, one or more numeric or character expressions that result in a value on which some decision depends.

constant

in SAS software, a number or a character string that indicates a fixed value.

data error

a type of execution error that occurs when a SAS program analyzes data that contains invalid values. For example, a data error occurs if you specify numeric variables in the INPUT statement for character data. SAS reports these errors in the SAS log but continues to execute the program.

data set

See SAS data set

data set option

a SAS language element that specifies actions that apply to only one particular SAS data set. For example, data set options enable you to rename variables, to select only the first or last n observations for processing, to drop variables from processing or from the output data set, and to specify a password for a SAS data set.

DATA step

in a SAS program, a group of statements that begins with a DATA statement and that ends with either a RUN statement, another DATA statement, a PROC statement, or the end of the job. The DATA step enables you to read raw data or other SAS data sets and to create SAS data sets.

data value

a unit of character, numeric, or alphanumeric information. This unit is stored as one item in a data record, such as a person's height being stored as one variable (namely, a column or vertical component) in an observation (row).

data view

See SAS data view

date and time format

instructions that tell SAS how to write numeric values as dates, times, and datetimes.

date and time informat

the instructions that tell SAS how to read numeric values that are represented as dates, times, and datetimes.

date constant

See SAS date constant

date value

See SAS date value

datetime constant

See SAS datetime constant

datetime value

See SAS datetime value

delimiter

a character that serves as a boundary that separates the elements of a text string.

descriptor information

information about the contents and attributes of a SAS data set. For example, the descriptor information includes the data types and lengths of the variables, as well as which engine was used to create the data. SAS creates and maintains descriptor information within every SAS data set.

directory

a named subdivision on a computer disk, used in organizing files and metadata about files.

DO group

a sequence of statements that starts with a simple DO statement and that ends with a corresponding END statement.

DO loop

a sequence of statements that starts with an iterative DO, DO WHILE, or DO UNTIL statement and that ends with a corresponding END statement. The statements are executed (usually repeatedly) according to directions that are specified in the DO statement.

double trailing at sign

a special symbol @@ that is used to hold a line of data in the input buffer during multiple iterations of a DATA step.

engine

a component of SAS software that reads from or writes to a file. Various engines enable SAS to access different types of file formats.

external file

a file that is created and maintained by a host operating system or by another vendor's software application. An external file can read both data and stored SAS statements.

field

the smallest logical unit of data in a file.

file reference

See fileref

fileref

a name that is temporarily assigned to an external file or to an aggregate storage location such as a directory or a folder. The fileref identifies the file or the storage location to SAS.

FIRST. variable

a temporary variable that SAS creates to identify the first observation of each BY group. The variable is not added to the SAS data set.

floating-point representation

a compact form of storing real numbers on a computer, similar to scientific notation. Different operating systems use different techniques for floating-point representation.

format modifier

a special symbol that is used in the INPUT and PUT statements and which enables you to control how SAS reads input data and writes output data.

formatted input

a style of input that uses special instructions called informats in the INPUT statement to determine how values that are entered in data fields should be interpreted.

function

See SAS function

informat

See SAS informat

input buffer

a temporary area of memory into which each record of data is read when the INPUT statement executes.

integer

a number that does not contain a decimal point.

interleaving

a process in which SAS combines two or more sorted SAS data sets into one sorted SAS data set based on the values of the BY variables.

keyword

See SAS keyword

LAST. variable

a temporary variable that SAS creates to identify the last observation of each BY group. This variable is not added to the SAS data set.

length variable

a numeric variable that is used with the $VARYING. informat or format to specify the actual length of a character variable whose values do not all have the same length.

library member

any of several types of SAS file in a SAS library. A library member can be a data set, a view, a catalog, a stored program, or an access descriptor.

library reference

See libref

libref

a SAS name that is associated with the location of a SAS library. For example, in the name MYLIB.MYFILE, MYLIB is the libref, and MYFILE is a file in the SAS library.

line-hold specifier

a special symbol used in INPUT and PUT statements that enables you to hold a record in the input or output buffer for further processing. Line-hold specifiers include the trailing at sign (@) and the double trailing at sign (@@).

list input

a style of input in which names of variables, not column locations, are specified in the INPUT statement. List input scans input records for data values that are separated by at least one blank or by some other delimiter.

LISTING output

SAS procedure output that is in a monospace font. All text in listing output has the same font size, and no special font styles are applied to it.

log

See SAS log

logical data model

a framework into which engines fit information for processing by SAS. This data model is a logical representation of data or files, not a physical structure.

logical operator

an operator that is used in expressions to link sequences of comparisons. The logical operators are AND, OR, and NOT.

match-merging

a process in which SAS joins observations from two or more SAS data sets according to the values of the BY variables.

member name

a name that is assigned to a SAS file in a SAS library.

merging

the process of combining observations from two or more SAS data sets into a single observation in a new SAS data set.

missing value

a type of value for a variable that contains no data for a particular row or column. By default, SAS writes a missing numeric value as a single period and a missing character value as a blank space.

modified list input

a style of input that uses special instructions called informats and format modifiers in the INPUT statement. Modified list input scans input records for data values that are separated by at least one blank (or by some other delimiter), or in some cases, by multiple blanks.

numeric format

a set of instructions that tell SAS to use a specific pattern for writing the values of numeric variables.

numeric informat

a set of instructions that tell SAS to use a specific pattern for reading numeric data values.

numeric value

a value that usually contains only numbers, which can include numbers in E-notation and hexadecimal notation. A numeric value can sometimes contain a decimal point, a plus sign, or a minus sign. Numeric values are stored in numeric variables.

observation

a row in a SAS data set. All of the data values in an observation are associated with a single entity such as a customer or a state. Each observation contains either one data value or a missing-value indicator for each variable.

one-level name

a one-part name for a SAS file in which the libref (library reference) is omitted and only the filename is specified. When you specify a one-level name, the default temporary library, Work, is assumed.

one-to-one matching

the process of combining observations from two or more data sets into one observation, using two or more SET statements to read observations independently from each data set.

one-to-one merging

the process of using the MERGE statement (without a BY statement) to combine observations from two or more data sets based on the observations' positions in the data sets.

operand

any of the variables and constants in an expression that contain operators, variables, and constants.

operator

See SAS operator

page size

the number of bytes of data that SAS moves between external storage and memory in one input/output operation. Page size is analogous to buffer size for SAS data sets.

PDV

See program data vector

permanent SAS data library

a SAS library that is not deleted when a SAS session ends, and which is therefore available to subsequent SAS sessions.

permanent SAS file

a file in a SAS library that is not deleted when the SAS session or job terminates.

physical filename

the name that an operating system uses to identify a file.

pointer

in the DATA step, a programming tool that SAS uses to keep track of its position in the input or output buffer.

pointer control

the process of instructing SAS to move the pointer before reading or writing data.

PROC

See SAS procedure

PROC step

a group of SAS statements that call and execute a SAS procedure. A PROC step usually takes a SAS data set as input.

procedure

See SAS procedure

procedure output file

an external file that contains the result of the analysis that a SAS procedure performs or the report that the procedure produces. Most SAS procedures write output to the procedure output file by default. Reports that are produced by SAS DATA steps, using PUT statements and a FILE statement along with a PRINT destination, also go to this file.

program compilation

the process of checking syntax and translating a portion of a program into a form that the computer can execute.

program data vector

the temporary area of computer memory in which SAS builds a SAS data set, one observation at a time. The program data vector is a logical concept and does not necessarily correspond to a single contiguous area of memory. Short form: PDV.

programming error

a flaw in the logic of a SAS program that can cause the program to fail or to perform differently than the programmer intended.

raw data

in statistical analysis, data (including data in SAS data sets) that has not had a particular operation, such as standardization, performed on it.

SAS catalog

a SAS file that stores many different kinds of information in smaller units called catalog entries. A single SAS catalog can contain different types of catalog entries.

SAS data file

a type of SAS data set that contains data values as well as descriptor information that is associated with the data. The descriptor information includes information such as the data types and lengths of the variables, as well as the name of the engine that was used to create the data.

SAS data set

a file whose contents are in one of the native SAS file formats. There are two types of SAS data sets: SAS data files and SAS data views. SAS data files contain data values in addition to descriptor information that is associated with the data. SAS data views contain only the descriptor information plus other information that is required for retrieving data values from other SAS data sets or from files whose contents are in other software vendors' file formats.

SAS data set option

an option that appears in parentheses after a SAS data set name. Data set options specify actions that apply only to the processing of that SAS data set.

SAS data view

a type of SAS data set that retrieves data values from other files. A SAS data view contains only descriptor information such as the data types and lengths of the variables (columns) plus other information that is required for retrieving data values from other SAS data sets or from files that are stored in other software vendors' file formats. Short form: data view.

SAS date constant

a string in the form 'ddMMMyy'd or 'ddMMMyyyy'd that represents a date in a SAS statement. The string is enclosed in quotation marks and is followed by the character d (for example, '6JUL01'd, '06JUL01'd, '6 JUL2001'd, or '06JUL2001'd).

SAS date value

an integer that represents a date in SAS software. The integer represents the number of days between January 1, 1960, and another specified date. For example, the SAS date value 366 represents the calendar date January 1, 1961.

SAS datetime constant

a string in the form 'ddMMMyy:hh:mm:ss'dt or 'ddMMMyyyy:hh:mm:ss'dt that represents a date and time in SAS. The string is enclosed in quotation marks and is followed by the characters dt (for example, '06JUL2001:09:53:22'dt).

SAS datetime value

an integer that represents a date and a time in SAS software. The integer represents the number of seconds between midnight, January 1, 1960, and another specified date and time. For example, the SAS datetime value for 9:30 a.m., June 5, 2000, is 1275816600.

SAS Enterprise Guide

a software application with a point-and-click interface that gives users access to the functionality of many components of SAS software. Interactive dialog boxes guide users through data analysis tasks and reporting tasks, and users can easily export the results of those tasks to other Windows applications or servers. SAS Enterprise Guide provides access not only to SAS data files, but also to data that is in a wide variety of other software vendors' formats and in other operating system formats.

SAS expression

a type of macro expression consisting of a sequence of operands and arithmetic operators that form a set of instructions that are evaluated to produce a numeric value, a character value, or a Boolean value. Examples of operands are constants and system functions. SAS uses arithmetic expressions in program statements to create variables, to assign values, to calculate new values, to transform variables, and to perform conditional processing.

SAS file

a specially structured file that is created, organized, and maintained by SAS. A SAS file can be a SAS data set, a catalog, a stored program, an access descriptor, a utility file, a multidimensional database file, a financial database file, a data mining database file, or an item store file.

SAS function

a type of SAS language element that can be used in an expression or assignment statement to process zero or more arguments and to return a value. Examples of SAS functions are MEAN and SUM. Short form: function.

SAS informat

a type of SAS language element that applies a pattern to or executes instructions for a data value to be read as input. Types of informats correspond to the data's type: numeric, character, date, time, or timestamp. The ability to create user-defined informats is also supported. Examples of SAS informats are BINARY and DATE. Short form: informat.

SAS keyword

a literal that is a primary part of the SAS language. For example, SAS keywords include DATA, PROC, RUN, names of SAS language elements, names of SAS statement options, and system variables.

SAS library

one or more files that are defined, recognized, and accessible by SAS and that are referenced and stored as a unit. Each file is a member of the library.

SAS log

a file that contains a record of the SAS statements that you enter, as well as messages about the execution of your program.

SAS name

a name that is assigned to items such as SAS variables and SAS data sets. For most SAS names, the first character must be a letter or an underscore. Subsequent characters can be letters, numbers, or underscores. Blanks and special characters (except the underscore) are not allowed. However, the VALIDVARNAME= system option determines what rules apply to SAS variable names. The maximum length of a SAS name depends on the language element that it is assigned to.

SAS operator

in a SAS expression, any of several symbols that request a comparison, a logical operation, or an arithmetic calculation.

SAS procedure

a program that provides specific functionality and that is accessed with a PROC statement. For example, SAS procedures can be used to produce reports, to manage files, or to analyze data. Many procedures are included in SAS software.

SAS program

a group of SAS statements that guide SAS through a process or series of processes in order to read and transform input data and to generate output. The DATA step and the procedure step, used alone or in combination, form the basis of SAS programs.

SAS session

the activity between invoking and exiting a specific SAS software product.

SAS statement

a string of SAS keywords, SAS names, and special characters and operators that instructs SAS to perform an operation or that gives information to SAS. Each SAS statement ends with a semicolon.

SAS system option

an option that affects the processing of an entire SAS program or interactive SAS session from the time the option is specified until it is changed. Examples of items that are controlled by SAS system options include the appearance of SAS output, the handling of some files that are used by SAS, the use of system variables, the processing of observations in SAS data sets, features of SAS initialization, and the way SAS interacts with your host operating environment.

SAS time constant

a string in the form 'hh:mm:ss't that represents a time in a SAS statement. The string is enclosed in quotation marks and is followed by the character t (for example, '09:53:22't).

SAS time value

an integer that represents a time in SAS software. The integer represents the number of seconds between midnight of the current day and another specified time value. For example, the SAS time value for 9:30 a.m. is 34200.

SAS variable

a column in a SAS data set or in a SAS data view. The data values for each variable describe a single characteristic for all observations (rows).

Sashelp library

a SAS library supplied by SAS software that stores the text for Help windows, default function-key definitions and window definitions, and menus.

Sasuser library

a default, permanent SAS library that is created at the beginning of your first SAS session. The Sasuser library contains a PROFILE catalog that stores the customized features or settings that you specify for SAS.

statement

See SAS statement

statement option
a word that you specify in a particular SAS statement and which affects only the processing that that statement performs.

syntax checking
the process by which SAS checks each SAS statement for proper usage, correct spelling, proper SAS naming conventions, and so on.

syntax error
an error in the spelling or grammar of a SAS statement. SAS finds syntax errors as it compiles each SAS step before execution.

system option
See SAS system option

target variable
the variable to which the result of a function or expression is assigned.

taskbar
the bar at the bottom of the Windows desktop that displays active applications. The taskbar enables you to easily switch between applications and to restore, move, size, minimize, maximize, and close applications.

temporary SAS file
a SAS file in a SAS library (usually the Work library) that is deleted at the end of the SAS session or job.

temporary SAS library
a library that exists only for the current SAS session or job. The most common temporary library is the Work library.

text output
another term for listing output.

time constant
See SAS time constant

time value
See SAS time value

title
a heading that is printed at the top of each page of SAS output or of the SAS log.

toggle
an option, parameter, or other mechanism that enables you to turn on or turn off a processing feature.

trailing at sign
a special symbol @ that is used to hold a line of input or output so that SAS can read from it or write to it in a subsequent INPUT or PUT statement.

variable
See SAS variable

variable attribute

any of the following characteristics that are associated with a particular variable: name, label, format, informat, data type, and length.

variable label

up to 256 characters of descriptive text that can be printed in the output by certain procedures instead of, or in addition to, the variable name.

variable length

the number of bytes used to store each of a variable's values in a SAS data set.

variable type

the classification of a variable as either numeric or character. Type is an attribute of SAS variables.

Work library

a temporary SAS library that is automatically defined by SAS at the beginning of each SAS session or SAS job. Unless you have specified a User library, any newly created SAS file that has a one- level name will be placed in the Work library by default and will be deleted at the end of the current SAS session or job.

Index

SAS® Publishing Delivers!

SAS Publishing provides you with a wide range of resources to help you develop your SAS software expertise.
Visit us online at **support.sas.com/bookstore**.

SAS® PRESS

SAS Press titles deliver expert advice from SAS® users worldwide. Written by experienced SAS professionals,
SAS Press books deliver real-world insights on a broad range of topics for all skill levels.

support.sas.com/saspress

SAS® DOCUMENTATION

We produce a full range of primary documentation:

- Online help built into the software
- Tutorials integrated into the product
- Reference documentation delivered in HTML and PDF formats—free on the Web
- Hard-copy books

support.sas.com/documentation

SAS® PUBLISHING NEWS

Subscribe to SAS Publishing News to receive up-to-date information via e-mail about all new SAS titles,
site features.

support.sas.com/spn

RSATION!

Visit our Web site for links to our pages on Facebook,
odcasts, and RSS feeds, too.

support.sas.com/socialmedia

§sas. | THE POWER TO KNOW.